Changing Climate

Report of the Carbon Dioxide Assessment Committee

Board on Atmospheric Sciences and Climate
Commission on Physical Sciences,
Mathematics, and Resources
National Research Council (U.S.). Carbon Dioxide Assessment Committee.

NATIONAL ACADEMY PRESS
Washington, D.C. 1983

National Academy Press • 2101 Constitution Avenue NW • Washington DC 20418

NOTICE The project that is the subject of this report was approved by the Governing Board of the National Research Council, whose members are drawn from the councils of the National Academy of Sciences, the National Academy of Engineering, and the Institute of Medicine. The members of the Committee responsible for the report were chosen for their special competences and with regard for appropriate balance.

This report has been reviewed by a group other than the authors according to procedures approved by a Report Review Committee consisting of members of the National Academy of Sciences, the National Academy of Engineering, and the Institute of Medicine.

The National Research Council was established by the National Academy of Sciences in 1916 to associate the broad community of science and technology with the Academy's purposes of furthering knowledge and of advising the federal government. The Council operates in accordance with general policies determined by the Academy under the authority of its congressional charter of 1863, which establishes the Academy as a private, nonprofit, self-governing membership corporation. The Council has become the principal operating agency of both the National Academy of Sciences and the National Academy of Engineering in the conduct of their services to the government, the public, and the scientific and engineering communities. It is administered jointly by both Academies and the Institute of Medicine. The National Academy of Engineering and the Institute of Medicine were established in 1964 and 1970, respectively, under the charter of the National Academy of Sciences.

Library of Congress Catalog Card Number 83-62680
International Standard Book Number 0-309-03425-6

Printed in the United States of America

Carbon Dioxide Assessment Committee

William A. Nierenberg, Scripps Institution of Oceanography, Chairman
Peter G. Brewer, Woods Hole Oceanographic Institution (on leave with the National Science Foundation)
Lester Machta, Air Resources Laboratory, National Oceanic and Atmospheric Administration
William D. Nordhaus, Yale University
Roger R. Revelle, University of California, San Diego
Thomas C. Schelling, Harvard University
Joseph Smagorinsky, Princeton University
Paul E. Waggoner, Connecticut Agricultural Experiment Station
George M. Woodwell, Marine Biological Laboratory, Woods Hole, Massachusetts

Consultants

David A. Katcher, Chevy Chase, Maryland
Gary W. Yohe, Wesleyan University

Staff

John S. Perry, Executive Secretary
Jesse H. Ausubel, Staff Officer
James A. Tavares
Adele King Malone (to 5/82)

Detection and Monitoring Working Group

Gunter Weller, University of Alaska, Chairman
D. James Baker, Jr., University of Washington
W. Lawrence Gates, Oregon State University
Michael C. MacCracken, Lawrence Livermore Laboratory,
Syukuro Manabe, Geophysical Fluid Dynamics Laboratory, National Oceanic and Atmospheric Administration
Thomas H. Vonder Haar, Colorado State University

John S. Perry, Executive Secretary

Climate Research Committee

Joseph Smagorinsky, Princeton University, Chairman
Tim P. Barnett, Scripps Institution of Oceanography
Harry L. Bryden, Woods Hole Oceanographic Institution
Isaac Held, Geophysical Fluid Dynamics Laboratory, National Oceanic and Atmospheric Administration
Frederick T. Mackenzie, Northwestern University
James C. McWilliams, National Center for Atmospheric Research
Norman Phillips, National Oceanic and Atmospheric Administration
Veerabhadran Ramanathan, National Center for Atmospheric Research
Jagadish Shukla, Goddard Space Flight Center, National Aeronautics and Space Administration
Thomas H. Vonder Haar, Colorado State University
John M. Wallace, University of Washington
Ferris Webster, University of Delaware
Gunter E. Weller, University of Alaska

Board on Atmospheric Sciences and Climate

Commission on Physical Sciences, Mathematics, and Resources

Foreword

The Energy Security Act of 1980, while focused on the development of synthetic fuels, also called for examination of some of the environmental consequences of their development. One such consequence perceived by the Congress was the buildup of carbon dioxide (CO_2) in the atmosphere, and the National Academy of Sciences (NAS) and the Office of Science and Technology Policy (OSTP) of the Executive Office of the President were requested to prepare an assessment of its implications.

Concern about the atmosphere's carbon dioxide and its influence on climate dates back to the last century. In the 1970s, however, with recognition of a growing world population and increasing per capita use of energy, attention markedly heightened. In 1977 the National Research Council issued a report, Energy and Climate,[1] prepared by a panel chaired by Roger Revelle, calling for an intensified program of research on CO_2. At around this time, the federal government began expanding its concern with CO_2, primarily through a research and assessment program in the Department of Energy. In a congressional symposium on CO_2 and energy policy in 1979 some scientists expressed the fear that atmospheric CO_2 could double by the first decade of the twenty-first century if coal and fossil-based synthetic fuels were vigorously exploited.

Such concerns and the increasing volume of research results led the Congress and the Executive to ask the NAS to consider anew various aspects of the issue. In July 1979, a brief preliminary statement about CO_2 and energy policy was released by the Academy, and later in that same summer a Panel of the Climate Research Board chaired by the late Jule Charney undertook an evaluation of the models being used to estimate likely effects of CO_2 on climate.[2] In the following winter and spring, a committee chaired by Thomas C. Schelling and including several other members of the current Committee considered

[1] Geophysics Study Committee (1977). Energy and Climate. National Academy Press, Washington, D.C.

[2] National Research Council (1979). Carbon Dioxide and Climate: A Scientific Assessment. National Academy Press, Washington, D.C.

some of the economic and social aspects of increase in CO_2.[3] At the same time, in April 1980, the Senate Committee on Energy and Natural Resources convened a hearing on the issue.

In this climate of concern, the Energy Security Act of 1980 was passed, calling upon OSTP to request a study by the Academy that would deal with, among others, the following issues:

- A comprehensive assessment of CO_2 release and impacts of CO_2 increase;
- Development of an international research and assessment program and definition of the U.S. role;
- Analysis of domestic resource requirements for international and domestic programs;
- Evaluation of the U.S. government CO_2 program; and
- Assessment of the need for periodic reports and a long-term assessment program.

Annex 3 of this report contains the relevant subtitle of the legislation.

As congressional interest was mounting, the Climate Research Board requested one of its members, William A. Nierenberg, to monitor developments and to advise the Board on appropriate actions. In response to the congressional mandate, the Carbon Dioxide Assessment Committee (CDAC) was formed under his leadership to develop a plan to accomplish the requested study.

With support from OSTP, the Committee developed a preliminary plan, which was provided to OSTP for comment in January 1981. The change of administration at that time slowed the discussions of the study's scope and objectives. However, in the summer of 1981 a plan of action was agreed on by which the NAS and OSTP could respond to the congressional request for an independent and comprehensive assessment. The report would comprise two major parts: (1) an overview or synthesis representing the views of the Committee as a whole on the issue and (2) a group of papers each addressing a specific topic or problem area and prepared by an individual committee member or a specialist group.

The Committee began its work in September 1981. Over the next 2 years, the Committee members met four times (September 28-29, 1981, Washington, D.C.; March 25-26, 1982, La Jolla, California; September 20-21, Berkeley Springs, West Virginia; January 13-14, 1983, Washington, D.C.) to monitor progress on their specific topics and to develop their collective views.

A number of considerations went into the design of the Committee and the selection of additional experts to contribute to its work. Competence was sought in each of the major subject areas of the question,

---

[3] Letter report of the Ad Hoc Study Panel on Economic and Social Aspects of Carbon Dioxide Increase, T. C. Schelling, Chairman. April 18, 1980. Climate Research Board, National Academy of Sciences, Washington, D.C.

as well as experience with assessment of long-range issues. A balance of viewpoints about environmental issues was sought. Finally, continuity was maintained with previous, related NRC efforts (Energy and Climate, Energy in Transition,[4] the Schelling report). Additional experience and skills were provided by consultants to the Committee and by ad hoc workshops and groups convened with the assistance of other NRC units.

Before the reader turns to the contents of this report, the question might be raised: Why another CO_2 report, apart from the legislative request? The CO_2 issue has been probingly addressed by many individuals and groups in recent years. One could mention, for example, Carbon Dioxide Review: 1982,[5] Carbon Dioxide from Coal Utilization,[6] On the Assessment of the Role of CO_2 on Climate Variations and Their Impact,[7] The CO_2-Climate Connection,[8] The Long-Term Impacts of Increasing Atmospheric Carbon Dioxide Levels,[9] Interactions of Energy and Climate,[10] as well as the NRC's own Energy and Climate, Charney report, and Schelling report. There are at least two reasons to contribute to the growing volume of literature on the CO_2 issue. One is that research developments are occurring at a rapid rate, as we see in many sections of the report that follows. A second reason is that the focus and perspectives of the reports on the CO_2 issue differ. A major distinguishing feature of Changing Climate is that it represents a sustained attempt by a group with a wide range of expertise to achieve a comprehensive and internally consistent assessment.

On behalf of the Board and the Academy, I wish to express our appreciation to Professor Nierenberg, the members of the Committee, and the many other participants in this study for their individual and collective contributions to this report. As described in the Historical Note (Annex 2), the carbon dioxide issue has been with us for a long time,

---

[4] National Research Council (1979). Energy in Transition 1985-2010. Final Report of the Committee on Nuclear and Alternative Energy Systems (CONAES). W. H. Freeman, San Francisco.

[5] W. C. Clark, ed. (1982). Carbon Dioxide Review: 1982. Oxford U. Press, New York, 469 pp.

[6] I. M. Smith (1982). Carbon Dioxide from Coal Utilization. Technical Information Service, International Energy Agency, Paris.

[7] WMO/ICSU/UNEP (1981). On the assessment of the role of CO_2 on climate variations and their impact. (Based on meeting of experts, Villach, Austria, November 1980.) World Meteorological Organization, Geneva, January 1981.

[8] G. B. Tucker (1981). The CO_2-Climate Connection. Australian Academy of Science, Canberra.

[9] G. J. MacDonald, ed. (1982). The Long-Term Impacts of Increasing Atmospheric Carbon Dioxide Levels. Ballinger, Cambridge, Mass., 263 pp.

[10] W. Bach, J. Pankrath, and J. Williams, eds. (1980). Interactions of Energy and Climate. Reidel, Dordrecht, The Netherlands.

and it will undoubtedly maintain a prominent place on our agenda for a long time to come. We continue to need well-coordinated programs of research, productively interdisciplinary in character and broadly international in scope. As Professor Nierenberg indicates in his Preface, this report is best viewed as one stepping-stone on a long pathway into the future. We are confident, however, that it will prove to be a solid step on that pathway toward more complete understanding of this complex issue.

Thomas F. Malone, Chairman
Board on Atmospheric Sciences and Climate

Preface

There is a broad class of problems that have no "solution" in the sense of an agreed course of action that would be expected to make the problem go away. These problems can also be so important that they should not be avoided or ignored until the fog lifts. We simply must learn to deal more effectively with their twists and turns as they unfold. We require sensible regular progress to anticipate what these developments might be with a balanced diversity of approaches. The payoff is that we will have had the chance to consider alternative courses of action with some degree of calm before we may be forced to choose among them in urgency or have them forced on us when other--perhaps better--options have been lost. Increasing atmospheric CO_2 and its climatic consequences constitute such a problem.

Research developments are taking place rapidly in this area. In the pages that follow we report our understanding of the status of a number of selected, critical aspects and comment on how well we think the overall attack on this complex matter is proceeding. Our stance is conservative: we believe there is reason for caution, not panic. Since understanding and proof of what is happening to climate as a result of practices that load the atmosphere with CO_2 may come too late to allow for corrective action, we may not be able to wait to make certain there is a best course. Thus, we must proceed in a manner that keeps open our major options on energy development and use, on water management, agricultural adjustment, and other relevant activities, as we move from one set of uncertainties to another. We make an effort in this report to point the way as we see it today.

A range of approaches was employed in developing the Committee's report. For example, in the study of possible future CO_2 emissions, a review of earlier research was commissioned and a new model was constructed to remedy some of the shortcomings of previous work. The carbon cycle was addressed through individual reviews of its oceanic, atmospheric, and biotic components, together with model-based sensitivity analyses. In the area of agriculture a survey was undertaken, and several outside experts were convened in a small informal workshop to address the relationship of climatic change to crop yield. A group was also convened, in Athens, Georgia, in May 1982, in conjunction with a meeting organized by the American Association for the Advancement of

Science on direct effects of CO_2 on plants. Small, informal workshops were also convened in the areas of hydrology and land surface processes (La Jolla, March 16-17, 1982), Antarctic Ice (La Jolla, March 18-19, 1982), and Arctic Ice (Philadelphia, June 1-2, 1982). Questions of sea-level rise were explored with assistance from experts from the Samenwerkende Instellingen ten Behoeve van beleidsanalytische Studies (SIBAS), Delft, The Netherlands. The general area of scenario construction and evaluation benefited from the participation of three Carbon Dioxide Assessment Committee (CDAC) members in a workshop on this topic at the International Institute for Applied Systems Analysis in Laxenburg, Austria, in July 1982.

The Committee recognized early that the central issue of CO_2 effects on climate would require intensive review. Thus, the Board's Climate Research Committee was asked to re-examine and update the work of the 1979 Charney panel. A panel chaired by Joseph Smagorinsky, a member of the CDAC and Chairman of the Climate Research Committee (CRC), undertook this task and issued its report, *Carbon Dioxide and Climate: A Second Assessment*, in 1982. This document should be considered an integral part of the present study; its conclusions are reproduced in this volume together with brief supplementary comments. The CRC was also asked to consider the detection of climatic change induced by CO_2 and the monitoring of climatic variables. With the advice of the CRC, a group of experts, including a number of CRC members, was asked to contribute to the report. These experts met several times during 1982 under the leadership of Gunter Weller, and the results of their conferring form Chapter 5 of this volume.

The scope of the Committee's report is broad, and I believe it meets as fully as a small group could the congressional request for a "comprehensive" assessment of the CO_2 issue. A truly complete assessment of the CO_2 issue might involve most or all of a very wide range of elements, including identification of various risks and prospective changes; estimation of probabilities of their occurrence; linkage of such events with various environmental and social consequences; and evaluation of the risks by comparison with costs, with other risks, with benefits, with alternative ways of reducing risks, or with risks of substitute activities. In this report the Committee does attempt to shed a little light on all of these aspects of the CO_2 issue: CO_2 emissions and concentrations are projected; possible climatic changes are assessed; implications of increasing CO_2 and climatic change for agriculture and water, sea level, and other selected areas are examined, including possible impacts that might have a low probability but a high cost; and possible consequences are evaluated against historical experiences and other current and future problems.

However, the report certainly does not exhaust the issue. A number of possibly important problems are left for future investigators. Among such problems are the effects of non-CO_2 greenhouse gases on climate, effects of altered climate on agriculture and water outside the United States, and the feasibility of alternative nonfossil energy strategies. Furthermore, as interest in synfuels development diminished, the CDAC chose to place less emphasis on this aspect of the

Contents

EXECUTIVE SUMMARY 1

1 SYNTHESIS 5
Carbon Dioxide Assessment Committee

1.1 INTRODUCTION, 5
1.2 THE OUTLOOK, 9
1.2.1 Future CO_2 Emissions, 9
1.2.2 Future Atmospheric CO_2 Concentrations, 14
1.2.3 Changing Climate with Changing CO_2, 27
1.2.4 Detection of CO_2-Induced Changes, 32
1.2.5 Agricultural Impacts, 36
1.2.6 Water Supplies, 40
1.2.7 Sea Level, Antarctic, and Arctic, 40
1.3 SERIOUSNESS OF PROJECTED CHANGES, 44
1.3.1 Specifiable Concerns, 45
1.3.1.1 Agriculture and Water Resources, 45
1.3.1.2 Rising Sea Level, 48
1.3.2 More Speculative Concerns, 48
1.3.3 The Problem of Unease about Changes of This Magnitude, 50
1.4 POSSIBLE RESPONSES, 55
1.4.1 Defining the Problem, 55
1.4.2 The Organizing Framework, 57
1.4.3 Categories of Response, 57
1.4.4 Reprise, 61
1.5 RECOMMENDATIONS, 61
1.5.1 Can CO_2 Be Addressed as an Isolated Issue?, 61
1.5.2 Actual and Near-Term Change of Policies, 62
1.5.3 Energy Research and Policy, 64
1.5.4 Synfuels Policy and CO_2, 65
1.5.5 Applied Research and Development, 66
1.5.6 Basic Research and Monitoring, 67
1.5.6.1 General Research Comments, 68
1.5.6.2 The International Aspect, 70

1.5.6.3 Projecting CO_2 Emissions, 72
1.5.6.4 Projecting CO_2 Concentrations, 72
1.5.6.5 Climate, 74
1.5.6.6 Detection and Monitoring, 76
1.5.6.7 Impacts, 78
1.6 CONCLUDING REMARKS, 81
REFERENCES, 81

2 FUTURE CARBON DIOXIDE EMISSIONS FROM FOSSIL FUELS 87

2.1 FUTURE PATHS OF ENERGY AND CARBON DIOXIDE EMISSIONS, 87
William D. Nordhaus and Gary W. Yohe
2.1.1 Overview, 87
2.1.2 Detailed Description of the Model, Data, and Results, 99
2.1.2.1 The Model, 100
2.1.2.2 The Data, 111
2.1.2.3 Results, 130
References, 151
2.2 A REVIEW OF ESTIMATES OF FUTURE CARBON DIOXIDE EMISSIONS, 153
Jesse H. Ausubel and William D. Nordhaus
2.2.1 Introduction, 153
2.2.2 Projections Based on Extrapolations, 155
2.2.3 Energy System Projections, 156
2.2.3.1 Perry-Landsberg (NAS), 157
2.2.3.2 IIASA, 157
2.2.3.3 Rotty et al., 159
2.2.3.4 Nordhaus, 160
2.2.3.5 Edmonds and Reilly, 161
2.2.3.6 Other Projections, 162
2.2.4 Projections with CO_2 Feedback to the Energy System, 164
2.2.4.1 Nordhaus, 164
2.2.4.2 Edmonds and Reilly, 165
2.2.4.3 CEQ, 166
2.2.4.4 A. M. Perry et al., 167
2.2.4.5 General Comments, 168
2.2.5 A Note on the Biosphere, 169
2.2.6 Projections of Non-CO_2 Trace Gases, 170
2.2.7 Findings, 171
2.2.7.1 The State of the Art, 171
2.2.7.2 Likely Future Outcomes, 173
2.2.8 Conclusion, 181
References, 181

3 PAST AND FUTURE ATMOSPHERIC CONCENTRATIONS OF CARBON DIOXIDE 186

3.1 INTRODUCTION Peter G. Brewer, 186
3.2 CARBON DIOXIDE AND THE OCEANS Peter G. Brewer, 188
3.2.1 Introduction, 188
3.2.2 The Cycle of Carbon Dioxide within the Oceans, 189
3.2.3 The Deep Circulation, 189
3.2.4 Biological Activity, 190
3.2.5 Deep Decomposition of Organic Matter, 191
3.2.6 Calcium Carbonate, 191
3.2.7 The Chemistry of CO_2 in Seawater, 194
3.2.8 Measurements of Ocean CO_2, 199
3.2.9 Models of Ocean CO_2 Uptake, 202
3.2.10 Future Studies and Problems, 209
3.2.11 Summary, 211
References, 211
3.3 BIOTIC EFFECTS ON THE CONCENTRATION OF ATMOSPHERIC CARBON DIOXIDE: A REVIEW AND PROJECTION George M. Woodwell, 216
3.3.1 Introduction, 216
3.3.2 How Much Carbon is Held in the Biota and Soils?, 217
3.3.2.1 The Biota, 217
3.3.2.2 The Soils, 217
3.3.2.3 Total Carbon Pool under Biotic Influences, 217
3.3.3 Metabolism and the Storage of Carbon in Terrestrial and Aquatic Ecosystems, 218
3.3.3.1 The Production Equation, 218
3.3.3.2 A Basis in the Metabolism of Forests for the Oscillation in Atmospheric CO_2 Concentration, 219
3.3.3.3 Factors Affecting Global Net Ecosystem Production, 221
3.3.4 Changes in Area of Forests of the World, 225
3.3.5 The Biota in the Context of the Global Carbon Balance, 229
3.3.6 A Projection of Further Releases from Biotic Pools, 234
3.3.7 Summary and Conclusions, 234
References, 236
3.4 THE ATMOSPHERE Lester Machta, 242
3.4.1 Introduction, 242
3.4.2 Changes in Atmospheric CO_2 Growth Rate with Time and Space, 243
3.4.2.1 Shorter-Term Variation and Its Possible Cause, 243
3.4.2.2 Longer-Term Variations, 246
3.4.2.3 Change in Annual Cycle, 247
3.4.2.4 Spatial Distribution, 248
3.4.2.5 Isotopic Content of Atmospheric CO_2, 249
3.4.3 Conclusions, 250
References, 250

3.5 METHANE HYDRATES IN CONTINENTAL SLOPE SEDIMENTS AND INCREASING ATMOSPHERIC CARBON DIOXIDE Roger R. Revelle, 252
3.5.1 Methane in the Atmosphere, 252
3.5.2 Formation of Methane Clathrate in Continental Slope Sediments, 253
3.5.3 Effect of Carbon Dioxide-Induced Warming on Continental Slope Clathrates, 256
3.5.4 Future Rate of Methane Release from Sedimentary Clathrates, 257
References, 260
3.6 SENSITIVITY STUDIES USING CARBON CYCLE MODELS Lester Machta, 262
3.6.1 Comparison among Different Models, 262
3.6.2 Comparison of Parameters within a Single Model, 263
3.6.3 Deforestation as a Source of CO_2, 264
3.6.4 Conclusion, 265
References, 265

4 EFFECTS ON CLIMATE 266

4.1 EFFECTS OF CARBON DIOXIDE Joseph Smagorinsky, 266
4.1.1 Excerpts from "Charney" and "Smagorinsky" Reports, 266
4.1.2 Epilogue, 277
References, 282
4.2 EFFECTS OF NON-CO_2 GREENHOUSE GASES Lester Machta, 285
References, 291

5 DETECTION AND MONITORING OF CO_2-INDUCED CLIMATE CHANGES 292
Gunter Weller, D. James Baker, Jr., W. Lawrence Gates, Michael C. MacCracken, Syukuro Manabe, and Thomas H. Vonder Haar

5.1 SUMMARY, 292
5.2 HAVE CO_2-INDUCED SURFACE TEMPERATURE CHANGES ALREADY OCCURRED?, 297
5.2.1 Introduction, 297
5.2.2 Requirements for Identifying CO_2-Induced Climate Change, 301
5.2.2.1 Climatic Data Bases, 303
5.2.2.2 Causal Factors, 306
5.2.2.3 Relating Causal Factors and Climatic Effects, 311
5.2.3 Attempts to Identify CO_2-Induced Climate Change, 313
5.2.3.1 Carbon Dioxide as a Causal Factor, 318
5.2.3.2 Volcanic Aerosol as a Causal Factor, 320
5.2.3.3 Solar Variations as a Causal Factor, 322
5.2.3.4 Combinations of Causal Factors, 324
5.2.4 Steps for Building Confidence, 327
5.3 A STRATEGY FOR MONITORING CO_2-INDUCED CLIMATE CHANGE, 330
5.3.1 The "Fingerprinting" Concept, 330

5.3.2 Considerations in Climate Monitoring, 331
5.3.2.1 Statistical Variability and Expectations of Change, 331
5.3.2.2 Initial Selection of Parameters, 332
5.3.2.3 Revision and Application of a Monitoring Strategy, 332
5.3.3 Candidate Parameters for Monitoring, 333
5.3.3.1 Causal Factors, 335
5.3.3.2 Atmospheric Parameters, 351
5.3.3.3 Cryospheric Parameters, 361
5.3.3.4 Oceanic Parameters, 367
5.3.4 Conclusions and Recommendations, 370
5.3.4.1 Priority of Parameters to be Monitored, 370
5.3.4.2 Measurement Networks, 371
5.3.4.3 Modeling and Statistical Techniques, 372
5.3.4.4 Objective Evaluation of Evidence, 372
REFERENCES, 373

6 AGRICULTURE AND A CLIMATE CHANGED BY MORE CARBON DIOXIDE 383
Paul E. Waggoner

6.1 INTRODUCTION, 383
6.1.1 Concentrating on a Critical, Susceptible, and Exemplary Subject, 383
6.1.2 Agriculture and Past Changes in the Weather, 384
6.1.3 The Range of Change in the Atmosphere, 386
6.2 EFFECTS OF CO_2 ON PHOTOSYNTHESIS AND PLANT GROWTH, 388
6.2.1 Photosynthesis, 388
6.2.1.1 Rate of Photosynthesis, 389
6.2.1.2 Duration of Photosynthesis, 390
6.2.1.3 Fate and Partitioning of Photosynthate, 390
6.2.2 Drought, 391
6.2.3 Nutrients, 391
6.2.3.1 Nitrogen Metabolism, 392
6.2.3.2 Organic Matter and Rhizosphere Association, 392
6.2.4 Phenology, 393
6.2.5 Weeds, 393
6.2.6 Direct Effects of CO_2 on Yield, 394
6.3 PREDICTING THE CHANGES IN YIELD THAT WILL FOLLOW A CHANGE TO A WARMER, DRIER CLIMATE, 396
6.3.1 History, 396
6.3.2 Simulation, 403
6.3.3 Summary, 405
6.4 PATHOGENS AND INSECT PESTS, 405
6.5 IRRIGATION IN A WARMER AND DRIER CLIMATE, 407
6.6 ADAPTING TO THE CHANGE TO A WARMER, DRIER CLIMATE, 409
6.6.1 Breeding New Varieties, 409
6.6.2 Adapting to Less Water, 411
6.7 CONCLUSION, 413
REFERENCES, 413

7 EFFECTS OF A CARBON DIOXIDE-INDUCED CLIMATIC CHANGE ON WATER SUPPLIES IN THE WESTERN UNITED STATES 419
Roger R. Revelle and Paul E. Waggoner

7.1 EMPIRICAL RELATIONSHIPS AMONG PRECIPITATION, TEMPERATURE, AND STREAM RUNOFF, 419
7.2 EFFECTS OF CLIMATE CHANGE IN SEVEN WESTERN U.S. WATER REGIONS, 421
7.3 THE COLORADO RIVER, 425
7.4 CLIMATE CHANGE AND WATER-RESOURCE SYSTEMS, 431
REFERENCES, 432

8 PROBABLE FUTURE CHANGES IN SEA LEVEL RESULTING FROM INCREASED ATMOSPHERIC CARBON DIOXIDE 433
Roger R. Revelle

8.1 THE OBSERVED RISE IN SEA LEVEL DURING PAST DECADES, 433
8.2 THE FUTURE RISE IN SEA LEVEL, 435
8.2.1 Melting of Greenland Ice Cap and Alpine Glaciers, 436
8.2.2 Heating of the Upper Oceans, 437
8.2.3 Possible Disintegration of the West Antarctic Ice Sheet, 441
REFERENCES, 447

9 CLIMATIC CHANGE: IMPLICATIONS FOR WELFARE AND POLICY 449
Thomas C. Schelling

9.1 INTRODUCTION, 449
9.1.1 Uncertainties, 451
9.1.2 The Time Dimension, 451
9.1.3 Discounting, Positive or Negative, 452
9.1.4 Perspective on Change, 453
9.1.5 Prudential Considerations, 454
9.1.6 Variation in Human Environments, 455
9.2 A SCHEMA FOR ASSESSMENT AND CHOICE, 456
9.2.1 Five Categories, 463
9.2.2 Background Climate and Trends, 465
9.2.3 Production of CO_2, 466
9.2.4 Removal of CO_2, 467
9.2.5 Modification of Climate and Weather, 468
9.2.6 Adaptation, 470
9.2.7 Breathing CO_2, 471
9.2.8 Change in Sea Level, 472
9.2.9 Defenses against Rising Sea Level, 472
9.2.10 Food and Agriculture, 474
9.2.11 Global Warming and Energy Consumption, 476
9.2.12 Distributional Impact, 477
9.3 SUMMING UP, 477

Annex 1: Report of Informal Meeting on CO_2 and the Arctic Ocean
Roger R. Revelle 483

Annex 2: Historical Note
Jesse H. Ausubel 488

Annex 3: Energy Security Act of 1980 492

Annex 4: Background Information on Committee Members 494

Executive Summary

1. Carbon dioxide (CO_2) is one of the gases of the atmosphere important in determining the Earth's climate. In the last generation the CO_2 concentration in the atmosphere has increased from 315 parts per million (ppm) by volume to over 340 ppmv. (Chapters 3, 4)

2. The current increase is primarily attributable to burning of coal, oil, and gas; future increases will similarly be determined primarily by fossil fuel combustion. Deforestation and land use changes have probably been important factors in atmospheric CO_2 increase over the past 100 years. (Chapters 2, 3)

3. Projections of future fossil fuel use and atmospheric concentrations of CO_2 embody large uncertainties that are to a considerable extent irreducible. The dominant sources of uncertainty stem from our inability to predict future economic and technological developments that will determine the global demand for energy and the attractiveness of fossil fuels. We think it most likely that atmospheric CO_2 concentration will pass 600 ppm (the nominal doubling of the recent level) in the third quarter of the next century. We also estimate that there is about a 1-in-20 chance that doubling will occur before 2035. (Chapters 2, 3)

4. If deforestation has been a large net source of CO_2 in recent decades, then the models that we are using to project future atmospheric concentrations are seriously flawed; the fraction of man-made CO_2 remaining airborne must then be lower, and CO_2 increase will probably occur more slowly than it otherwise would. (Chapter 3)

5. Estimates of effects of increasing CO_2 on climate also embody significant uncertainties, stemming from fundamental gaps in our understanding of physical processes, notably the processes that determine cloudiness and the long-term interactions between atmosphere and ocean. (Chapter 4)

6. Several other gases besides CO_2 that can affect the climate appear to be increasing as a result of human activities; if we project

increases in all these gases, climate changes can be expected significantly earlier than if we consider CO_2 alone. (Chapter 4)

7. From climate model simulations of increased CO_2 we conclude with considerable confidence that there would be global mean temperature increase. With much less confidence we infer other more specific regional climate changes, including relatively greater polar temperature increase and summer dryness in middle latitudes (e.g., the latitudes of the United States). (Chapter 4)

8. Results of most numerical model experiments suggest that a doubling of CO_2, if maintained indefinitely, would cause a global surface air warming of between 1.5°C and 4.5°C. The climate record of the past hundred years and our estimates of CO_2 changes over that period suggest that values in the lower half of this range are more probable. (Chapters 4, 5)

9. By itself, CO_2 increase should have beneficial effects on photosynthesis and water-use efficiency of agricultural plants, especially when other factors are not already limiting growth. (Chapters 3, 6)

10. Analysis of the effects of a warmer and drier climate on rainfed agriculture in the United States suggests that over the next couple of decades negative effects of climate change and positive effects from CO_2 fertilization both will be modest and will approximately balance. The outlook is more troubling for agriculture in lands dependent on irrigation. Longer-term impacts are highly uncertain and will depend strongly on the outcome of future agricultural research, development, and technology. (Chapter 6)

11. Changes in temperature and rainfall may be amplified as changes in the annual discharge of rivers. For example, a 2°C warming could severely reduce the quantity and quality of water resources in the western United States. (Chapter 7)

12. (a) If a global warming of about 3 or 4°C were to occur over the next hundred years, it is likely that there would be a global sea-level rise of about 70 cm, in comparison with the rise of about 15 cm over the last century. More rapid rates could occur subsequently, if the West Antarctic Ice Sheet should begin to disintegrate. (Chapter 8)

(b) Such a warming might also bring about changes in Arctic ice cover, with perhaps a disappearance of the summer ice pack and associated changes in high-latitude weather and climate. (Annex 1)

13. Because of their large uncertainties and significant implications, it is important to confirm the various predictions of climate changes at the earliest possible time and to achieve greater precision. This can best be done through carefully designed monitoring programs of long duration emphasizing the ensemble of variables believed to influence climate or to reflect strongly the effect of CO_2. (Chapter 5)

14. The social and economic implications of even the most carefully constructed and detailed scenarios of CO_2 increase and climatic consequences are largely unpredictable. However, a number of inferences seem clear:

(a) Rapid climate change will take its place among the numerous other changes that will influence the course of society, and these other changes may largely determine whether the climatic impacts of greenhouse gases are a serious problem.

(b) As a human experience, climate change is far from novel; large numbers of people now live in almost all climatic zones and move easily between them.

(c) Nevertheless, we are deeply concerned about environmental changes of this magnitude; man-made emissions of greenhouse gases promise to impose a warming of unusual dimensions on a global climate that is already unusually warm. We may get into trouble in ways that we have barely imagined, like release of methane from marine sediments, or not yet discovered.

(d) Climate changes, their benefits and damages, and the benefits and damages of the actions that bring them about will fall unequally on the world's people and nations. Because of real or perceived inequities, climate change could well be a divisive rather than a unifying factor in world affairs. (Chapter 9)

15. Viewed in terms of energy, global pollution, and worldwide environmental damage, the "CO_2 problem" appears intractable. Viewed as a problem of changes in local environmental factors--rainfall, river flow, sea level--the myriad of individual incremental problems take their place among the other stresses to which nations and individuals adapt. It is important to be flexible both in definition of the issue, which is really more climate change than CO_2, and in maintaining a variety of alternative options for response. (Chapter 9)

16. Given the extent and character of the uncertainty in each segment of the argument--emissions, concentrations, climatic effects, environmental and societal impacts--a balanced program of research, both basic and applied, is called for, with appropriate attention to more significant uncertainties and potentially more serious problems. (Chapter 1)

17. Even very forceful policies adopted soon with regard to energy and land use are unlikely to prevent some modification of climate as a result of human activities. Thus, it is prudent to undertake applied research and development--and to consider some adjustments--in regard to activities, like irrigated agriculture, that are vulnerable to climate change. (Chapters 1, 9)

18. Assessment of the CO_2 issue should be regarded as an iterative process that emphasizes carry over of learning from one effort to the next. (Chapter 1)

19. Successful response to widespread environmental change will be facilitated by the existence of an international network of scientists

conversant with the issues and of broad international consensus on facts and their reliability. Sound international research and assessment efforts can turn up new solutions and lubricate the processes of change and adaptation. (Chapter 1)

20. With respect to specific recommendations on research, development, or use of different energy systems, the Committee offers three levels of recommendations. These are based on the general view that, if other things are equal, policy should lean away from the injection of greenhouse gases into the atmosphere.

(a) Research and development should give some priority to the enhancement of long-term energy options that are not based on combustion of fossil fuels. (Chapters 1, 2, 9)

(b) We do not believe, however, that the evidence at hand about CO_2-induced climate change would support steps to change current fuel-use patterns away from fossil fuels. Such steps may be necessary or desirable at some time in the future, and we should certainly think carefully about costs and benefits of such steps; but the very near future would be better spent improving our knowledge (including knowledge of energy and other processes leading to creation of greenhouse gases) than in changing fuel mix or use. (Chapters 1, 2, 9)

(c) It is possible that steps to control costly climate change should start with non-CO_2 greenhouse gases. While our studies focused chiefly on CO_2, fragmentary evidence suggests that non-CO_2 greenhouse gases may be as important a set of determinants as CO_2 itself. While the costs of climate change from non-CO_2 gases would be the same as those from CO_2, the control of emissions of some non-CO_2 gases may be more easily achieved. (Chapters 1, 2, 4, 9)

21. Finally, we wish to emphasize that the CO_2 issue interacts with many other issues, and it can be seen as a healthy stimulus for acquiring knowledge and skills useful in the treatment of numerous other important problems. (Chapter 1)

1 Synthesis

Carbon Dioxide Assessment Committee

1.1 INTRODUCTION

For more than a hundred years scientists have been suggesting that slight changes in the chemical composition of the atmosphere could bring about major climatic variations. Since the turn of the century, the focus has been particularly on worldwide release of carbon dioxide (CO_2), as a result of burning of coal, oil, and gas and changes in land use that release CO_2 from forests and soils.* In recent decades many aspects of the argument that enough CO_2 will be released to bring about unwanted and unwonted changes in climate have been filled out and strengthened; at the same time, new questions about segments of the argument have arisen, and possible benefits have been identified, including directly favorable implications for plant growth from increasing CO_2.

At this stage in the history of the CO_2 question, many readers are familiar with its basic aspects, so we have limited this introduction to two fundamental points. The first is that CO_2, along with water vapor, ozone, and a variety of other compounds, is a key factor in determining the thermal structure of the atmosphere. These so-called "greenhouse" gases do not strongly absorb incoming radiation for most of the shortwave solar spectrum, but they are more effective absorbers of the long-wavelength (infrared) radiation of the Earth's surface and atmosphere (see Figure 1.1). The mix and distribution of the gases account in no small part for the generally hospitable climate of Earth and the inhospitable climate of other planets. Concern arises about human activities that release greenhouse gases because important absorption bands for CO_2 and other atmospheric gases are far from saturation; increasing the concentration of the gases will continue to affect the net emission or absorption of energy from a given layer of the atmosphere and thus the climate. The second fundamental point is that the atmospheric concentration of CO_2 is rising. Figure 1.2 shows an exceptionally accurate and reliable record of measurements

---

*See "Annex 2, Historical Note," for the early history of the CO_2 issue.

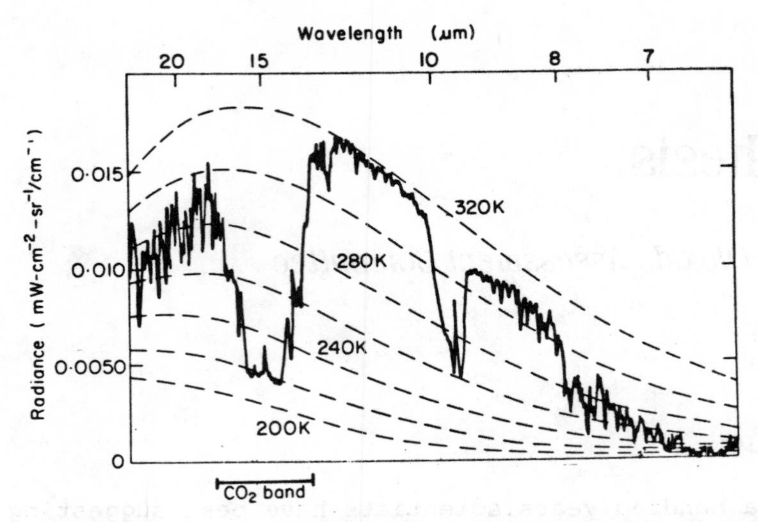

FIGURE 1.1 Infrared spectrum of the Earth as taken by a scanning interferometer on board the Nimbus-4 satellite over the North African desert. Also shown (dotted lines) are the blackbody radiances that would be observed at various temperatures. Thus, in the 10-13-µm region, the atmosphere is transparent and the radiance corresponds closely to that expected from the hot desert surface (at 320 K or 47°C). In the CO_2 band, however, the radiance is from the stratosphere at a temperature of 220 K, and energy from the Earth's surface is blocked. Other important infrared-absorbing trace gases include water vapor, nitrous oxide, methane, the chlorofluorocarbons, and ozone (in the troposphere). (From Paltridge and Platt, 1976, after Hanel et al., 1972).

starting in 1958. In short, there is a strong physical basis for attention to the CO_2 question in both theory and measurement.

Another, quite different, aspect of the CO_2 issue that requires introduction is that the time horizons of the subject and, therefore, of this report are very long. We talk about American agriculture in the year 2000, global energy use to 2100, and possible changes in sea level over the next three to five centuries. Is it meaningful to talk of such remote times? We think it is, and we have tried to devise approaches that take the time dimension seriously, although much of the report had to be speculative. Is it necessary to look so far into the future? Again, we think the answer is yes. Once the CO_2 content of the atmosphere rises significantly, it is likely to remain elevated for centuries; so from a physical point of view one must consider the long run. From the perspective of human activities, the time periods to be considered are also necessarily long. It takes many decades to replace the capital and infrastructure associated with a particular form of energy, and the time to develop large-scale water supply systems can be equally great. No policy, no matter how forceful, will make the issue of climate change disappear for at least decades to come. Finally, in considering the environment, one must think in terms of long-term

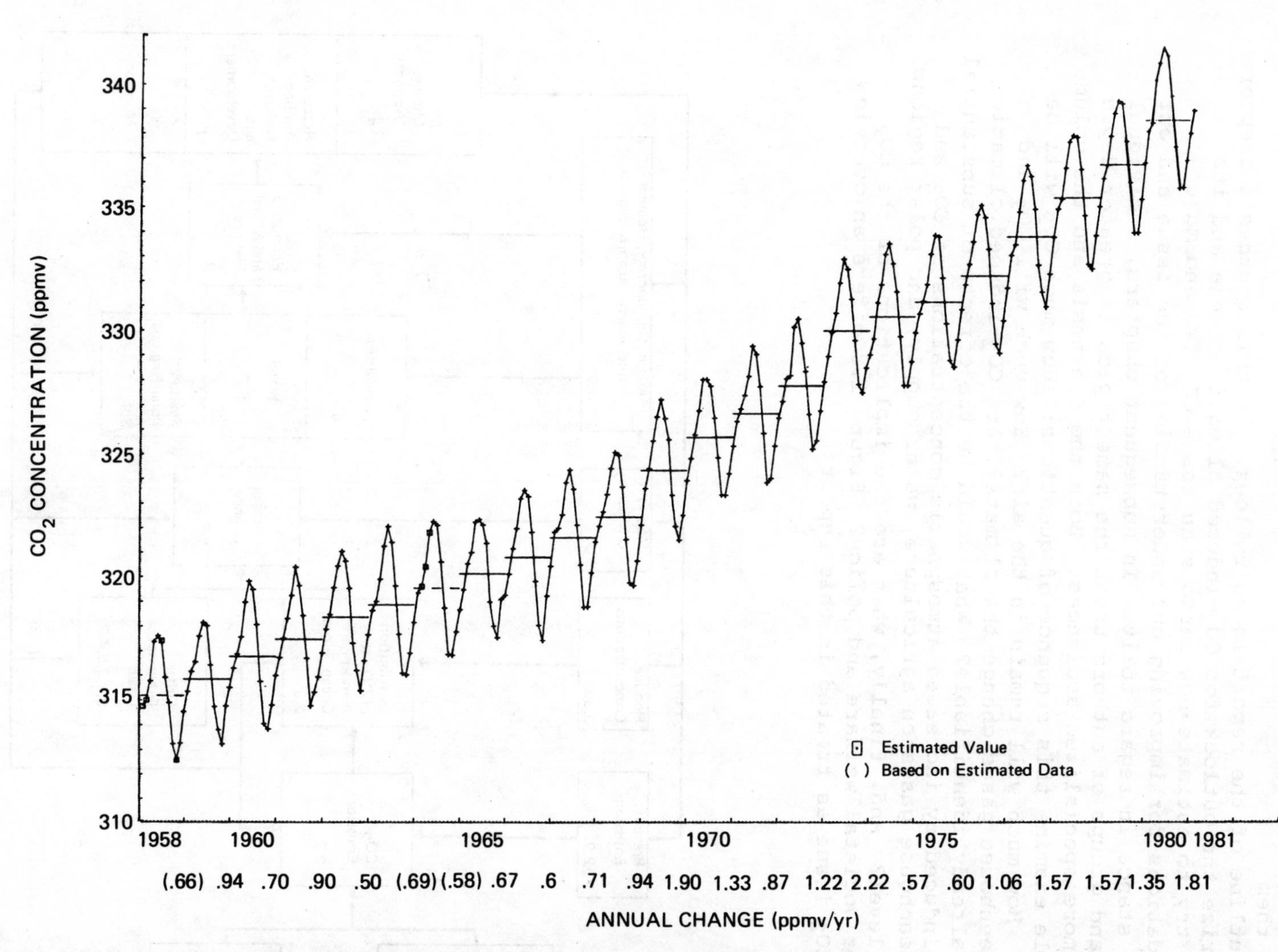

FIGURE 1.2 Mean monthly concentrations of atmospheric CO_2 at Mauna Loa. The yearly oscillation is explained mainly by the annual cycle of photosynthesis and respiration of plants in the northern hemisphere. See Section 1.2.2 for discussion of annual cycle. (Source: Geophysical Monitoring for Climate Change, National Oceanic and Atmospheric Administration.)

sustainable strategies. While adverse consequences of 100 years from now are obviously less pressing than those of next year, if they are also of large magnitude and irreversible, we cannot in good conscience discount them.

The outline of the report is as follows. In this Synthesis chapter we summarize the outlook for CO_2-induced climatic change and its effects, try to estimate how serious an issue CO_2 is, and make recommendations for improving our understanding of the issue and our societal stance in regard to it. In subsequent chapters, individual authors and groups of authors treat the same topics in greater depth and for more specialized audiences. Both the Synthesis and the volume as a whole examine this sequence of questions: How much CO_2 will be emitted? How much will remain in the air? How much will CO_2 and other greenhouse gases change the climate? Are CO_2-induced climatic changes already identifiable? What would be the effects of substantial warming induced by increased atmospheric concentrations of CO_2 and other greenhouse gases on agriculture, water supply, and polar regions and sea level? And, finally, what are the implications of the CO_2 issue for societal welfare and policy? Figure 1.3 offers an overview of the CO_2 issue as treated in this report.

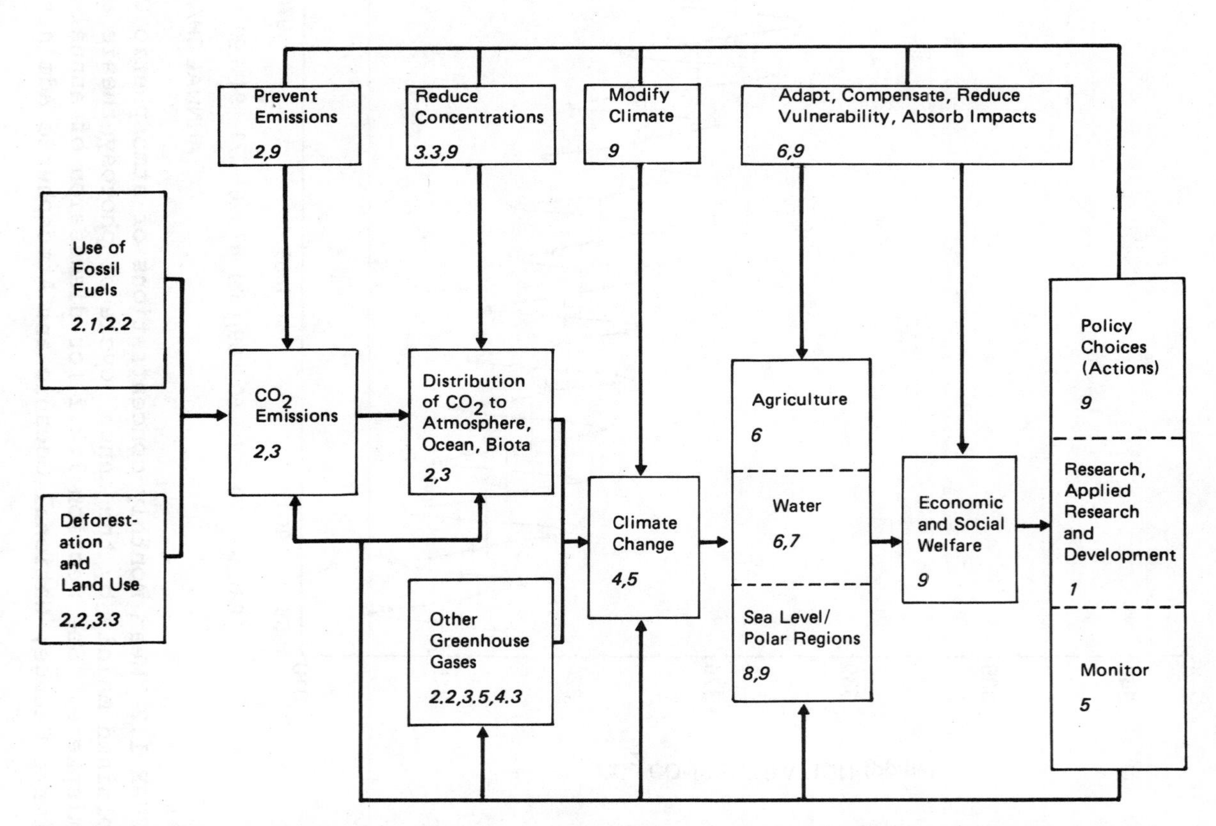

FIGURE 1.3 An overview of the CO_2 issue. Numbers refer to chapters or sections that focus on topic in box.

1.2 THE OUTLOOK

1.2.1 Future CO_2 Emissions

By far the largest potential sources of man-made CO_2 emissions are the fossil fuels, especially the abundant supplies of coal. Current annual fossil fuel emissions are estimated at about 5 x 10^9 tons of carbon (Gt of C) ± 10% (Marland and Rotty, 1983). In 1981, emissions came about 44% from oil, 38% from coal, and 17% from gas. The United States accounts for about one quarter of worldwide fossil fuel emissions, as do Western Europe and Japan, the Soviet Union and Eastern Europe, and developing countries.

To estimate future emissions of CO_2 from fossil fuels, Nordhaus and his co-authors adopted two approaches. One was to review previous global, long-range energy studies and use the range of projections as a guide to the uncertainty of scientific judgment (Ausubel and Nordhaus, this volume, Chapter 2, Section 2.2). The second approach (Nordhaus and Yohe, this volume, Chapter 2, Section 2.1), developed for this assessment, explicitly allows estimation of future emissions and their uncertainty based on a range of values for key parameters.

Review of previous energy studies shows that almost all studies applicable to estimation of CO_2 emissions project a continued marked growth of energy demand. For example, projections of energy demand in the year 2030 generally range from about 2-1/2 to 5 times the recent rate of energy use of 8 terawatt (TW, 10^{12} W) years per year. The studies vary so widely in quality, approach, level of detail, time horizon, data base, and geographic aggregation that strict comparisons are generally inappropriate. However, some generalizations may be ventured. Most studies looking beyond the year 2000 project average energy growth between about 2% and slightly above 3% per year, rates reasonably consistent with the 2.2% global annual average increase in primary energy consumption* that has prevailed over the past 120 years. Of course, the absolute range of projections spreads as the time horizon is extended, as a result of compounding the varied annual rates of increase. To illustrate, the range embracing almost all of the more detailed projections increases from 14-21 TW yr/yr in A.D. 2000 to 20-40 TW yr/yr a generation later. There are no strong signs of convergence toward a single, widely accepted projection or set of assumptions, although generally estimates have been lower in the last few years than in the 1970s. Figure 1.4 summarizes past global energy consumption and most of the long-range projections.

Combining estimates of energy demand and the mix of fuels leads to projections of CO_2 emissions. When the mix includes a large share of fossil energy, the projections show relatively high levels of CO_2 emissions. Figure 1.5 shows paths of CO_2 emissions derived from about a dozen long-range energy projections. Average annual rates of

---

*The 2.2%/yr figure includes wood and noncommercial energy sources (Marchetti and Nakicenovic, 1979; Nakicenovic, 1979).

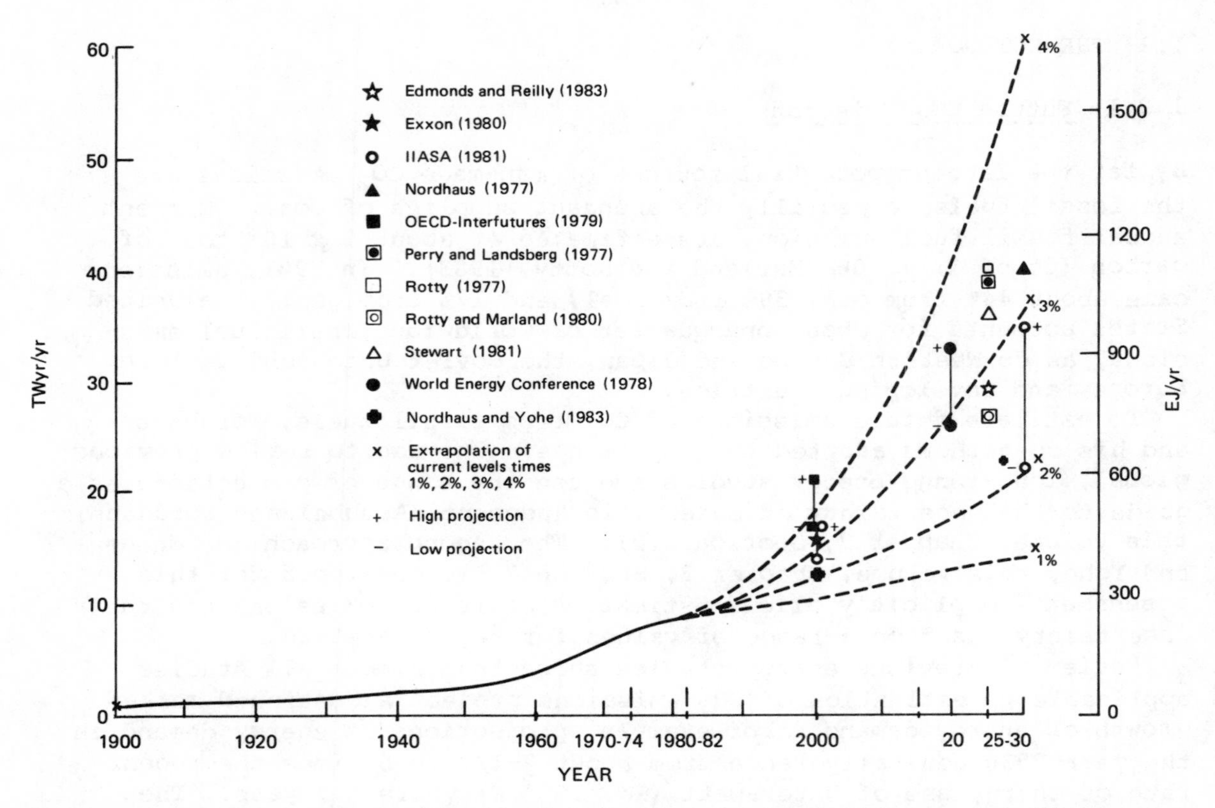

FIGURE 1.4 Past and projected global energy consumption. See Chapter 2, Section 2.2 for discussion of projections included here.

increase in CO_2 emissions to 2030 generally range from about 1 to 3.5%.* Estimated annual emissions range between about 7 and 13 Gt of C in the year 2000 and, with a few exceptions, between about 10 and 30 Gt of C in 2030. Thus, based on a review of past efforts, one might infer that energy consumption 50 years hence could differ by at least a factor of 2 and associated CO_2 emissions by a factor of 3 or more.

For purposes of understanding future outcomes and weighing policy choices, the past efforts reviewed leave open important questions. They generally do not allow a judgment as to the accuracy with which a forecast is made. It is of central importance in many policy problems to know not only the best judgment about an event (such as the time when atmospheric CO_2 will pass a certain level) but also to be able to estimate the degree of precision or approximation about that judgment. Some studies have approached the difficulties of forecasting by

*This range contrasts with the 4.3% figure for past and projected growth in CO_2 emissions that prevailed for several years in the literature on the CO_2 issue. The mean growth rate of fossil fuel CO_2 emissions over the past 120 years has more recently been estimated at about 3.5% per year (Elliott, 1983).

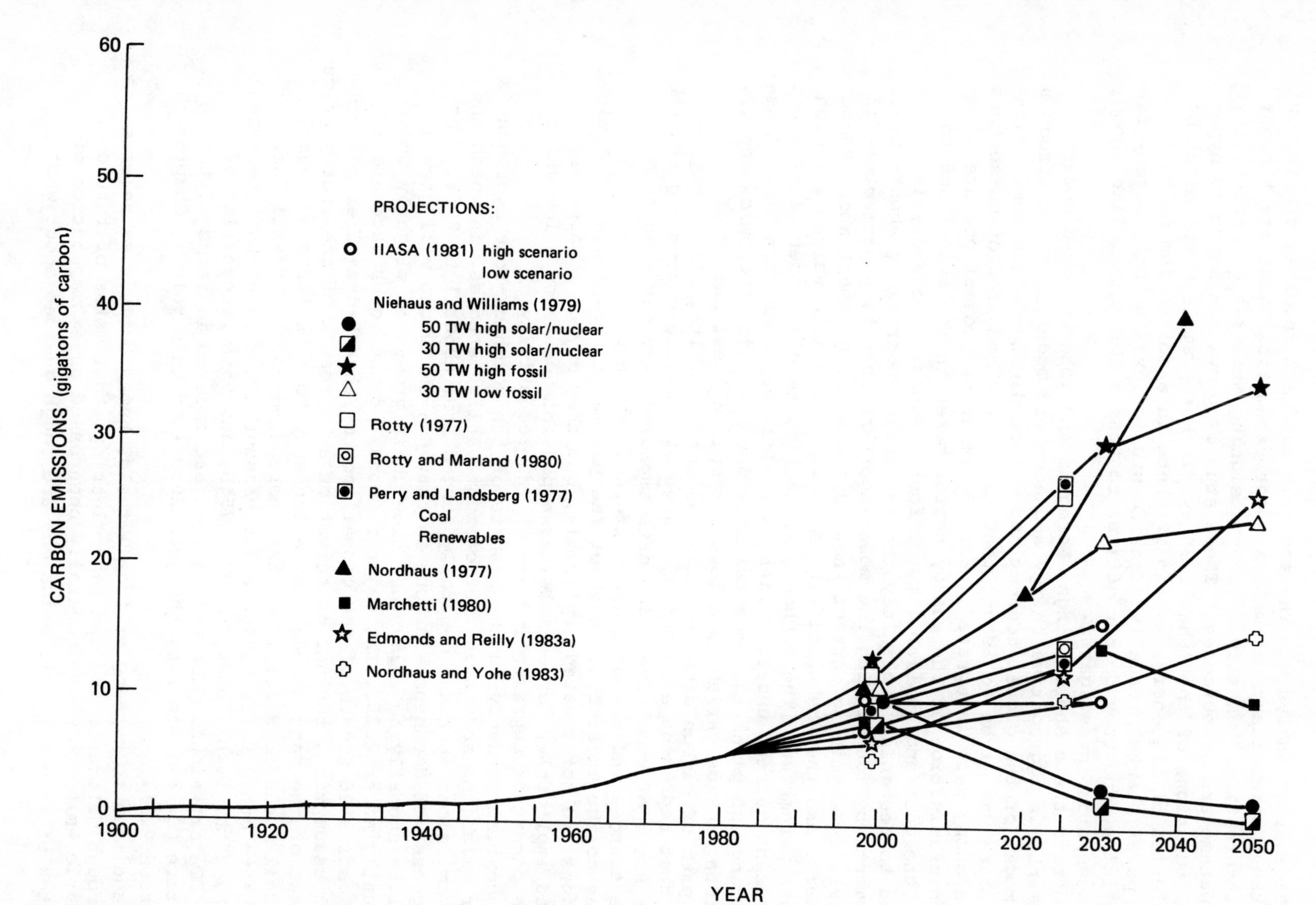

FIGURE 1.5 CO_2 emissions derived from long-range energy projections and historic production from fossil fuels. See Chapter 2, Section 2.2 for discussion of projections included here.

use of "scenario" analysis. The scenario approach involves tracing out time paths for important parameters under assumptions that are thought to be "interesting," usually without assigning measures of probability to the parameters or outcomes. These studies provide answers to hypothetical questions of the "what if?" type. For example, What might be the evolution of the energy system if there is a moratorium on building nuclear power plants? Usually scenario studies examine only a very few possibilities, and they do not attempt to assess the actual likelihood of the scenarios investigated.

To address these shortcomings, Nordhaus and Yohe employed modern developments in aggregative energy and economic modeling to construct a simple model of the global economy and carbon dioxide emissions. Particular care was given to assure that the energy and production sectors of the economy were integrated (most CO_2 emission projections are based on examination of the energy sector taken in isolation) and to respect the cost and availability of fossil fuels. The analysis attempts to recognize explicitly the intrinsic uncertainty about future developments by identifying the most important uncertain parameters of the model, by examining current knowledge and disagreement about these parameters, and then by specifying a range of possible values for each uncertain parameter. The emphasis was not to resolve uncertainties but to represent current uncertainties as realistically as possible. A use of the range of paths and uncertainties for the major economic, energy, and carbon dioxide variables allows not only a "best guess" of the future path of carbon dioxide emissions but also alternative trajectories that represent a reasonable range of possible outcomes given the current state of knowledge. The data employed were gathered from diverse sources and are of quite different levels of precision; judgments as to the uncertainties about the parameters are rough. Political conditions are not treated explicitly, but they may be regarded as included implicitly, for example, as a possible cause of a low value for the parameter representing growth in productivity.

The central tendency in the results of the Nordhaus-Yohe approach is a lower emissions rate than that of most earlier studies, in which the annual emissions increase generally ranged from about 1 to 3.5%. The "best guess" of Nordhaus and Yohe is that CO_2 emissions will grow at about 1.6% annually to 2025, then slow their growth to slightly under 1% annually after 2025. The major reasons for the lower rate are a slower estimated growth of the global economy than had earlier been the general assumption, further conservation as a result of the energy price increases of the past deacde, and a tendency to substitute nonfossil for fossil fuels as a result of the increasing cost of fossil fuels relative to other fuels. Figure 1.6 presents five paths that represent the 5th, 25th, 50th ("best guess"), 75th, and 95th percentiles of annual CO_2 emissions. The percentiles are indexed in terms of the cumulative CO_2 emissions by the year 2050 (see this volume, Chapter 2, Section 2.1).

In addition to burning of fossil fuels, human activities release CO_2 through deforestation and land clearing. Estimates of future biospheric emissions have generally been based on extrapolation of estimates of recent biotic emissions and rough guesses about what

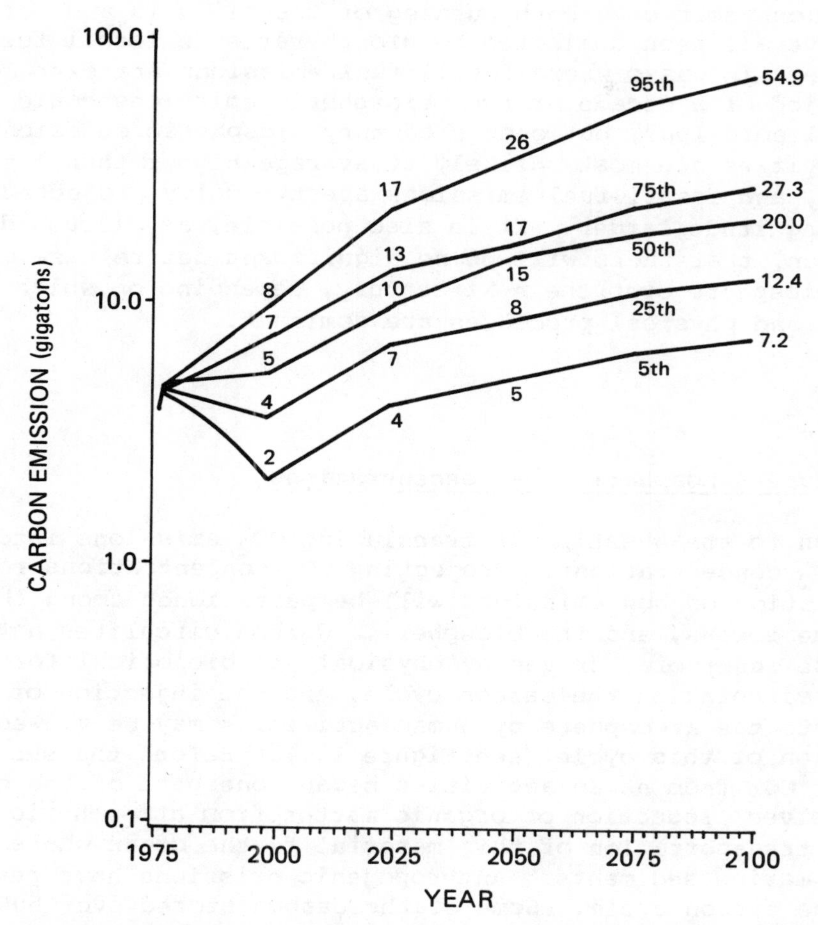

FIGURE 1.6 Carbon dioxide emissions from a sample of 100 randomly chosen runs. The 5th, 25th, 50th, 75th, and 95th percentile runs for yearly emissions, with emissions for years 2000, 2025, 2050, and 2100 indicated. See Chapter 2, Section 2.1, and Figure 2.17 for further detail.

proportion of carbon in the biosphere might be subject to human influence. Models of the carbon in forests and soils are now being developed that might provide projections with a stronger theoretical basis. An estimate for the maximum possible future addition from all biospheric sources is 240 Gt of C (Revelle and Munk, 1977), and Woodwell (this volume, Chapter 3, Section 3.3) offers a similar projection. Baumgartner (1979) estimates that clearing of all tropical forests might contribute about 140 Gt of C. The total carbon content of the Amazon forest is estimated at about 120 Gt of C (Sioli, 1973). Chan et al. (1980) develop a high deforestation scenario in which total additional transfer of carbon from the biosphere to the atmosphere by the year 2100 is about 100 Gt of C. The World Climate Programme (1981) group of experts adopted a range of 50 to 150 Gt of C for biospheric emissions in the 1980 to 2025 period. Projections of future atmospheric CO_2

concentrations embracing both burning of fossil fuels and terrestrial sources have all been dominated by growth rates in fossil fuel emissions, except in cases where fossil fuel emissions are extremely low. Over a period of a decade or two, biospheric emissions could rival fossil fuel emissions, but over a century biospheric emissions from human activities are most unlikely to average higher than 1 to 3 Gt of C per year, and fossil fuel emissions are typically projected to be an order of magnitude larger. It is also possible, as discussed in the next section, that there will be no significant net release of carbon from the biosphere over the next century, depending on which human activities and physical processes are dominant.

1.2.2 Future Atmospheric CO_2 Concentrations

We now turn to the question of translating CO_2 emissions into atmospheric CO_2 concentrations. Projecting CO_2 concentrations requires a determination of how emissions will be partitioned among the atmosphere, the oceans, and the biosphere. Carbon circulates naturally among these reservoirs driven by physical and biological forces; we term this circulation the carbon cycle, and the injection of carbon dioxide into the atmosphere by human activities may be viewed as a perturbation of this cycle (see Figure 1.7). Before the substantial release of CO_2 from human activities began, one part of the carbon cycle involved production of organic matter from atmospheric CO_2 and water and transportation of this material to the ocean where it was buried in marine sediments. Anthropogenic emissions have reversed this part of the carbon cycle. Some of the carbon stored over 500 million years in marine sediments is now returning to the atmosphere in a few short generations.

By means of quantitative models of the carbon cycle, we can estimate the effects of postulated rates of carbon dioxide injections on the pools and fluxes of carbon, in particular on the concentration of carbon in the atmosphere. Moreover, we can also assess the degree to which uncertainties in our understanding of one or another factor influence our forecasts of future atmospheric CO_2 levels.

The atmosphere forms a thin film over the Earth. Its composition is in large part the result of biological activity, and it is powerfully shaped by interaction with the oceans that cover some 70% of the globe. The concentration of CO_2 in the atmosphere has varied over the ages; for example, there is evidence that it may have been about 200 ppm during the last ice age 18,000 years ago. Reliable instrumental records are available only since 1957. Since then, the concentration has increased from about 315 ppm to slightly above 340 ppm and at an average rate of about 0.4% per year over the last decade (Figure 1.2). As discussed by Machta (this volume, Chapter 3, Section 3.4), carbon dioxide is well mixed in the atmosphere; measurements from the global network of sampling sites show relatively small spatial and temporal variations that are explainable largely in terms of fossil fuel sources

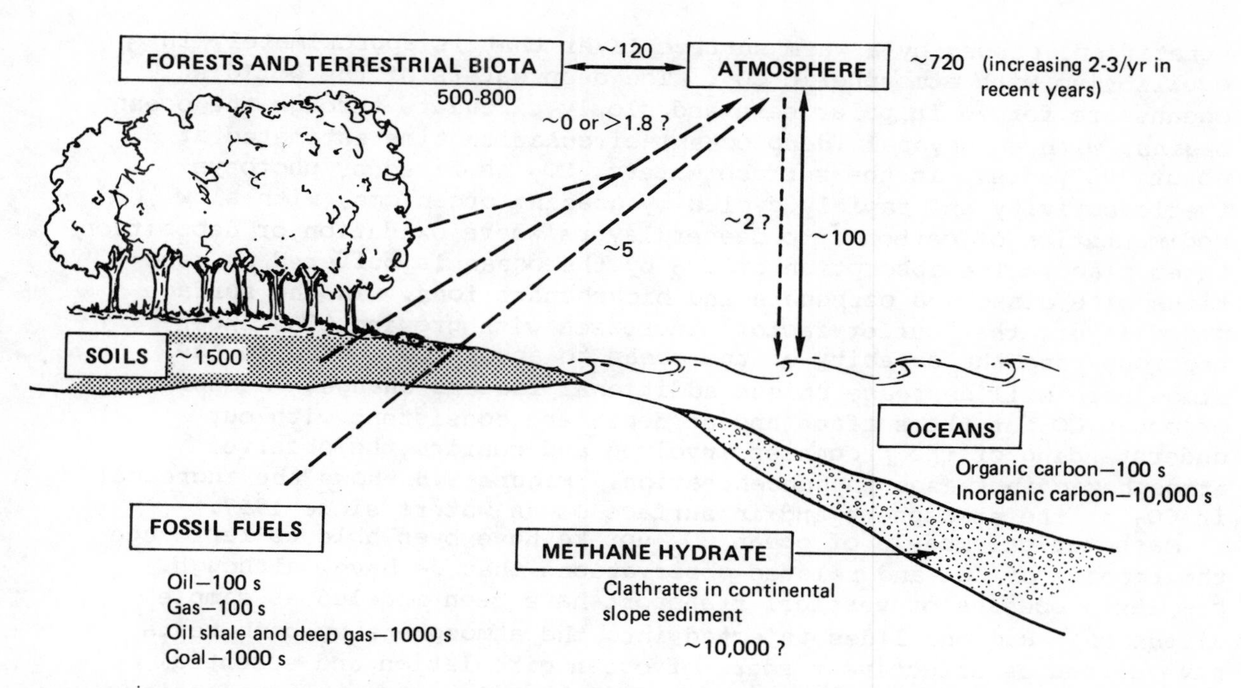

FIGURE 1.7 Some global carbon pools and annual fluxes. Estimated sizes of pools and fluxes are in gigatons of carbon. Estimates are rounded from figures given in Chapters 2 and 3 and from Clark, 1982. Pools that are only broadly measured are assessed here in order of magnitude, e.g., hundreds (100s). Dashed arrows represent additional fluxes due to human activities.

and biological, atmospheric, and oceanic processes, such as the annual cycle of the terrestrial biota and the quasi-periodic Southern Oscillation. Recently noted increases in the amplitude of the annual cycle may be related to changes in the cycling of CO_2 through photosynthesis and respiration. The year-to-year increases in atmospheric concentrations are generally becoming larger with time, roughly in step with emissions of CO_2 from fossil fuel combustion. Indeed, according to Machta (this volume, Chapter 3, Section 3.4), most of the atmospheric variations during the past 20 years are more easily accounted for when we consider a growing fossil fuel source alone than when we consider any significant additional source, such as deforestation.

Comparison of fossil fuel CO_2 releases and growth in atmospheric concentrations shows that a quantity somewhat larger than half the fossil fuel CO_2 has remained in the atmosphere. The rest must have been transferred to some other reservoir, primarily to the ocean. As described by Brewer (this volume, Chapter 3, Section 3.2), the capacity of the ocean as a sink for CO_2 is a function of its chemistry and biology, and the rate at which its capacity can be brought into play is a function of its physics. The low- and mid-latitude oceans are stably stratified, capped by a warm surface layer that is approximately in equilibrium with atmospheric CO_2. The deep waters of the world's oceans are formed in polar seas and slowly circulate through the ocean basins, with an abyssal (deep ocean) circulation time estimated at about 500 years. In the surface waters, CO_2 is fixed by photosynthetic activity and rapidly cycled by grazing organisms, with slow sedimentation of carbon into deeper layers where oxidation or deposition takes place. The absorption of CO_2 by the ocean is buffered by reactions with dissolved carbonate and bicarbonate ions. In the surface mixed layer, the "buffer factor" increases with growing CO_2 concentrations, and the capacity of the ocean to absorb CO_2 added to the atmosphere will decrease unless additional factors change. Measurements of ocean CO_2 at the surface and at depth are consistent with our understanding of the processes involved and confirm the observed atmospheric increases in concentration. Figure 1.8 shows the increase in CO_2 in the atmosphere and in surface ocean waters since 1957.

Mathematical models of ocean CO_2 uptake have been able to reproduce the records of CO_2 and related observations that we have, although complex processes of vertical transport have been modeled as simple diffusion. Radionuclides injected into the atmosphere by bomb tests have served as effective tracers of ocean circulation and essential empirical calibrators of these ocean models. However, the expression of all oceanic physics as a single unrealistic process hardly inspires confidence in the models' ability to deal with altered climatic regimes in the future. Moreover, potentially significant processes, such as changing riverine fluxes of terrestrial carbon and nutrients to the sea, may not yet have been adequately evaluated and represented. Nevertheless, at least for the short run, present-day models appear satisfactory to answer the question of how much anthropogenic CO_2 is taken up by the ocean. Brewer (this volume, Chapter 3, Section 3.2) reports several estimates that give a value at the present time of about 2 Gt of C/yr, or 40% of fossil fuel emissions.

The terrestrial biota and soils contain about three times as much carbon as the atmosphere, and their changes could influence the atmospheric burden. The most active and vulnerable portion of the biota is in forests, which probably contain between about 260 and 500 Gt of C (Olson and Watts, 1982). As discussed by Woodwell (see Figure 1.9 and Chapter 3, Section 3.3), the net flux of carbon between the atmosphere and any ecosystem depends on the balance between photosynthetic production by green plants and respiration by both plants and other organisms. Woodwell observes that on land photosynthesis is more susceptible to disturbance than respiration, so disturbance of the biota tends to release carbon into the atmosphere in the period following disturbance. Subsequently, over a period of years to decades or longer, the balance may shift through recovery due to succession and the slow migration of plants in response to a new, stable environment. Increased CO_2 enhances photosynthesis but does not necessarily lead to increased storage of carbon; on the other hand, increased temperature tends to increase respiration. Extension of growing seasons and extension or reduction of biomes with climate change can also affect the terrestrial carbon balance, particularly on longer time scales. It is thus

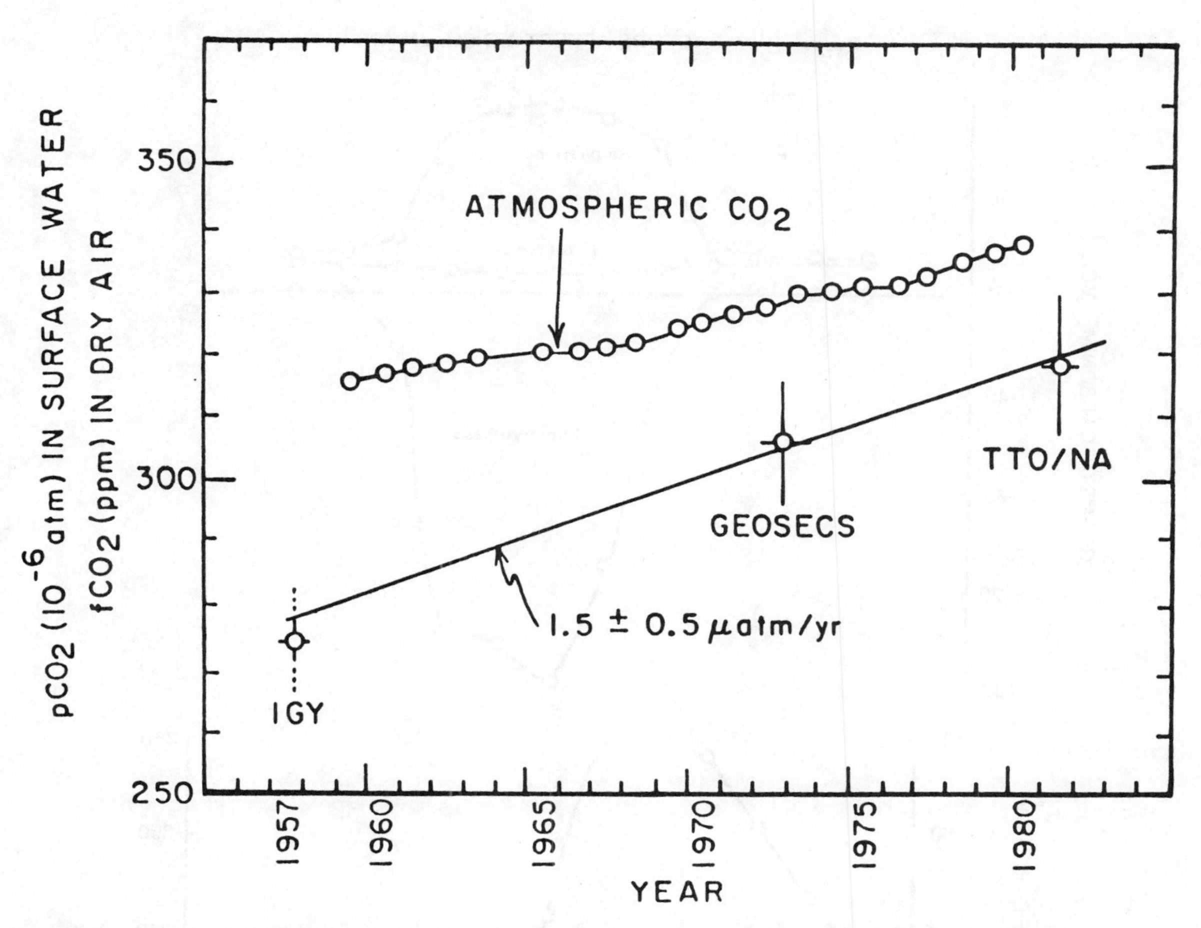

FIGURE 1.8 Mean Sargasso gyre surface water pCO_2 versus time. The atmospheric CO_2 concentration is expressed in mole fraction CO_2 in dry air at Mauna Loa, Hawaii. The IGY was the International Geophysical Year; GEOSECS is the Geochemical Ocean Sections Study; TTO/NA refers to the program of observation of transient tracers in the North Atlantic. (Source: Takahashi et al., 1983.) See Brewer, Chapter 3, Section 3.2 for further explanation.

plausible that future changes in the atmosphere could lead to a significantly increased net biotic flux of carbon to or from the atmosphere.

The possible sources of biotic CO_2 emissions--deforestation and land disturbance--are poorly documented. There are several indirect approaches to estimating what this biotic contribution may have been, largely based on correlation between growth in human population and the rate of conversion of forest for agriculture. The main direct data sources on deforestation have been the production yearbooks of the Food and Agriculture Organization (FAO) of the United Nations, published since 1949, and these may be inaccurate. All examinations of the biota conclude that there has been a marked reduction in storage of carbon over the past century or so; however, the timing and amount are the

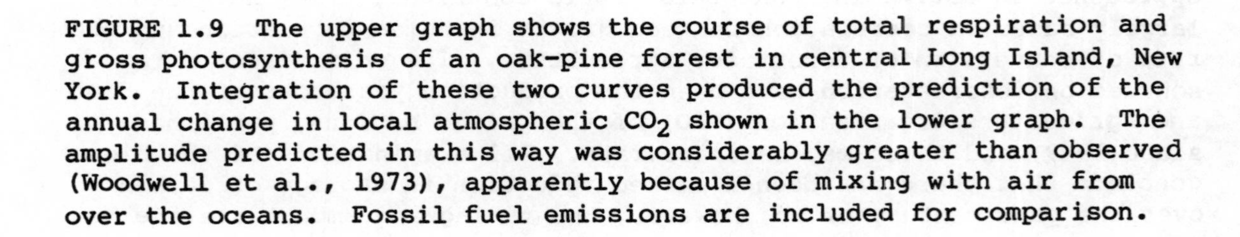

FIGURE 1.9 The upper graph shows the course of total respiration and gross photosynthesis of an oak-pine forest in central Long Island, New York. Integration of these two curves produced the prediction of the annual change in local atmospheric CO_2 shown in the lower graph. The amplitude predicted in this way was considerably greater than observed (Woodwell et al., 1973), apparently because of mixing with air from over the oceans. Fossil fuel emissions are included for comparison.

subject of substantial dispute. Based on a model of the terrestrial carbon cycle, Woodwell (this volume, Chapter 3, Section 3.3) reports a range of 1.8 to 4.7 Gt of C per year at present. Other researchers have suggested lower figures (see Table 3.3). All but the lowest values in Woodwell's range are impossible to reconcile with atmospheric observations and present-day ocean models. A net biotic carbon source of about 2 Gt/yr suggests that about 40% of all (fossil fuel plus biotic) CO_2 emissions have been remaining in the atmosphere, whereas a net biotic source near zero suggests about 60% of all anthropogenic CO_2 emissions remain airborne.

There is also disagreement about whether significant regrowth of forests in some areas and stimulation of plant growth ("fertilization") by increased atmospheric CO_2 are taking place, perhaps countering losses from deforestation. While the increasing amplitude of the seasonal cycle may be interpreted as an indication of regrowth and stimulation, there is as yet no direct evidence of regrowth and stimulation of net ecosystem storage of carbon sufficient to balance the apparent effects of deforestation. At the plant level, there are arguments for increased growth; but at the community level, especially in forests, the relevant mechanisms may be limited by other factors (see Chapter 3, Section 3.3, and Chapter 6). Growth of much biomass may be limited by availability of land, solar radiation, water, and nutrients other than carbon. Inventories of biota and other measurements (e.g., width of tree rings) are not sufficient at present to resolve the debate.

Finally, there is dispute about the level of CO_2 in the atmosphere in the last century before anthropogenic sources began to raise it. The preindustrial (circa 1850) concentration probably lay in the range 250-295 ppm, with 260-280 ppm a preferred interval (see Machta, this volume, Chapter 3, Section 3.4). Backward extrapolation of contemporary observations based solely on estimated fossil fuel CO_2 injections and a constant airborne fraction leads to an estimate of about 290 ppm at the turn of the century. Chemical and other measurements made at that time span a range about the same value. Inferences from air trapped in glacial ice and from deep-ocean measurements indicate concentrations in the vicinity of 265 ppm in the middle of the last century. The discrepancy between mid- and late-nineteenth century data, if real, might partly be accounted for by emissions from the terrestrial biosphere.

For the period since about 1950, we have records of CO_2 emissions from fossil fuels that are reliable within 10-15%. For the period before this, the record of fossil fuel emissions is less reliable.

In view of the uncertainty and conflicting evidence about the preindustrial concentration, fossil fuel emissions, the biotic contribution, and ocean uptake, one reasonable approach is to use models to project different outcomes based on different assumptions about the various reservoirs and mechanisms (see Machta, this volume, Chapter 3, Section 3.6). In the last few years, carbon cycle models, like energy models and climate models, have become more sophisticated. There are now several dynamic, process-oriented models that are making progress in representing, for example, accumulation and decay of dead vegetation; processing of carbon in soils and humus; and the biology, chemistry, and physics of the oceans. Published models have been calibrated to agree well with what is known about recent CO_2 trends, but no model has been properly validated against all trends and all data on emission rates.

The uncertainty of future projections may be examined by comparing different models or by varying parameters within a single model. Study of a group of models, each individually plausible, shows substantial agreement in projections for a fixed scenario of CO_2 input to the atmosphere. Maximum deviation of the lowest and highest concentration from the average among five models is less than 10%. Variation of parameters within plausible ranges in a single model shows at most about 30% variation from the mean (see Table 1.1). Thus, if current carbon-cycle models are accepted as valid representations of reality, reasonable variations in their parameters do not significantly affect predictions of future concentrations of CO_2; research simply to refine these parameters may not be effective in reducing uncertainties.

On the other hand, if net releases of CO_2 from the biosphere comparable to those from fossil fuels are now in progress and have been for the past several decades, the question of carbon-cycle modeling is different. If, for example, CO_2 released annually from deforestation were in the upper part of the range that Woodwell suggests, the current models would fail to reproduce the observed atmospheric CO_2 growth

TABLE 1.1 Sensitivity Study Using a Box Model of the Carbon Cycle (Keeling and Bacastow, 1977) and the Nordhaus-Yohe 50th Percentile CO_2 Emissions Estimate

Variation in Parameter	Range[a] (ppmv)	Range[b] (%)
Rate of exchange between air and sea		
2X and 0.5X standard rate of exchange	2	0.3
Rate of exchange between mixed layer of the ocean and the deep ocean		
2X and 0.5X standard rate of exchange	70	9
Both of above taken together	74	10
Biospheric uptake due enhanced atmospheric CO_2[c]		
No uptake and a β value of 0.266	229	29
Buffer factor		
Constant (10) and variable according to predicted oceanic chemistry change	61	8

[a]The Range is the higher minus the lower predicted by the changes in arithmetic number used for the parameter in the year 2100.
[b]Range divided by 784 ppmv, the predicted value for the year 2100, times 100.
[c]This is the CO_2 fertilization effect. 0.26 is the standard value for the so-called β-factor.

after 1958. The models would most likely have to be modified, since no reasonable adjustment of the parameters will allow a good fit of predictions to observations after 1958. The airborne fraction--the ratio of atmospheric increase in a year to the net (fossil and biotic) amount added to the atmosphere--would drop to about 0.3 from the value of almost 0.6 that is consistent with a net CO_2 release from the biosphere near zero. Increases of predicted concentration would be accordingly slower.

While the sophistication of carbon-cycle models has been increasing, their predictive capability may diminish markedly as we depart from current CO_2 concentrations, reservoir sizes, and climate conditions. For example, the terrestrial biotic reservoir of carbon may increase or decrease in response to climate change as a result of warming, longer growing seasons, and change in rainfall patterns, for example. No model contains a satisfactory long-term treatment of climate feedbacks to the biosphere. For a decade, most carbon-cycle models in estimating biotic response have depended on the so-called Beta (β) factor, a measure of how much plant growth increases as a result of increase in atmospheric CO_2 concentration. The use of the β factor needs to be replaced by separate analyses of effects of changes in the area of forests and potential changes in net ecosystem production caused by both increased atmospheric CO_2 and changes in climate. In sum, the current generation of carbon-cycle models appears to be satisfactory for forecasting concentrations in the next few decades, but credibility of the models fades as concentrations rise.

Keeping in mind the state of the art of projecting CO_2 emissions and their partitioning among different reservoirs, we now report the estimates of Nordhaus and Yohe (this volume, Chapter 2, Section 2.1) and Machta (this volume, Chapter 3, Section 3.6) on possible future atmospheric concentrations and factors that affect them. Perhaps the most useful graph to study is Figure 1.10, which shows the percentiles of CO_2 concentrations for the different Nordhaus-Yohe emission trajectories. To calculate concentrations, Nordhaus and Yohe use an estimate of 0.47 as the fraction of current emissions that remains airborne during the first year after emission. (This amount is consistent with a historic average annual contribution of CO_2 from the biosphere of about 1 Gt of C.) For a quarter or half century, the inertia built into the world economy and carbon cycle leaves an impression of relative certainty about outcomes. After the early part of the next century, however, the degree of uncertainty becomes extremely large. The time at which CO_2 concentrations are assumed to pass 600 ppm, the conventional "doubling" of the concentration representative of the beginning of the twentieth century, can be shown as follows:

Percentile	Doubling Time
5	After 2100
25	2100
50	2065
75	2050
95	2035

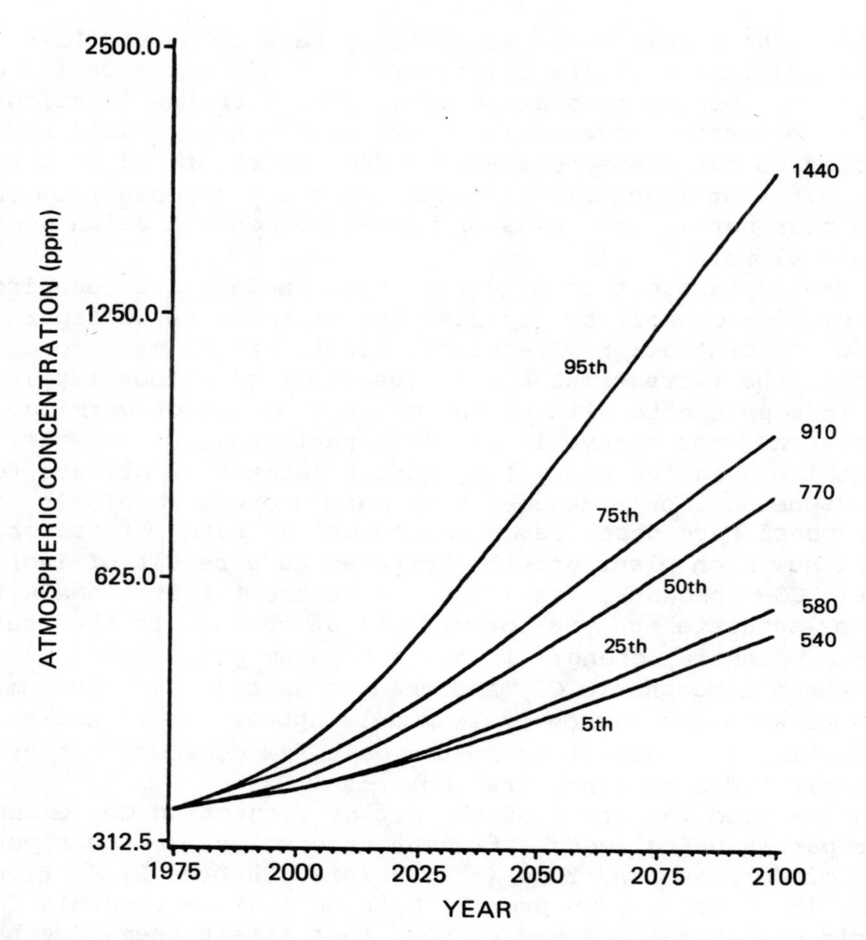

FIGURE 1.10 Atmospheric concentration of carbon dioxide in parts per million. The indicated percentile runs for concentrations; the numbers on the right-hand side indicate concentrations in the year 2100 for each run. See Nordhaus and Yohe, Chapter 2, Section 2.1, and Figure 2.18 for further detail.

On this result, Nordhaus and Yohe base a central conclusion: Given current knowledge, odds are even whether the doubling of carbon dioxide will occur in the period 2050-2100 or outside that period. It is a 1-in-4 possibility that doubling will occur before 2050 and a 1-in-20 possibility that doubling will occur before 2035. The median estimate for passing 600 ppm is 2065. For the year 2000, the most likely concentration is 370 ppm, with an upper limit of about 400 ppm.

Nordhaus and Yohe also address the question of the relative importance of different uncertainties in making concentration projections. Table 1.2 displays the contribution to overall uncertainty made by individual variables or parameters, calculated as the uncertainty induced when a parameter takes its full range of uncertainty and all other parameters are set equal to their most likely values.

TABLE 1.2 Indices of Sensitivity of Atmospheric Concentration in 2100 to Uncertainty about Key Parameters[a] (100 = Level of Effect of Most Important Parameter[b])

	Marginal Variance from Most Likely Outcome
Ease of substitution between fossil and nonfossil fuels	100
General productivity growth	79
Trends in real costs of producing energy	73
Ease of substitution between energy and labor	70
Airborne fraction for CO_2 emission	62
Extraction costs for fossil fuels	56
Population growth	36
Fuel mix among fossil fuels	24
Trends in relative costs of fossil and nonfossil fuels	21
Total resources of fossil fuels	5

[a]For full explanation of parameters see Chapter 2, Section 2.1.2.
[b]Value of sensitivity is scaled at 100 for the parameter that has the highest marginal variance.

The ranking of the importance of uncertainties shown in Table 1.2 contains several surprises. The most important parameters are those relating to future production trends, and the ease with which it is possible to substitute nonfossil sources of energy (e.g., uranium) for fossil sources (e.g., coal) is at the top of the list; several of these parameters have rarely been noted as factors affecting future CO_2 trends. Another surprise concerns two parameters that have been extensively discussed in the CO_2 literature: the extent of world resources of fossil fuels and the carbon cycle (based on a range for the airborne fraction of between 0.38-0.59).* The Nordhaus-Yohe estimates indicate that in projecting future CO_2 concentrations uncertainty about resource inventories is trivial. Uncertainty about the airborne fraction is of intermediate significance.

*The airborne fraction for the last 25 years may have been less than 0.38 if the higher figures reported by Woodwell in Chapter 3, Section 3.3, are accepted, and it could be greater than 0.59 in the future if the figures Woodwell reports are significant overestimates and the buffering capacity of the ocean declines; however, this range is a treatment of uncertainty roughly comparable with that used in Nordhaus-Yohe for the other parameters, which could also have values outside the range employed (see Chapter 2, Section 2.1).

Machta (this volume, Chapter 3, Section 3.6) employs a carbon-cycle model to consider CO_2 from deforestation as a possible real and important source of atmospheric CO_2. If some reasonable amounts of future CO_2 from deforestation are added to CO_2 from future fossil fuel combustion, the error that would be introduced by the omission of the future deforestation CO_2 would be small in the year 2100 assuming, say, a 2% per year growth rate in fossil fuel CO_2 after 1980. To give an extreme example, oxidizing 300 Gt of C, or about half the terrestrial biota, would result in an increase of perhaps 75 ppmv in a predicted value of about 1,000 ppmv in the year 2100 or about 12% of the total increase. If the rate of emissions from fossil fuels is slower, then biotic emissions could account for a somewhat more significant share of overall increase.

Nordhaus and Yohe have also made extremely tentative estimates of the effect of energy-sector policies designed to reduce the burning of fossil fuels, in particular the imposition of fossil fuel taxes, set for illustrative purposes at $10 per ton of coal equivalent. The taxes lower emissions noticeably during the period in which they are in place, but their effect on concentrations at the end of the twenty-first century is small. These examples suggest that use of carbon dioxide taxes (or their regulatory equivalents) will have to be very forceful to have a marked effect on carbon dioxide concentrations.

A review of several studies (Chapter 2, Section 2.4), all quite tentative, shows that if fossil fuel growth rates are 1 or 2%/yr and concentrations of 400-450 ppm are judged acceptable, there is little urgency for reductions in CO_2 emissions below an uncontrolled path before A.D. 1990. The review suggests that if a limit in the vicinity of 450-500 ppm is desirable steps to reduce emissions below an uncontrolled path would need to be initiated around A.D. 2000.

Along with considering the climatic implications of increased CO_2, it is important also to take into account other possible man-made changes in atmospheric composition (see Machta, this volume, Chapter 4, Section 4.3). Reliable measurements have now shown that background concentrations of several radiatively active gases besides CO_2 have increased worldwide in the 1970s. These include the chlorofluorocarbons CF_2Cl_2 and $CFCl_3$, N_2O (nitrous oxide), CH_4 (methane), and ozone (in the troposphere). Since these gases also absorb and emit thermal radiation, their effects on climate may add to those of CO_2.

Chlorofluorocarbons. This class of gases originates from industrial activities and has been emitted to the atmosphere during the past 50 years. These gases are increasing in the atmosphere approximately as expected from their growth in emissions. CFC-11, CFC-12, and CFC-22, the three most abundant ones, all have long residence times in the air (tens of years) so that they can accumulate. Both the sources and sinks of the chlorofluorocarbons are believed to be known. The emissions from industrial production and products (such as aerosol propellants) represent the only source of any consequence. Photochemical destruction, mainly in the stratosphere, and very slow uptake by the oceans are the only known significant sinks. Theoretically, chlorofluorocarbons are implicated as potential destroyers of stratospheric ozone, the destruction of which in turn could result in damage to human health and the environment from increased ultraviolet radiation. Since emissions of these gases may be increasingly restricted, an extrapolation of current or past growth rates of chlorofluorocarbons to predict future atmospheric concentrations may be misleading at this time.

Nitrous oxide. It is likely that most nitrous oxide in the air has come from denitrification in the natural or cultivated biosphere. One would therefore expect to find the largest part of atmospheric nitrous oxide to be derived from nature, unrelated to human activity. Recent, careful measurements have suggested a small growth rate of the concentration of nitrous oxide in ground-level air at remote locations. The source of the small increase is unknown, but prime candidates are the continued expanded use of nitrogen fertilizers around the world to improve agricultural productivity and high temperature combustion in which atmospheric nitrogen is oxidized. If such fertilizers are the source, the current slow increase is likely to continue into the foreseeable future because the demand for food will grow with population size.

Methane. The most abundant hydrocarbon, methane, often called natural gas, is increasing in the atmosphere. It is thought to be a natural constituent of the air arising as it does from many biological processes and from seepage out of the Earth. Measurements in the 1950s and 1960s were imprecise, and there were spatial differences so that the observed temporal variability was not viewed as an upward trend. However, in the late 1970s several investigators using gas chromatography have unequivocally demonstrated an upward trend.

Increase in the number of ruminant farm animals and expansion of rice production might well explain, at least qualitatively, the atmospheric methane growth. Other biological activities, such as termite destruction of wood, and possible leakage from man's mining and use of fossil methane might also contribute to methane in the air, but their contribution to its increase is less clear. The higher concentrations far north of the equatorial region suggest that the termite source may be minor. The relatively rapid recent increase with time, about as fast as for CO_2, combined with the uncertainty as to its origin, are both intriguing features of the methane growth in air.

There is no reason to expect the upward trend in atmospheric methane concentration to stop soon since the most likely sources of methane are related to population size. In the long run, those sources that are dependent on the size of a biospheric feature (e.g., cows or rice paddies) will ultimately be limited by space. Thus, the growth in atmospheric concentration from these sources might continue for many decades but perhaps not for many centuries.

Methane also forms another link in the question of future atmospheric composition. As discussed by Revelle (this volume, Chapter 3, Section 3.5), large amounts of methane are believed to be stored in methane hydrates in continental slope sediments. Methane hydrate is a type of clathrate in which methane and smaller amounts of ethane and other higher hydrocarbons are trapped within a cage of water molecules in the form of ice. Methane hydrates are stable at low temperatures and relatively high pressures. With a rise in ocean-bottom temperatures, the uppermost layers of ocean sediments would also become warmer; and methane hydrates would become unstable in the upper limit of their depth range, that is, about 300 m in the Arctic and about 600 m at low latitudes. The quantity of clathrates that will be released from sediments under the seafloor as a result of ocean warming depends on the distribution of clathrates with depth and on their abundance in the sediments. Estimates of total amount by different authors differ by a factor of 500, from 10^3 to 5×10^5 Gt of C. Revelle, assuming a warming induced by CO_2 and other trace gases released by human activities and a stock of about 10^4 Gt of C, estimates that the resulting increase in atmospheric methane toward the latter part of the twenty-first century could be two thirds to four thirds of the current amount; the uncertainties in the series of methane calculations are so great that the result cannot be thought of as a projection for the future, but it is equally obvious that we must be attentive to the possibility of new, important feedbacks affecting the chemical composition of the atmosphere.

Tropospheric ozone. Tropospheric ozone was originally believed to be primarily a consequence of transport from stratospheric ozone by air motions. It can also be created within the troposphere by man and nature. Locally, as in the Los Angeles Basin, large amounts of ozone are derived from reactions among oxides of nitrogen, hydrocarbons, and sunlight. Few scientists believe that these local sources of pollution can increase the upper-tropospheric concentrations of ozone, because ozone is so reactive that its lifetime in the lower atmosphere is no more than a few days. Nevertheless, an analysis of a limited number of measurements suggests an upward trend. It has been suggested that increase of mid- and upper-troposphere ozone concentration in the northern hemisphere may result from photochemical reactions of the oxides of nitrogen and hydrocarbons emitted by high-flying jet aircraft. Since the lifetime of an ozone molecule in the upper troposphere is also relatively short, little accumulation takes place. An increase in concentration must therefore reflect a continual increase in aircraft emissions, if they are the source.

Some other gases. Several other gases being measured show upward trends and may have absorption lines in the infrared window of the electromagnetic spectrum, making them potential greenhouse gases, for example, carbon tetrachloride (CCl_4) and methyl chloroform (CH_3CCl_3). Very likely both of these gases have both natural and man-made sources. On the other hand, measurements at the Mauna Loa Observatory exhibit no or insignificant increases in carbon monoxide (CO). It is likely that the list of atmospheric gases studied for their trends and potential greenhouse effects will grow in years to come: the study of greenhouse gases other than CO_2 is still in its infancy.

The atmospheric concentrations of these trace gases are not all independent of one another. Complicated chemical reactions among them, as well as with other gases not particularly radiatively active, can affect their concentrations. In addition to chemical reactions among today's atmospheric components, there are likely to be new climate-chemistry interactions in the future. As the atmospheric composition changes, the expected higher atmospheric water-vapor content will further affect the atmospheric chemistry.

Unlike CO_2, which generally does not undergo chemical changes in the air, these trace gases frequently do. Not only can the mean concentration be affected by other chemicals and sunlight, but distribution particularly in the vertical can be influenced (ozone is a prime example). To estimate future concentrations will require more than estimates of natural and man-made emission rates, fundamental though those rates will be.

1.2.3 Changing Climate with Changing CO_2

With projections given of rising atmospheric concentration of CO_2 and other greenhouse gases, the next question to be addressed is that of possible effects on climate. The main tools for evaluating the possible role of greenhouse gases are numerical models of the climate system. Simplified models permit economically feasible analyses over a wide range of conditions, but they are limited in the information they can provide, for example, on regional climate change. More detailed inferences may be obtained primarily from three-dimensional general circulation models (GCMs), which represent the global atmospheric circulation, as well as the oceans, the land, and ice (Figure 1.11). Comparisons of simulated time means of a number of climatic variables with observations show that modern climate models provide a reasonably satisfactory simulation of the present large-scale global climate and its average seasonal changes. However, the capabilities of even the most advanced current models remain severely limited; for example, the three-dimensional GCMs are generally deficient in the treatment of ocean heat transport and dynamics and feedback between the ocean and the atmosphere.

The adequacy and results of climate model studies in the CO_2 context were examined in 1979 by an NRC panel chaired by the late Jule Charney, and again in connection with the present study by a panel led by Joseph Smagorinsky. Comprehensive reviews of modeling methods and results have been carried out by Schneider and Dickinson (1974) and international groups (Gates, 1979). The Smagorinsky panel also evaluated empirical approaches to assessing climate sensitivity. The earlier NRC panel reports form one basis of our assessment of the climatic implications of increasing CO_2. A second basis is the search for a "CO_2 signal" in the recent climatic record; this detection approach is discussed below in Section 1.2.4 and by Weller et al. in Chapter 5.

The Carbon Dioxide Assessment Committee did not extend the systematic treatment of uncertainty adopted in several sections of this report to its analysis of climate modeling results. However, Smagorinsky (this volume, Chapter 4, Section 4.2) notes that such approaches are now being initiated by climate researchers.

The primary effect of an increase of CO_2 is to cause more absorption and re-radiation of thermal radiation from and to the Earth's surface and thus to increase the air temperature in the lower troposphere. A strong positive feedback mechanism is the likely accompanying increase of moisture, which is an even more powerful

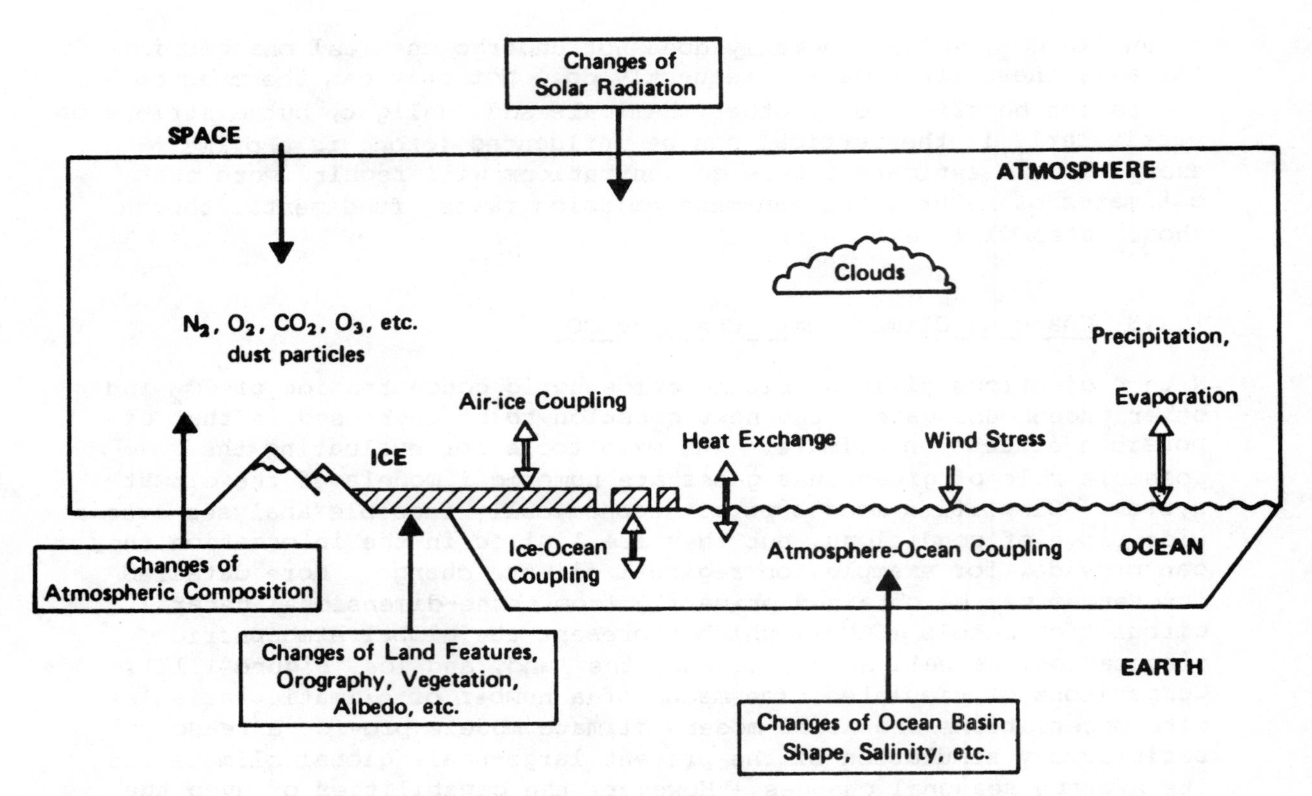

FIGURE 1.11 Schematic illustration of the components of the coupled atmosphere-ocean-ice-earth climatic system. The solid arrows are examples of external processes, and the open arrows are examples of internal processes in climatic change. Source: U.S. Committee for the Global Atmospheric Research Program (1975).

absorber of terrestrial radiation. None of the known potential negative feedback mechanisms, such as increase in the areal extent of low or middle cloud amount, can be expected to vitiate the principal conclusion that there will be appreciable warming, since they do not appear in most current models to be as strong as the positive moisture feedback. There is, however, always the possibility of some overlooked or underestimated factor.

When it is assumed that the CO_2 content of the atmosphere is doubled and statistical thermal equilibrium is achieved, all models predict a global surface warming. None of the calculations with more physically comprehensive models predicts negligible warming. Calculations with the three-dimensional, time-dependent models of the global atmospheric circulation indicate global warming due to a doubling of CO_2 from 300 ppm to 600 ppm to be in the range between about 1.5 and 4.5°C, as suggested in 1979 by the Charney panel. Simpler models that appear to contain the main physical factors give similar results.

Dissenting inferences suggesting negligible CO_2-induced climate change have been drawn in recent years, but careful review shows these studies to be based on incomplete and misleading analysis (National

Research Council, 1982; Luther and MacCracken, 1983). Investigations with a variety of climate models continue to show a broad range of estimates of appreciable warming.

A major uncertainty has to do with the transfer of increased heat into the oceans, which serve as a thermal regulator for the planet. The heat capacity of the oceans is potentially great enough to slow down the response of climate to increasing CO_2 by several decades and to cause important regional differences in response. The influences of clouds, aerosols, and land-surface processes on sensitivity of climate to increased CO_2 also remain poorly understood. Table 1.3 indicates the possible contribution of various processes to the sensitivity of simple climate models. For example, including the feedback provided by reduction in the extent of snow and ice (Model 5) increases the equilibrium surface temperature rise for doubled CO_2 in the model by a factor of 1.3-1.4.

TABLE 1.3 Equilibrium Surface Temperature Increase Due to Doubled CO_2 (300 ppm → 600 ppm) in One-Dimensional Radiative-Convective Models[a,b,c]

Model	Description[d]	ΔT_s (°C)	f	F (W m^{-2})
1	FAH, 6.5LR, FCA	1.2	1	4.0
2	FRH, 6.5LF, FCA	1.9	1.6	3.9
3	Same as 2, except MALR replaces 6.5LR	1.4	0.7	4.0
4	Same as 2, except FCT replaces FCA	2.8	1.4	3.9
5	Same as 2, except SAF included[e]	2.5-2.8	1.3-1.4	
6	Same as 2, except VAF included[f]	3.5	1.8	

[a]Source: National Research Council (1982).
[b]Data from Hansen et al. (1981).
[c]Model 1 has no feedbacks affecting the atmosphere's radiative properties. The feedback factor f specifies the impact of each added process on model sensitivity to doubled CO_2. F is the equilibrium thermal flux into the planetary surface if the ocean temperature is held fixed (infinite heat capacity) when CO_2 is doubled; this is the flux after the atmosphere has adjusted to the radiative perturbation within the model constraints indicated but before the surface temperature has increased.
[d]FRH, fixed relative humidity; FAH, fixed absolute humidity; 6.5LR, 6.5°C km^{-1} limiting lapse rate; MALR, moist adiabatic limiting lapse rate; FCA, fixed cloud altitude; FCT, fixed cloud temperature; SAF, snow-ice albedo feedback; VAF, vegetation albedo feedback.
[e]Based on Wang and Stone (1980).
[f]Based on Cess (1978).

Warming of the lower troposphere will be accompanied by regional shifts in the geographical distribution of the various climatic elements, such as temperature, rainfall, evaporation, and soil moisture. Indeed, variations with latitude, longitude, and season are likely in many cases to be more striking than globally averaged changes. Unfortunately, we cannot at present predict the magnitude and locations of regional climate changes with much precision or confidence. Although current models are not sufficiently realistic to provide reliable predictions in the detail desired, they do suggest scales and ranges of temporal and spatial variations. Along with global warming, the main conclusions of the model studies, discussed in greater detail in the report of the Smagorinsky panel (National Research Council, 1982) and in Chapter 4 of this report, may be summarized as follows:

- A cooling of the stratosphere with relatively small latitudinal variation is expected.
- Global-mean rates of both evaporation and precipitation are projected to increase.

With less confidence it is concluded that:

- Increases in surface air temperature would vary significantly with latitude and over the seasons:

  (a) Warming at equilibrium would be 2-3 times as great over the polar regions as over the tropics; warming would probably be significantly greater over the Arctic than over the Antarctic.

  (b) Temperature increases would have large seasonal variations over the Arctic, with minimum warming in summer and maximum warming in winter. In lower latitudes (equatorward of 45° latitude) the warming has smaller seasonal variation.

- Some qualitative inferences on hydrological changes averaged around latitude circles may be drawn from model simulations of large, fixed CO_2 increase in equilibrium:

  (a) Annual-mean runoff increases over polar and surrounding regions.

  (b) Snowmelt arrives earlier and snowfall begins later.

  (c) Summer soil moisture decreases in large regions in middle and high latitudes of the northern hemisphere.

  (d) The coverage and thickness of sea ice over the Arctic and circum-Antarctic oceans decrease.

As CO_2 slowly increases, land and ocean will both warm but at different rates. Eventually, with a steady CO_2 concentration, a new equilibrium would be reached with a climate different from today's. However, we should not expect that the detailed distribution and timing of changes during the intervening years can be obtained solely by interpolation. The differences between rapidly heated land and slowly warming water, coupled with the irregular distribution of continents and oceans and continually changing CO_2 concentrations, may bring varying "transient" patterns of climate change.

While the climatic effects of CO_2 have been explored with a wide variety of climate models, most of the estimates of the climate effects of other greenhouse gases have been based on simple one-dimensional radiative-convective models. Typically the calculation involves doubling a reference concentration of the gas (for the chlorofluorocarbons, increases from 0 to 1 or 2 ppb are used) while other constituents are held constant. Table 1.4 gives some estimates of the change in surface temperature due to either a doubling of their concentration or an increase from 0 to 1 ppb for the halocarbons. The table was adapted from Table 2a in the report of the World Meteorological Organization (1983). There are other published values, but they generally do not disagree by more than about ±30% with the figures given here.

The models used to obtain these results generally gave a sensitivity to doubled CO_2 between 2 and 3°C. None of the changes of individual gases by itself approaches CO_2, but it is clear that the summation of all of these potential changes could be of the same magnitude as CO_2. It is worth noting that because the concentration of each of these gases is small enough for their radiative effect to be treated as optically thin, the temperature effect is linearly proportional to

TABLE 1.4 Some Estimates of Surface Temperature Change Due to Changes in Atmospheric Constituents Other Than CO_2

Constituent	Mixing Ratio Change (ppb) From	To	Surface Temperature Change (°C)	Source[a]
Nitrous oxide (N_2O)	300	600	0.3-0.4	1,3
Methane (CH_4)	1500	3000	0.3	3,4
CFC-11 ($CFCl_3$)	0	1	0.15	1,5
CFC-12 (CF_2Cl_2)	0	1	0.13	1,5
CFC-22 (CF_2HCl)	0	1	0.04	7
Carbon tetrachloride (CCl_4)	0	1	0.14	1,5
Carbon tetrafluoride (CF_4)	0	1	0.07	2
Methyl chloride (CH_3Cl_3)	0	1	0.013	1,5
Methylene chloride (CH_2Cl_2)	0	1	0.05	1,5
Chloroform ($CHCl_3$)	0	1	0.1	1,5
Methyl chloroform (CH_3CCl_3)	0	1	0.02	7
Ethylene (C_2H_4)	0.2	0.4	0.01	1
Sulfur dioxide (SO_2)	2	4	0.02	1
Ammonia (NH_3)	6	12	0.09	1
Tropospheric ozone (O_3)	F(Lat,ht)	2 F(Lat,ht)	0.9	4,6
Stratospheric water vapor (H_2O)	3000	6000	0.6	1

[a]Sources: 1, Wang et al. (1976); 2, Wang et al. (1980); 3, Donner and Ramanathan (1980); 4, Hameed et al. (1980); 5, Ramanathan (1975); 6, Fishman et al. (1979); 7, Hummel and Reck (1981).

their concentration, whereas the CO_2 effect depends logarithmically on the concentration.

The implication of the prospective increases in trace gases is that climatic changes of the character expected for elevated CO_2 concentrations may be encountered sooner than if CO_2 were the only cause of change, or, alternatively, that estimates solely of a CO_2 effect may be conservative.

At present little or nothing can be estimated about changes in extreme conditions that might accompany changes in mean climatic conditions and more generally about the weather of future climate. We do not know, for example, whether the climate will show more or less year-to-year variability under generally warmer planetary conditions. Variability of climate is one of its most important features. Food production, human settlements, and numerous aspects of the environment are strongly influenced by occasional extreme episodes. An area of particular interest is that of severe storms. The frequency, severity, and track of hurricanes and other severe storms are likely to be affected by CO_2-induced climatic changes, such as warming of ocean waters. Neither our current knowledge of storm genesis nor the current capabilities of climate models are great enough to allow convincing linkages at this time.

Besides numerical modeling approaches, past climates and recent climate fluctuations have been studied empirically to discern possible regional patterns of climatic variation associated with elevated CO_2 levels or warmer mean temperatures. These studies can be useful in exploring the sensitivity of climate to various factors, in evaluating how well climate models perform, and in exploring the kinds of regional patterns of change that are possible. However, the search for a historical analogue to CO_2-induced climatic change is hampered by inadequacies in data and by the absence of close parallels of cause and effect. Maps of a warmer earth derived from analogue approaches should not be viewed as predictions of regional effects of CO_2-induced changes.

1.2.4 Detection of CO_2-Induced Changes

Given the historic increase in atmospheric CO_2 and the results of climate models concerning the effects of increasing CO_2, it is appropriate to ask whether climatic records tend to confirm model estimates. Weller et al. examine this question at length in Chapter 5 of this report. Observational verification of model-based predictions is also important in calibrating climate models so that we can attach more confidence to their predictions of future changes.

The most clearly defined change expected from increasing atmospheric CO_2 is a large-scale warming of the Earth's surface and lower atmosphere. Thus, a number of investigators have examined trends in globally or hemispherically averaged surface temperature for evidence of CO_2-induced changes. Although differing in detail because of varying data sources and analysis methods, the records of large-scale average temperatures reconstructed by a number of investigators are in general agreement for the period of instrumental records, i.e., about the last 100 years. Northern hemisphere temperatures, mostly measured over land in the 20-70° latitude zone, increased from the late nineteenth century to the 1940s, decreased until the mid-1970s, and have apparently increased again in recent years (Figure 1.12). If one selects the 1970s to compare with the 1880s, one finds that the mean temperature of the recent decade was about 0.5°C warmer; other selective examinations of the time series of northern hemisphere temperatures could show different results. To the extent that one can judge from scanty data, southern hemisphere temperatures have increased more steadily than in the north by about the same total amount. In view of the relatively large and inadequately explained fluctuations over the last century, we do not believe that the overall pattern of variations in hemispheric-mean or global-mean temperature or associated changes in other climatic variables either confirms or contradicts model projections of temperature changes attributable to increasing atmospheric CO_2 concentration.

Factors other than CO_2--such as atmospheric turbidity, solar radiation, and albedo--also influence climate. Attempts have been made to account for such influences on the temperature record and thereby make the sought-for CO_2 signal stand out more clearly. Unfortunately, only indirect sources of historical data are available. For example, stratospheric turbidity has been inferred primarily from volcanic activity, and solar radiance from phenomena such as sunspots. The quantitative reliability of these inferences is unknown.

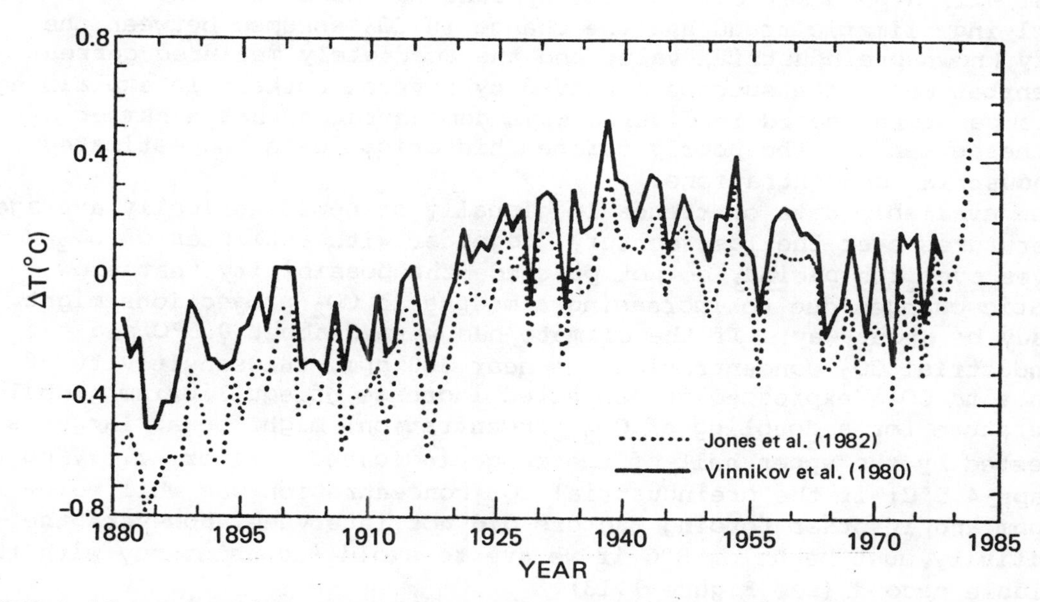

FIGURE 1.12 Comparison of the reconstructions of annual surface air temperature anomalies for the northern hemisphere from Jones and Wigley (1980) and Vinnikov et al. (1980). Figure from Clark (1982), but data for 1981 added to Jones and Wigley (Jones, 1981). See Weller et al., Chapter 5, for further discussion.

Despite these difficulties, a number of investigators, employing various combinations of data and methodology, have related the global or hemispheric mean temperature record to indices of turbidity and solar radiance and to estimates of the effect of increasing CO_2. Although good agreement between modeled and observed variations has been obtained in some of these studies, it is clear that considerable uncertainties remain. When attempts are made to account for climatic influences of such other factors as volcanic and solar variations, an apparent temperature trend consistent with the trend in CO_2 concentrations and simulations with climate models becomes more evident. However, uncertainties preclude acceptance of such analyses as more than suggestive. The studies done to date have been most helpful in raising questions, suggesting relationships, and identifying gaps in data and observations.

In essence, the problem of detection is to determine the existence and magnitude of a hypothesized CO_2 effect against the background of quasi-random climatic variability, which may be in part due to internal processes in the atmosphere and ocean and in part explainable in terms of fluctuations in other external factors. A reasonable approach is to assume that the record of some climatic parameter, e.g., temperature, is the sum of a hypothesized "natural" value, a perturbation due to CO_2, and a random component. The "natural" value may be taken as a constant long-run preindustrial mean or perhaps that mean corrected for the factors discussed above. The random component will have statistical characteristics different from simple "white noise" and will be difficult to model. It is clear that the magnitude of the derived CO_2 signal will depend markedly on the hypothesis chosen for the unperturbed underlying climatic trend and the change in CO_2 assumed between the poorly known preindustrial value and the accurately measured current concentrations. The success achieved by several workers in explaining the temperature record in diverse ways demonstrates that a number of hypotheses can fit the poorly defined historical data and estimated preindustrial concentrations.

The available data on trends in globally or hemispherically averaged temperatures over the last century, together with estimates of CO_2 changes over the period, do not preclude the possibility that slow climatic changes due to increasing atmospheric CO_2 projections might already be under way. If the climate has warmed about 0.5°C and the preindustrial CO_2 concentration was near 300 ppm, the sensitivity of climate to CO_2 (expressed as projected increase of equilibrium global temperature for a doubling of CO_2 concentration) might be as large as suggested by the upper half of the range indicated earlier, i.e., up to perhaps 4.5°C; if the preindustrial CO_2 concentration was well below 300 ppm and if other forcing factors did not intervene, however, the sensitivity must be below 3°C if we are to avoid inconsistency with the available record (see Figure 1.13).

If, as expected, the CO_2 signal gradually increases in the future, then the likelihood of perceiving it with an appropriate degree of statistical significance will increase. Given the inertia created by the ocean thermal capacity and the level of natural fluctuations, achieving statistical confirmation of the CO_2-induced contribution to

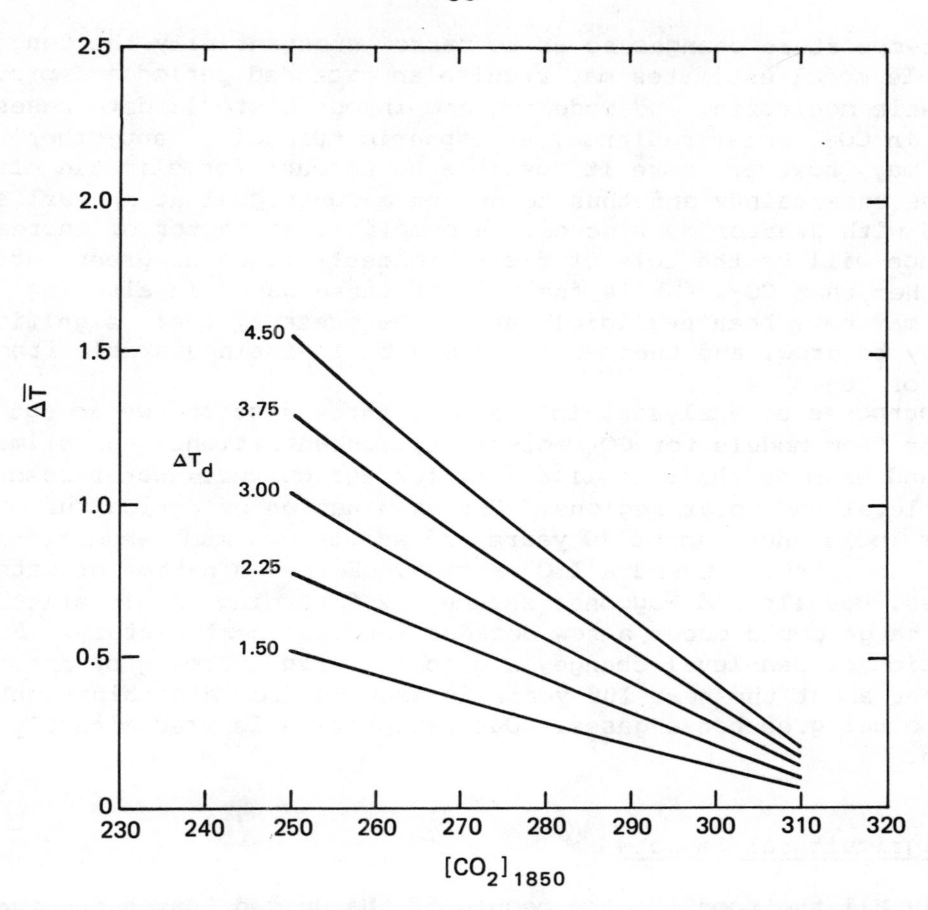

FIGURE 1.13 Relationship between CO_2 change, temperature change, and climate sensitivity assuming no other factors intervene. The abscissa represents a range of values for the preindustrial (1850) concentration of CO_2. The ordinate represents the increase ($\Delta\bar{T}$) in global mean equilibrium surface temperature between 1850 and the period 1961-1980. The response is calculated for a range of values of ΔT_d, the change of global mean equilibrium temperature for a doubling of CO_2 concentration (assumed independent of initial CO_2 concentration), and assumes that the temperature range is logarithmically related to the change in CO_2 concentration (Augustsson and Ramanathan, 1977). An ocean response time (mean thermal lag) of 15 years is used. The concentration of CO_2 was assumed in each case to increase linearly from the indicated value in 1850 to 310 ppm in 1950, and then linearly from 1950 to 340 ppm in 1980. Note that if the temperature increase from 1850 to the interval 1961-1980 is taken to be 0.5°C, then for consistency, ΔT_d may be as large as 4.5°C only if nineteenth century CO_2 concentrations were about 300 ppm, whereas ΔT_d may be as small as about 1.5°C if nineteenth century CO_2 concentrations were as low as 250 ppm. For ocean response times shorter than 15 years, the isolines slide upward and move in the opposite direction for longer ocean lag time. Varying the time of the start of the increase in CO_2 concentrations from 1850 to 1920 has little effect. See Weller et al., Chapter 5, for further discussion.

global temperature changes so as to narrow substantially the range of acceptable model estimates may require an extended period. Improvements in climatic monitoring and modeling and in our historic data bases for changes in CO_2, solar radiance, atmospheric turbidity, and other factors may, however, make it possible to account for climatic effects with less uncertainty and thus to detect a CO_2 signal at an earlier time and with greater confidence. A complicating factor of increasing importance will be the role of rising concentrations of greenhouse gases other than CO_2. While the role of these gases in altering climate may have been negligible up to the present, their significance is likely to grow, and their effects may be indistinguishable from the effects of CO_2.

For purposes of analysis, in the next three sections we accept the estimates from models for CO_2 emissions, concentrations, and climate change and examine their implications for agriculture, water resources, and sea level and polar regions. For examination of agriculture, Waggoner looks ahead about 20 years and adopts maximum assumptions of change: about 400 ppm and a 1°C warming. For examination of water resources, Revelle and Waggoner assume a 2°C warming in midlatitudes; such a change could occur a few decades into the next century. For examination of sea-level change, a global average warming of about 3-4°C over about the next 100 years is assumed from a combination of CO_2 and other greenhouse gases. Our perspective is predominantly American.

1.2.5 Agricultural Impacts

Virtually all the food for the people of the United States and the feed for our animals grow on a third of a billion acres (1.35×10^{12} m^2) of cropland and vast rangelands and pastures, exposed to the annual lottery of the weather and climate. Rather than seek to examine all aspects of agriculture that might be affected by CO_2-induced climatic changes, we concentrate on a critical, susceptible, and illustrative aspect of agriculture: American crop production. These crops are critical to Americans because they feed us and bring $40 billion of our foreign exchange. They are also critical to others. For example, in 1979 the United States provided 42% of the wheat and 19% of the rice traded between the nations of the world, and fully 43% of the world's corn crop is American. These crops may be susceptible and illustrative because most are grown in latitudes from 35 to 49°, within a zone that climate researchers expect will experience a substantial change in weather and climate as CO_2 increases. Waggoner, in the detailed analysis in Chapter 6, concentrates particularly on wheat, corn, and soybeans, which outdistance in value any other American crop.

From past experience we know that:

- Farmers fit husbandry and crops to weather and climate.
- Rapid environmental change disrupts agriculture.
- Colder as well as warmer and wetter as well as drier conditions can damage crops.

- Pests can amplify effects of bad weather.
- The very soil can be changed by atmospheric conditions.
- Farm and range animals are affected by weather and climate.
- Occasional extremes destroy agriculture.
- Impact of changed weather is sharp in marginal climates.

Against this background it is possible to calculate or speculate on the changes in crops that would follow hypothetical changes in the atmosphere. The hypothetical change that Waggoner considers is at the high end of the range estimated for the year 2000: an increase in CO_2 to about 400 ppmv, a mean warming of about 1°C in the northern United States with a growing season about 10 days longer, and more frequent drought in the United States caused by somewhat less rain and slightly more evaporation. Regional studies in other areas or for other times might well begin by assuming different changes.

Carbon dioxide is a major substrate for photosynthesis and, therefore, can directly affect plant growth if a lack of CO_2 rather than a shortage of water or some other nutrient is the limiting factor. Since the current 340 ppmv appears to be limiting in many cases, a rise in atmospheric CO_2 should increase photosynthesis. However, most effects of CO_2 on photosynthesis and plant growth have been studied and measured during short periods when other factors such as light, water, temperature, and nutrients were adjusted to an optimal level. In addition, growth habits and adaptations to different environments might alter the effects of changing CO_2 concentration.

Increased CO_2 affects photosynthetic rate and duration, as well as the fate and partitioning of photosynthate. Increased CO_2 may also improve the hydration of plants, because it influences the opening of the stomates, the pores through which plants gain CO_2 and transpire water. Increased production of photosynthates may improve the availability of nutrients by encouraging growth of nitrogen-fixing symbionts, enlarging the pool of soil organic matter and increasing soil nitrogen levels. Changed environment not only affects crops but also weeds, pests, and their interrelationships, sometimes benefiting the crop, sometimes its competitors and predators.

Although it is difficult to predict the changes in yield in real crops that might follow a rise in atmospheric CO_2, a survey of experiments in growth chambers and greenhouses leads to the conclusion that CO_2 enrichment to 400 ppmv by A.D. 2000 may increase yields of well-tended crops by, say, 5% (see Table 1.5). In comparison, yield of Illinois corn has quadrupled in about half a century. Although the quantitative evidence for changes in yield under growing conditions in which factors other than CO_2 are limiting is equivocal, some increase in yield--even in poor circumstances--is indicated.

Two orderly means are at hand to calculate how the yields of corn, soybeans, and wheat will change if rising CO_2 makes the climate warmer and drier. In one method, statistical regression, history is distilled to obtain the change in yield for a specified change in weather, such as the decrease in wheat yield in Kansas after a 10% decrease in March precipitation. In the other method, the physiology of, e.g., wheat and the physics of evaporation are assembled in a computer program or

TABLE 1.5 Changes in Yields of Crops in Optimum and Stressful Environments Anticipated from Atmospheric Enrichment to 400 ppmv of CO_2

Crop	Change in Yield (%)	Component Harvested	Yield Increment/ CO_2 Increment (%/ppmv of CO_2)	Yield Change by Enrichment (%/60 ppmv of CO_2)	Reference
Optimum Environments					
Barley	0.9[a]	Grain	0.18	11	Gifford et al. (1973)
Corn	0.28	Young shoots	0.03	1.9	Wong (1979)
Cotton	0.6	Lint	0.34	20	Mauney et al. (1978)
Soybean	0.4[a]	Grain	0.04	2	Hardman and Brun (1971)
Wheat	0.4	Grain	0.13	8	Gifford (1979)
Wheat	0.3	Grain	0.07	4	Sionit et al. (1980)
Wheat	0.6	Grain	0.13	8	Sionit et al. (1981)
Stressful Environments					
Corn (1/3 normal N)	0.28	Young shoots	0.03	1.9	Wong (1979)
Wheat (water limited)	0.6	Grain	0.44	26	Gifford (1979)
Wheat (one H_2O stress cycle)	0.5	Grain	0.10	6	Sionit et al. (1980)
Wheat (two H_2O stress cycles)	0.2	Grain	0.05	3	Sionit et al. (1980)
Wheat (1/8 normal nutrient)	0.1	Grain	0.02	1	Sionit et al. (1981)

[a]Calculated from shoots only.

"simulator" of wheat to calculate the change in wheat yields per change in environment. Waggoner employs both these methods, comparing and verifying them as well as making predictions. Despite a long list of qualifications and warnings, a clear conclusion is obtained (see Table 1.6). If we assume no significant adaptation of inputs and limited geographic mobility, the warmer and drier climate assumed to accompany the increased CO_2 will decrease yields of the three great American food crops over the entire grain belt by 5 to 10%, tempering any direct advantage of CO_2 enhancement of photosynthesis.

Sometimes a change in the weather that has only a modest direct effect on a crop is amplified into a disaster by a third party, a pest. Although pests will change, we cannot predict how.

Turning briefly to other countries and longer times, one can make some extrapolations. The direct benefit of more CO_2 to photosynthesis is universal and will continue for a long time. Crops in northern nations will benefit from warming, and tropical crops will be less

TABLE 1.6 Climate Change and Agricultural Productivity[a]

Crop and Region/State	Present Yield (quintals/hectare)	Estimated Change for 1°C Temperature Increase and 10% Precipitation Decrease	
		Amount (quintals/hectare)	Percentage Change (%)
Spring Wheat			
Red River Valley	18.2	-1.32	-7
North Dakota	14.9	-1.77	-12
South Dakota	12.0	-1.36	-11
Winter Wheat			
Nebraska	21.3	-1.04	-5
Kansas	21.3	-1.04	-5
Oklahoma	19.7	-0.37	-2
Soybeans			
Iowa	23.6	-1.55	-7
Illinois	21.9	-0.82	-4
Indiana	22.0	-1.25	-6
Corn			
Iowa	72.7	-2.36	-3
Illinois	68.8	-1.72	-3
Indiana	65.3	-2.80	-4

[a]Examples of the effect of a hypothetical climate change on crop yields, if we assume no significant adaptation of inputs and limited geographic mobility. Results shown are based on statistical multiple regression analysis of observed crop and weather data and are calculated for a nominal 1°C increase in temperature and 10% decrease in precipitation for each season or monthly period used as input to the analysis (see Chapter 6).

affected, if, as indicated by climate models, temperatures change little there. Where rainfall is now meager, an increase will have great benefit and a decrease can be tragic. Adaptation will be easier in countries that span several climatic zones and have sufficient wealth, ingenious farmers, and capable scientists with a practical outlook. Although adaptation will continue if the rise in CO_2 continues for generations, an accompanying and continuing desiccation could pass the ability to adapt.

Returning to American crops, one sees in the end that the effects on plants of the changes in CO_2 and climate foreseen for A.D. 2000 are modest, some positive and some negative. The best forecast of yield for the next few decades in the United States, therefore, seems a continuation of the incremental increases in production accomplished in the past generation as scientists and farmers adapt crops and husbandry to an environment that is slowly changing with the usual annual fluctuations around the trend.

1.2.6 Water Supplies

As discussed above in connection with agriculture, CO_2-induced climate changes would involve changes in precipitation, temperature, and their seasonal characteristics. Such changes must be expected to have consequences for rivers and thus for the availability of water for personal use, industry, inland navigation, and irrigation. Of the rain that falls on a given watershed, a large part is eventually evaporated and transpired; the remaining runoff feeds the streams, rivers, and aquifers that drain the region. Despite the current imprecision in predictions of climate changes, it thus seems useful to consider their implications for runoff available for cities and irrigation.

To assess the effects on the water resources of the United States of probable climate change, Revelle and Waggoner (Chapter 7) use the empirical relationships found by Langbein et al. (1949) among mean annual precipitation, temperature, and runoff. The catchments studied by Langbein and his colleagues were distributed over climates from warm to cold and from humid to arid, but Revelle and Waggoner focus on the relations among runoff, temperature, and precipitation only for relatively arid areas. From Langbein's data, they observe that for any given annual precipitation, runoff diminishes rapidly with increasing temperature. Similarly, for any given temperature, the proportion of runoff to precipitation increases rapidly with increasing precipitation.

For any particular region, the relations derived are rather crude approximations because many physical factors, including geology, topography, size of drainage basin, and vegetation, may alter the effect of climate on runoff. Revelle and Waggoner believe, nevertheless, that these relations can be used without serious error to describe the effects of relatively small changes in average temperature and precipitation on mean annual runoff. Table 1.7 shows the approximate percentage decrease in runoff to be expected for a 2°C increase in temperature. Table 1.8 shows the approximate percentage decreases in runoff for a 10% decrease in precipitation. From these exploratory investigations it is evident that in arid and semiarid lands relatively small changes in temperature or precipitation can produce amplified changes in runoff, river flow, and hence the availability of water for irrigation.

1.2.7 Sea Level, Antarctic, and Arctic

Many processes can cause an apparent change in sea level at any particular location. They include local or regional uplift or subsidence of the land; changes of atmospheric pressure, winds, or ocean currents; changes in the volume of the ocean basins owing to volcanic activity, marine sediment deposition, isostatic adjustment of the Earth's crust under the sea, or changes in the rate of seafloor spreading; changes in the mass of ocean water brought about by melting or accumulation of ice in ice sheets and alpine glaciers; and thermal expansion or contraction of ocean waters when these become warmer or colder. Only the last two processes are of primary interest in con-

TABLE 1.7 Approximate Percentage Decrease in Runoff for a 2°C Increase in Temperature[a]

Initial Temperature (°C)	Precipitation (mm yr-1) 200	300	400	500	600	700
- 2	26	20	19	17	17	14
0	30	23	23	19	17	16
2	39	30	24	19	17	16
4	47	35	25	20	17	16
6	100	35	30	21	17	16
8		53	31	22	20	16
10		100	34	22	22	16
12			47	32	22	19
14			100	38	23	19

[a]Source: Revelle and Waggoner, Chapter 7. Computed from data on runoff as a function of precipitation and temperature (Table 7.1) taken from Langbein et al. (1949).

sidering worldwide changes in sea level resulting from climate change, such as the warming that may be induced by increasing greenhouse gases in the atmosphere. (Melting or formation of sea ice and floating ice shelves have no effect on sea level--a glass of ice water filled to the brim does not overflow while the ice melts.) But the other processes contribute to the "noise" that afflicts all sea-level records and may make their interpretation over periods of a few decades difficult or impossible.

For orientation, it is useful to keep in mind that sea level has risen 150 m in the 150 centuries since the peak of the last glacial period. Hence, the present rate of 10-20 cm per century is small compared with the average rate of 1 m per century over the past 15 millennia and very much smaller than the inferred maximum rise of perhaps 5 m per century immediately following the glacial period. Indeed, the present is a time of quiet sea level compared with the violent oscillations that occurred during most of the last 100,000 years.

The projected climatic warming from increasing atmospheric CO_2 and other greenhouse gases will lead to an increased transfer of water mass to the sea from continental (Greenland and Antarctic) and alpine glaciers. As shown by Revelle in Chapter 8, the resulting rise in sea level could be about 40 cm over the next century. Increased downward infrared radiation will also lead to a warming and, therefore, expansion of the upper ocean waters, which can contribute another 30 cm for a total of 70 cm. Assuming the correctness of the figure of 4 W m^{-2} for the increased downward infrared flux with a doubling of CO_2

TABLE 1.8 Approximate Percentage Decrease in Runoff for a 10% Decrease in Precipitation[a]

Temperature	Initial Precipitation (mm yr-1)				
(°C)	300	400	500	600	700
- 2	12	16	17	18	18
0	14	16	17	19	19
2	15	16	19	19	20
4	17	19	19	21	21
6	23	23	21	21	21
8	30	24	24	22	22
10		24	27	23	23
12		40	30	25	25
14			34	30	27
16			50	36	29

[a]Source: Revelle and Waggoner, Chapter 7. Computed from data on runoff as a function of precipitation and temperature (Table 7.1) taken from Langbein et al. (1949).

(higher concentrations of other infrared absorbing gases might further increase this flux), the estimates for both ice melting and ocean thermal expansion still have large uncertainty--at least ±25%. These are due to our uncertainty over the causes of the current rise in sea level, our inability to predict whether changes in atmospheric circulation will cause more or less snow to fall on the ice caps, our ignorance of the conditions for advance or retreat of alpine glaciers, and our lack of understanding of the physical processes associated with the flux of heat to the ocean.

Of even greater uncertainty is the potential disintegration of the West Antarctic Ice Sheet, most of which now rests on bedrock below sea level. This could cause a further sea-level rise of 5 to 6 m in the next several hundred years.

West of the Transantarctic Mountains (approximately from the Meridian of Greenwich, across the Antarctic Peninsula to 180° W), most of the Antarctic Ice Sheet rests on bedrock below sea level, some of it more than 1000 m beneath the sea surface. In its present configuration, this "marine ice sheet" is believed to be inherently unstable; it may be subject to rapid shrinkage and disintegration under the impact of a CO_2-induced climatic change (Mercer, 1978). Events in the fairly recent geologic past suggest that rapid disappearance of West Antarctic ice has occurred before. However, evidence from radar soundings of flow lines extending across the Ross Ice Shelf indicates that the remaining West Antarctic Ice Sheet has been relatively stable for the last 1000-2000 years. Indeed, various lines of evidence suggest that the mass balance of the entire Antarctic Ice Sheet may be positive, i.e., ice may be accumulating.

If the West Antarctic Ice Sheet were to "collapse" (slide into the sea), it would release about 2 million cubic kilometers of ice before the remaining half of the ice sheet began to float.

The rate at which the West Antarctic Ice Sheet could disappear under the impact of a CO_2-induced warming has recently been examined by Bentley (1983). He concluded that rates of discharges and removal of icebergs might make disappearance barely possible, although unlikely, in 200 years, but only after removal of the ice shelves.

If the time required for the ice shelves to disappear is 100 years, Bentley's analysis would not be incompatible with a minimum time of 300 years for disintegration of the West Antarctic Ice Sheet. The corresponding average rate of rise of sea level would be slightly less than 2 m/100 years, beginning about the middle of the next century. Bentley's "preferred" minimum time of about 500 years would give a rate of sea-level rise of 1.1 m/100 years, which is, as pointed out earlier, about the mean rate for the last 15,000 years. To either of these figures we must add a rise of 70 $\pm$ 18 cm between 1980 and 2080, which Revelle has shown is likely to result from ocean warming and ice ablation in Greenland and Antarctica, plus a possible retreat of alpine glaciers. These processes may well continue in later centuries.

Like the Antarctic ice, Arctic ice has been a stable climate feature (see Annex 1). There is quite good evidence for persistence of the ice cover all year round for the last 700,000 years and perhaps for the past 3,000,000 years, although there is debate about whether the Arctic may have been open in summer from 700,000 to 3,000,000 years ago. The existence of glacial marine sediments in the Arctic basin shows that ice rafting occurred during the past 5,000,000 years. Longer ago than 5,000,000-15,000,000 years, the Arctic may have been open year round. Global cooling patterns are such that an initial freeze-up of the Arctic may have occurred 15,000,000 years before the present, although there is no direct evidence. The physical reasons for the persistence of the Arctic ice are not well understood; but they may reflect both dynamic and thermodynamic processes, such that when little (excess) ice exists, correspondingly more (less) ice is produced the next winter.

Studies on whether the Arctic sea ice will completely melt in summer, and if so, whether the ice will remain melted in winter, as suggested by Flohn (1983), have produced ambiguous results. Given the apparent long-term stability of Arctic ice, one must be cautious in projecting a melting due to prospective warming from increasing CO_2 concentrations. A number of climate and ice models suggest that the Arctic ice may melt in summer with a warming of about the magnitude that may be induced by a doubling of CO_2 and increase of other greenhouse gases, but this conclusion must be viewed as still tentative. The representations of the Arctic in energy balance and most climate models that have melted Arctic ice with a CO_2 warming usually do not include changes in cloud cover, ice dynamics, or the effects of open leads and salinity stratification.

Owing to dynamic and thermodynamic processes, thickness of ice may respond more readily to temperature increases than extent of ice. However, verification of ice extent and thickness estimates from climate models is not yet adequate.

Oceanographic studies are also quite limited for the case of an open Arctic. There is now a very strong, salinity-induced, density stratification, the causes of which are not fully understood. If this stratification can be broken and does not reform, then the Arctic might be able to remain open through the winter. This possibility is not considered likely.

Finally, there have been few studies of the effect of less ice or no ice in summer on atmospheric circulation. While atmospheric effects of reduction in Arctic ice remain highly speculative, some poleward shift of storm tracks seems likely and most significant climatic effects may occur during transition seasons.

1.3 SERIOUSNESS OF PROJECTED CHANGES

In assessing the seriousness of the changes projected in the preceding sections, there are two enormous sources of uncertainty. One source is the contents of the above outlook itself: uncertainties about sources and uses of energy, which in turn embody uncertainties about population, per capita income, energy-using and energy-producing technologies, density and geographic distribution of populations, and the distribution of income; a multitude of uncertainties about the carbon cycle; uncertainties in translating a growth curve for CO_2 in the atmosphere into appropriately time-phased changes in climate in all the regions of the globe; uncertainties about whether human activities other than release of CO_2 will be affecting the climate and what "natural" climatic trends will be; and, finally, uncertainties about effects on plant growth, water supplies, sea level, and other factors.

The second source is uncertainty about the kind of world the human race will be inhabiting as the decades go by, through the coming century, and beyond. This source overlaps the uncertainties just mentioned; per capita income both influences the use of fossil fuels and affects how readily the world's population can afford, or can adapt to, changes in climate. And for both purposes the distribution of income--the income disparities among different parts of the world, within countries as well as between countries--affects the calculations. Similarly, the structures people inhabit, the ways people and goods are transported, the foods people eat, the ways countries defend themselves, and the geographical distributions of populations within and among countries, all affect land use and the kinds and amounts of energy used and hence the production of CO_2; but they also affect the ways that climate impinges on living and earning, even on what climates are preferred. The mobility of people, capital, and goods--the readiness with which people can migrate, goods can be traded, and capital for infrastructure and productive capacity can flow among regions and countries--would also determine how much difference the changes in climate would make. The location and significance of national boundaries and various international and supranational institutions would have much to do with whether adverse climatic effects in parts of the world could be offset, in a welfare assessment, by improvements in other places. Different individuals and groups will interpret these uncertainties in different ways, depending on their culture, training, societal vantage point, and other factors.

While emphasizing that the uncertainties just described must be kept in mind, we now discuss the areas of concern we have been able to identify in our discussions of the CO_2 issue. First we address areas of specifiable concern, then of more speculative concern, and, finally, of poorly defined but potentially serious concern.

1.3.1 Specifiable Concerns

1.3.1.1 Agriculture and Water Resources

The outlook for American agriculture over the next couple of decades based on a foundation of physiology and history (summarized above and presented in detail by Waggoner in Chapter 6) tells how a CO_2-induced change in climate would change the yields of crops if farmers, ignoring the weather, persisted in planting the same varieties of the same species in the same way in the same place. The safest prediction of any made by Waggoner is that farmers will adapt to a change in climate, exploiting it and, probably, proving our predictions to be pessimistic. If the climate changes, farmers will move themselves, change the crops, modify varieties, and alter husbandry. The loss of acreage to the margin of the desert, for example, may be replaced by yield and acres at the cold margin. Seeking higher yields and more profit, farmers will correct their course annually, and they may even adapt to a slowly changing climate unconsciously and successfully. Thus, we do not regard the hypothesized CO_2-induced climate changes as a major direct threat to American agriculture over the next few decades. Of course, shifts at the margins may be easy for the nation but not for those involved.

While the effects of increased CO_2 plus climatic warming on agriculture might be relatively small in the United States, such effects might be much larger in countries that do not have or build a good agricultural research infrastructure or the agricultural flexibility that come from relatively large capital investments in agriculture, rural transportation, farm credit and crop insurance, and marketing systems. Through trade linkages and political and social awareness, climate-induced agricultural problems in other parts of the world will become America's concern. Longer-term agricultural impacts, as global climatic conditions continue to shift, perhaps at an increasing rate, might also conceivably be much more serious. Will these be offset by the benefits of CO_2 for photosynthesis and water-use efficiency? At present we lack tools with which to evaluate in a credible way the very long-run prospects.

While on balance U.S. agriculture as a whole may not suffer significantly, irrigated agriculture is both important and susceptible. Its importance derives from its expanse and the value of its crops. Fully 50 million acres, or about 1 in 7 American acres, of cropland are irrigated. The quarter trillion cubic meters of irrigation water withdrawn from American streams and groundwater represent about half of all withdrawals of this natural resource. Averaging wheat yields over all American fields, humid as well as arid, one sees an average of 1.9 tons/ha on unirrigated versus 3.7 tons/ha on irrigated land. Valuable crops are grown on irrigated fields because irrigation reduces variability of water and produces consistently high yields. Thus, most of the irrigated cropland (44 million acres) occurs on only 12% of the farms, but these farms produce fully 40% of the market value of the crops from all American cropland (Jensen, 1982).

Irrigation is susceptible, because it is such a heavy user of water. In seven U.S. water regions examined by Revelle and Waggoner (Chapter 7), the share of total water withdrawals for irrigation ranged from 68% to 95%. For these regions, Revelle and Waggoner draw on the efforts of Stockton and Boggess (1979) and perform a corroborating study of the Colorado River to estimate the effects of a climatic change on water.

Revelle and Waggoner (Chapter 7) show that the effect of climatic change on water must be considered separately for different regions. Following Stockton and Boggess (1979), they make several simplifying assumptions, the most important of which are: 1) variations in annual runoff are predominantly influenced by climate, although other factors, such as geology, topography, and vegetation, have effects, and 2) that evapotranspiration is controlled only by temperature. A warmer and drier climate would severely affect the seven water regions: the Missouri, Texas Gulf, Rio Grande, Arkansas-White-Red, Upper Colorado, Lower Colorado, and California. All are in the western United States; and although they cover about half the country, they have less than 15% of the runoff. A 10% decrease in precipitation, combined with a 2°C warming, would decrease runoff in these regions between 40 and 76% (see Table 1.9). The impact would be especially severe in the Missouri, Rio Grande, Upper Colorado, and Lower Colorado regions where even current water requirements would exceed the supplies after climatic change by between 20 and 270%. Local shortages and a general deterioration of water quality would occur in the Arkansas-White-Red, Texas Gulf, and California regions. Much of the irrigated area might have to be abandoned unless water could be imported from other regions with more abundant supplies, such as the Pacific Northwest or the Upper and Lower Mississippi. Major additions to reservoirs would be required in several regions to maintain a safe yield of water during drought, even after a reduction of irrigated area and the maximum practicable rise in the efficiency of the use of water in irrigation.

Except for the Rio Grande, the Colorado River is more intensively used than any other major stream in the United States. Half the estimated "normal" river flow of 18 billion cubic meters per year has been allocated by interstate compact, confirmed by federal law, to the "lower basin" states of Arizona and California, with minor amounts going to Nevada, although nearly all of the runoff originates from snow in the high mountains of western Colorado, southwestern Wyoming, and eastern Utah. Revelle and Waggoner examined mean annual precipitation, temperature, and river flow for 1931 to 1976 and found a high correlation between variations in precipitation and temperature, on the one hand, and runoff on the other. A rise of 2°C in average temperature from 4.2 to 6.2°C would reduce runoff by 29 $\pm$ 6%, and a 10% decrease in precipitation would cause a further reduction of about 11 $\pm$ 1.4% in

TABLE 1.9 Comparison of Water Requirements and Supplies for Present Climatic State and for a 2°C Increase in Temperature and 10% Reduction in Precipitation in Seven Western U.S. Water Regions[a]

Water Region[b]	Present Climate						Warmer and Drier Climate		
	Area ($10^{10}/m^2$)	Mean Annual Runoff (10^{10} m^3 yr^{-1})	(mm)	Mean Annual Supply (10^{10} m^3 yr^{-1})	Mean Annual Requirements[c] (10^{10} m^3 yr^{-1})	Ratio of Requirement to Supply	Mean Annual Supply (10^{10} m^3 yr^{-1})	Percent Change in Supply	Ratio of Requirement[d] to Supply
Missouri	132.4	8.50	64	8.50	3.63	0.43	3.07	-63.9	1.18
Arkansas-White-Red	63.2	9.35	148	9.35	1.67	0.18	4.32	-53.8	0.39
Texas Gulf	44.9	4.92	110	4.92	1.74	0.35	2.47	-49.8	0.70
Rio Grande	35.2	0.74	21	0.74	0.67	0.91	0.18	-75.7	3.72
Upper Colorado	29.6	1.64[e]	55	1.64	1.63[f]	0.99	0.99	-39.6	1.65
Lower Colorado	40.1	0.38	10	1.15[g]	1.37	1.19	0.50[g]	-56.5	2.68
California	42.9	9.56	222	10.18[g]	4.22	0.41	5.71[g]	-43.9	0.74
For the 7 regions together	388.3	35.09	90.4	35.09[h]	14.93	0.43	16.53[h]	-53	0.90

[a]Source: Stockton and Boggess (1979) and calculations from Revelle and Waggoner, this volume, Chapter 7.
[b]As defined by the U.S. Water Resources Council (1978).
[c]Projected through A.D. 2000.
[d]Assuming no increase in requirement because of increased evapotranspiration from irrigated farms or reservoirs.
[e]Average "virgin flow" of the Colorado River at Lee Ferry from 1931 to 1976.
[f]Includes allocation to Lower Basin states, California included, of 0.93×10^{10} m^3 yr^{-1}.
[g]Includes water received from Upper Colorado Basin, but not mined groundwater.
[h]Total is less than sum of the column because of flow of Lower Colorado derived from Upper Colorado (g).

runoff. A rise in temperature--even without a decrease in precipitation--would seriously affect entire states.

A 2°C warming and a 10% reduction in precipitation would probably not have serious effects on water supplies in the humid regions east of the 100th meridian. Neither would effects be severe in the water-rich Pacific Northwest and the Great Basin (parts of Nevada, Utah, and Idaho), where demand is relatively small and groundwater reserves are large.

Loss in irrigated yield accompanying a change in climate can be envisaged in two ways. For grain, one might simply and roughly say that some areas can no longer be irrigated, and on those acres the yield will be reduced by at least 50% because the subsequent dryland crops will be grown in alternate years. For produce, a decrease in water and irrigated area for truck croplands could reduce yields to zero on many acres and thus decrease the supply of fresh vegetables in the supermarkets, especially in the winter. Of course, decline of irrigation systems, often caused by increasing salinity and rising water table when drainage facilities are inadequate, is not a new experience for mankind.

1.3.1.2 Rising Sea Level

If one accepts the projections of warming, then certain physical consequences seem inescapable. One of these is the slow rise in global sea level. As explained by Revelle, melting of land ice and thermal expansion of the ocean may lead to a rise of about 70 cm in global sea level over the next 100 years, continuing thereafter. Many shoreline problems (for example, coastal erosion, storm surges, and salinity of groundwater) are sensitive to sea-level changes on the order of decimeters, and 70 cm, though modest-sounding on a calm day at the seashore, could effect a variety of unwelcome changes. We discuss the question of larger and continuing sea-level rise in the section that follows.

1.3.2 More Speculative Concerns

Our more speculative concerns center on the West Antarctic Ice Sheet (WAIS) and sea-level rise, the Arctic, and human health. These concerns are more speculative in that both the scientific uncertainties are greater and the potential effects are more distant.

Resuming the question of sea level, we concluded above that with a postulated warming of about 3 or 4°C from CO_2 and other greenhouse gases a gradual rise is probable over the next 100 years as a result of thermal expansion of the ocean, ablation of the Greenland and Antarctic ice caps, and retreat of alpine glaciers. We have also mentioned that, because of events in Antarctica, a much larger rate of rise is not unlikely during the following several centuries. Rates of sea-level rise could reach 1-2 m per 100 years. A complete collapse of the WAIS would produce a worldwide rise in sea level of between 5 and 6 m.

How serious would such rates and levels of sea-level rise be? As Schelling discusses in Chapter 9, there are three principal ways that human populations can adapt to a rising sea level: retreat and abandonment, construction of dams and dikes, or building on piers and landfill. The basic division is between abandonment and defense.

Defense against sea-level rise has received little attention in this country. It is therefore worth emphasizing that there are ways to defend against rising sea levels. For built-up and densely populated areas, defenses could be cost effective for a rise of as much as 5 or 6 m. Even where defending against 5 m would not be cost effective, defending against a meter or two could make sense for a century or two. Defense is not an empty hypothetical or purely speculative option.

The economics of dikes and levees depends on the availability of materials (sand, clay, rock); on the configuration of the area to be protected; on the differential elevation of sea level and internal water table; on the depth of the dike where it encloses a harbor or estuary; on the tide, currents, storm surges, and wave action that it must withstand; and on the level of security demanded for contingencies such as extreme ocean storms, extreme internal flooding, earthquakes, military action, sabotage, and uncertainties in the construction itself. On the economics of diking, it is worth remembering that the Dutch for centuries have found it economical to reclaim the bottom of the sea, at depths of several meters, for agricultural, industrial, and residential purposes.

The situation is totally different for an area like the coast of Bangladesh. Defense would be extremely costly for a region with a huge coastal area subject to inundation, rather than a concentration of capital assets that could be enclosed by a few miles of dikes. Such an area would be so susceptible to internal flooding with freshwater that levees required to protect the country would be many times greater than the length of the shoreline.

Where defense is not practicable, retreat is inevitable, at least selectively. In urban concentrations, where buildings may last a century, good 100-year predictions of sea-level change (including likely erosion and storm damage) might permit the orderly evacuation and demolition of buildings without excessive write-off of undepreciated assets.

Changes in Arctic weather and climate would have both practical and noneconomic implications. Open seas and easier ice conditions would have bearing on long-term strategies of use for northern seas and channels with respect to both navigation and seafloor development. Oil and gas exploration, drilling, production, and transportation could become easier and less expensive. The old dream of a "Northwest Passage" might become a reality. An ice-free Arctic Ocean in summer and a less hostile environment in North American, Russian, and Scandinavian Arctic regions would also have implications for military strategy and tactics, if technology does not shift military issues to other spheres entirely. For example, surface and aircraft-carrying fleets could operate in the Arctic during the summer months, as they do now in other oceans. One effect of warming should be a change in the stability and distribution of permafrost. This change would, in turn, suggest design changes for overland vehicles, construction equipment, pipelines, and buildings. On a different plane, concern arises about possible loss of habitats and the conservation of nature; polar regions are among the wilder and more pristine environments remaining.

In contrast to polar and sea-level change, not much consideration has been given by those who study increasing CO_2 and climate change to any possible direct effect on human health or the animal population from CO_2 in the air we breathe. The natural a priori concern with the health effects of a doubling or quadrupling of an important gas in the air we breathe--the substance that actually regulates our breathing rate--is relieved by the observation that for as long as people have been living indoors, not to mention burning fuel to heat themselves, they have been spending large parts of their lives--virtually entire lives in the case of people who work indoors and travel in enclosed vehicles--in an atmosphere of elevated CO_2. Doubling or even quadrupling CO_2 would still present a school child with a lesser concentration during outdoor recess than the child faces in today's average classroom.

There is, furthermore, no documented evidence that CO_2 concentrations of five or ten times the normal outdoor concentration damage human or animal tissue, affect metabolism, or interfere with the nervous system. Nor is there a theoretical basis for expecting direct effects on health from the kinds of CO_2 concentrations anticipated.

But even though this answer is reassuring, the question has to be faced. It will occur to people who hear about changes in the atmosphere that their grandchildren are going to breathe. And experiments have not been carried out with either people or large animals whose whole lives, including prenatal life, were spent in an environment that never contained less than, say, 700 ppmv of CO_2. So the question deserves attention, even though there is no known cause for alarm.

Probably more serious is the effect of elevated temperatures on health and welfare. If a 3 or 4°C increase in average temperatures occurs, as might be expected in different parts of the United States with a CO_2 doubling, extreme summer temperatures in warm years might rise by an equal amount. Excess human death and illness are already characteristic of summer "hot spells," and these might be worsened by much higher extreme summer temperatures. And, climatic shifts may change the habitats of disease vectors or the hosts for such vectors.

1.3.3 The Problem of Unease about Changes of This Magnitude

Enveloping our specific and more speculative concerns about impacts of climatic change on water resources, sea level, and other areas discussed is a profound uneasiness about inducing environmental changes of the magnitude envisaged with major increases in atmospheric CO_2 and other greenhouse gases.

To establish a context, consider, for example, the most frequently quoted index--change in global average surface temperature. This crude measure of climate tells us little about what temperature change to expect for specific regions and nothing about the type of climate that would be experienced. Global average surface temperature has come to such prominence in large part because it represents a relative measure of CO_2 effects among climate models. Indeed, for many models it is the only result with much scientific validity. Nevertheless, changes in average surface temperature may suggest well the nature of our unease.

Increasing CO_2 is expected to produce changes in global mean temperature that, in both magnitude and rate of change, have few or no precedents in the Earth's recent history. Consider the ranges of temperature experienced in various periods in the past (Figure 1.14). A range of less than a degree was experienced in the last century, less than 2°C in the last thousand years, and only 6 or 7°C in the last million years. The development of civilization since the retreat of the last glaciation has taken place in a global climate never more than 1°C warmer or colder than today's. Despite the modest decline of time-averaged global-mean temperatures since the 1940s, we are still in an unusually warm period in the Earth's history. Indeed, according to one source (Jones, 1981), 1981 was the warmest year on record. Thus, the temperature increases of a couple of degrees or so projected for the next century are not only large in historical terms but also carry our planet into largely unknown territory. Increasing CO_2 promises to impose a warming of unusual magnitude on a global climate that is already unusually warm.

Furthermore, the question of threshold responses arises. It is possible that a change in the central tendency of climate will come about smoothly and gradually. It is also possible that discontinuities will occur. For example, Lorenz (1968) and others have suggested the possibility of more than one climatic equilibrium.

As Schelling (Chapter 9) points out, our calm assessment of the CO_2 issue rests essentially on the "foreseeable" consequences of climatic change. Less well-seen aspects remain troubling. We have mentioned the possible release of methane clathrates from ocean sediments. We have also mentioned melting of the central Arctic sea ice. Disappearance of the permanent Arctic ice would result in a marked increase in the thermal asymmetry of the planet, with only one pole still glaciated. Such asymmetric conditions could produce further, unanticipated climatic changes (Flohn, 1982). Warming amplified at high-latitude regions could also affect major features of the oceanic circulation, and these too could lead to unexpectedly different climatic conditions, as well as changes in the capacity of the oceans to absorb CO_2. At the level of ecosystems, surprising changes may also result from climatic shifts.

We are not complacent about global-average temperature changes that sound small; very serious shifts in the environment could well be implied. There is probably some positive association between what we can predict and what we can accommodate. To predict requires some understanding, and that same understanding may help us to overcome the problem. What we have not predicted, what we have overlooked, may be what we least understand. And when it finally forces itself on our attention, it may appear harder to adapt to, precisely because it is not familiar and well understood. There may yet be surprises. Antici-

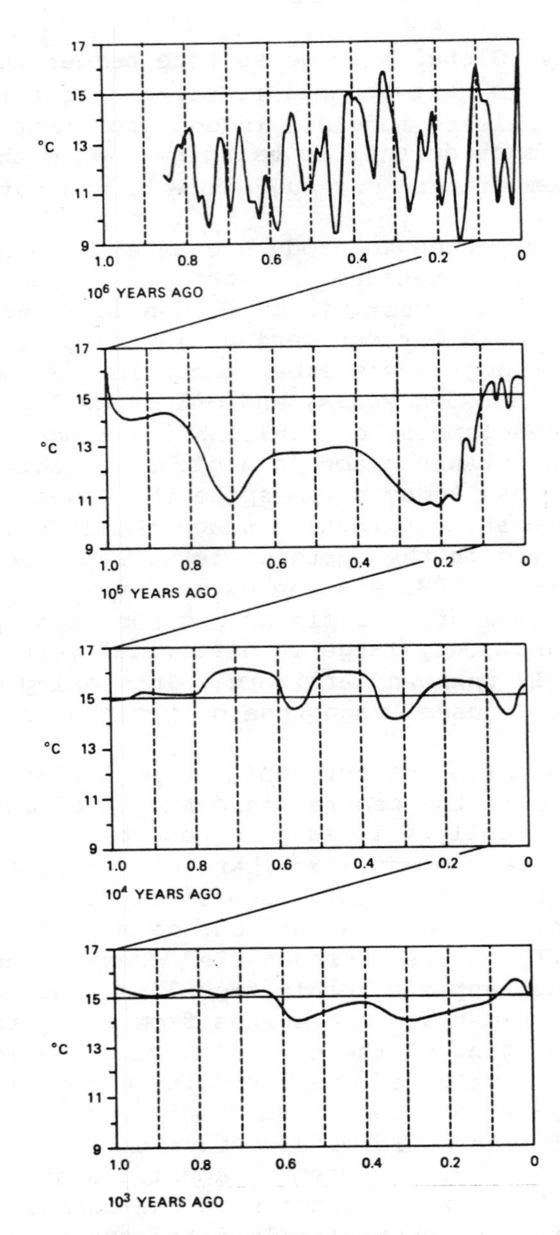

FIGURE 1.14 An approximate temperature history of the northern hemisphere for the last 850,000 years. The panels are at the same vertical scale. The top panel shows the past million years, the second panel amplifies the past 100,000 years, the third panel the past 10,000 years, and the bottom panel the past 1000 years. The horizontal line at 15°C is included simply for reference. Considerable uncertainty attaches to the record in each panel, and the temperature records are derived from a variety of sources, for example, ice volume, as well as more direct data. Spatial and temporal (e.g., seasonal) variation of data sources is also considerable. From Clark (1982). Original data from Matthews (1976), Mitchell (1979), and National Research Council (1975).

pating climate change is a new art. In our calm assessment we may be overlooking things that should alarm us.

At the same time, one might observe that--barring the kind of surprises mentioned above--the climate changes under consideration are not large in comparison with the climate changes individuals and social groups have undergone historically as a result of migration. Table 1.10 shows U.S. population for 1800, 1860, 1920, and 1980, distributed according to the climatic zones in Figure 1.15. These data have been transformed into a series of maps of the United States in which the areas of our various climatic zones are drawn so as to be proportionate to their populations at various times (see Chapter 9). The maps seemingly depict massive climate change; formerly empty, thus small, climatic zones become heavily populated and grow large. But it is not that deserts have expanded or that the climate has changed from permafrost to rain forest, or from prairie to Mediterranean west coast, or to places where it gets cold but does not quite freeze from where it got a little colder and did freeze. People have moved, and to all climates, to places of enormous extremes like the Dakotas and places of little change like Puerto Rico. People have moved from the seacoast to the prairie, from the snows to the Sun Belt.

Not only have people moved, but they have taken with them their horses, dogs, children, technologies, crops, livestock, and hobbies. It is extraordinary how adaptable people can be in moving to drastically

TABLE 1.10 U.S. Population by Climatic Zone[a,b,c]

Climatic Zone[c]	Description	Population 1800	1860	1920	1980
Aw	Tropical wet and dry (Savannah)	0	2,996 (1)	129,741 (1)	2,793,140 (1)
BS and BS_k	Semiarid and steppe	0	64,018 (1)	4,291,664 (4)	21,000,465 (9)
BW_h	Tropical and subtropical desert	0	28,029 (1)	743,263 (1)	4,955,742 (2)
Caf	Humid subtropical (warm summer)	2,034,536 (42)	9,426,517 (32)	32,360,561 (29)	71,932,014 (32)
Cb	Marine (cool summer)	0	39,246 (1)	1,795,406 (2)	4,447,811 (2)
Cs	Dry-summer subtropical (Mediterranean)	0	202,420 (1)	1,636,597 (2)	8,675,763 (4)
Daf	Humid continental (warm summer)	2,348,030 (49)	16,074,866 (54)	59,811,474 (54)	90,882,262 (40)
Dbf	Humid continental (cool summer)	435,665 (9)	3,586,555 (12)	9,394,792 (8)	13,710,636 (6)
H	Undifferentiated highlands	0	184,896 (1)	1,559,963 (1)	9;147,733 (4)

[a]Source: U.S. Census Bureau, 1800, 1860, 1920, 1980. Data compiled by Clark University Cartographic Service.
[b]Figures in parentheses are percentage of total population in that climate zone.
[c]Climatic zones shown in Figure 1.15.

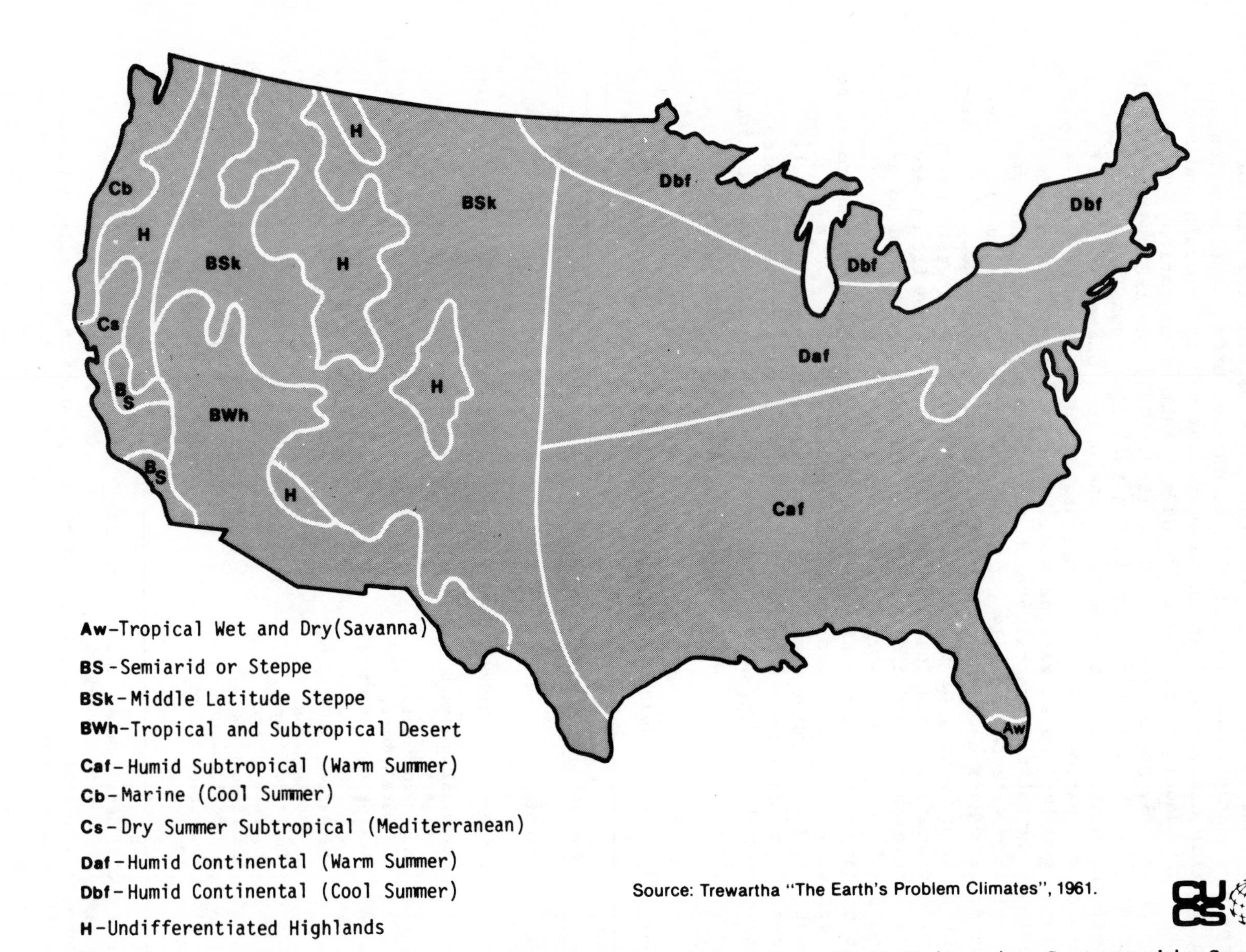

FIGURE 1.15 Climatic zones of the United States. Prepared by Clark University Cartographic Service.

different climates. That adaptability may suggest that if climates change only by shifting familiar climates around the world, it is not altogether different from leaving the climates alone and moving the people around. Of course, when people moved from England to Massachusetts or from the East Coast to the Great Plains, there were substantial difficulties in adapting; and if the climate changes and people stay, they may also have substantial difficulties. But it appears that a change in the climates where people live may not be altogether different from people moving to another climate. It may be that what we have to look forward to is not quite so historically unusual as a human experience as the descriptions from the paleoclimatic record would suggest. We have really become accustomed to marked climate change. For the individual, in contrast to the environment, the idea of climate change in a generation or two is far from novel.

While people may be able to adapt readily to climatic change, they may be unwilling to accept climatic changes imposed on them involuntarily by the decisions of others. Thus, in trying to clarify our unease about CO_2-induced climatic change, it is necessary to point out the potentially divisive nature of the issue. It is important to recognize the distribution of incentives for, and effects of, human-induced climatic changes. Although it might be in the interest of the world economy to restrict, at some cost, the use of fossil fuels, it is probably not in the interest of any single region or nation to incur on its own the cost of reduction in global CO_2. For example, countries that view heavy rains as disasters and countries that view them as water for their crops would have different preferences about which, if any, rains to avoid or restore and whether they or another country should forgo (or burn) fossil fuels to help effect the change. The marginal effects of climatic change on the distribution of wealth may range from quite positive to quite negative. In short, CO_2-induced climatic changes, and more generally weather and climate modification, may be a potent source of international conflict.

1.4 POSSIBLE RESPONSES

So far we have developed an outlook for CO_2-induced climate change and made some tentative evaluations of the seriousness of possible changes in prospect. In the preceding discussions we have occasionally referred to potential societal responses, for example, taxes on CO_2 emissions, agricultural adjustments, and migration. Now we discuss possible responses in a more systematic fashion and offer two sets of comments. One set relates to flexibility in defining the issue, the other to specific categories of response.

1.4.1 Defining the Problem

As Schelling points out in Chapter 9, how one defines a problem or issue often governs or biases the search for solutions and sometimes in a way that puts emphasis on more difficult or less attractive solutions.

The protagonist of this study has been CO_2. Recent research reflected in this report has been largely motivated, first, by the observation that atmospheric CO_2 is increasing as the use of fossil fuels expands and, second, by the known potential for a "greenhouse effect" that could generate quantitatively significant changes in climate worldwide. The members of the group responsible for the report are known as the Carbon Dioxide Assessment Committee, the work was authorized by an Act of Congress concerned with carbon-intensive fossil fuels, and the agency principally charged with managing the research is the Department of Energy. The topic is generally referred to as "the carbon dioxide problem," a global challenge to the management of energy resources.

However, there are good reasons for taking climate change itself as the main perspective rather than CO_2 or energy. One reason is that over the span of time that this report has to cover there could be changes and fluctuations in climate not due to human activity. A second reason is that CO_2 is not the only climate-affecting substance that society releases to the atmosphere. Not only must the impact of CO_2 be assessed in conjunction with other climate-changing activities, but any policy response needs a focus broader than CO_2. A third reason is that there is a natural tendency to define a problem by reference to the agent of change and to seek solutions in the domain suggested by the naming of the problem, e.g., "fossil fuels" or "CO_2."

There is a legitimate presumption that where the Earth's biosphere is concerned any drastic change may produce mischief. There is also a widespread methodological preference for preventive over alleviating programs, and for dealing with causes rather than symptoms. But it may be wrong to commit ourselves to the principle that, if fossil fuels and CO_2 are where the problem lies, they must also be where the solution lies. Although a precautionary attitude toward any drastic changes in world climates would be prudent, definition of the problem requires investigation not only of what changes in climates may occur but also of what damages or blessings the changes may bring.

To illustrate the point, while for any expected adverse consequences of CO_2, conserving fossil fuel is an obvious policy option at the outset, the parallel importance of water supply and conservation emerges only later. Defining the issue as "the CO_2 problem" can focus attention too exclusively on energy and fossil fuels and divert it from rainfall or irrigation or, more even-handedly, the broad issue of climate change.

Something else is illustrated about the character of the issue. If the solution has to be reduced CO_2 emissions, both the problem and the solution are global in a severe sense. A ton of CO_2 produced anywhere in the world has the same effects, for good or ill, as a ton produced anywhere else. Any nation or locality that attempts to mitigate prospective changes in climate through a unilateral program of conservation, fuel switching, biomass enhancement, or scrubbing of CO_2 from smokestacks, in the absence of some global fuel rationing or compensation arrangement, pays alone the cost of its program while sharing the consequences with the rest of the world. In contrast, water resources are usually regional or local. Worldwide agreements involving some of the main consumers or producers of fossil fuels would be essential to programs for reducing CO_2 emissions; in contrast, water development and conservation are national in scope or involve a few neighboring countries.

1.4.2 The Organizing Framework

If we accept that the issue is climate, then it follows that the organizing framework for welfare and policy implications of atmospheric CO_2 should also be built around climate change, not around CO_2.

As Schelling argues, the framework ought to be comprehensive. It should include theoretical possibilities that may be of no contemporary significance, because we have to think about choices as they evolve through the next century. The framework should make room for imagination, not just for options that currently look cost effective.

The framework should lend itself to different levels of universality. While atmospheric CO_2 is a global condition, its consequences and many of its policy implications will be regional and local. Governments will assess consequences and choose policies according to the climatic impacts on their own populations and territory. At the same time, some national governments, including ours, will need a framework for assessing worldwide consequences and policy options that are international in scope.

Just as governments will assess differently the implications of climate change for their own countries, some perceiving gains and others losses, so will interests be divided within countries. Not only are some countries, like our own, large enough to have diverse climates subject to different kinds of change, but people in the same climate are affected differently according to how they live and earn their living, their age and health, what they eat, and how they take their recreation. Our framework has to be susceptible of disaggregation.

The framework should be construed as moving through time. The changes take time; the uncertainties unfold over time; policies and their effects have lead times, lag times, and growth rates. Governments and people will attach different discounts to events and conditions at different distances in the future. And a country that appears to be victim or beneficiary of a climate-change forecast for the next 75 years would not be helped or hurt the same amount, or necessarily even in the same direction, by an additional 75 years of the same scenario.

1.4.3 Categories of Response

Schelling (Table 1.11) develops a framework consisting of four categories of response, arrayed against background climate and trends. Category 1 is prevention, containing options for affecting the production of CO_2. Category 2 is removal: if you cannot help producing too much CO_2, can you remove some? Category 3 consists of policies deliberately intended to modify climate and weather: if too much CO_2 is produced and not enough can be removed, so that concentration is

TABLE 1.11 CO_2-Induced Climatic Change: Framework for Policy Choices

Possibly Changing Background Factors	Policy Choices for Response[a] (1) Reduce CO_2 Production	(2) Remove CO_2 from Effluents or Atmosphere	(3) Make Countervailing Modifications in Climate, Weather, Hydrology	(4) Adapt to Increasing CO_2 and Changing Climate
Natural warming, cooling, variability			Weather Enhance precipitation Modify, steer hurricanes and tornadoes	Environmental controls heating/cooling of buildings, area enclosures Other adaptations habitation, health, construction, transport, military
Population global, distribution: nation, climate zone, elevation (sea level), density				Migrate--internationally, intranationally
Income global average distribution				Compensate losers--intranationally, internationally

Governments				
Industrial emissions non-CO_2 greenhouse gases particulates			Climate Change production of gases, particulates Change albedo ice, land, ocean Change cloud cover	
Energy per capita demand fossil versus nonfossil	Energy management Reduce energy use Reduce role of fossil energy Increase role of low-carbon fuels	Remove CO_2 from effluents Dispose in ocean, land Dispose of byproducts in land, ocean		
Agriculture, forestry, land use, erosion Farming and other dust Agricultural emissions (N_2O, CH_4)	Land use Reduce rate of deforestation Preserve undisturbed carbon-rich landscapes	Reforest Increase standing stock, fossilize trees		Change agricultural practices: cultivation, plant genetics Change demand for agricultural products, diet Direct CO_2 effects Change crop mix Alter genetics
Water supply, demand, technology, transport, conservation, exotic sources (icebergs, desalinization)			Hydrology Build dams, canals Change river courses	Improve water-use efficiency

[a]Responses may be considered at individual, local, national, and international levels.

going to increase and climate is going to change in systematic fashion, can we do something about climate? Finally, Category 4 is adaptation, consisting of all the policies or actions taken in consequence of anticipated or experienced climate change.

In Category 1, production of CO_2, there are two main subdivisions: energy and land use. Energy breaks down into three main subdivisions: reduction in total energy use, reduction in the fossil fuel component, and switching to less carbon-intensive fossil fuels. Land use cuts across categories 1 and 2. There is no way to "unuse" fuel that has been burned, but forests can be grown or cut, and the net effect can go either way. Preserving a growing forest rather than cutting it can be thought of as producing less CO_2 or removing it from the atmosphere. The relevant land use encompasses more than forests and other living biomass. What happens to forests affects the release of carbon from the exposed soil, and so does what happens to unforested land through cultivation, erosion, and other disturbances or changes.

Category 2, removal of CO_2, shares with production the characteristic that it affects the global carbon inventory. Removal can be subdivided into processes that take CO_2 out of the atmosphere at large and those that "scrub" the CO_2 or otherwise remove it directly from the products of combustion, i.e., from stack gases and other exhausts. With respect to using photosynthesis and reforestation to reduce atmospheric CO_2 one conclusion is inescapable, irrespective of a hundred years' technological change: increasing the standing stock of trees can be no great part of any solution to the growing CO_2 problem. That does not mean that a strategy for the use of lands and forests should ignore CO_2, only that the role of trees, standing or fossilized, will be modest. "Scrubbing" from stacks and "washing" by the oceans offer the possibility of yielding to technological advance.

Category 3, modification of climate and weather, can be summarized in four points. (1) From study of CO_2 we know that, in principle, modification of climate and weather is feasible; the question is what kinds of advances in climate and weather modification will emerge over the coming century? (2) Interest in CO_2 may generate or reinforce a lasting interest in national and international means of climate and weather modification; once generated, that interest may flourish independently of whatever is done about CO_2. (3) Climate and weather modification may be more a source of international tension than a relief. (4) CO_2 may not dominate the subject of anthropogenic climate change as it does now; emission of, or reduction of emission of, non-CO_2 greenhouse gases may become increasingly important to policy on climate change.

Category 4, adaptation, policies, or actions taken in consequence of anticipated or experienced climatic change, will consist of a multitude of largely decentralized, unconnected actions. Adaptation can be undertaken by units of all sizes, families, firms, ministries and departments, cities, states, nations, and international organizations. Impacts of climatic change could, of course, be numerous and diverse, affecting agriculture and water supply, ecosystems, and location of industry, for example, and adaptive response could thus take numerous forms, from writing assessment reports (studying the problem), to developing markets for water and incentives for water conservation, to enlarging buffer stocks, to strengthening financial institutions (e.g., insurance), air conditioning and central heating, and educational and training activity. Qualitatively, the basic adaptive responses would seem to be learning new skills and relocation or migration (Meyer-Abich, 1980).

1.4.4 Reprise

Overall, we find in the CO_2 issue reason for concern, but not panic. Although the prospect of historically unprecedented climatic changes is troubling, the problems that may be associated with it are of quite uncertain magnitude, and both climate change and increased CO_2 may also bring benefits. There are theory and evidence for each link in the chain of causal inference that we have described, but it could be that emissions will be low, or that concentrations will rise slowly, or that climatic effects will be small, or that environmental and societal impacts will be mild. Thus, we make some tentative suggestions about actual and near-term changes of policies, firmer recommendations about applied research and development with regard to the possibility of a CO_2-induced climatic change, and strong recommendations about acquiring more knowledge of various aspects of the CO_2 question. In our judgment, the knowledge we can gain in coming years should be more beneficial than a lack of action will be damaging; a program of action without a program for learning could be costly and ineffective. In the words of one reviewer of the manuscript of this report, our recommendations call for "research, monitoring, vigilance, and an open mind."

1.5 RECOMMENDATIONS

1.5.1 Can CO_2 Be Addressed as an Isolated Issue?

Before discussing actions and policies, we raise the question of whether CO_2 should be treated jointly with other issues or as a separable, isolated issue.

If one chooses to isolate the CO_2 issue, it can be judged in basically two ways. One is the approach of welfare economics: to measure the potential costs of a CO_2 buildup against the potential costs of controlling CO_2 emissions (or other societal reponses). For example, Nordhaus (1980) has proposed an optimal control strategy for concentrations, in which strategies are judged by the effects they generate on paths of consumption. "Consumption" here is interpreted in a broad way, including not only conventional items, such as food, clothing, and shelter, but also intangibles, such as enjoying the environment. The purpose of economic policy--and CO_2 control--is to enhance total consumption to the greatest extent. The central result of an analysis along these lines at present is that, given current knowledge, we are highly uncertain about the appropriate direction and stringency of CO_2 controls. The key uncertainties in the analysis are (1) the economic and social impact of elevation of CO_2 concentrations, (2) the economic costs of controlling CO_2 emissions, and (3) the relevant value judgments (e.g., discount rates) that we should apply in weighing alternative paths. In this approach the best single investment strategy for coping with the CO_2 issue is more research.

A second approach that treats the CO_2 issue in isolation compares it with other areas in which societal investments might be made. Is CO_2 more or less important than nuclear war, or economic depression, or population growth, or drug addiction? If CO_2 comes out far down the agenda, the implication is that little or no investment, even of research, need be made in it. Meyer-Abich (1980) has argued that CO_2 is "chalk on a white wall" or a "particular darkness in the night." He proposes that the world already faces such serious problems of energy, agriculture, water, and land use that additional problems from climatic change will be trivial; CO_2 will be a marginal, probably negligible, factor.

If the significance of CO_2 is not isolated as an issue, the approach is to stress its ties to other social, economic, and environmental problems. For example, use of fossil fuels generates not only CO_2 but other problems for air and water quality. Deforestation creates not only CO_2 but problems of soil erosion. And, similarly, responses that might be useful with respect to CO_2-induced climatic changes--such as reducing water demand or increasing water supply in the Great Plains of the United States--might be appropriate responses to other problems, like depletion of the Ogallala Aquifer. Attitudes toward CO_2 may be different if one treats it jointly with other issues; policies judged expensive for one issue might seem more affordable as responses to a combination of issues. Advocates of action deriving from potential CO_2 problems generally rely on this argument. A single problem, like CO_2 or acid rain or soil erosion, may not weigh heavily in the cost-benefit calculus, but combinations of problems may weigh heavily indeed. It is the essence of the political process to come to terms with arrays of problems.

1.5.2 Actual and Near-Term Change of Policies

We recommend caution in undertaking any major changes in current behavior and policies solely on account of CO_2. It is probably wiser not to act aggressively to "solve the CO_2 problem" right now when we really do not know the future consequences or context of CO_2 increase. In trying to consider the world of 50 or 100 years from now, we cannot be sure that we can tell the difference between solutions and problems. It is instructive to look back at, say, 1905 to see if even the best guesses made at that time about accumulating world problems and their solutions were actually valid or useful as planning guides for the twentieth century. Life has changed since then in many unexpected ways; penicillin and air transport are vivid examples.

It is not easy to anticipate now what will appear as correct decisions with respect to CO_2 50 or 100 years hence. For example, we might anticipate that climatic warming in the north would permit seaborne commerce through the Northwest Passage. However, it is quite possible that transport will be so different in generations hence that surface ice cover in the Arctic will be largely irrelevant, and expenditures to develop Arctic transport as we can currently envision it would be wasteful.

Allowing for this general caution, we nevertheless recommend that activities and planning that involve long time scales (i.e., on the order of decades or more)--particularly concerning agriculture and water resources--explicitly incorporate the assumption that the climate of the future is unlikely to resemble the climate of the recent past. In fact, because of trace gases besides CO_2, future climate is likely to diverge increasingly from our recent climate whether we control CO_2 or not. Planners of hydroelectric or irrigation systems, agricultural infrastructure, forestry, hazardous-waste disposal, nature conservation, and other areas might ask whether current policies are vulnerable to climatic change and whether alternative policies can be designed that will both serve our needs and be robust in the face of climatic change.

The need to design and manage on the basis of likely future climates as well as past climates is probably greatest in the area of water resources. Experience has shown that the planning and construction of water-resource developments in major river basins can take several decades. It is not too soon to begin to think of ways in which the planned use of water could alleviate potential effects of climatic changes or even take advantage of them. Several possible measures come to mind: changes in legislation that would allow water to be transferred from one river basin to another; improved efficiency in the use of water for irrigation; conservation of waste water and of municipal water supplies; limitations on the size of irrigated areas; increases in crop yields per unit volume of applied water; and enhancement of the recharge of aquifers (Revelle, 1982).

In fact, greater efficiency in water use is a prudent goal to pursue even if we neglect CO_2 concerns. Opportunities for conservation of agricultural water through improved conveyance and farm distribution systems, application methods, scheduling, crop selection, and cropping practices are great. So, too, are the opportunities to halt and reverse the degradation of irrigated lands beset by waterlogging and salinization (White, 1983).

In fact, as Waggoner (this volume, Chapter 6) shows, we already know how to reduce greatly the harm of drought by storing more precipitation and getting more yield from the stored water. Fewer weeds and tillages decrease the loss of soil moisture, while more stubble captures more precipitation. Other means of increasing storage include barriers that catch snow, leveling and terracing to decrease runoff, and harvesting water from nearby acreage. Matching irrigation to need can decrease pumping; but if most of the former excess was returned to the groundwater supply, the saving of water will not be great. Getting more yield per acre increases the yield of marketable product per unit of water consumed. Changing to shorter growing-season crops can also increase water-use efficiency.

Concern about climatic changes may foster readjustments in agriculture and water management that will be beneficial in any event. For example, some of the same measures that would help us to prepare for a permanently drier average climate--such as encouragement of water conservation or provision of additional carry-over water storage--would make current agriculture and water use more resistant to passing droughts.

1.5.3 Energy Research and Policy

The major impetus to this report was concern about the projected impact on atmospheric CO_2 of fossil fuel combustion, coal conversion, and related synthetic fuels activities authorized in the Energy Security Act of 1980 (PL-96-294, Title VII, subtitle B, appended). During 1979 many scientists expressed concern that U.S. energy policy and its changing emphasis would exacerbate the CO_2 problem. They recommended, in part, conservation of fossil fuels and choice among fossil fuels and conversion processes based on CO_2 emissions, implying a bias against coal and carbon-based synthetic fuels.

Our study does not support the extent of concern expressed about the importance of decisions planned at that time about synthetic fuels in relation to the CO_2 issue (see the following section for more detail). Analysis of the contribution of various factors to uncertainty about future emissions of CO_2 shows that other factors besides the choice among fossil fuels are much more important in determining future emissions. If one wishes to reduce emissions through decisions relating to energy policy, placing a moratorium or limitation on development of synthetic fuels, especially in one country, even a large one like the United States, is likely to be a poor choice. By itself, shifting the fuel mix within fossil fuels is highly unlikely to achieve a reduction of CO_2 concentrations by more than a few parts per million by the year 2100, even if accomplished globally. A global policy banning carbon-based synfuels might lead to a reduction in CO_2 concentration of only 10-20 ppm by the year 2100, because of the large alternative supply of coal available. Suggestions that a near-term shift to carbon-based synfuels could advance the time of CO_2 doubling by decades are much exaggerated. Our study does, however, suggest that considerations of the ease of substitution between fossil and nonfossil fuels and between energy and labor inputs in the economy, along with extraction costs of fossil fuels, and the bias of technological change in the energy sector are of great importance in determining future emissions, which will account for considerable variance in atmospheric concentration over the next century.

Thus, we conclude that

1. The possibilities that concerns about the CO_2 issue will become more serious provide strong arguments for stimulating research on nonfossil energy sources. We may find that emissions are rising rapidly, that the fraction remaining airborne is high, that climate is very sensitive to CO_2 increase, or that the impacts of climate change are costly and divisive. In such a case, we want to have an enhanced ability to make a transition to nonfossil fuels.

2. The potential disruptions associated with CO_2-induced climatic change are sufficiently serious to make us lean away from fossil fuel energy options, if other things are equal. However, our current assessment of the probability of an alarming scenario justifies primarily increased monitoring and vigilance and not immediate action to curtail fossil fuel use.
3. Analysis of prospective CO_2 emissions does not offer a strong argument for making choices among particular patterns of fossil fuel use at this time.

It should be kept in mind that important uncertainties about future emissions stem from variables such as rates of productivity and population growth, as well as from decisions centered in the energy sector. Several of these variables that affect future emissions, like productivity growth, population growth, or technological change, strongly resist adjustment by policymakers.

1.5.4 Synfuels Policy and CO_2

By synfuels policy we refer to the development and use of new carbon-based fuels derived from fossil sources to replace a portion of the existing fuel mix. Synfuels would include oil and gas from coals and shales. At present, synfuels are a negligible part of world energy supply.

Overall, some skepticism is in order about the relation between the CO_2 issue and encouragement of synfuels. The causal links are complex and numerous; the direction of the effect is ambiguous; the importance of the effect has not been convincingly demonstrated. As suggested above, it would probably be more effective to address the issue of CO_2 emissions through policies other than the highly uncertain mechanism of slowing or speeding synfuels development.

Synfuels policy might be separated into three components:

1. Short-run emergency preparedness (for example, stockpiles or temporary surge capacity).
2. Tax or subsidy arrangements for research and development (R&D) on synfuels or for the use of synfuels.
3. Less specifically focused policies that might have an effect on synfuels and CO_2 emissions (mass transit, hydrogen R&D, nuclear- or solar-power policy).

A range of interactions can be envisaged between the different kinds of policies and CO_2. From short-run policies in one or a few nations, such as strategic stockpiles, there is unlikely to be significant impact on CO_2. More widespread, enduring tax and subsidy arrangements are likely to have more important effects on CO_2, but the impacts are not obvious. Of course, taxes on carbon-based fuels are the most predictable in their emission-reducing impact.

The most likely kind of synfuels policy is to encourage R&D and commercialization of synfuels; this was the major goal of the 1980 Energy

Security Act. An example of such a policy is an interest guarantee, which acts as an implicit subsidy to production and as a subsidy to learning about the technology. The effect of such a subsidy is subtle. If in fact the technology is viable, the subsidy will speed up its introduction, but it is unlikely to have a major impact on CO_2 emissions in the long run. If the technology is not viable, it will, like commercial supersonic transport, tend to disappear even with subsidies.

A second subtlety arises because the impact of early introduction of synfuels on CO_2 emissions is ambiguous and could either speed up or slow down CO_2 emissions. For example, if the price of the synfuel turns out to be high and the availability of other fuels is restricted, demand for energy may fall, and CO_2 emissions may decline. If, on the other hand, the new synfuel is attractive economically, it may lead to greater CO_2 emissions. Another ambiguity arises because the synfuel may replace either a carbon-based or noncarbon-based fuel. In the former case, it might have a small impact on CO_2 emissions; in the latter, a larger one. It is impossible to determine a priori the net effect of such influences, but using energy models might provide a range of answers.

Probably the most important determinant of the role of synfuels will be their interaction with other policies. Thus, a policy that encourages mass transit or solar energy could discourage carbon-based synfuels. A nuclear moratorium or heavy regulation of nuclear power could encourage synfuels. A stringent environmental policy, banning surface mining or growth in emissions of air pollutants, could discourage synfuels. These more subtle influences are probably important in determining CO_2 emissions from synfuels, more important than synfuels policy itself.

1.5.5 Applied Research and Development

The prospect of climate change clearly lends urgency to applied research and development in two areas besides energy: agriculture and water resources. Although no detailed timetable of climatic change is yet available, we have some notion of the general character of the climatic challenges that may be ahead of us. America is fortunate in possessing widespread and effective research networks in both agriculture and water resources. The tasks are to build and to maintain a strong and flexible national capability to adapt to changing climate and indeed to exploit new opportunities that changed climates may offer.

Modern agriculture possesses great flexibility and adaptability to change. For example, changing crops can be swift. And changing the variety of a crop by planting a different strain can be even swifter, because little in the process needs to be altered, from the dealer who supplies chemicals, to the farmer who must finance equipment, to the consumer who may be scarcely aware of the change. The question is whether breeders can develop new varieties to adapt to climate as fast as it changes. A complete breeding cycle is approximately a decade from the beginning of inbreeding to the marketing of a product; the objectives can be shifted during the first 5 years, and new hybrids emerging are those adapted to the final 3 years. Ideally, one would identify the critical environmental changes for a given crop and engineer appropriate genetic changes to cope with the new conditions. This process is as yet beyond our capabilities, but research may bring it within our grasp. In any event, there is ample reason to believe that steady work by plant breeders will continue to produce varieties of the life-sustaining major crops which are adapted to a changing environment.

There are also numerous avenues to pursue if there is less rainfall and runoff. White (1983) emphasized two sets of advanced techniques with potentially great impact and broad application: those relating to groundwater exploration and extraction and those relating to re-use of water. Improved seismic and geological surveys, well drilling, and pumping methods are opening up a huge volume of water previously ignored or inaccessible. Where this water exists in aquifers that are easily rechargable, it represents a potentially permanent addition to water supplies. White pointed out that the techniques have been of major importance in developing countries that can use them to gain access to previously untapped supplies without building elaborate storage and conveyance works. As a result of advances in treatment methods and in system planning, the re-use of water is also beginning to be viewed as a practical measure in both urban and agricultural settings. Other more tested alternatives deserve more widespread appraisal as well; these include water-pricing policies, leak detection, water-conserving devices, canal lining when seepage from the canal enters a saline-water aquifer, water application scheduling, drip irrigation, and choice of water-efficient or salt-tolerant crops.

Water technology needs to be developed complementary to changes in social and economic institutions. Research and activities should not focus exclusively on water control but on the combination of water development, land-use management, and economic and social adjustment.

1.5.6 Basic Research and Monitoring

The CO_2 issue involves virtually every branch of science and impinges on virtually every area of human activity. Whether motivated by concern for CO_2 or some other problem, or simply by scientific interest, research is in progress on nearly every identifiable topic that could contribute to our understanding. Research is funded by several federal agencies and coordinated through the National Climate Program and the lead agency for CO_2-related research, the Department of Energy (DOE). The DOE Carbon Dioxide Research and Assessment Program itself supports a broad program of investigation specifically aimed at the periodic development of detailed and comprehensive technical assessments drawing on the work of a large community. Indeed, in developing some portions of this report, we have worked closely with the DOE program and drawn on its participants, research results, and periodic coordination and review activities.

As we have emphasized throughout this report, the uncertainties regarding the CO_2 issue are numerous. Some areas of uncertainty are unlikely to diminish rapidly. The issue, and research directed at its illumination, will be with us for a long time. While we have tried to carry out some exemplary analyses in this report, we can propose no fresh and efficient route to the knowledge we would like to have. Rather than stress the need to attack any particular link in the CO_2 argument, we stress the need for balanced attention to the major components: emissions, concentrations, climate change, and environmental and social impacts and responses. A plethora of research recommendations exists, and several areas, for example, the carbon cycle and climatic effects, are being pursued with considerable vigor. Other areas may require more concentrated effort. For example, detection of the effects of increasing CO_2 may require better focus and increased interaction among climatologists, statisticians, investigators from several other disciplines, and those responsible for design and operation of monitoring programs. We offer several general comments about research with regard to CO_2 assessment before addressing specific subject areas.

1.5.6.1 General Research Comments

1. A broad, healthy program of basic research in the physical, biological, and social sciences is an indispensable foundation for our efforts to understand the CO_2 issue, which offers vivid evidence of the indivisibility of basic and applied research. The kinds of knowledge that we would like to have for analyzing the CO_2 issue--knowledge, for example, about the role of clouds and the ocean in the climate system or about the behavior of ice sheets--are unlikely to be produced on a procurement schedule defined by contracting agencies. While rates of learning appear to be rapid in most of the areas of concern for the CO_2 issue, fundamental, difficult questions are involved, and in some areas we simply do not know whether or when insights will be forthcoming.

2. The importance of geophysical and biospheric monitoring must be stressed. Over the past decade or so we have seen a series of climate-related issues become prominent and subjected to costly analysis. These include, for example, the Sahelian drought, the impacts of stratospheric flight on the ozone layer, and now CO_2. In each case, data bases are deficient, whether about the variability of climate in the Sahel, or on concentrations of gases in the stratosphere, or of carbon inventories, global temperatures, and CO_2 concentrations. With sound, stably supported, long-term geophysical and biospheric monitoring programs we can have the capability for dealing with such issues and with new ones that will undoubtedly arise in the future.

Certain kinds of routine data collection are both expensive and uninteresting and may not appeal to the most gifted scientists, yet their cumulative significance over time can be important. There is a problem of how to ensure that an adequate investment is made in this kind of data collection and that it is done well with methods that are validated by the most knowledgeable people. Historically, this kind of deficiency has been serious with respect to research programs in many environmental areas (Brooks, 1982).

3. More systematic setting of research priorities based on the relative contributions of various problems to the overall uncertainty of the CO_2 issue should be considered. For example, the sensitivity studies of Nordhaus and Yohe (this volume, Chapter 2, Section 2.1) and Machta (this volume, Chapter 3, Section 3.6) are suggestive about specific research priorities in certain areas. We cannot, it should be emphasized, move directly from estimates of sources of uncertainties to a budget allocation for research funds on CO_2. It may be easier, for example, to reduce uncertainties about the "depletion factor" for carbon fuels than about future "productivity growth," even though the latter appears to be a greater source of uncertainty. The problem is to provide incentives that will induce the most gifted researchers to shift to the most significant problems, i.e., to those areas contributing the most to uncertainties regarding future impact of CO_2. But these researchers must be convinced not only by the social importance of the problem but also by the perception of genuine scientific opportunities to contribute to its resolution.

4. The question of overall program balance in the CO_2 area also needs to be considered. We argued above that the way an issue is defined is fundamental to what research and policies are considered and that the main perspective here should probably be climate change, not simply CO_2. A CO_2 program geared heavily toward estimation of CO_2 concentrations and associated warming will probably be less useful than one with strong interests in, for example, other greenhouse gases, water resources, and sea-level rise, as well. Indeed, it may be wise for the government to anticipate the evolution of concern in this area into a broader program on human activities-greenhouse gases-climate change-adaptation.

5. Human-induced climatic change and the CO_2 issue more generally are not problems for which traditional paradigms of policy analysis and methods of assessment, like those taken from economics, engineering, and decision analysis, have had much success (Glantz et al., 1982). It is naive, indeed mistaken and misleading, to expect a "definitive" assessment of the CO_2 issue here, or from other groups or individuals in the foreseeable future. What is required is a sustained approach, probably no larger than the sum of efforts currently under way, which emphasizes carry over of learning from one effort to the next.

A healthy reaction to CO_2 would be one in which there is a steady production of knowledge, and in which every few years, either in the United States or in another country or in a combination of countries, developed or developing, east or west, some group undertakes a thoughtful synthesis. As we become generally better at global geophysical and biospheric modeling and at thinking one or two or more generations into the future, assessment of the CO_2 issue will deepen and strengthen. Each assessment must be regarded as part of an iterative process, consisting of efforts to assemble relevant knowledge, to assess the problem based on the assembled knowledge, to estimate the marginal value of additional knowledge to reduce uncertainties in the assessment, and to diffuse results to individuals and groups with power to act (Mar, 1982).

6. While the CO_2 issue remains--appropriately so, in our view--largely in the research community, it is important to consider the possibility that the issue may become prominent in political arenas. It is quite possible that, as a result of weather conditions or bad harvests in one part or another of the world, CO_2 will abruptly rise nearer to the top of many national and international agendas, regardless of the scientific basis for concern. How long it might hold such a position one can only speculate; most issues star for only a short while. And it is unlikely that prominence for one session of the United Nations or the United States Congress would bring about policy changes that would be of lasting importance with respect to CO_2. It is important that we fashion perspectives and programs that can be sustained through periods of excessive attention or inattention to the issue.

1.5.6.2 The International Aspect

Should it be desirable to control CO_2 emissions, it would be natural to suppose that they could be controlled in a way similar to conventional pollutants, but this supposition would be optimistic. Most externalities, or side effects of economic activity, are at least internal to nations; thus a government can weigh the costs and benefits of a control program and decide that, on balance, it is in the interests of its citizens. The CO_2 problem is different from conventional pollutants because it is an externality across so much space and time. Thus, just as we as individuals have little reason to curtail our emissions, we as a nation have little incentive to curb CO_2 emissions. By curbing our CO_2 output, we make little contribution to the solution and do not know whether we will receive any benefits. With respect to a CO_2-induced climate change, there is little incentive to act alone. The problem is exacerbated by the long time period over which CO_2 can affect our society and the environment. Although politicians may have one eye on posterity, political systems tend to be myopic and to emphasize short-term rewards.

Given the need for widespread, long-term commitment, a CO_2 control strategy could only work if major nations successfully negotiated a global policy. While such an outcome is possible, there are few examples where a multinational environmental pact has succeeded, the nuclear test ban treaty being the most prominent. Other clearly recognized problems--whale fisheries, acid rain, undersea mining, the ozone layer--emphasize how time on the order of decades is required to achieve even modest progress on international management strategies.

With regard to CO_2 increase, the multilateral bargaining is severely complicated by the likelihood that some major countries will probably _benefit_, at least from a moderate rise. For example, it is sometimes conjectured that the Soviet Union and Canada would benefit from a warmer climate. Given that these two countries (and the former's allies) burn 25% of world coal and hold a larger share of carbon resources in the ground, it is hard to see how a CO_2 control strategy can succeed without them. Given the unlikelihood that the United States or other western nations will compensate the Soviet Union for participating, it is hard to see why the Soviet Union would participate. If a major nation or group of nations does not participate, it is difficult to envisage others, particularly developing countries, making a major sacrifice. Thus, differences in the experience and expectations of nations pose major obstacles to any international agreement for control of CO_2.

While we may not be optimistic that agreement could ever be reached among nations about a control strategy for CO_2 (should such a control strategy be the desirable response in the first place), it is essential that research and dialogue about the CO_2 issue be carried on internationally. Study of world energy supply and demand inherently necessitates data from many countries, and a variety of national and international views of the energy situation is likely to enrich our analysis substantially. Study of the carbon cycle is also inherently global, and significant capability in this area, moreover, resides outside the United States. The atmosphere and oceans are similarly global, and in many countries there are strong research communities investigating them. Finally, monitoring--of the climate, the biosphere, the oceans--requires international participation. Exchange between nations and international collaboration should therefore be pursued extensively with respect to study of the CO_2 issue.

In addition, no matter how outstanding the analysis coming out of a particular country might be, other countries will always view individual national studies with suspicion. With a potentially divisive issue like CO_2, it is critical that assessments be undertaken independently by several nations, as well as by relatively neutral international groups. We commend the cooperation that has been initiated among the International Council of Scientific Unions, the World Meteorological Organization, and the United Nations Environment Program to prepare a careful assessment of the CO_2 issue during the next few years. The United States should contribute energetically to this effort from both governmental and nongovernmental communities concerned with studying and responding to the prospect of a CO_2-induced climate change.

It is worth noting that on particular issues, like detection of a CO_2-induced climatic change and evaluation of climate model results, there may be considerable efficiency in having an international focal point. Qualified individuals and groups from the United States should participate in support of such centers and assist where possible and appropriate in sharing, comparing, and analyzing findings on important questions. We note that in several areas--for example, biogeochemical cycles, atmospheric monitoring, and climate research--international programs and organizations have been functioning well.

In general, diffusion of sound information and calm, thoughtful anticipation of the future may be the best international insurance against poor decisions about possible control of CO_2 emissions, against possible adverse consequences of climatic change, and against the issue's becoming a source of major conflict. Indeed, if approached in a constructive manner, the CO_2 issue offers an opportunity to strengthen international cooperation and capabilities in many highly desirable respects. Vigorous efforts should be made to prevent CO_2 from becoming politically divisive; instead the nations of the world should seek to benefit from it as a catalyst for learning how to treat common problems effectively.

1.5.6.3 Projecting CO_2 Emissions

The current modeling and knowledge of future CO_2 emissions appears marginally adequate today; we have a general idea of likely future trends and the range of uncertainty. It may be that further effort could increase the accuracy of our forecasts substantially. Given the large uncertainty that future energy growth and energy projections are contributing to the CO_2 issue, this area may well merit more research attention and support than it has received in the past. Future research efforts might be designed with four points in mind.

1. In general, the most detailed and theoretically based projections of CO_2 emissions have been a spillover from work in other areas, particularly energy studies. This fact suggests that continued support of energy modeling efforts will be of importance in pushing out the frontier of knowledge about future CO_2 emissions, as well as the interaction between possible CO_2 controls and the economy.
2. We have identified a serious deficiency in the support of long-run economic and energy models in the United States. There is not one U.S. long-range global energy or economic model that is being developed and constantly maintained, updated with documentation, and made usable to a wide variety of groups. This shortcoming is in contrast to climate or carbon-cycle models, where several models receive long-term support, are periodically updated, and can be used by outside groups. Another contrast is with short-run economic models, which are too plentiful to enumerate.
3. Most CO_2 projections have been primitive from a methodological point of view. Work on projecting CO_2 emissions has not drawn sufficiently on existing work in statistics, econometrics, or decision theory. There has been little attention to uncertainties and probabilities. Also, considerable confusion of normative and positive approaches exists in modeling of CO_2 emissions.
4. Application of models for analysis of policies, where there are, for example, feedbacks to the economy from climatic change or CO_2 control strategies, is just beginning. Efforts to evaluate the effectiveness for CO_2 control of energy policies of particular nations or groups of nations in a globally consistent framework have been lacking.

1.5.6.4 Projecting CO_2 Concentrations

Projecting CO_2 concentrations consists essentially of the application of our knowledge of the carbon cycle to projections of CO_2 emissions.

After more than a decade of intense research, findings on the roles of the different reservoirs of carbon, particularly the biosphere and the oceans, remain in obvious, unresolved conflict. Efforts to improve our understanding of the carbon cycle are desirable for many reasons, and monitoring of carbon in its various forms must be maintained and in some cases expanded. However, examination of uncertainties in the CO_2 issue introduced by different factors suggests that uncertainty about the airborne fraction is of less significance for the overall CO_2 issue than the extensive discussion in the CO_2 literature would suggest. Questions that may merit more attention in the carbon cycle area include the history of CO_2 concentrations and the future behavior of the oceans and biosphere beneath a high-CO_2 atmosphere.

With respect to the biosphere, surveys, field experiments, and dynamic models will be the means to achieve insights into the responses of the biota to increasing CO_2 concentrations and changing climate. To provide a better basis for this work, inventories of carbon in the biosphere should be improved; satellite surveillance may be particularly helpful in strengthening the empirical basis of biospheric research and in assessing biospheric changes on a regular basis. Clarifying the history of the size of the biotic pool over the past century is also useful; it may help corroborate data on the preindustrial concentration, reduce uncertainty about the fraction of emissions remaining airborne, and provide a check on the quality of carbon-cycle models.

Historical and geological data on CO_2 from records of the past such as ice cores have proven to be valuable, and an expanded effort to confirm and refine previous findings should be undertaken. Recent ice-core data have placed new constraints on what the atmospheric CO_2 levels could have been in past centuries. Isotopic studies of tree rings, lake sediments, and current air samples offer the potential to elucidate further the history and fate of atmospheric CO_2.

With respect to the atmosphere, the need to continue high-precision observations cannot be overemphasized. The atmospheric CO_2 data provide information beyond simple evidence that atmospheric CO_2 is increasing. Through careful measurements, one is able to derive valuable information from the temporal and spatial variability of CO_2. For example, the pattern of results so far is suggestive of a minimal contribution from sources of CO_2 other than fossil fuel over the past couple of decades or even that the biota are a net sink for CO_2, although the limited quantity of the data (and the possibility of alternative explanations) prevent any definitive statement today that excludes nonfossil fuel sources.

With respect to the oceans, it now appears to be quite possible to measure the changing CO_2 properties of the ocean over time by using modern techniques, though no ongoing program yet exists to do so. Previous programs have provided regional coverage in different years and seasons. We recommend initiation of a program with more consistency in space and time. Until quite recently, oceanic CO_2 system measurements contained substantial inaccuracies.

Although many oceanographers believe that current methods appear satisfactory to answer approximately the question of how much CO_2 the sea takes up, there is room for improvement. Models of ocean CO_2 uptake have depended greatly on tracer data, particularly natural and bomb-produced ^{14}C, tritium, and, more recently, halogenated hydrocarbons. Typically, these models represent some features of ocean chemistry quite well, even though they represent vertical transport by a simple diffusion coefficient. The models treat only the CO_2 perturbation and do not yet adequately mimic the natural and complex CO_2-oxygen nutrient biogeochemical cycles within the ocean. As our ocean data base grows, the current generation of one-dimensional models will become increasingly inadequate, and incorporation of the CO_2 and tracer data into new models will be required. Warming accompanying atmospheric CO_2 rise will also affect the ocean. Storage of heat in the upper layers will postpone, but not prevent, climate change. Models of this heat storage generally treat it as passive uptake, not affecting water mass formation and vertical circulation. If changes occur in ocean transport and dynamics, they may affect ocean CO_2 uptake in significant ways. For example, changes in formation of oceanic bottom waters in high latitudes may affect the rate of transfer of dissolved CO_2 to the deep ocean. Progress in incorporating such processes and features into more advanced ocean models is anticipated and should be encouraged. Such models will require complex global measurements from ocean research ships, ocean-scanning satellites, and other sources.

With respect to the methane hydrate clathrates, we recommend further consideration of the probable effects of a rise in ocean-bottom temperatures on the stability of the clathrates. We also recommend a sediment sampling program on continental slopes to determine the depth, thickness, and distribution of methane hydrate clathrates, especially where oceanfloor temperatures and depths are such that methane release is possible from ocean warming during the next century.

Finally, more attention should be given to interactions among the biogeochemical cycles of carbon, sulfur, nitrogen, and phosphorus (Bolin et al., 1983). These interactions may offer another example of an area where there are important, as yet unforeseen, feedbacks. To illustrate, lower atmospheric CO_2 concentrations 10,000 to 20,000 years ago may have resulted from changes in oceanic biologic production, perhaps related to larger quantities of ocean nitrogen and phosphate.

1.5.6.5 Climate

The width of the range of projections associated with a given CO_2 increase and the desire for more detailed information on prospective climate change warrant continuing support for climate research. Special attention should be given to the role of the oceans and clouds, model comparison and validation, extremes, and non-CO_2 greenhouse gases.

The heat capacity of the upper ocean is potentially great enough to delay by decades the response of climate to increasing atmospheric CO_2, as modeled without it; and the lagging ocean thermal response may cause important regional differences in climatic response to increasing CO_2. The role of the ocean in time-dependent climatic response deserves special attention in future modeling studies, stressing the regional nature of oceanic thermal inertia and atmospheric energy-transfer mechanisms. Progress in understanding the ocean's role must be based on a broad program of research and ocean monitoring. Particular attention should be paid to improving estimates of mixing time scales in the main thermocline.

Cloud amounts, heights, optical properties, and structure may be influenced by CO_2-induced climatic changes. In view of the uncertainties in our knowledge of cloud parameters and the crudeness of cloud prediction schemes in existing climate models, it is premature to draw conclusions regarding the influence of clouds on climate sensitivity to increased CO_2, particularly on a regional basis. Empirical approaches, including satellite-observed radiation budget data, are an important means of studying the cloudiness-radiation problem, and they should be pursued.

Simplified models permit economically feasible analyses over a wide range of conditions. Although they can provide only limited information on local or regional effects, simplified models are valuable for focusing and interpreting studies performed with more complete and realistic models.

In addition to improvement and validation of models, we recommend more research into the question of statistical properties of a warmer, CO_2-enriched atmosphere. Particular attention should be given to possible changes in the character and frequency of extreme conditions and to severe storms. Variance studies with general circulation models (GCMs) are one potentially useful approach. Insights into the question of extremes may also be obtained from research in other fields, such as hydrology.

With respect to non-CO_2 greenhouse gases, improved, sustained monitoring is called for. Available instrumentation and methods are probably sufficient for high-quality data collection.

Equally important is the goal of obtaining an understanding of the mechanisms by which the gases increase. For the chlorofluorocarbons and several other potentially significant trace gases, only industrial production statistics are needed to establish annual emissions. For CH_4 and N_2O, biological sources probably dominate. Microbial processes in soils, water bodies, and living organisms produce and release these gases and a number of others. (Release of CH_4 and higher hydrocarbons from clathrates may become significant in the future.) Key sites for emissions and processes have been identified, but a variety of field and laboratory measurements requires implementation. Similarly, there is need for more elucidation of the mechanisms that control levels of ozone in the troposphere, where increases of ozone can have a warming effect. Interactions among the gases and with changing climate must be considered.

There is also need to make careful new projections of future emissions of non-CO_2 greenhouse gases. Projections of future emissions of these non-CO_2 gases are generally at a more primitive stage than are CO_2 projections. Projections have typically been derived from simple assumptions of linear increase or exponential growth based on a short segment of recent years. The times in the future vary to which the relevant studies of agricultural production, smelting, industrial use of chemicals, and other activities extend, and the assumptions employed vary as well. It would be desirable to have studies of the greenhouse effect that use assumptions consistently in generating both CO_2 emissions and emissions of other infrared-absorbing trace gases. In human activities--like those relating to emissions of the chlorofluorocarbons--where rapid technological change is occurring and where less inertia is imposed by a large and expensive capital stock than in the energy system, projections for several decades are especially hazardous.

The spectroscopic parameters of several of these gases are not well known, and even the band strengths of some have not been measured. The spectral transmittance and total band absorptions also need to be more closely determined. These improvements will help in developing more accurate radiative-transfer models and in answering questions about band overlap between constituents and with water vapor. Such information is needed for defining more accurate parameters in climate models.

1.5.6.6 Detection and Monitoring

In view of the importance of verifying the theoretical results about climatic effects of CO_2, a careful, well-designed program of monitoring and analysis must proceed. The information obtained will help us not only in detecting CO_2-induced changes as early as possible but in improving, validating, and calibrating the climate models employed for prediction of future changes.

If, as expected, the CO_2 signal gradually increases in the future, then the likelihood of perceiving it with an appropriate degree of statistical significance will increase. Given the inertia created by the ocean thermal capacity and the level of natural fluctuations, we expect that achieving statistical confirmation of the CO_2-induced contribution to global temperature changes so as to narrow substantially the range of acceptable model estimates may require an extended period. Improvements in climatic monitoring and modeling and in our historic data bases for changes in CO_2, solar radiance, atmospheric turbidity, and other factors may, however, make it possible to account for climatic effects with less uncertainty and thus to detect a CO_2 signal at an earlier time and with greater confidence.

A complicating factor of increasing importance will be the role of rising concentrations of greenhouse gases other than CO_2. While the role of these gases in altering climate may have been negligible up to the present, their significance is likely to grow. It will be difficult to distinguish between the climatic effects of CO_2 and those of other radiatively active trace gases. Their expected relative contributions to climatic change will have to be inferred from model calculations and precise monitoring of radiation fluxes.

A monitoring strategy should focus on parameters expected to respond strongly to changes in CO_2 (and other greenhouse gases) and on other factors that may influence climate. Candidate parameters may be identified, their variability estimated, and their evolution through time predicted by means of climate model simulations. Through analysis of past data, continued monitoring, and a combination of careful statistical analysis and physical reasoning, the effects of CO_2 may eventually be discerned.

Monitoring parameters should include not only data on the CO_2 forcing and the expected climate system responses but also data on other external factors that may influence climate and obscure CO_2 influences. Climate modeling and monitoring studies already accomplished provide considerable background for the selection of these parameters. Since fairly distinct climate changes are expected to become evident only over one or more decades, monitoring for both early detection and more rapid model improvement should be carried out for an extended period. Parameters may be selected for early emphasis on the basis of the following criteria:

1. Sensitivity. How do the effects exerted on climate by the variables or the changes experienced by the variables on decadal time scales compare with those associated with corresponding changes in CO_2?
2. Response characteristics. Are changes likely to be rapid enough to be detectable in a few decades?
3. Signal-to-noise ratio. Are the relevant changes sufficiently greater than the statistical variability to be measured accurately?
4. Past data base. Are data on the past behavior of the variable adequate for determining its natural variability?
5. Spatial coverage and resolution of required measurements.
6. Required frequency of measurements.
7. Feasibility of technical systems. Can we make the required measurements?

Initial application of these criteria leads to this list of recommended variables for monitoring:

Monitoring Causal Factors by Measuring Changes in	Monitoring Climatic Effects by Measuring Changes in
CO_2 concentrations	Troposphere/surface temperatures (including sea temperatures)
Volcanic aerosols	Stratospheric temperatures
Solar radiance	Radiation fluxes at the top of the atmosphere
"Greenhouse" gases other than CO_2	Precipitable water content (and clouds)
Stratospheric and tropospheric ozone	Snow and sea-ice covers
	Polar ice-sheet mass balance
	Sea level

In the above list, evaluated more thoroughly in Chapter 5, emphasis has been given to parameters that may contribute, either directly or through model improvements, to detection of CO_2 effects at the earliest possible time. Over the long run, it is important to build up a relatively complete data base of possible causes and effects of climate change and characteristics of climate variability, not simply for detection but to assist in research on and calibration of models of the climate system. Once we become convinced that climate changes are indeed under way, we will seek to predict their future evolution with increasing urgency and with increasing emphasis on parameters of societal importance (e.g., sea level and rainfall). We should thus anticipate that a detection program will gradually evolve into a more comprehensive geophysical monitoring and prediction program. It should be emphasized that the strategy proposed here is a simple tentative step in what must be an iterative process of measurement and study.

Collection of the desired observations will require a healthy global observing system, of which satellites will be a major component. Satellites can provide or contribute to long-term global measurements of radiative fluxes, planetary albedo, snow/ice extent, ocean and atmospheric temperatures, atmospheric water content, polar ice-sheet volume, sea level, aerosols, ozone, and trace atmospheric components; a well-designed and stable program of space-based environmental observation is essential if we are to monitor the state of our climate. Requirements and technical systems for monitoring high-priority CO_2 variables are summarized in Table 5.1.

We will also have to continue to improve climate models to reduce the uncertainties in predictions of climate effects and to validate the models against observations, although we believe that current climate models are sufficiently sound and detailed to enable us to identify a set of variables that could form the basis for an initial monitoring strategy. Statistical techniques for assessing the significance of observed changes may have to be improved so as to deal with the characteristics of the monitored variables. In the end, confidence that we have detected the effect of CO_2 will have to rest on a combination of both statistical testing and physical reasoning.

1.5.6.7 Impacts

1.5.6.7.1 Sea Level and Antarctica

With respect to sea-level rise, we recommend research to confirm the estimate of a possible 70-cm rise over the next century (Revelle, this volume, Chapter 8) and to explore the implications of a variety of other more gradual and more rapid rates of warming for sea-level rise. High priority should also be given to strengthening our understanding of the possibilities of disintegration of the WAIS and the rate at which this might occur. One promising approach to these questions is through examination of the morphology of reef corals at different depths in terraces formed during the last interglacial period about 125,000 years ago, when the ice mass in question may previously have disappeared.

Other studies and monitoring programs should also be undertaken. In doing so, five problems deserve special emphasis: possible changes in the mass balance of the Antarctic Ice Sheet; interaction between the Ross and Filchner-Ronne ice shelves and adjacent ocean waters; ice-stream velocities and mass transport into the Amundsen Sea from Pine Island and Thwaites Glaciers; modeling of the ice-sheet response to CO_2-induced climate change; and deep coring of the West Antarctic Ice Sheet to learn whether it did in fact disappear 125,000 years ago.

1.5.6.7.2 The Arctic Environment

A number of research efforts should bring progress in understanding effects of a greenhouse warming on the Arctic. Specifically, efforts should be made to

1. Improve GCMs and other models (sea ice, Arctic stratus, ocean dynamics, radiation balance, for example) and use them in studies focused on Arctic response. Proper handling of cloud cover in the Arctic merits special attention as do sensitivity studies using improved sea-ice models.
2. Study stability of the Arctic Ocean density stratification and the potential for its destruction.
3. Obtain long central Arctic sediment cores that could improve the record of variations in the Arctic Ocean for the period from 10,000 to 15,000,000 years ago.

1.5.6.7.3 Agriculture

Basic research on agriculture in relation to CO_2 assessment falls into two broad categories: (1) effects of CO_2 on photosynthesis and plant growth and (2) predicting the changes in yield that will follow a change to a warmer, drier (or other forecast) climate. With respect to the first category, research should be pursued in five obviously related areas: rate of photosynthesis, duration of photosynthesis, and fate and partitioning of photosynthate; drought and transpiration; relationship of increasing CO_2 to demand for and availability of nitrogen and other nutrients; phenology; and weeds. With respect to the second category, the relationship of climate to agriculture should be explored through both historical (e.g., regression) studies and through simulation. A difficult area that deserves consideration is the relationship of changing climate to insect pests and pathogens. Finally, attention should be given to analyzing effects of concurrent changes in climate and atmospheric composition on agriculture, as we make progress on the individual aspects. While our assessment is that near-term effects of CO_2 and climate change on U.S. agriculture will be modest, the evaluation needs to be extended to other regions and, if sound methods are available, to more distant times.

1.5.6.7.4 Ecosystem Response

The response of ecosystems to projected atmospheric conditions remains largely unexplored. Research is called for on ecosystem character and net ecosystem production in relation to increasing atmospheric CO_2 and climatic change. While there is accumulating evidence of effects of increasing CO_2 in increasing the growth of well-watered, fertilized plants, there is a question as to whether these effects extend to natural communities. The question arises particularly with respect to forests, where plants live in conditions of extreme competition for light, water, nutrients, space, and, probably, CO_2 during daylight. The question of redistributing forests with respect to climate change also needs to be addressed. Effects of increasing temperature on respiration of plants have received inadequate attention; by comparison, effects of other factors on respiration may be small. Finally, distribution of ecosystems over the globe induced by CO_2 and climate change may also affect global distribution of albedo; this relationship and its feedback to climate should be explored.

1.5.6.7.5 Water Resources

There is need to develop further the conceptual basis of analysis for all river basins; but the relationships between climate and water resources are complex and unique to each river basin, so that basin-by-basin studies are also needed. Priority should be given to regions with large commitments to irrigated agriculture and for basins where scanty or overabundant flow is already a problem. The roles of extreme events and interannual variability should be kept in mind. More specifically research and analysis is needed on

1. Relations among temperature, rainfall, runoff, and groundwater recharge rates; relations during past decadal or longer climatic excursions may be indicative of future possibilities.
2. Regional, seasonal, and interannual characteristics (including extremes) of rainfall and evaporation that might occur with a greenhouse warming.
3. Societal response to variations in water supply of different duration; water management, especially rates of development of water-resource systems and institutional change.
4. Geomorphologic changes during the last 5000 years from the perspective of changes in water regime.

1.5.6.7.6 Human Health

The need for research relating rising CO_2 concentrations to possible effects on human health arises primarily from a lack of biological data from which it would be possible to project thresholds of CO_2 concentrations hazardous over long periods of time. We know that high doses, 5000 ppm, for example, have measurable effects. There is evidence that levels approachable as a result of the activities discussed in this report will never be dangerous (Clark et al., 1982, p. 43, fn. 56). However, the literature and supporting experimentation in the area are scant (U.S. Department of Energy, 1982). Given the continuing difficulty of identifying health effects of much less subtle changes in the atmosphere, we are skeptical about whether meaningful or interpretable results could be available soon in this area and whether the research would be cost effective. However, it does deserve further consideration. The area of heat stress under climatic conditions and the relationship of climate to disease and disease vectors is less difficult and also merits attention.

1.6 CONCLUDING REMARKS

The CO_2 issue has been with us for over a century, and impacts of increasing CO_2 will be experienced in the century to come. Buildup of CO_2 in the atmosphere is one of many issues arising from our growth in numbers and power on a planet of finite size and limited--though large--resources. Indeed, an argument for continuing attention to the CO_2 issue is that it reminds us of the need to solve intellectual and societal problems, which are important to solve for other, perhaps more immediate, reasons. The skills called for to provide better analysis of and response to the CO_2 issue are similar to the skills we would like to have to tackle other problems. With more insight into the long-term evolution of the economy and technology, the carbon cycle, the oceans and the atmosphere and the ice, the responses of agriculture and ecosystems to environmental change, and why systems and societies collapse or adapt well, and with more extensive cooperation among the carbon-rich nations, many other problems besides CO_2 might yield to solution. The CO_2 issue has proven to be a stimulus to communication across academic disciplines and to cooperation among scientists of many nations. While it may be a worrisome issue for mankind, it is in some respects a healthy issue for science and for people. It is conceivable that CO_2 could serve as a stimulus not only for the integration of the sciences but for increasingly effective cooperative treatment of world issues.

REFERENCES

Augustsson, T., and V. Ramanathan (1977). A radiative-convective model study of the CO_2 climate problem. J. Atmos. Sci. 34:448-451.

Baumgartner, A. (1979). Climatic variability and forestry. In Proceedings of the World Climate Conference. World Meteorological Organization, Geneva, Switzerland.

Bentley, C. R. (1983). The West Antarctic Ice Sheet: diagnosis and prognosis. In Carbon Dioxide, Science and Consensus. Proceedings: Carbon Dioxide Research Conference, Sept. 19-23, 1983. Dept. of Energy CONF-820970, NTIS, Springfield, Va.

Bolin, B., P. J. Crutzen, P. M. Vitousek, R. G. Woodmansee, E. D. Goldberg, and R. B. Cook (1983). The Biogeochemical Cycles and Their Interactions. SCOPE 24. Wiley, New York.

Brooks, H. (1982). Science indicators and science priorities. Science, Technology, and Human Values 7:14-31.

Cess, R. D. (1978). Biosphere-albedo feedback and climate modeling. J. Atmos. Sci. 35:1765.

Chan, Y.-H., J. Olson, and W. Emanuel (1980). Land use and energy scenarios affecting the global carbon cycle. Environ. Internat. 4:189-206.

Charney, J. (1979). Carbon Dioxide and Climate: A Scientific Assessment. Report of the Ad Hoc Study Group on Carbon Dioxide and Climate, J. Charney, chairman. National Research Council, National Academy Press, Washington, D.C.

Clark, W. C., ed. (1982). Carbon Dioxide Review: 1982. Oxford U. Press, New York.

Clark, W. C., K. H. Cook, G. Marland, A. M. Weinberg, R. M. Rotty, P. R. Bell, L. J. Allison, and C. L. Cooper (1982). The carbon dioxide question: a perspective for 1982. In Carbon Dioxide Review: 1982, W. C. Clark, ed. Oxford U. Press, New York.

Donner, L., and V. Ramanathan (1980). Methane and nitrous oxide: their effect on the terrestrial climate. J. Atmos. Sci. 37:119-124.

Edmonds, J. A., and J. M. Reilly (1983). Global energy and CO_2 to the year 2050. Institute for Energy Analysis, Oak Ridge, Tenn. Submitted to the Energy Journal.

Elliott, W. P. (1983). A note on the historical industrial production of carbon dioxide. Clim. Change 5:141-144.

Exxon Corporation (1980). World Energy Outlook. Exxon Corp., New York, December.

Fishman, J., V. Ramanathan, P. Crutzen, and S. Liu (1979). Tropospheric ozone and climate. Nature 282:818-820.

Flohn, H. (1982). Climate change and an ice-free Arctic Ocean. In Carbon Dioxide Review: 1982, W. C. Clark, ed. Oxford U. Press, New York, pp. 145-177.

Gates, W. L., ed. (1979). Report of the JOC Study Conference on Climate Models: Performance, Intercomparison and Sensitivity Studies, Vols. I and II. GARP Publ. Ser. No. 22. Joint Planning Staff, Global Atmospheric Research Programme, World Meteorological Organization, Geneva, Switzerland, 1049 pp.

Gifford, R. M. (1979). Growth and yield of CO_2-enriched wheat under water-limited conditions. Aust. J. Plant Physiol. 6:367-378.

Gifford, R. M., P. M. Bremner, and D. B. Jones (1973). Assessing photosynthetic limitation to grain yield in a field crop. Aust. J. Agric. Res. 24:297-307.

Glantz, M. H., J. Robinson, and M. E. Krenz (1982). Climate-related impact studies: a review of past experiences. In Carbon Dioxide Review: 1982, W. C. Clark, ed. Oxford U. Press, New York, pp. 57-93.

Hameed, S., R. Cess, and J. Hogan (1980). Response of the global climate to changes in atmospheric composition due to fossil fuel burning. J. Geophys. Res. 85:7537-7545.

Hanel, R. A., B. J. Conrath, V. G. Kunde, C. Prabhakara, I. Revah, V. V. Salomonson, and G. Wolford (1972). The Nimbus 4 infrared spectroscopy experiment, 1. Calibrated thermal emission spectra. J. Geophys. Res. 11:2629-2641.

Hansen, J., D. Johnson, A. Lacis, S. Lebedeff, P. Lee, D. Rind, and G. Russell (1981). Climate impact of increasing atmospheric carbon dioxide. Science 213:957.

Hardman, L. E., and W. A. Brun (1971). Effect of atmospheric CO_2 enrichment at different developmental stages on growth and yield components of soybeans. Crop. Sci. 11:886-888.

Hummel, J. R., and R. A. Reck (1981). The direct thermal effect of $CHClF_2$, CH_3CCl_3 and CH_2Cl_2 on atmospheric surface temperatures. Atmos. Environ. 15:379-382.

Interfutures Project (1979). Facing the Future. Organization for Economic Cooperation and Development (OECD), Paris.

International Institute for Applied Systems Analysis (1981). Energy in a Finite World: A Global Systems Analysis. Ballinger, Cambridge, Mass.

Jensen, M. E. (1982). Overview-irrigation in U.S. arid and semiarid lands. Prepared for the Office of Technology Assessment, Water-Related Technologies for Sustaining Agriculture in Arid and Semiarid Lands.

Jones, P. D. (1981). Summary of climatic events during 1981. Clim. Mon. 10:113-114.

Jones, P. D., and T. M. L. Wigley (1980). Northern hemisphere temperatures, 1881-1979. Climate Monitor 9:43-47.

Keeling, C. D., and R. B. Bacastow (1977). Impact of industrial gases on climate. In Energy and Climate. Geophysics Study Committee, National Research Council, National Academy Press, Washington, D.C.

Langbein, W. B., et al. (1949). Annual Runoff in the United States. U.S. Geological Survey Circular 5. U.S. Dept. of the Interior, Washington, D.C. (reprinted, 1959).

Lorenz, E. N. (1970). Climatic change as a mathematical problem. J. Appl. Meteorol. 9(3):325-329.

Luther, F. M., and M. C. MacCracken (1983). Empirical Development of a Climate Response Function: An Analysis of the Propositions of Sherwood B. Idso. Report UCID-19792 (draft). Lawrence Livermore Laboratory, Livermore, Calif., March 1983, 56 pp.

Mar, B. (1982). Commentary. In Carbon Dioxide Review: 1982, W. C. Clark, ed. Oxford U. Press, New York, pp. 96-98.

Marchetti, C. (1980). On energy systems in historical perspective. International Institute for Applied Systems Analysis, Laxenburg, Austria.

Marchetti, C., and N. Nakicenovic (1979). The dynamics of energy systems and the logistic substitution model. RR-79-13. International Institute for Applied Systems Analysis, Laxenburg, Austria.

Marland, G., and R. M. Rotty (1983). Carbon dioxide emissions from fossil fuels: a procedure for estimation and results for 1950-1981. DOE/NBB-0036. NTIS, Springfield, Va.

Matthews, S. W. (1976). What's happening to our climate? Nat. Geographic Mag. 150:576-615.

Mauney, J. R., K. E. Fry, and G. Guinn (1978). Relationship of photosynthetic rate to growth and fruiting of cotton, soybean, sorghum, and sunflower. Crop Sci. 18:259-263.

Mercer, J. H. (1978). West Antarctic Ice Sheet and CO_2 greenhouse effect: a threat of disaster. Nature 271:321-325 (January 1978).

Meyer-Abich (1980). Chalk on the white wall: on the transformation of climatological facts into political facts. In Climatic Constraints and Human Activities, J. Ausubel and A. K. Biswas, eds. Pergamon Press, Oxford.

Mitchell, J. M. (1979). Some considerations of climatic variability in the context of future CO_2 effects on global-scale climate. In Workshop on the Global Effects of Carbon Dioxide from Fossil Fuels, W. P. Elliott and L. Machta, eds. CONF-770385. U.S. Dept. of Energy, Washington, D.C., pp. 91-99.

Nakicenovic, N. (1979). Software Package for the Logistic Substitution Model. Report RR-79-12. International Institute for Applied Systems Analysis, Laxenburg, Austria.

National Research Council (1975). Understanding Climatic Change: A Program for Action. Report of the U.S. Committee for the Global Atmospheric Research Program. National Academy of Sciences, Washington, D.C., 239 pp.

National Research Council (1982). Carbon Dioxide and Climate: A Second Assessment. Report of the CO_2/Climate Review Panel, J. Smagorinsky, chairman. National Academy Press, Washington, D.C.

Niehaus, F., and J. Williams (1979). Studies of different energy strategies in terms of their effects on the atmospheric CO_2 concentration. J. Geophys. Res. 84:3123-3129.

Nordhaus, W. D. (1977). Strategies for the control of carbon dioxide. Cowles Foundation Discussion Paper No. 443. Yale U., New Haven, Conn.

Nordhaus, W. D. (1980). Thinking about carbon dioxide: theoretical and empirical aspects of optimal control strategies. Cowles Foundation Discussion Paper No. 565. Yale U., New Haven, Conn.

Olson, J. S., and J. A. Watts (1982). Carbon in land vegetation. In Carbon Dioxide Review: 1982, W. C. Clark, ed. Oxford U. Press, New York, pp. 435-437.

Paltridge, G. W., and C. M. R. Platt (1976). Radiative Processes in Meteorology and Climatology. Elsevier, New York, 318 pp.

Perry, H., and H. H. Landsberg (1977). Projected world energy consumption. In Energy and Climate. Geophysics Study Committee, National Academy of Sciences, Washington, D.C.

Ramanathan, V. (1975). Greenhouse effect due to chlorofluorocarbons: climatic implications. Science 190:50-52.

Revelle, R. R. (1982). Carbon dioxide and world climate. Sci. Am. 247(2):36-43.

Revelle, R., and W. Munk (1977). The carbon dioxide cycle and the biosphere. In Energy and Climate. Geophysics Study Committee, National Research Council, National Academy of Sciences, Washington, D.C.

Rotty, R. (1977). Present and future production of CO_2 from fossil fuels. ORAU/IEA(O)-77-15. Institute for Energy Analysis, Oak Ridge, Tenn.

Rotty, R., and G. Marland (1980). Constraints on fossil fuel use. In Interactions of Energy and Climate, W. Bach, J. Pankrath, and J. Williams, eds. Reidel, Doredrecht, pp. 191-212.

Schilling, H. D., and R. Hildebrandt (1977). Primärenergie-Elektrische Energie. Clückauf, Essen, FRG.

Schneider, S. H., and R. E. Dickinson (1974). Climate modeling. Rev. Geophys. Space Phys. 12:447-493.

Sioli, H. (1973). Recent human activities in the Brazilian Amazon region. In Tropical Forest Ecosystems in Africa and South America, B. J. Meggers et al., eds. Smithsonian Institution, Washington, D.C.

Sionit, N., H. Helmers, and B. R. Strain (1980). Growth and yield of wheat under CO_2 enrichment and water stress. Crop Sci. 20:687-690.

Sionit, N., D. A. Mortensen, B. R. Strain, and H. Helmers (1981). Growth response of wheat to CO_2 enrichment and different levels of mineral nutrition. Agron. J. 74:1023-1027.

Stewart, H. (1981). Transitional Energy Policy 1980-2030. Pergamon, Oxford.

Stockton, C. W., and W. R. Boggess (1979). Geohydrological implications of climate change on water resource development. U.S. Army Coastal Engineering Research Center, Fort Belvoir, Va.

Takahashi, T., D. Chipman, and T. Volk (1983). Geographical, seasonal, and secular variations of the partial pressure of CO_2 in surface waters of the North Atlantic Ocean: the results of the North Atlantic TTO Program. In Proceedings: Carbon Dioxide Research Conference: Carbon Dioxide, Science and Consensus. U.S. Dept. of Energy Report CONF 820970, Part II, pp. 123-145.

U.S. Committee for the Global Atmospheric Research Program (1975). Understanding Climatic Change: A Program for Action. National Academy of Sciences, Washington, D.C., 239 pp.

U.S. Department of Energy (1982). Effects of CO_2 on mammalian organisms. Report of a workshop, June 5-6, 1980, Bethesda, Maryland. CONF-8006249. NTIS, Springfield, Va.

U.S. Water Resources Council (1978). The Nation's Water Resources, The Second National Water Assessment. U.S. Govt. Printing Office, Washington, D.C.

Vinnikov, K. Ya., G. V. Gruza, V. F. Zakharov, A. A. Kirillov, N. P. Kovyneva, and E. Ya. Ran'kova (1980). Current climatic changes in the northern hemispere. Sov. Meteorol. Hydrol. 6:1-10.

Wang, W. C., and P. H. Stone (1980). Effect of ice-albedo feedback on global sensitivity in a one-dimensional radiative-convective climate model. J. Atmos. Sci. 37:545.

Wang, W. C., Y. Yung, A. Lacis, T. Mo, and J. Hansen (1976). Greenhouse effects due to man-made perturbation of trace gases. Science 194:685-690.

Wang, W. C., J. P. Pinto, and Y. Yung (1980). Climatic effects due to halogenated compounds in the earth's atmosphere. J. Atmos. Sci. 37:333-338.

White, G. F. (1983). Water resource adequacy: illusion and reality. Natural Resources Forum 7(1):11-21.

Woodwell, G. M., R. A. Houghton, and N. R. Tempel (1973). Atmospheric CO_2 at Brookhaven, Long Island, New York: patterns of variation up to 125 meters. J. Geophy. Res. 78:932-940.

Wong, S. C. (1979). Elevated atmospheric partial pressure of CO_2 and plant growth. Oecologia 44:68-74.

World Climate Programme (1981). On the assessment of the role of CO_2 on climate variations and their impact. Report of a WMO/UNEP/ICSU meeting of experts in Villach, Austria, November 1980. World Meteorological Organization, Geneva, Switzerland.

World Energy Conference (1978). World Energy Resources 1985-2020, An Appraisal of World Coal Resources and Their Future Availability; World Energy Demand (Report to the Conservation Commission). IPC Science and Technology Press, Guildford, U.K.

World Meteorological Organization (1983). Report of the Meeting of Experts on Potential Climatic Effects of Ozone and Other Minor Trace Gases. Report No. 14. WMO Global Ozone Research and Monitoring Project, Geneva, Switzerland, 38 pp.

2 Future Carbon Dioxide Emissions from Fossil Fuels

2.1 FUTURE PATHS OF ENERGY AND CARBON DIOXIDE EMISSIONS

William D. Nordhaus and Gary W. Yohe

This section deals with the uncertainty about the buildup of CO_2 in the atmosphere. It attempts to provide a simple model of CO_2 emissions, identify the major uncertain variables or parameters influencing these emissions, and then estimate the best guess and inherent uncertainty about future CO_2 emissions and concentrations. Section 2.1.1 is a self-contained overview of the method, model, and results. Section 2.1.2 contains a detailed description of sources, methods, reservations, and results.

2.1.1 Overview

There is widespread agreement that anthropogenic carbon dioxide emissions have been rising steadily, primarily driven by the combustion of fossil fuels. There is, however, enormous uncertainty about the future emission rates and atmospheric concentrations beyond the year 2000; and even greater uncertainty exists about the extent of climatic change and the social and economic impacts of possible future trajectories of carbon dioxide. Yet, if the appropriate decisions are to be made, the balance of future risks and costs must be weighed, and producing best possible estimates of future emission trajectories is therefore imperative.

Many of the early analyses of the carbon dioxide problem have produced estimates of future emissions and concentrations from extrapolative techniques (see Section 2.2, the accompanying survey by Ausubel and Nordhaus). For the purposes of understanding future outcomes and policy choices, these techniques leave important questions unanswered. First, they do not allow an assessment to be made about the degree of precision with which the forecast has been constructed. Moreover, little information is generated about the underlying structure that produced the reported trajectories. It is, therefore, hard to know how changing economic structures might alter the pattern of CO_2 emissions. But information about precision and sensitivity are sometimes of critical importance to policy makers. It is critical to know not only the best scientific assessment of an event but also the extent to which that judgment is precisely or vaguely known. Particularly in cases where policy decisions are irreversible, for example, the best decision in the face of great uncertainty might be simply to gather more information. But that decision would be extremely difficult to reach without some notion of the extent of the uncertainty surrounding the projections.

In an attempt to address uncertainties, a second generation of studies, employing scenario analysis, has arisen. These studies, notable among them Limits to Growth (Meadows et al., 1972), CONAES (Modeling Resource Group, 1978), and IIASA (1981), have traced time paths for important variables with a well-defined model and specified sets of assumptions. The studies, which we call "nonprobabilistic scenario analysis," represent a marked improvement over earlier efforts. They still fall short of providing the policymaker with a precise notion of the likelihood that a particular combination of events might occur: Is the "high" scenario 1 in 10, 1 in 100, or what?

In this section an effort is made to put more definite likelihoods on alternative views of the world. The technique, called probabilistic scenario analysis, extends the scenario approach to include modern developments in aggregate energy and economic modeling in a simple and transparent model of the global economy and carbon dioxide emissions. Particular care is given not only to assure that the energy and production sectors are integrated but also to respect the cost and availability of fossil fuels.

In addition, the analysis presented here attempts to recognize the intrinsic uncertainty about future economic, energy, and carbon cycle developments. This is done by specifying the most important uncertain parameters of the model, by examining current knowledge and disagreement about these parameters, and then by specifying a range of possible outcomes for each uncertain variable or parameter. The emphasis is not to resolve uncertainties but to represent current uncertainties as accurately as possible and integrate them into the structure in a consistent fashion. The result of the entire process is the generation of a range of paths and uncertainties for major economic, energy, and carbon dioxide variables--projections of not only a "best guess" of the future paths of important variables but also a set of alternative trajectories and associated probabilities that quantify the range of possible outcomes on the basis of the current state of knowledge about the underlying uncertainty of the parameters.

It is reasonable, even at this early stage, to ask why such an elaborate effort to quantify uncertainty should be undertaken. Cannot prudent policy be written on the basis of the "best-guess" trajectories of important variables? In general, the answer is "no." To limit analysis to the best-guess path is to limit one's options. Similarly, to consider some possible path with no assessment of its likelihood relative to other possible paths is a formula for frustration, leading to endless arguments about which path should be taken most seriously. But to consider a full range of possibilities, along with each one's likelihood, allows a balanced weighing of the important and the unimportant in whatever way seems appropriate.

Armed with probabilistic scenarios, in other words, the policymaker will be able to evaluate a new dimension of his problem. He can assess not only a policy along a most likely trajectory but also along other trajectories that cannot be ruled out with some degree of statistical significance. With some knowledge of the range of uncertainty, he might decide to ask for more information to narrow the range, particularly if a policy seems to be warranted only by a few selected outcomes. Alternatively, he might choose to minimize the risk of proceeding along an undesirable set of possible paths. And finally, he might undertake a policy based simply on an expected value. Any one of these options might prove to be prudent, but none of them is possible without a quantified range of possibilities. It is toward providing such a range that this section is directed.

The plan of the overview is this. We first sketch the model that is used to relate the different variables and project future carbon dioxide emissions. We then describe the data sources and some adjustments that we have made to the data. Finally, we describe the results. It should be noted that a full description of the methods is contained in Section 2.2.

The economic and energy model is a highly aggregative model of the world economy and energy sector. It is based on the idea of a multi-input production function that represents the relationship between world Gross National Product (GNP) (the output), on the one hand, and labor, fossil fuels, and nonfossil fuels (the inputs), on the other. In addition, to reflect the likelihood that economic efficiency will continue to improve in the future, various technological parameters are included to describe the rate of growth of economic efficiency in general, as well as the extent to which that growth is more or less rapid in the energy sectors than in the nonenergy sectors.

A further important feature is the explicit incorporation of both the extent to which it is relatively easy or difficult to substitute nonenergy inputs (insulation or radial tires) for energy inputs (heating oil or gasoline) and the extent to which it is easy or difficult to substitute nonfossil energy (nuclear or solar-derived electricity or hydrogen) for fossil energy (coal-fired electricity or gasoline).

The prices of different inputs play a central role in reflecting scarcity and driving the relative quantities of different inputs. We thus introduce a cost function for fossil fuels that relates their price to their degree of exhaustion or abundance. On the other side of the market we generate an economically consistent derived demand for energy from the structure of the production function as the response of economic agents to changes in relative prices of different fuels and other inputs. Thus, if fossil fuels are scarce and costly, the system will economize on this input and use relatively more nonfossil fuels and labor inputs.

Finally, we recognize that there are a number of important uncertainties about the model and future trends. We thus incorporate 10 key uncertainties in the model. These relate to variables such as the rate of population growth, the availability and cost of fossil fuels, the rate of growth of productivity, the extent to which productivity growth will be relatively more rapid in fossil fuels versus nonfossil fuels, or in energy versus nonenergy, and so forth. A complete list of the uncertainties, with their relative importance in determining the total uncertainty is presented later in this overview.

The data are gathered from diverse sources and are of quite different levels of precision. In general, we have surveyed the recent literature on energy and economic modeling to determine what are commonly held views of such variables as future population growth or productivity growth. Other variables, such as the ease of substitution or differential productivity growth, were ones that are not the subject of common discourse; for these, we examined recent trends or results generated by disaggregated studies.

A much more difficult data problem arose from the need to estimate the uncertainty about key parameters or variables. Our starting assumption was to view the dispersion in results of published studies as a reflection of the underlying uncertainty about the variable studied. This starting point was modified in two respects. First, it is commonly observed that even trained analysts tend to cluster together excessively--that is, they tend to underestimate the degree of uncertainty about their estimates. To account for this tendency to move toward the current consensus, we have spread out some of the distributions of observations by slightly less than 50%. Second, we have also imposed our own judgments about the uncertainties in those cases where no external data or disagreement existed or where the disagreement was so small as to convince us that the predictions were not independent. We must emphasize that these judgments about the uncertainties in our understanding of the variables are only rough judgments; the methods of estimation are difficult, and they could be quite far from the estimates that would come from a more thorough study. On the other hand, however, several validation exercises revealed that our results were within the statistical realm of reason, i.e., they fell within reasonable bounds of uncertainty that could be deduced by other means on the basis of historical experience.

Using the model and data just sketched, we investigated the range of outcomes for economic, energy, and carbon dioxide variables. There were ultimately 10 uncertain variables, each of which was discretized into high, medium, and low values in such a way that the variance of the discretized variable was equal to the variance of the continuous variable. We thus ended up with 3^{10} (=59,049) different possible outcomes. Rather than do a complete description, we settled on sampling 100 or 1000 of the different possible outcomes. The results reported below, then, should be interpreted as samples of the underlying distributions (although the sampling errors are known to be quite low).

We now turn to brief descriptions of the major results of this study, with the promise that a more complete description can be found in Section 2.1.2. The first set of results pertains to the central estimates and range of estimates of the central variables, carbon dioxide emissions and concentrations. Our central estimate here is taken as the sample mean of 1000 runs.* Carbon dioxide emissions are projected to rise modestly to the end of our time horizon, the year 2100. We estimate that carbon dioxide emissions will grow at about 1.6% annually to 2025, then slow to slightly under 1% annually after 2025. Atmospheric concentrations in the average case are expected to hit the nominal doubling level (600 ppm) around the year 2070.

These results show a considerably slower emissions rate and carbon dioxide buildup than many of the earlier studies (see Ausubel and Nordhaus, Section 2.2) for two major reasons. First, the expected growth of the global economy is now thought to be slower than had earlier been generally assumed; our work includes this new expectation. Perhaps more importantly, we also include the tendency to substitute nonfossil for fossil fuels as a result of the increasing relative prices of fossil fuels. This is an important effect that has frequently been ignored.

The next result concerns our attention on the degree of uncertainty about future carbon dioxide emissions and concentrations. The range of uncertainty is shown in Figures 2.1 to 2.4. Figures 2.1 and 2.2 show 100 randomly chosen outcomes for carbon dioxide emissions and concentrations. These are shown mainly to give a visual impression of the range of outcomes. Figures 2.3 and 2.4 present the five runs that represent the 5th, 25th, 50th, 75th, and 95th percentiles of outcomes--where we measure the outcomes in terms of the cumulative carbon dioxide emissions by the year 2050. It should be emphasized that the percentiles are derived from the actual sampling distribution. They reflect, therefore, the distribution of outcomes derived from the interaction of expert opinion on the underlying random parameters and the economic model; they reflect judgment not an objectively derived distribution.

Perhaps the most useful graph to study is Figure 2.4, which shows the percentiles of carbon dioxide concentrations. For a quarter or half century, the inertia built into the economy and the carbon cycle leave an impression of relative certainty about the outcomes. After the early part of the next century, however, the degree of uncertainty becomes extremely large. In terms of our conventional doubling time, note the time at which carbon dioxide concentrations are assumed to hit 600 ppm:

Percentile	Doubling Time
5	After 2100
25	2100
50	2065
75	2050
95	2035

---

*More precisely, if x_j were the value assumed by run j, whose underlying sample of the 10 random variables gave it a probability of P_j, then the central estimate would be $\sum_{j=1}^{1000} P_j x_j$.

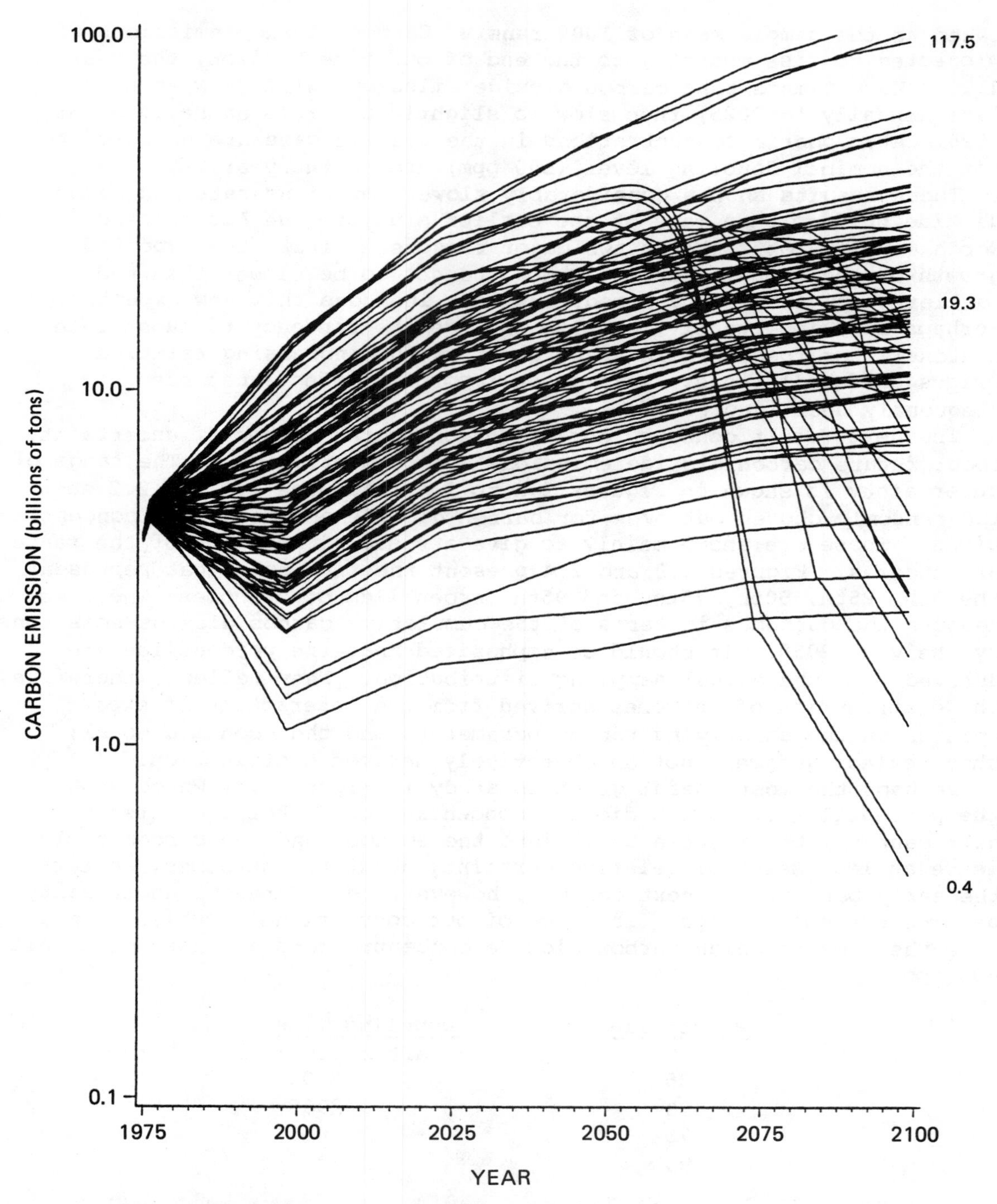

FIGURE 2.1 Carbon dioxide emissions for 100 randomly drawn runs (billions of tons of carbon per year). Outcomes of 100 randomly chosen runs; the numbers on the right-hand side indicate the mean projected yearly emission for the year 2100 and the extreme high and low outcomes.

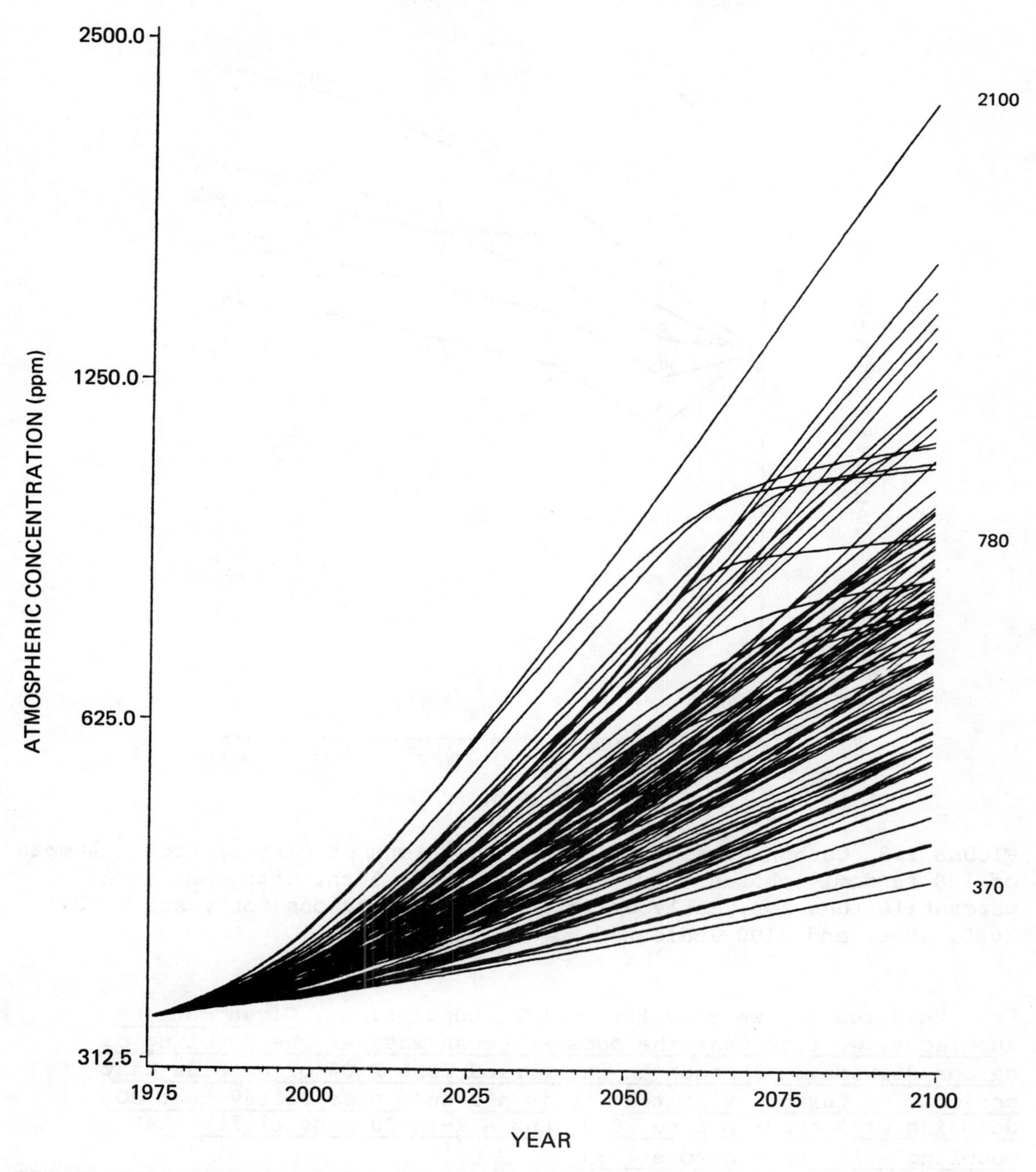

FIGURE 2.2 Atmospheric concentration of carbon dioxide (parts per million). Outcomes of 100 randomly selected runs; the numbers on the right-hand side indicate the mean concentration for the year 2100 and the extreme high and low outcomes.

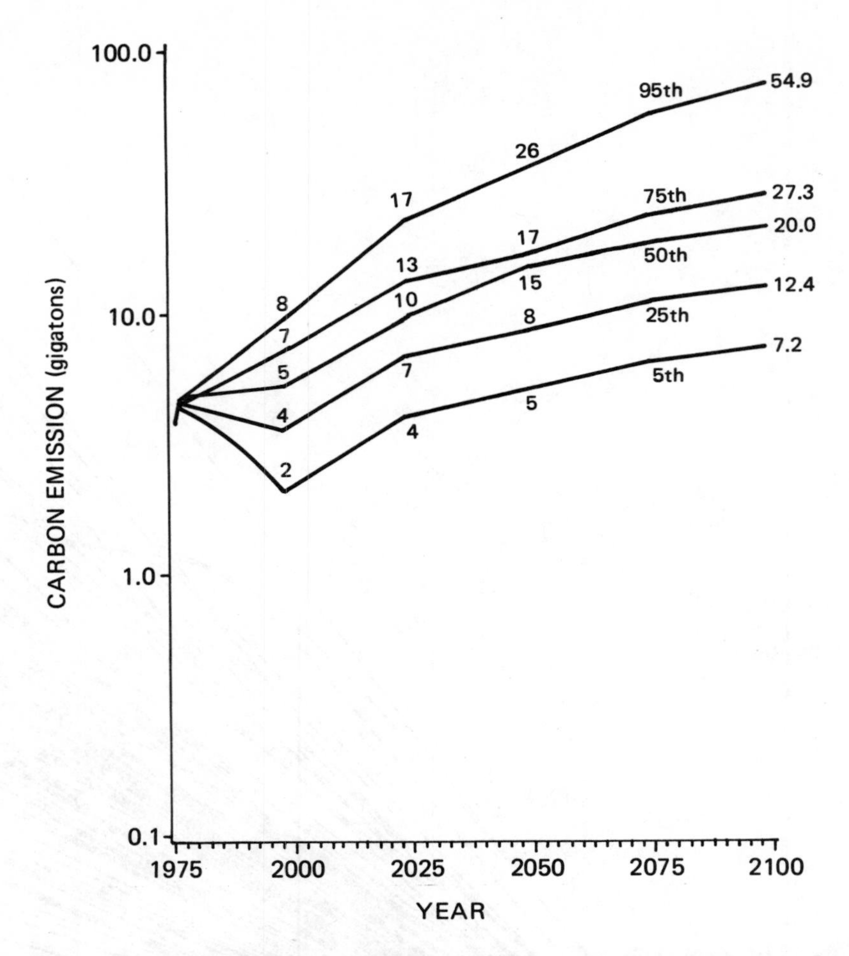

FIGURE 2.3 Carbon dioxide emissions (gigatons of carbon) from a sample of 100 randomly chosen runs. The 5th, 25th, 50th, 75th, and 95th percentile runs for yearly emissions, with emissions for years 2100, 2025, 2050, and 2100 indicated.

From this result, we make the central conclusion: Given current knowledge, we find that the odds are even whether the doubling of carbon dioxide will occur in the period 2050-2100 or outside that period. We further find that it is a 1-in-4 possibility that CO_2 doubling will occur before 2050, and a 1-in-20 possibility that doubling will occur before 2035.

The next issue addressed is the question of the relative importance of different uncertainties. We have computed by two different techniques the relative importance of the ten uncertain variables discussed above, and the results are shown in Table 2.1. This table calculates the contribution to the overall uncertainty that is made by each variable taken by itself.

In one case, shown in column (2), the contribution is calculated as the uncertainty introduced when a variable takes its full range of uncertainty and all other variables are set equal to their most likely

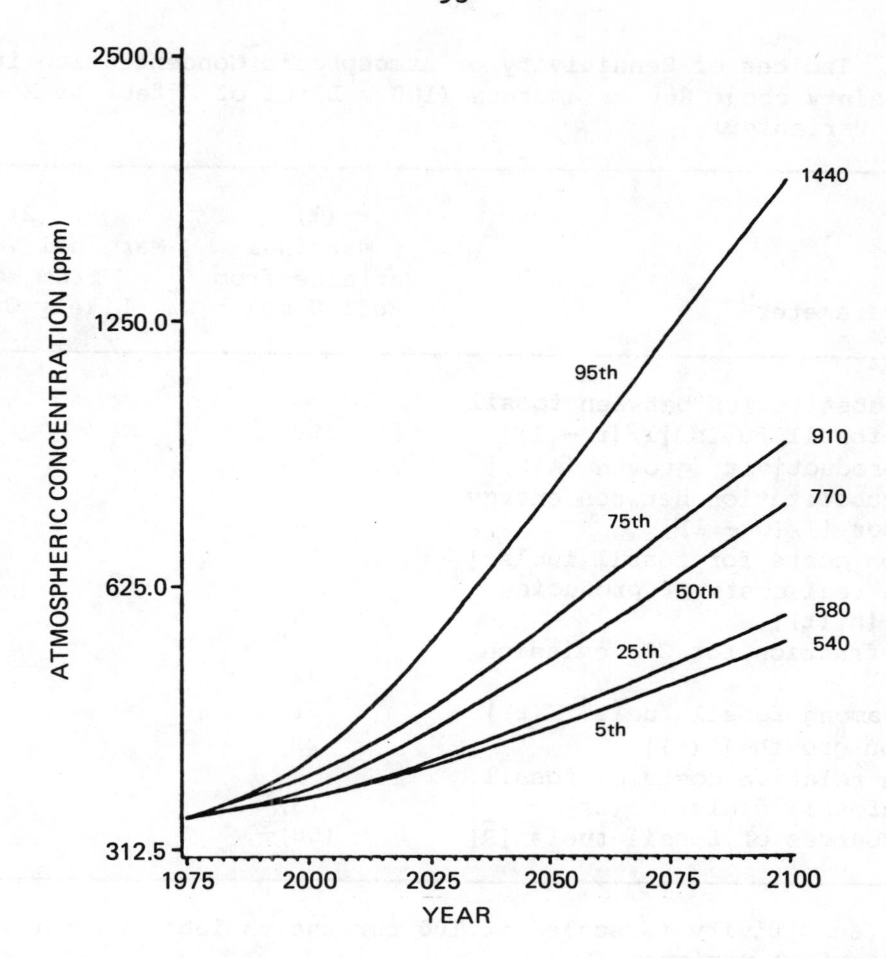

FIGURE 2.4 Atmospheric concentration of carbon dioxide (5th, 25th, 50th, 75th, and 95th percentiles; parts per million). The indicated percentile runs for concentrations; the numbers on the right-hand side indicate concentrations in the year 2100 for each run.

values. In the first column, the uncertainty is calculated from the run in which the full panoply of uncertainties is deployed--that is, when all other nine uncertain variables are allowed to take their full ranges of outcomes. In both cases, we have created index numbers with the variable that induces the most uncertainty set equal to 100, and other variables are scaled by their ratio of uncertainty added to that of the variable with the largest contribution.

The two indices are used because they convey different information. The variable in column (1) is more relevant to uncertainty reduction in the real world (because other variables do indeed have uncertainty); but the calculations in column (1) are dependent on the uncertainties assumed for the whole range of variables. The numbers in column (2) are, for that reason, more robust to misspecification in the uncertainty of other variables.

TABLE 2.1 Indices of Sensitivity of Atmospheric Concentration in 2100 to Uncertainty about Key Parameters (100 = Level of Effect of Most Important Variable[a])

Parameter[b]	(1) Marginal Variance from Full Sample[c]	(2) Marginal Variance from Most Likely Outcome[d]
Ease of substitution between fossil and nonfossil fuels [1/(r - 1)]	100	100
General productivity growth [A(t)]	76	79
Ease of substitution between energy and labor [1/(q - 1)]	56	70
Extraction costs for fossil fuels [g_1]	50	56
Trends in real costs of producing energy [$h_1(t)$]	48	73
Airborne fraction for CO_2 emission [AF(s)]	44	62
Fuel mix among fossil fuels [Z(t)]	31	24
Population growth [L(t)]	22	36
Trends in relative costs of fossil and nonfossil fuels [$h_2(t)$]	(3)[e]	21
Total resources of fossil fuels [$\bar{R}$]	(50)[e]	5

[a]Value of sensitivity is scaled at 100 for the variable that has the highest marginal variance.
[b]Notation in the square brackets refers to variable notation in the model presented formally in Section 2.2.
[c]"Marginal variance" from full sample equals (1) the variance in the base case (i.e., with all variables varying according to their full range of uncertainty) minus (2) the variance with listed variable set at its most likely value (but all nine other variables varying according to their full uncertainty). Note that no resampling occurs.
[d]"Marginal variance from most likely outcome" calculated as the variance when the listed variable assumes its full range of uncertainty and all nine other variables are set equal to their most likely value.
[e]Parentheses indicate that the marginal variance is negative.

The ranking of the importance of uncertainties shown in Table 2.1 contains several surprises. First, note that an unfamiliar production parameter ranks at the top of both columns--the ease of substitution between fossil and nonfossil fuels. While some studies have included substitution parameters in their model specifications (notably Edmonds and Reilly, 1983), the sensitivity of concentration projections to assumptions about substitution has not been noted earlier. A second set of variables on the list include those that have been exhaustively discussed in the carbon dioxide literature--the world resources of fossil fuels and the carbon cycle ("airborne fraction"). Our estimates indicate that these are of modest significance in the uncertainty about future carbon dioxide concentrations.

Table 2.1 is also extremely suggestive about research priorities in the carbon dioxide area. We cannot, it should be emphasized, move directly from the source of uncertainties to a budget allocation for research funds on carbon dioxide. It may be much easier, for example, to resolve uncertainties about the "depletion factor" for carbon fuels than about future "productivity growth." As a result research funds might therefore be more fruitfully deployed in the first prior area than in the second.

On the other hand, the results suggest that considerably more attention should be paid to some uncertainties that arise early in the logical chain from combustion to the carbon cycle, particularly better global modeling of energy and economy. It is striking, for example, to note that the United States supports considerable work on global carbon cycle and global general circulation (climate) models, but much less attention in the United States has been given to long-run global economic or energy modeling (see Section 2.2 for a further discussion.

Finally, it is possible to explore more fully the ramifications of Figure 2.4--the figure that indicated 5th, 25th, 50th, 75th, and 95th percentile trajectories for atmospheric concentrations of carbon dioxide. One might ask, what parameters are most influential in determining whether the concentration path deviates from the median in either direction; Table 2.1 provides the answer. If uncertainty in a parameter is significant in its effect on the overall variance of the outcome, then it follows that movement in that parameter away from its projected median would be significant in its effect on the outcome variable. Clearly, therefore, an increase (decrease) in the ease of substitution out of fossil fuel as it becomes more expensive would significantly increase the likelihood that the concentration trajectory would be lower (higher) than the 50th percentile. Similarly, slower (more rapid) productivity growth would cause slower growth in energy consumption and produce a significant lowering (raising) of the concentration trajectory.

As a final set of experiments, we have used our procedure to make extremely tentative estimates of the effect of energy-sector policies that are designed to reduce the burning of fossil fuels. The particular policy we investigate is the imposition of fossil fuel taxes, set, for illustrative purposes, at $10 per ton of coal equivalent and at a more stringent level. Taxes were not chosen for any reason other than modeling ease. Any type of emissions restraint can be represented analytically by its equivalent tax. These runs use the most likely outcome as a base case. Figure 2.5 shows the trajectory of taxes that we have investigated, while Figures 2.6 and 2.7 show the effects of the different tax policies on the level of carbon dioxide emissions and on carbon dioxide concentrations. In general, the taxes lower emissions noticeably during the period in which the taxes are in place. The effect on concentrations at the end of the twenty-first century of the $10 tax are quite modest. These examples are included here only to illustrate the nature of a problem that deserves much more attention.

They suggest, as does some other work in the literature (Edmonds and Reilly, 1983), that the use of carbon taxes (or their regulatory equivalents) will have to be quite forceful to have a marked effect on carbon concentrations, even if they are imposed worldwide. Unilateral regulations would, of course, have to be substantially more restrictive.

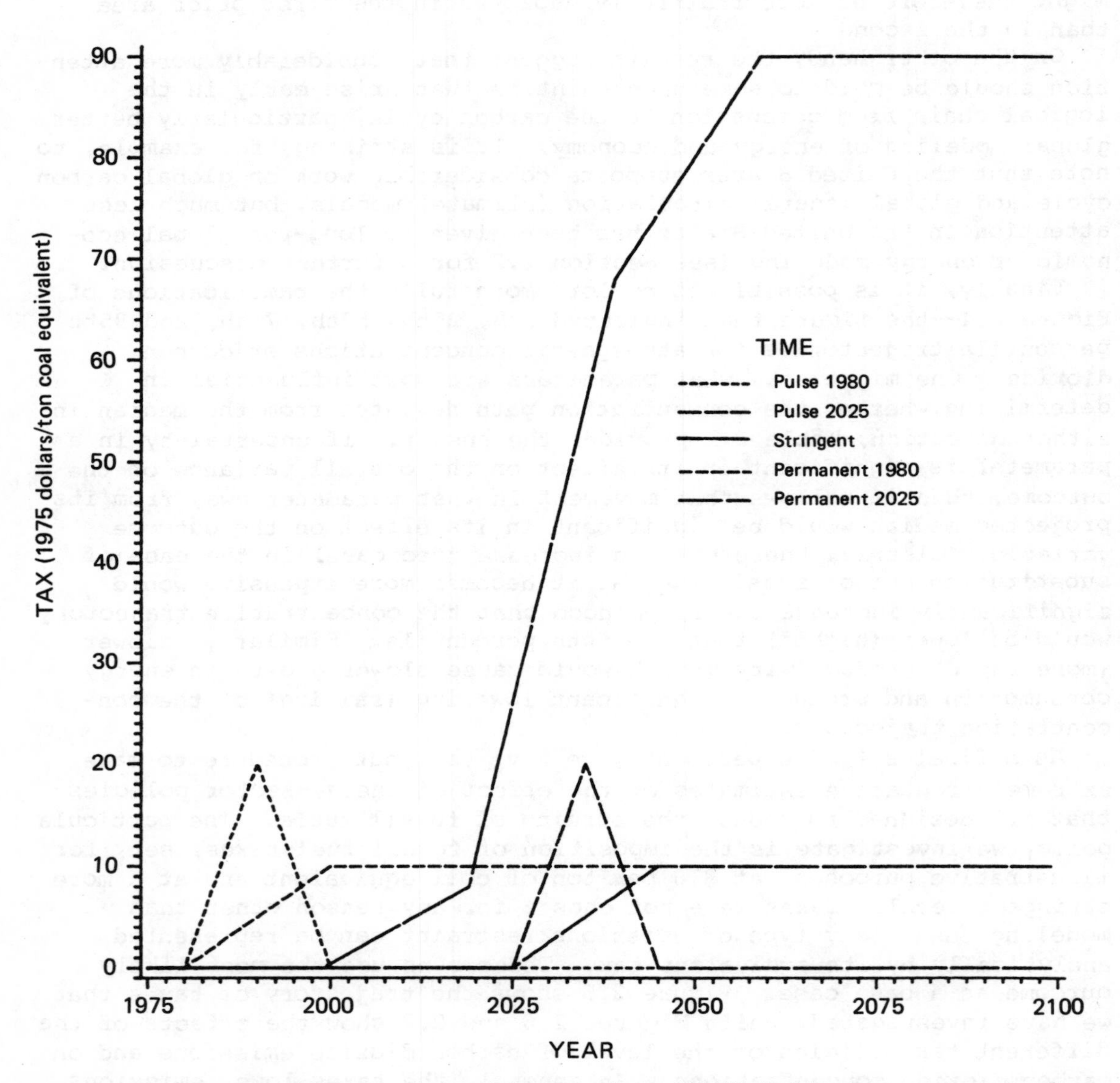

FIGURE 2.5 Taxation on carbon fuel price (1975 dollars per ton coal equivalent). The time tracks of a stringent tax and four alternative $10 per ton of coal equivalent taxes; the temporary taxes peak at $20 to accommodate the model.

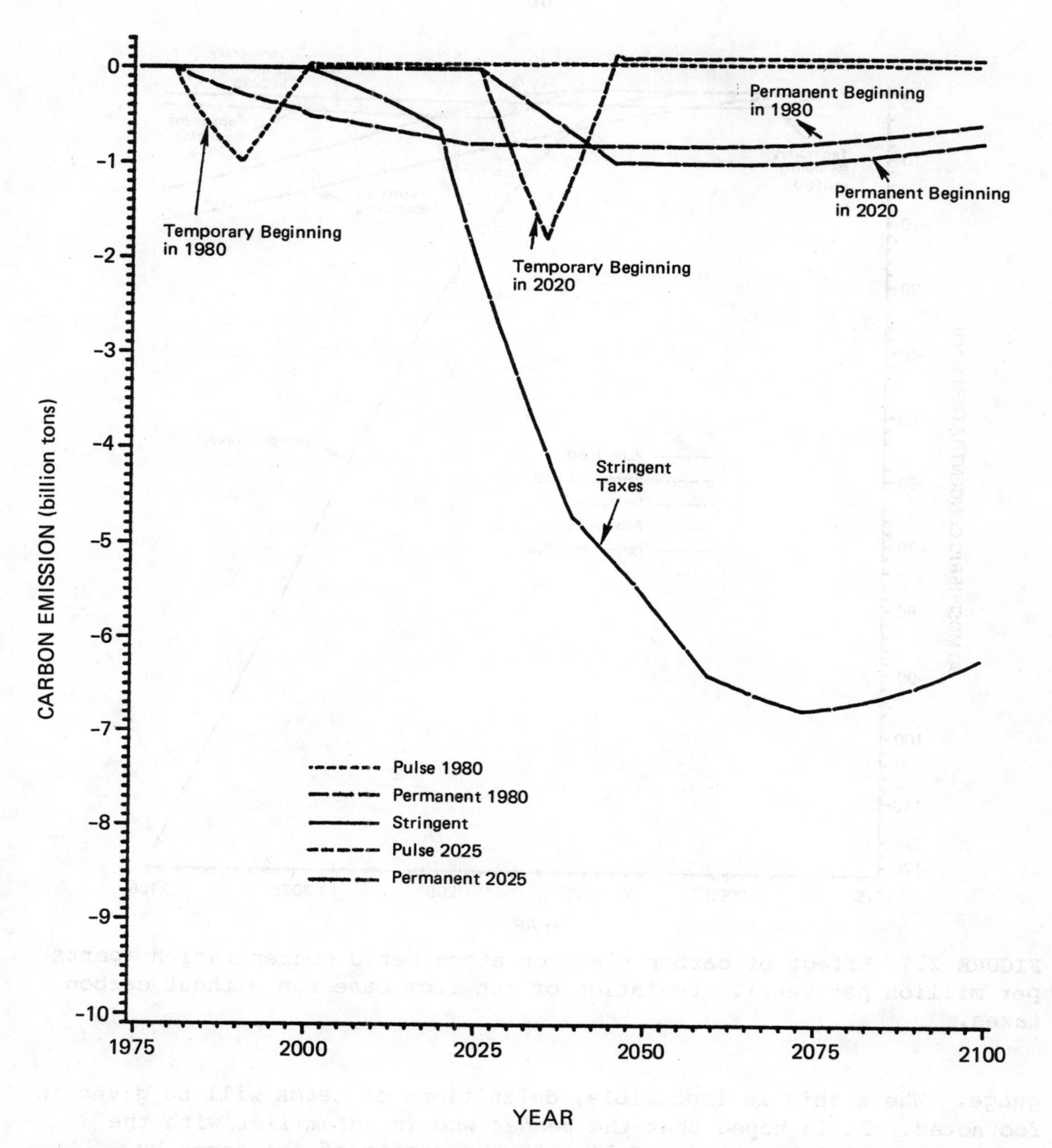

FIGURE 2.6 Plot of carbon emission versus time for taxed runs. Deviation in emissions from the base run for various taxes.

2.1.2 Detailed Description of the Model, Data, and Results

The preceding section provided an overview of the paper--its motivation, its methodology, and its results. This section will provide a more complete description of the procedures. Throughout, an attempt will be made to refrain from using economic jargon and overly technical lan-

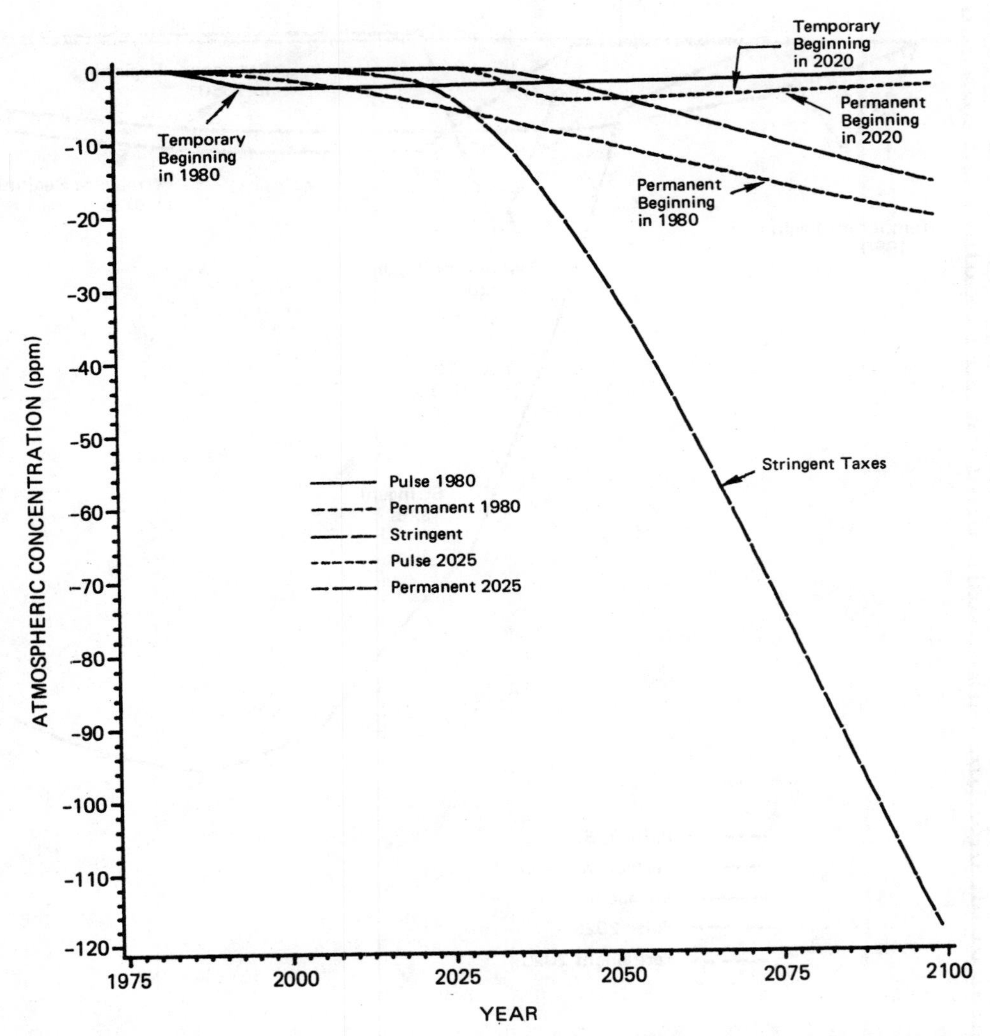

FIGURE 2.7 Effect of carbon taxes on atmospheric concentration (parts per million per year). Deviation of run from base run without carbon taxes.

guage. Where this is impossible, definitions of terms will be given in footnotes. It is hoped that the reader who is unfamiliar with the terminology will be able to follow the reasoning of the paper by referring to these notes.

2.1.2.1 The Model

2.1.2.1.1 Methodological Summary

Before we launch into a detailed description of the data and methods used in the present study, it may be useful to give a brief symbolic overview of the events to follow. We start by denoting the variables:

x_t = endogenous variables (those determined within the system), as they unfold over time;

z_t = exogenous variables (those determined outside the system), as they unfold over time;

k = parameters of the system, assumed to be constant over time;

G = a functional relation, mapping the exogenous variables and parameters into the endogenous variables.

The present study is concerned with the future evolution of CO_2 emissions and concentrations as the endogenous variables--these are the x_t variables. The key exogenous variables (or z_t) are economic "events," such as population growth or fossil fuel reserves. Parameters that relate the variables (the k's) are ones such as the airborne fraction or the emissions per unit energy produced.

We thus write our system symbolically as

$$x_t = G(z_t, k).$$

The pages that follow describe how we derive the relational function, G; how we estimate future trajectories of the exogenous variables, z_t; and how we estimate the parameters, k. The central difficulty with studies of this kind is that the system is imperfectly known. We are not able to forecast the z_t with accuracy; indeed, we may not know which are the important exogenous variables. The parameters, also, are imperfectly known.

The technique that follows uses a procedure that we have denoted probabilistic scenario analysis. We start with a simple representation of the system (the function G). We then estimate future trajectories and subjective probability distributions [denoted by g(.) and h(.)] of the exogenous variables and parameters z_t and k:

$$g(z_t),\ h(k) \text{ (judgmental probability distributions on } z_t, k).$$

These then map through the G function to give us a conditional probability distribution--f(.)--on the variables of concern, the x_t:

$$G{:}[g(z_t),\ h(k)] \rightarrow f(x_t).$$

All of this is, unfortunately, much more easily described than accomplished. The major issues that arise are these: First, the G function is not known in advance and may be extremely complex. Second, neither the exogenous variables nor the parameters are well known. The scientific and economic literature can be used to illuminate the "best guess" about these variables or parameters, however. Third, the judgmental probability distributions are ignored in most of the applied scientific literature. Attempting to determine these distributions is the hardest part of our task. And finally, there is no established methodology for developing probabilistic scenarios. The text that follows outlines one attempt to overcome these difficulties.

An aggregate world production function sets the stage.* We chose the simplest conventional form that would allow explicit parameters for both the share of GNP devoted to energy and the ease of substituting between fossil and nonfossil fuels †:

$$X(t) = A(t)L(t)^{d(t)}[bE^c(t)^r + (1 - b)E^n(t)^r]^{[1 - d(t)]/r}, \quad (1)$$

where

$X(t)$	=	world GNP at time t in constant 1975 U.S. dollars;
$L(t)$	=	world population at time t;
$A(t)$	=	level of labor productivity at time t in U.S. dollars of output per capita;
$(1 - d)$	=	the proportion of GNP devoted to paying for energy;
$E^c(t)$	=	consumption of fossil fuels at time t in metric tons of coal equivalent;
$E^n(t)$	=	consumption of nonfossil fuel at time t in metric tons of coal equivalent;
r	=	a parameter reflecting the ease of substitution between $E^c(t)$ and $E^n(t)$; and
b	=	a parameter reflecting the relative levels of use of $E^c(0)$ and $E^n(0)$.

Equation (1) is not so mysterious as it might first appear. It assumes its peculiar form because of well-established techniques in microeconomics. They mandate that if a production process with certain conventional properties is to be represented mathematically, then the researcher is locked into an equation of the general type exhibited in (1). A slightly simpler form exists (the straight Cobb-Douglas form) and could have been employed, but that would have restricted the degree of substitution between the two types of energy in an arbitrary and unacceptable manner. To preserve desired flexibility in the specification of the ease of substitution, Equation (1) is the simplest option available.

Turning now to a brief discussion of some of those desirable properties, it is important to note, first, that Equation (1) displays constant returns to scale; i.e., doubling $L(t)$, $E^c(t)$, and $E^n(t)$ from any level necessarily doubles output. As growth proceeds, this

---

*A production function is a mathematical representation of a production process that employs a variety of inputs in a variety of combinations to manufacture some type of output. General production functions allow for substitution between any of the inputs in response to changes in input prices so that the manufacturer can maximize profits.

†The estimation of production functions is a well-developed field in economics. There have been numerous surveys, of which that by Johnston (1972) is perhaps the most comprehensive. Also see Nerlove (1965) for a careful study of the identification and estimation of the particular production function that we use, the "Cobb-Douglas" version.

feature guarantees that payments to labor employed and for energy consumed exhaust output. It should be noted, at least in passing, that almost all production relationships that spring to mind easily, and almost all that are used in existing studies, display this constant returns-to-scale property. Imposing it on our production schedule did not drive us afield of conventional economic modeling.

Note as well that Equation (1) aggregates the value of all nonenergy inputs into labor--the third factor of production. This aggregation produces a simplification that allows the model to isolate the potential substitution both between the two types of energy and between energy and other inputs without being unnecessarily cluttered by an index of what those inputs might be. No assumptions about constant capital-labor ratios are being made, and any increased productivity that might be created by this secondary substitution is captured in the A(t) parameter. Nonetheless, the substitutions of critical importance in an energy-emissions model--substitution between the two major types of energy and substitution between energy and other inputs--are both identifiable and quantifiable. We will return to them shortly.

Before doing so, however, it is convenient to discuss the derived demand for energy implicit in the functional form of Equation (1) as it stands. Since energy demand is derived entirely from production, Equation (1) imposes some unavoidable structure on the demand for energy. The share of X(t) devoted to paying for energy at any point in time is, first of all, fixed; i.e., letting $P^n(t)$ represent the real price of $E^c(t)$ at time t in 1975 U.S. dollars and $P^c(t)$ represent the real price of $E^c(t)$ at time t, then the share devoted to energy can be expressed as

$$P^n(t)E^n(t) + P^c(t)E^c(t) = (1 - d)X(t) \qquad (2a)$$

for all t. Letting

$$E^n(t) + E^c(t) = E(t)$$

represent total energy demand and

$$[P^c(t)E^c(t)/E(t)] + [P^n(t)E^n(t)/E(t)] = P(t)$$

represent the weighted aggregate price of energy, Equation (2a) can immediately be rewritten in the more convenient form:

$$E(t) = (1 - d)X(t)/P(t). \qquad (2b)$$

It becomes clear, therefore, that Equation (1) imposes unitary price and income elasticities on the aggregate energy demand equation.*

---

*The price elasticity of demand is a reflection of the responsiveness of the quantity demanded to changes in the price. It is formally defined as the ratio between the percentage change in the quantity

(continued overleaf)

Additionally, the particular nested form of Equation (1) necessarily requires that the relative demand for $E^c(t)$ and $E^n(t)$ take the form

$$\frac{E^c(t)}{E^n(t)} = \left[\frac{(1 - b)}{b} \frac{P^c(t)}{P^n(t)}\right]^{1/r - 1} . \quad (3)$$

Implicit in Equation (1), then, is the condition that a 1% increase in the energy price ratio of fossil to nonfossil fuel must always generate a$[(r - 1)^{-1}]$% reduction in the ratio of fossil to nonfossil fuel use. Since the parameter r could be arbitrarily specified, however, Equation (3) is not nearly so restrictive as Equation (2b). Still, the point is this: by specifying the production function, we fully specify the underlying structure of demand for all three of our inputs--labor, fossil fuel, and nonfossil fuel.

Quantification of the ability to substitute between the two types of energy and between labor and energy now follows straightforwardly from the derived demand schedules just noted. To that end, let

$$s = \frac{d \ln[E^c(t)/E^n(t)]}{d \ln[P^c(t)/P^n(t)]}$$

represent the notion of the "elasticity of substitution" between the two types of energy; i.e., let s represent a measure of how responsive the ratio of fossil to nonfossil fuel consumed worldwide is to changes in the relative prices of the two fuels. Logarithmic differentiation of Equation (3) then reveals that

$$s = (r - 1)^{-1}.$$

If r were to equal 0.5, therefore, s would equal -2.0, indicating that any 1% increase in the relative price of carbon-based fuel would

---

(continued from overleaf)

demanded and the percentage change in the price that caused demand to change; i.e.,

$$\text{(price elasticity)} = d \ln[E(t)]/d \ln[P(t)].$$

Given Equation (2b), therefore, it is clear that the price elasticity of the derived demand for energy is (-1). The income elasticity of demand is similarly defined as the ratio between the percentage change in the quantity demanded and the percentage change in income. Since notationally the income elasticity of demand is $d \ln[E(t)]/d \ln[X(t)]$, it is equally clear that this elasticity must also equal 1 for the schedule listed in Equation (2b). In conclusion, therefore, the structure of the production schedule recorded in Equation (1) implies that (i) a 1% increase in the price of energy would always cause a 1% reduction in the demand for energy, while (ii) a 1% increase in world GNP would always cause a 1% increase in the demand for energy.

produce a 2% reduction in the carbon to noncarbon fuel consumption ratio. A similar computation meanwhile shows that the corresponding elasticity of substitution between either $E^c(t)$ or $E^n(t)$ and labor is unity. Thus, the unitary price and income elasticities of aggregate energy demand already noted from Equation (2b) were to be expected.

The lack of flexibility in this last elasticity was a source of concern. We were not anxious to be boxed into a structure of unitary elasticities in the demand for energy, but we were bound by a well-known result of economic theory: in maintaining the simple production structure that we felt was required to preserve the necessary transparency in the intertemporal model, we were forced to set the elasticity of substitution between energy and labor either to $s = (r - 1)^{-1}$ or to unity. Rather than loosen this theoretical binding by resorting to a more complicated production function, we chose instead to provide the desired flexibility by keeping Equation (1) as our fundamental production relationship and adjusting the share of world GNP used to pay for energy over time. To see how this was accomplished, let

$$X(t) = A(t)[mL(t)^q + (1 - m)E(t)^q]^{1/q} \qquad (1')$$

represent the next logical generalization of production. The parameter q here reflects the ease of substituting between energy and labor in the production process; it is the analog to the parameter r in Equation (1), and $(q - 1)^{-1}$ is the corresponding elasticity between energy now aggregated into one factor and labor. The resulting derived demands for labor and energy could then be combined to form the analog of Equation (3):

$$\frac{E(t)}{L(t)} = \left[\frac{(1 - m)}{m}\frac{P(t)}{w(t)}\right]^{1/(q - 1)}, \qquad (4)$$

where w(t) represents the unit price of labor. Multiplying both sides of Equation (4) by [P(t)/w(t)], a more convenient form emerges:

$$\frac{[1 - d(t)]}{d(t)} = \frac{P(t)E(t)}{w(t)L(t)} = kP(t)^{q/(q - 1)}, \qquad (4')$$

where $k = [(1 - m)/m]^{1/(q - 1)}$ and taking the approximation that w(t) = 1.* Notice that the first part of that equation simply states that the relative share of GNP devoted to paying for energy must equal the ratio of the energy bill of the world, P(t)E(t), and the global wage bill w(t)L(t). It makes the d(t) parameter defined here the precise

---

*Setting w(t) = 1 requires an approximation, as follows:

The model assumes that if the relative price of energy to labor, [P(t)/w(t)], is constant, then d (the share of labor) is constant. We are attempting to examine the effect of changes in P(t)/w(t) on d.

(continued overleaf)

analog of the d parameter recorded in Equation (1). Rearranging terms, then, the equation

$$d(t) = [kP(t)q/(q - 1) + 1]^{-1} \quad (5)$$

provides a means by which the share of world GNP paid to energy can be adjusted from period to period in a manner consistent with an elasticity of substitution between labor and energy equal to $(q - 1)^{-1}$. We were able, with this procedure, to approximate a more general schedule like Equation (1') with a series of simpler schedules of the type shown in Equation (1) by simply adjusting energy's share of GNP in a way that was consistent with the more complex structure that we needed. We are, in other words, out of our theoretical bind.

With this final step completed, we are able to set both the elasticity of substitution between fossil and nonfossil fuels and the elasticity of substitution between energy and labor equal to whatever the data suggested were appropriate values without overburdening the model with unnecessary complication.

To proceed from this point it is necessary to specify how A(t), L(t), and the prices of the two fuels are to be determined over time. Productivity and population are the easiest; they take the forms

$$A(t) = A_0 e^{a(t)t}$$

and

$$L(t) = L_0 e^{l(t)t},$$

where

A_0 = labor productivity at t = 0 (1975);
$a(t)$ = rate of growth of labor productivity at time t;
L_0 = world population at t = 0; and
$l(t)$ = rate of growth of population at time t;

The last three are taken to be exogenous.

(continued from overleaf)

If we were instead to set w(t) equal to its model solution, then Equation (4') could be written as

$$\lambda(t) = k\,\theta(t)^{\gamma}, \quad (4'')$$

where $\lambda = (1 - d)/d$, $\theta = P/w$, $\gamma = q/(1 - q)$. Thus the change in λ from one path to another would be

$$\Delta \ln \lambda(t) = \gamma[\Delta \ln P(t) - \Delta \ln w(t)].$$

For given L(t) and P(t), we can solve this for w(t), and $\Delta \ln w(t)$ is an order of magnitude smaller than $\Delta \ln P(t)$, because the share of E is about one tenth of the share of L. Our approximation thus misstates the change in d by about one tenth. Also note that the share of L changes only a little.

Energy prices are divided into production and distribution cost components (roughly the difference between wholesale and retail prices), and production costs are presumed to be subject to technological change. The price of noncarbon-based fuel is, more specifically, given by

$$P^n(t) = P^n_d + P^n_0 e^{[h_1(t) + h_2(t)]t}, \qquad (6a)$$

where

P^n_d = distribution costs in 1975 U.S. dollars per metric ton of coal equivalent;

P^n_0 = initial production costs in 1975 dollars per metric ton of coal;

$-h_1(t)$ = rate of technological change in the energy industry at time t; and

$-h_2(t)$ = bias of technological change toward noncarbon energy at time t

are all exogenous. The last two entries in the list may require a little explanation. The rate of technological change in the energy industry is the rate at which the efficiency in the industry is improving; conversely, it can be viewed as the inverse of the rate of change in the real price of energy. If, for example, the price of energy were decreasing at a rate of 1% per year, this would be consistent only with a rate of technological change of 1% per year. The bias in the rate of technological change reflects the possibility that technical change and innovation will not proceed at the same rate in both energy sectors. If innovation were more rapid in the nonfossil fuel sector, for instance, then the bias would favor that sector, and $h_2(t)$ would be positive.

The price equation for fossil fuel takes similar form. The only complicating element here is the implicit inclusion of a depletion factor--a reflection of the usual expectation that the price of fossil fuel should increase over time as the world's resources of fossil fuels are used up. We do not, here, necessarily include supply and demand effects.* The depletion factor represents, more accurately, the notion

---

*The model has one theoretical flaw from the point of view of the economics of exhaustible resources. This is that there are no rents charged to scarce fossil fuels. The economics and some estimates of such scarcity rents are provided in Nordhaus (1979) for a model without uncertainty.

There is a very great difficulty in the present model, however, in calculating the appropriate scarcity rents. This difficulty arises because the appropriate scarcity rents will be different in each of the 3^{10} possible trajectories. And the actual rent at each point of time will depend on the way that the uncertainties are revealed over time.

(continued overleaf)

that the price of fossil fuel must increase as the cheaper wells are drilled or mines are exhausted and as more expensive sources come on line. Depletion is represented as follows:

$$P^c(t) = P^c_d + \left(g_0 + g_1 \left\{ R(t)/[\bar{R} - R(t)] \right\}^{g_2} \right) e^{h_1(t)t} + T(t - \bar{t}), \quad (6b)$$

where

P^c_d = distribution costs in 1975 U.S. dollars per metric ton of coal equivalent;

g_0 = initial production costs in 1975 dollars per metric ton of coal equivalent;

R = a measure of the world's remaining carbon-based fuel reserves in metric tons of coal equivalent in 1975;

$T(t - \bar{t})$ = a tax policy parameter used to reflect taxation of fossil fuels;

$R(t)$ = $[E^c(0) + \ldots + E^c(t - 1)]$ = total carbon-based fuel consumed since 1975 in metric tons of coal equivalent; and

$g_i (i = 1,2)$ = depletion parameters.

In this list, of course, all but R(t) are taken to be exogenous.

At this point, then, only three parameters remain to be determined: b, A, and m. These are specified, given assumed values for s, r, q, d, L_0, $E^c(0)$, $E^n(0)$, $(p^n_0 + P^n_d)$, and $(g_0 + P^c_d)$, so that the entire set of parameters satisfied Equations (1), (3), and (5) at time zero. No further data are necessary.

Special mention needs to be made of the policy variable $T(t - \bar{t})$. It is included to reflect any policy that might be designed to reduce carbon dioxide concentrations either directly by taxes or indirectly by discouraging the consumption of fossil fuels. Since either type of policy would make it more expensive to burn these fuels, either would be captured by the tax $T(t - \bar{t})$ that increased the price of $E^c(\bar{t})$. The parameter t simply denotes the lag between the imposition of a CO_2 reduction policy and its effect on the day-to-day operations of fuel burners. The use of a tax to summarize even quantity-based restrictions is widespread in the economic literature and is supported by the following equivalence theorem: for any targeted quantity restriction on, for example, carbon-based fuels, there exists a tax to

---

(continued from overleaf)
After some thought about the best way to calculate the rents, we finally gave it up as hopelessly complicated.

In reality, it seems that, except for oil and gas, the scarcity rents are likely to be quite small for most of the time. This conclusion is based on a reading of the estimates from Nordhaus (1979). However, it should be noted that omission of the scarcity rent leads to a downward bias in the market price of fossil fuels and consequently in an upward bias in the estimate of CO_2 emissions and concentrations. We expect that this bias is likely to be on the order of 0 to 2% during the period under consideration.

be added to the price of carbon-based fuels such that consumers, in their own best interest, will undertake actions to lower their consumption to the prescribed target level.* Either tool, properly computed, can therefore achieve any arbitrary policy objective, and the generality of the tax approach is assured.

Some have noted that the two alternatives need not be equivalent, in terms of their efficiency, under uncertainty. A similar theorem exists, however, when the comparison is conducted between alternatives computed to generate the same expected result. Others have worried that the equivalent tax might, in practice, be difficult to compute. Whether that is true or not, of course, this purported difficulty does not damage the treatment in the present paper.

Returning now to the model, Equations (1) and (6) complete a simple, economically consistent vehicle with which to project the driving force of industrial CO_2 emissions. Only a link to the atmosphere is required; that link is represented by

$$C(t) = \left[Z_0 e^{z(t)t} \right] \left[E^c(t) \right],$$

where

$C(t)$ = emissions of carbon in gigatons per year;
Z_0 = the "emissions factor," equal to the initial ratio of carbon emissions to fossil fuel consumption in 1975; and
$z(t)$ = the rate of growth of the emissions factor.

The last two are, of course, exogenous. The ratio $z(t)$ is, moreover, presumed to increase over time because of a supply-induced change in the fuel mix (i.e., toward coal and shales). Nonfossil fuels are presumed to provide energy without adding to carbon emissions.

An airborne fraction approach to link emissions to atmospheric concentrations is finally employed to complete the model. Formally,

$$M(t) = M(t - 1) + AF(s)[0.471\ C(t)] - sM(t - 1), \tag{7}$$

where

s = a seepage factor reflecting the slow absorption of airborne carbon dioxide into the deep oceans;
$AF(s)$ = the marginal airborne fraction of carbon dioxide; and
$M(t)$ = carbon mass in the atmosphere in period t measured in parts per million.

---

*See Yohe (1979) for a survey of the literature on this point.

Equation (7) is a standard representation of the complex workings of the atmosphere, frequently used in the carbon cycle literature.* The coefficient 0.471 preceding C(t) simply converts gigatons of carbon into the appropriate atmospheric units of parts per million. The seepage factor is a subject of current debate among researchers (see Brewer, this volume, Chapter 3, Section 3.2); in separate work, we have shown that the maximum likelihood estimate of the airborne fraction is quite sensitive to the specification of s; thus, the AF(s) notation.

In summary, then, the model operates with the demands for fossil and nonfossil fuel being derived entirely from a production function that, for any year, assumes the form

$$X(t) = A(t)L(t)^{d(t)} [bE^{c}(t)^{r} + (1 - b)E^{n}(t)^{r}]^{[1 - d(t)]/r}. \quad (1)$$

They emerge summarized by

$$P^{n}(t)E^{n}(t) + P^{c}(t)E^{c}(t) = [1 - d(t)]X(t) \quad (2a)$$

and

$$\frac{E^{c}(t)}{E^{n}(t)} = \left[\frac{(1 - b)}{b}\,\frac{P^{c}(t)}{P^{n}(t)}\right]^{1/r - 1} \quad (3)$$

with

$$d(t) = [kP(t)q/(q - 1) + 1]^{-1}. \quad (5)$$

Furthermore, the neutral productivity growth factor and labor growth component of Equation (1) are given exogenously by

$$A(t) = A_{0}e^{a(t)t}$$

and

$$L(t) = L_{0}e^{l(t)t},$$

respectively. The supply conditions from fossil and nonfossil fuels are meanwhile determined by

---

*See Bolin (1981) for a complete discussion of the concentration model. Our modification of that work specifies a marginal airborne fraction--the fraction of period t emissions that remain in the atmosphere on the margin (i.e., in period t). Whenever the seepage factor is nonzero, the marginal fraction does not equal the average fraction that most of the previous studies have employed.

$$P^n(t) = P_d^n + P_0^n e^{[h_1(t) + h_2(t)]t} \tag{6a}$$

and

$$P^c(t) = P_d^c + \left(g_0 + g_1\left\{R(t)/[\bar{R} - R(t)]\right\}^{g_2}\right) e^{h_1(t)t} + T(t - \bar{t}) \tag{6b}$$

with $R(t) = E^c(0) + \ldots + E^c(t - 1)$.

Emissions are then recorded according to

$$C(t) = \left[z_0 e^{z(t)t}\right] E^c(t)$$

and atmospheric concentrations according to

$$M(t) = M(t - 1) + AF(s)[0.471\ C(t)] - sM(t - 1). \tag{7}$$

Figure 2.8 represents a geometric interpretation of this process.

2.1.2.2 The Data

Two kinds of data are required. Initial conditions are, first of all, required. Table 2.2 records the estimates for world GNP, world population, and world fossil and nonfossil fuel in 1975. Since these initial conditions are based on historical evidence, consensus is not difficult to achieve. Existing studies and comparison with published data are sufficient to generate consistent estimates for these parameters. Initial energy prices are a bit more problematical. We want aggregate prices based on the historical distribution of, for example, fossil fuels between coal, oil, and gas. Table 2.3 records both the necessary raw data and their sources. Table 2.4 produces the aggregates and illustrates the procedure that is employed in their construction. Table 2.5 registers the emissions ratios of the various types of fossil fuels from which the initial value for the aggregate emissions ratio is computed. Table 2.4 also records that aggregation procedure. Finally, an initial level of atmospheric carbon concentration is required; current measures set the 1975 value at 331 parts per million (ppm) (see Keeling et al. in Clark, 1982, Table 1, page 378).

Data are also required to set the long-term context of the study--a more difficult problem. Projections of various important parameters into the near and distant future were compared, but the uncertainties inherent in such projection made consensus impossible. Existing studies provide ranges for variables like world population growth, world productivity growth, energy prices, and the emissions factor, but no generally accepted paths emerge. The observed ranges are, however, viewed as more than spurious disagreement among researchers. They are, instead, viewed as a reflection of the inherent uncertainty about the variables.

A more precise description of the technique we use is the following: we assume that the published estimates for each of the random variables

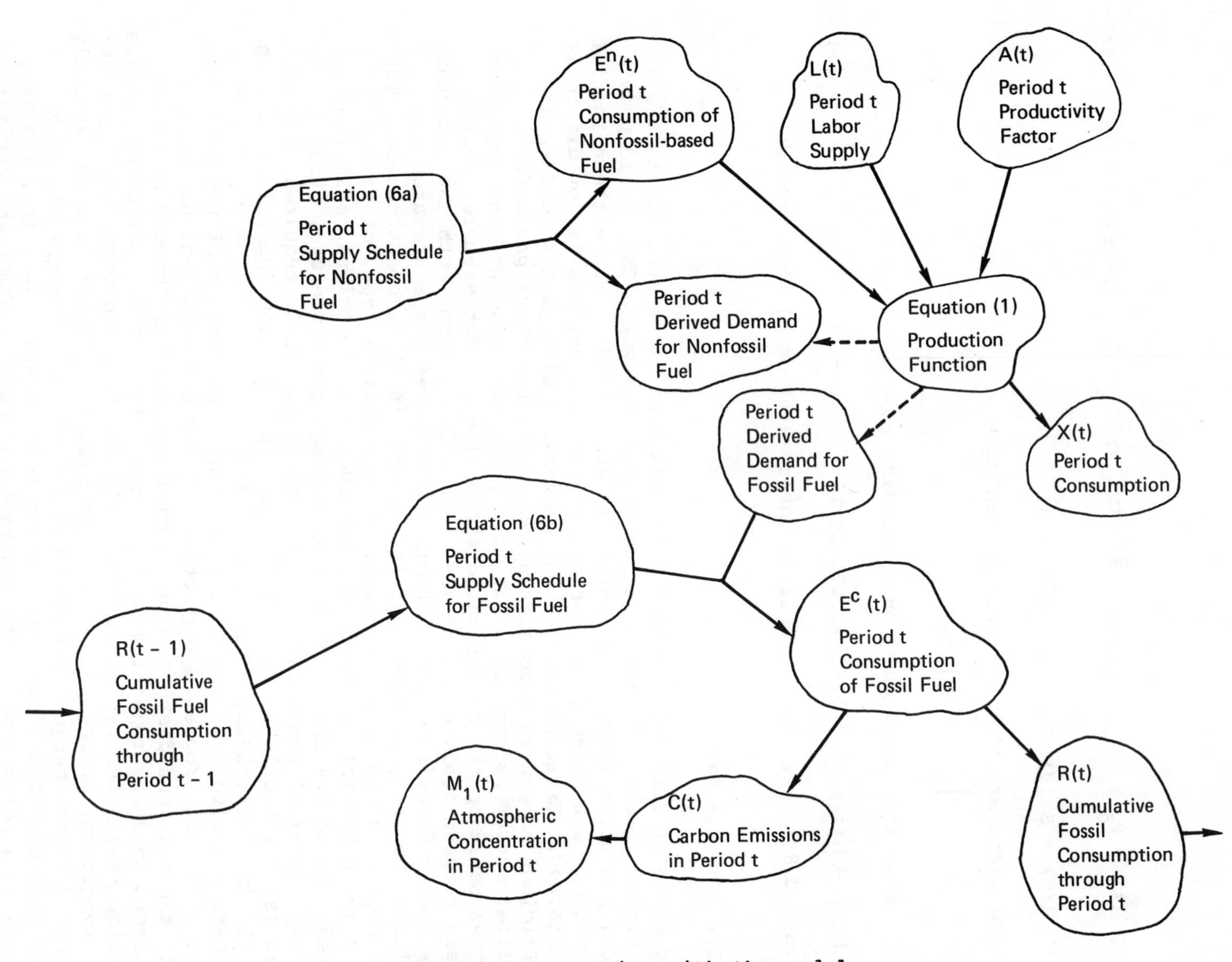

FIGURE 2.8 Iteration with the model.

TABLE 2.2 Initial Conditions for Population, GNP, and Aggregate Fuel Consumption

A. World Population in 1975	
IIASA (1981, p. 133)	4×10^9
Ridker and Watson (1980, p. 45)	3.8×10^9
Keyfitz (1982)	3.98×10^9
Value employed: L(0)	4×10^9
B. World GNP in 1975	
IIASA (1981, p. 457)	$\$ 6.4 \times 10^{12}$
Ridker and Watson (1980, p. 45)	$\$ 4.5 \times 10^{12}$
Report of the President (1980)	$\$ 6.53 \times 10^{12}$
Value employed: X(0)[a]	$\$ 6.4 \times 10^{12}$
C. Fossil Fuel Consumption in 1975	
IIASA (1981, p. 136)	8.127×10^9 mtce
Ridker and Watson (1980, p. 185)	8.114×10^9 mtce
Value employed: $E^c(0)$	8.1×10^9 mtce
D. Nonfossil fuel Consumption in 1975	
IIASA (1981, p. 136)	0.67×10^9 mtce
Ridker and Watson (1980, p. 185)	0.69×10^9 mtce
Value employed: $E^n(0)$	0.7×10^9 mtce

[a]The Ridker and Watson estimate was ignored because it was built around an estimate of per capita income that appeared to be low relative to other published data.

identified in our model is an unbiased, but not necessarily independent, estimate of that variable. We use the means and variances of those estimates as a basis for constructing _judgmental probability distributions_ for each variable. To obtain a manageable number of alternatives from which to sample, we assume that each judgmental probability distribution is normally distributed. We then take high, middle, and low values (corresponding to 25, 50, and 25%, respectively) that maintain the same means and variances as the estimated normal distributions.

To put it more intuitively, we have constructed discrete distributions to mirror the level of uncertainty at present surrounding the various analysts' published projections. In deference to the tendency of individuals to underestimate uncertainty, however, the procedure for reflecting uncertainty does not stop there. Particularly when the estimated variances declined over time, future variances are expanded beyond their computed ranges to correct for systematic underestimation of uncertainty. Section 2.1.2.2.1 is devoted to a thorough exploration of this second procedural phase; the remainder of this section will concentrate on applying the first phase to the critical parameters.

In passing, though, it should be noted that our procedure produces more than a purely subjective view of the future paths of some complicated variables. It produces a "judgmental" view that weighs the expert opinions of many researchers as expressed in their published

TABLE 2.3 Energy Price and Disaggregate Consumption Data

A. Consumption Patterns (1975)[a]		
I. Fossil Fuel		
Coal	2.42×10^9 mtce	30%
Oil	4.10×10^9 mtce	50%
Gas	1.61×10^9 mtce	20%
Total	8.13×10^9 mtce	100%
II. Nonfossil Fuel		
Hydro	0.53×10^9 mtce	81%
Nuclear	0.13×10^9 mtce	19%
Total	0.66×10^9 mtce	100%
B. Energy Prices (1975)[b]		
I. Primary[c]		
Coal	$ 15/mtce	
Oil	$ 54/mtce	
Gas	$ 19/mtce	
Electricity	$118/mtce	
II. Secondary[d]		
Coal	$ 23/mtce	
Oil	$ 98/mtce	
Gas	$ 28/mtce	
Electricty	$255/mtce	
C. Energy Prices (1981)[e]		
I. Primary[c]		
Coal	$ 15/mtce	
Oil	$108/mtce	
Gas	$ 86/mtce	
Electricty	$118/mtce	
II. Secondary[d]		
Coal	$ 23/mtce	
Oil	$171/mtce	
Gas	$116/mtce	
Electricty	$255/mtce	

[a]Source: IIASA (1981), p. 471.
[b]Measured in 1975 United States dollars. Source: Reilly et al. (1981).
[c]These are wholesale prices.
[d]These are retail prices.
[e]Measured in 1975 United States dollars. They are derived from the 1975 prices to reflect the impact of the 1979 Iran-Iraq war as follows. The June 10, 1982, issue of Blue Chip Indicators provided an estimate for the 1981 wholesale oil price of $34.00 (current dollars) per barrel. That translates into $22.50 per barrel in 1975 dollars, or $108/mtce. The btu equivalent price for gas was then computed to equal [0.8($108/mtce)] = $86/mtce. Coal and electricity (generated from nonfossil fuels) were assumed to remain constant over the 6-year period. The secondary prices for oil and gas were then computed by adding the reported consensus differences between wholesale and retail prices: $13/barrel or $63/mtce for oil and $30/mtce for oil. Thus, secondary prices of [$108 + $63] = $171/mtce and [$86 + $30] = $116/mtce were recorded, respectively.

TABLE 2.4 Aggregate Prices and Emissions

A. Energy Prices in 1975[a]	
I. Primary	
Fossil fuel[b]	$ 35/mtce
Nonfossil fuel[c]	$ 118/mtce
II. Secondary	
Fossil fuel[b]	$ 62/mtce
Nonfossil fuel[c]	$ 255/mtce
B. Energy Prices in 1981[a]	
I. Primary	
Fossil fuel[d]	$ 76/mtce
Nonfossil fuel[c]	$ 118/mtce
II. Secondary	
Fossil fuel[d]	$ 116/mtce
Nonfossil fuel[c]	$ 255/mtce
C. Emissions Ratio in 1975[e]: Z(0)	580 g of C/mtce

[a]Measured in constant 1975 dollars.
[b]Computed using the prices and weights recorded in Table 2.3. In 1975, for example, coal amounted to 30% of the total fossil fuel consumed and cost $15/mtce, oil amounted to 50% of the total and cost $54/mtce, and gas amounted to 20% at a cost of $19/mtce. Thus, the aggregate price of fossil fuel in 1975 was 0.3($15) + 0.5($54) + 0.2($19) = $35/mtce.
[c]The price of nonfossil fuel was taken to be the price of electricity generated from nonfossil sources.
[d]Computed using the weights and prices recorded in Table 2.3. The weights employed were the 1975 numbers because it was unlikely that major substitutions could have occurred in the 6 years from 1975 and 1981. This presumption is borne out by data published in the BP Statistical Review of the World Oil Industry, 1980, p. 16.
[e]The ratio of grams of carbon emitted per mtce of fuel consumed. The 1975 consumption weights of Table 2.3 were combined with the emission data of Table 2.5 to produce Z(0); i.e., Z(0) = 0.3(700) + 0.5(577) + 0.2(404) = 580 g of C/mtce.

work. The data for obtaining parameter estimates are not, in other words, anonymous and private. They are public and thus presumably derived with the care that scientists use in producing work attached to their names. And they are judgmental views about nonelemental variables--not variables like GNP growth or energy growth that depend

TABLE 2.5 Carbon Emissions from Fossil Fuels[a]

Fuel	kg of C/10^9 J	kg of C/mtce[b]
Petroleum	19.7	577
Gas	13.8	404
Coal	23.9	700
Shale oil[c]	41.8	1224

[a]Source: Marland (1982).
[b]Conversion based on 1 mtce = 29.29 x 10^9 J; kg of C, kilograms of carbon.
[c]Includes carbon dioxide emissions due to shale oil mining and extraction.

on a host of known and unknown effects, but variables like population growth and resource availability that depend on fewer things.*

Beginning once again with population numbers, Table 2.6 shows that a variety of growth projections have been made for at least the next 50 years. Each assumes no major catastrophes; and while the general trend in each calls for a steady decline in the rate of growth, there is some disagreement. Differences are to be expected, of course, but it is interesting to note that these differences found their source in the assumptions made about the less-developed countries; the historical experience of the LDCs has been so widely varied that a common expectation would, of course, have been surprising. The full effect of that disagreement is not reflected in the world projections, however, because the larger, more-developed countries have displayed low, stable growth rates over the past few decades.

Of further interest was the marked reduction in the variance of projections beyond the year 2025; most researchers predict that the world's population will stabilize sometime after the first third of the twenty-

*Two technical points might be raised:

First, are the estimates independent? It is likely that some of the figures depend on previous estimates--indeed, they might all go back to a single careful study. We have been unable to check for such an occurrence in every case, but in some we are confident that the outcomes are truly independent, even competitive.

Second, are the underlying judgmental probability distributions independent? While some lingering correlations probably exist, we took care to construct our random variables so that the correlations were low--that is, the variables are intended to be orthogonal. Thus, the rate of productivity growth in energy is thought to be independent of the difference in productivity growth between fossil and nonfossil fuels.

TABLE 2.6 Projected Trends in the Growth Rates of World Population[a]

Source of Estimate[b]	1975-2000	2000-2025	2025 and beyond
OECD	1.6%	--	--
IIASA[c]	1.7%	0.9%	d
RFF (high)	1.9%	1.52%	d
RFF (low)	1.4%	0.75%	d
Hudson Institute	2.0%	1.4%	--
Keyfitz[c]	1.6%	0.9%	0.3%
Mean	1.7%	1.1%	0.3%
Standard deviation	0.2%	0.36%	n.a.
Cell extremes	1.4%; 2.0%	0.6%; 1.6%	n.a.

[a]Estimates of the l(t) parameters of population growth equation.
[b]Sources: OECD Interfutures Project (1979), IIASA (1981), Ridker and Watson (1980), Kahn et al. (1976), and Keyfitz (1982).
[c]The IIASA projections were based on an earlier set of estimates by Keyfitz.
[d]The IIASA and RFF studies report the expectation of stable world population some time after the first third of the twenty-first century.

first century. Sometimes they have reached that conclusion because they believe that by then the volatile LDC behavior will have evolved into the predictable model of the developed countries; sometimes their predicted stability was based on some other presumption. In either case, their behavioral hypothesis was as much of a guess about the unknown future as any other growth path, and it is hard to see why uncertainty should diminish as time goes forward. The observed reduction in range of population growth projections beyond 2025 is thus a likely candidate for the adjustment discussed further in the next section.

As troublesome as the later estimates might have been, however, the earlier ranges provided excellent arenas for illustrating the summarizing procedure for the observed variation. The various estimates, ranging from 1.4% growth per year up to 2.0% for 1975-2000 are, for example, assumed to be observations drawn from an underlying normal distribution of the true uncertainty. These observations (X_i) are then used to compute estimates of the mean (μ_x) and variance (σ^2) of that distribution:

$$\mu_x = \bar{X} = \frac{1}{n}\left\{X_1 + \ldots + X_n\right\}$$

and

$$\sigma^2 \approx s^2 = \frac{1}{n-1}\left\{(X_1 - \bar{X})^2 + \ldots + (X_n - \bar{X})^2\right\},$$

where n represents the number of observations. To discretize the distribution defined by $\bar{X}$ and S^2 into three cells of probabilities 0.25, 0.50, and 0.25, therefore, $\bar{X}$ is assigned a probability of 0.5 and $(\bar{X} \pm s\sqrt{2})$ probabilities of 0.25. In this way, mean and variance $[s^2 = 0.25(2s^2) + 0.25(2s^2)]$ are both preserved. For the 1975-2000 range, therefore, 1.7% is the "middle" estimate, while 1.4% and 2.0% represent the extremes. Under this procedure, roughly 8% of the underlying probability is left beyond the extremes on both sides. Similarly, 1.0% is the middle estimate for the period 2000-2025, with 0.5% and 1.5% catching the 0.25 probability tails.

Estimates of growth in world productivity are recorded in Table 2.7. They are, for the most part, based on a somewhat surprising assumption about the growth of world trade over the next several decades. Each researcher found that the growth of the world economy will be bounded by growth in the largest markets--the developed countries. Many studies have identified productivity growth as a critical parameter for energy and carbon dioxide projections. The common presumption about the growth of world trade, ironically, has caused otherwise independent studies to project estimates of output growth that converge over time. Of particular note is the decline in the variance in projected growth rates beyond the year 2025. It may have been caused more by a dearth of estimates than anything else, but its range includes only the lower tail of the long-run historical experience of the United States, and it misses the Japanese experience completely. These ranges, too, are subject to revision later.

TABLE 2.7 Projections of the Rate of Growth of World Productivity[a]

Source of Estimate[b]	1975-2000	2000-2025	2025 and beyond
OECD (high)	3.4%	--	--
OECD (mid)	2.8%	--	--
OECD (mid)	1.9%	--	--
OECD (low)	2.7%	--	--
IIASA (high)	2.3%	0.9%	--
IIASA (low)	1.2%	1.9%	--
Hudson	2.8%	1.4%	1.2%
RFF (high)	2.4%	2.1%	--
RFF (low)	1.6%	1.8%	--
Hudson (low)	--	--	0.75%
Mean	2.3%	1.6%	1.0%
Standard deviation	0.7%	0.5%	0.3%
Cell extremes	1.2%; 3.4%	0.9%; 2.3%	0.5%; 1.5%

[a]Estimates of the a(t) parameter in the productivity growth expression.
[b]Sources: OECD Interfutures Project (1979), IIASA (1981), Ridker and Watson (1980), and Kahn et al. (1976).

TABLE 2.8 Projections of the Rate of Growth of Noncarbon Energy Prices[a]

Source of Estimate[b]	1975-2000	2000-2025	2025 and beyond
IEA	0.0%	0.0%	0.0%
RFF (DH NU)[c]	1.0%	0.6%	--
RFF (DHP1)[c]	0.7%	-0.1%	--
RFF (DHP2)[c]	1.0%	-0.4%	--
Mean	0.6%	0.0%	0.0%
Standard deviation	0.5%	0.4%	n.a.
Cell extremes	-0.1%; 1.3%	-0.5%; 0.5%	n.a.

[a]Estimates of the $h_1(t)$ and $h_2(t)$ parameters of the energy price (supply) equations.
[b]Sources: Reilly et al. (1981), Ridker and Watson (1980).
[c]The difference between these three scenarios is essentially a difference in the assumption about solar and nuclear development. The particulars are not so important, for our purpose, as the spread of uncertainty.

Table 2.8 records projected future adjustments in the primary real price of noncarbon-based energy--electricity not derived from burning carbon-based fuel. These trends can, however, be interpreted as the inverse of the rate of technological change in the energy sector, i.e., in the notation of the previous section, $h_1(t)$. Since technological change can continue in the fossil fuel sector as well, these estimates are also used to frame the difference in the rate of advance between the two sectors $[h_2(t)]$. These estimates, then, are clearly dependent not only on growth assumptions (and thus the need for new technology) but also about the future contributions of sources like nuclear, fusion, and solar-generating facilities. Despite the obvious uncertainties involved in projecting either factor into the twenty-first century, the estimates recorded in Table 2.8 again converge. The Reilly et al. (1982) view of constant real prices is therefore included as the middle case, and the ultimate variation around that case expanded.

Estimates of the emissions factor are based both on the unit emissions for each source recorded in Table 2.5 and on projected mixes of carbon-based fuels in the future. For each case, the mix of oil, gas, coal, and shale oil is computed and used to weight the unit emissions in computing an aggregate. The procedure has already been illustrated in the calculation of Z(0) for Table 2.4. Results of the other computations are noted in Table 2.9. The summarizing procedure is applied across the ranges of emissions for each period to produce high, medium, and low trends. Notice that the high trend includes a 31% contribution from shales (as projected by RFF) by the year 2050. The lower two paths stabilized at 100% coal, or 700 g of C/mtce by 2075. The three possibilities are illustrated in Figure 2.9.

TABLE 2.9 Aggregate Carbon Emissions

A. Fuel Proportions

Year	Source[a]	Proportions Oil	Gas	Coal	Shale	Carbon Emissions
1975	Table 2.3	0.50	0.20	0.30	0.00	580
2025	IIASA (high)	0.33	0.29	0.38	0.00	580
	IIASA (low)	0.34	0.23	0.43	0.00	590
	IEA	0.28	0.18	0.54	0.00	612
	RFF	0.28	0.23	0.37	0.06	607
2050	IIASA (high)	0.25	0.10	0.65	0.00	604
	IIASA (low)	0.28	0.19	0.53	0.00	598
	IEA	0.26	0.08	0.66	0.00	644
	RFF	0.02	0.02	0.65	0.31	854

B. Projected Growth in Emissions--Z(t)

Year	Mean	Emissions Std.Dev.	Extremes	Growth Mean	Extremes
1995-2025	597	15	582; 612	0.05%	0.0%; 0.1%
2025-2050[b]	627	25	602; 799	0.2%	0.1%; 1.1%
2050-2075[c]	n.a.	n.a.	700; 700; 854	0.4%	0.6%; 0.3%
2075-2100[c]	n.a.	n.a.	700; 700; 854	0.0%	0.0%; 0.0%

[a]Sources: IIASA (1981), Reilly et al. (1982), Ridker and Watson (1980).
[b]The mean and standard deviation reported here excludes the shale estimate to generate the low extreme and the middle growth paths. The higher extreme includes the shale estimate in both computations.
[c]The two lower runs exclude shale; the high extreme converges to a 30% shale share of carbon-based fuel.

Elasticities of substitution between energy and labor, on the one hand, and between the two types of energy, on the other, are estimated on the basis of the literature on price elasticities of demand. In the former case, for example, it is noted that many would put the overall price elasticity of demand for energy somewhere in the inelastic range, i.e., they would expect a 1% price increase to reduce consumption by something less than 1%. A range for $s' = (q - 1)^{-1}$, the elasticity of substitution between E(t) and L(t), that select -0.4 and -1.2 for the 25% probability extremes and -0.7 for the mean is therefore employed. Similar reasoning puts the extremes for $s = (r - 1)^{-1}$, the elasticity between $E^c(t)$ and $E^n(t)$, at -0.5 and -2.0 around the middle run of -1.2.

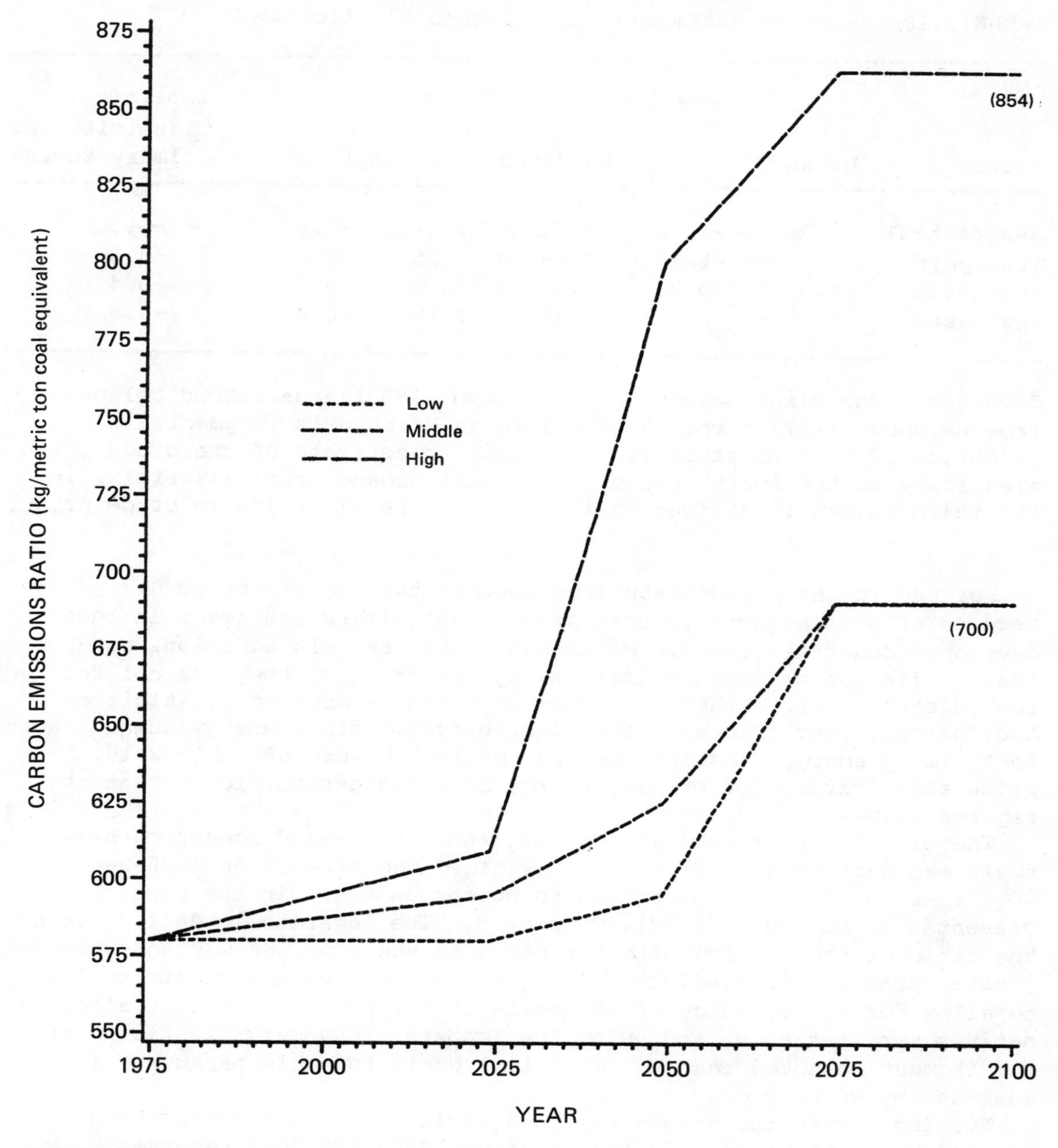

FIGURE 2.9 Carbon emissions ratio.

There are by now a wide variety of studies of the elasticity of substitution between energy and nonenergy inputs. To a first approximation, this parameter is equal to the price elasticity of the derived demand for energy [this is shown in Nordhaus (1980a)]. Table 2.10, drawn from Nordhaus (1980a), gives central estimates and ranges for the price elasticity of the demand for energy.

TABLE 2.10 Range of Estimated Final Demand Elasticities[a]

Sector	Hogan	Nordhaus	Best Guess	Implicit Elasticity for Primary Energy
Residential	-0.28 to -1.10	-0.71 to -1.14	-0.9	-0.3
Transport	-0.22 to -1.30	-0.36 to -1.28	-0.8	-0.2
Industrial	-0.49 to -0.90	-0.30 to -0.52	-0.7	-0.4
Aggregate	--	-0.66 to -1.15	-0.8	-0.3

[a]Sources: The first column is from Hogan (1980); the second column, from Nordhaus (1977); the third column is Nordhaus's judgmental weighting of various studies. To obtain an estimate of the crude price elasticity in the fourth column, the final demand price elasticity in the third column is divided by the ratio of retail price to crude price.

For use in the present study we lowered the elasticity to 0.7 because of some suggestion that price elasticities are lower in less-developed countries than in developed countries. In addition, note that, while the price elasticities may appear high, they are not for two reasons. First, they are long-run rather short-run elasticities. And, second, they relate to the elasticity for final energy demand, not for primary energy. As is shown in the last column of Table 2.10, price elasticities for primary energy are considerably lower than the figures we use.

The elasticity of substitution between carbon- and noncarbon-based fuels was derived as follows. We examined the effects of different CO_2 taxes on the ratio of carbon to noncarbon fuels in the runs presented in Nordhaus (1979), Chapter 8. The logarithmic derivative of the ratio of the two fuels to the ratio of their prices was somewhat greater than 1.5 in absolute value. We reduced the elasticity to 1.2 to allow for the tendency of LP models to "overoptimize." The alternatives were set above and below the important boundary elasticity of 1. It must be noted that the empirical basis for this parameter is as weak as any we rely on.

Turning now to the parameters in Equation (6b), estimates for g_1, g_2, and R are required. Estimates of world fossil fuel reserves vary widely according to the assumptions that are made about economic feasibility. Table 2.11 registers the variety from which our estimates were drawn. The low range includes only proven reserves that will certainly become economically feasible in the foreseeable future. The middle range captures a large increment of reserves that most researchers think will become feasible in that time span; it quadruples the low range by including difficult oil deposits and extensive use of cleansed coal. The upper range adds a small percentage of potential shale availability to the supply and puts world resources well beyond quantities that will be consumed over the span of our study--the next 125 years.

TABLE 2.11 World Resources of Fossil Fuels[a]

A. Certain Economic Feasibility (low R)[b]	
IIASA	2.7×10^{12} mtce
WAES	3.2×10^{12} mtce
Value employed for low R	3×10^{12} mtce
B. Probable Economic Feasibility (middle R)	
IIASA	11×10^{12} mtce
IEA (high)	12.2×10^{12} mtce
IEA (low)	12.0×10^{12} mtce
Value employed for middle R	12×10^{12} mtce
C. Including Shale Estimate (high R)	
Total deposits--Duncan and Swanson	144×10^{12} mtce
Value employed for high R (= middle R + 0.07 shale)	22×10^{12} mtce

[a]Sources: IIASA (1981); Energy: Global Prospects, Workshop on Alternative Energy Strategies (WAES) (1977); Reilly et al. (1982); Duncan and Swanson (1965).
[b]These numbers are consistent, component by component, with other incomplete data found in Moody and Geiger (1975), and World Energy Conference (1978).
[c]The inclusion of shale allows for incredible availability of fossil fuel. The 7% utilization rate, chosen rather arbitrarily, generated a resource constraint that was always nonbinding through the year 2100.

The procedure for computing g_1 and g_2 is more involved. For simplicity, first of all, g_2 is set equal to 1; manipulating g_1 provides more than enough flexibility. A range of prices for fossil fuel in some future time after an arbitrary R_1 mtce of fossil fuel had been consumed, is then constructed. Denoting those prices by P_j and the various reserve estimates cited above by $\bar{R}_k$, a collection of g_1 values, now clearly dependent on both P_j and $\bar{R}_k$, are computed according to

$$P_j = g_0 + g_1(j,k)[R_1/(\bar{R}_k - R_1)], \tag{8}$$

where j and k index high, middle, and low values for P_j and $\bar{R}_k$, respectively. Figure 2.10 shows that this procedure generated three possible paths for each of the three $\bar{R}_k$; i.e., nine separate specifications of the $g_1(j,k)$ and thus nine specifications of Equation (6b). Table 2.12 meanwhile records the prices estimated by several studies for $R_1 = 1100 \times 10^9$ mtce. It is a value chosen because of the availability of these price projections, and the table shows how the necessary aggregate prices are computed. Several other studies cited either prices without aggregation weights or consumption mixes without prices, so they were of little use. It is, nonetheless,

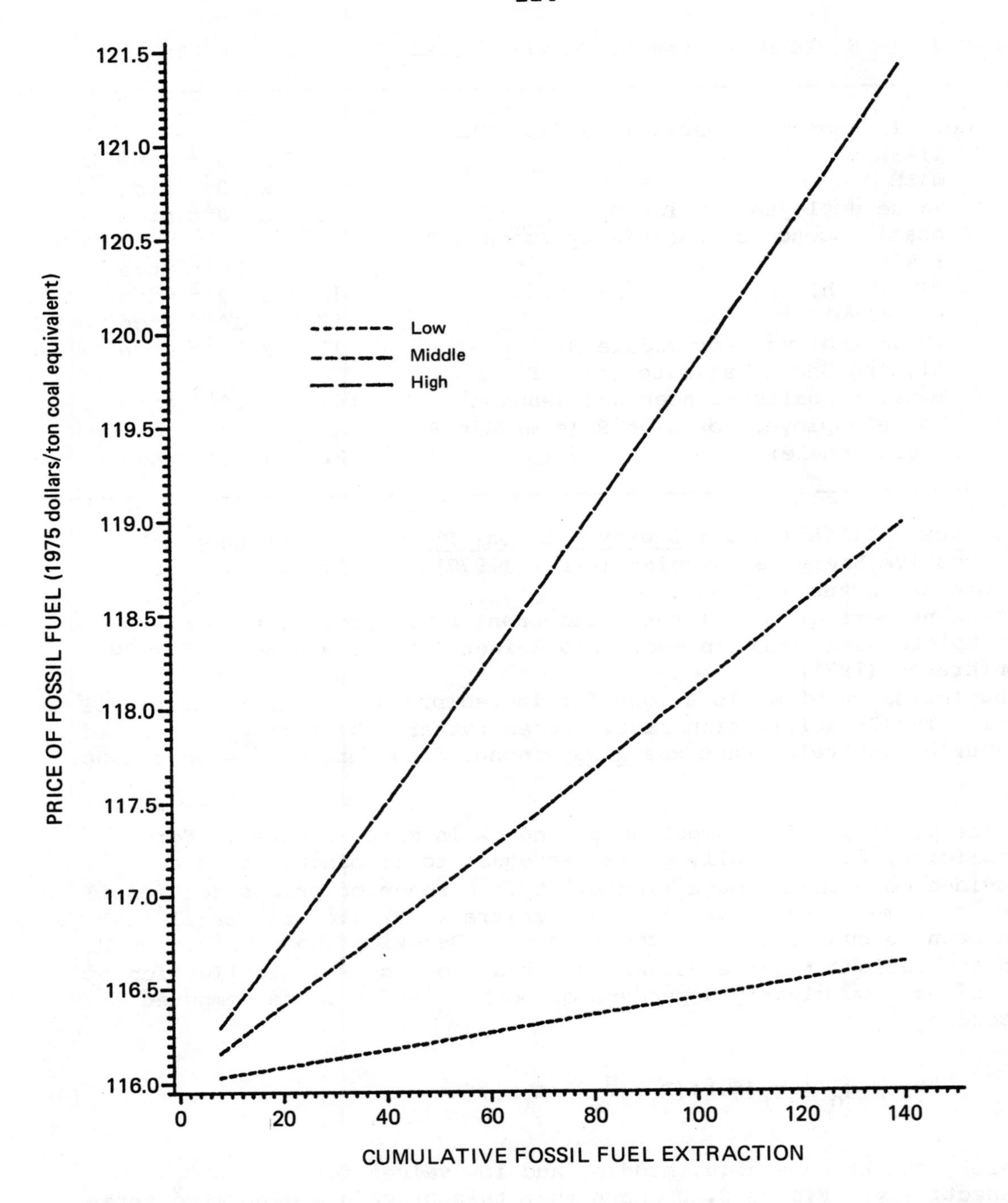

FIGURE 2.10 Secondary (retail) price of fossil fuel as a function of cumulative fossil fuel extraction. Prices are 1975 dollars per metric ton of coal equivalent. Cumulative extraction is measured from 1975 in billion metric tons of coal equivalent. In the terms of this section, these are price paths for a given $\bar{R}_k$ and $P_j = (P_1, P_M, P_H)$ for R_1.

TABLE 2.12 Primary (Wholesale) Fossil Fuel Prices when Cumulative Extraction from 1975 (R_1) = 1100 x 10^9 mtce

Source of Projection[a]	Oil[b]	Gas[b]	Coal[b]	Aggregate Price[c, d]
IIASA (high)	$130(0.33)	$70(0.29)	$25(0.39)	$73/mtce
IIASA (low)	$150(0.34)	$65(0.23)	$25(0.43)	$76/mtce
IEA (low)	$140(0.28)	$65(0.18)	$25(0.54)	$64/mtce
IEA (high)	$140(0.34)	$65(0.18)	$25(0.47)	$71/mtce
RFF	$123(0.38)	$75(0.27)	$40(0.35)	$81/mtce
Mean				$73/mtce
Standard deviation				$5.7/mtce
Cell extremes				$65; $81/mtce

[a]Sources: IIASA (1981), Reilly et al. (1982).
[b]The proportions of each source in the total consumption are given in the parentheses.
[c]Aggregate prices are computed as weighted sums of the components, with the weights being the proportions noted in the parentheses. Thus, for IIASA (high) (0.33)($130) + (0.29)($70) + (0.39)($25) = $73/mtce.
[d]Many studies did not provide a full range of necessary data. A compilation of 23 surveys of projected oil prices produced during the 1981 Stanford International Energy Workshop did, however, provide a good sample of oil price expectations through the year 2000. Extrapolating those data through 2030 [the year when most studies see R(t) = 1100 x 10^9 mtce] using the range of weights listed here produced a much wider range extending from $43/mtce to $106/mtce. These are rough estimates, of course, but lead to some widening later.

interesting to note that interpolating between the estimated oil prices used in this study for 2025 (and contained in Table 2.12) and the 1981 oil prices provides us a benchmark for comparison. This benchmark fell in the third decile of 23 studies (i.e., lower end) compiled during the Stanford International Energy Workshop of 1981 (Energy Modeling Forum, Stanford University).

Even with these data collected, our work on the price equation for fossil fuel is not completed. Section 2.1.2.1 outlines an approximation procedure that allowed both the simplicity of the nested production function recorded in Equation (1) and the flexibility of being able to vary the elasticity of substitution between energy and labor across time. It was an adjustment made necessary by a desire to incorporate a source of dynamic uncertainty into the model that could well loom large in the balance of this century. Much in the same spirit, we now need two similar adjustments in the fossil fuel equation. The first is designed to preserve the structure of Equation (3) even as the short-term effects of restricted oil supplies were recognized. The second gives society enough foresight to prepare for the imminent exhaustion of fossil fuel supplies.

The need for the first adjustment can be seen by looking at the very recent past. The model, as presented above, allows instantaneous substitution into and out of aggregate energy and between its carbon and noncarbon components each year in response to changes in relative input prices. While this is a conventional assumption for long-range growth models in which people are presumed to predict price movements accurately and plan accordingly, it does not conform well to the uncertain world that has confronted energy consumers since 1975. The dramatic disruption in world oil supplies caused by the advent of OPEC in 1973, the events in Iran, the oil glut, and the decontrol of gas and oil prices in the United States are all examples of factors that have contributed to the uncertainty; and their net effect has been to increase the primary price of carbon-based fuel from $35/mtce in 1975 to $76/mtce in 1981. Investment decisions taken in 1975 were, however, made in response to 1975 prices and 1975 expectations. Much of the world's present capital stock was, in fact, put into place before the oil shocks of 1973. The decisions that produced these investments were clearly not made with the type of accurate foresight required in the model. Nor can it be presumed that instantaneous substitution would have brought all the existing capital up to date relative to current energy prices. Thus, there exists a need to provide a longer reaction time at the beginning of the model to reflect the difficulty faced by most consumers in responding to such enormous price changes.

One possible adjustment would involve making alterations in the production function, but that course is again rejected to avoid complexity. Rather than produce the complications of more complex intertemporal substitution, we modify the early fossil fuel prices against which consumption decisions would be made. For the first 25 years of each run, in particular, a linear combination of projected current fossil fuel prices (computed from 1981 prices) and the lower 1975 prices is employed to slow the rate of growth of fossil fuel prices; the result is a reduction in the reaction to higher fossil fuel prices mandated by Equation (3). After the year 2000, however, this delayed reaction is stopped and decisions are assumed to be made on the basis of prevailing fuel prices.

More specifically, the 1975 primary price of fossil fuel ($35/mtce) is used in conjunction with the price ranges computed for $R_1 = 1100 \times 10^9$ mtce to compute the appropriate g_1 coefficients. They are recorded here in Table 2.13. This computation, with the R_1 price range expanded to $43, $73, and $103 per mtce to reflect the larger dispersion of the Stanford estimates of oil prices, is appropriate because the price estimates on which the range was based were made under 1975 expections. Nonetheless, the primary price of fossil fuel did reach $76/mtce by 1981, and distribution costs did rise by $40/mtce from 1975 through 1981. These figures, therefore, are used as initial conditions for the long-term supply equation; i.e., the equation

$$P^c(t) = \left(76 + g_1\left\{R(t)/[\bar{R} - R(t)]\right\}^{g_2}\right)\exp[h_1(t)t] + 40 \quad (9)$$

fully specifies the long-term price equation for fossil fuel. Still, the point of this adjustment is that imposing these inflated prices in

TABLE 2.13 The g_1 Parameters of $P^c(t)$[a]

R	Price at R_1[b]	g_1	Probability
3200 x 10^9 mtce	$43/mtce	13.8	0.06
3200 x 10^9 mtce	$73/mtce	65	0.13
3200 x 10^9 mtce	$103/mtce	118	0.06
11000 x 10^9 mtce	$43/mtce	72	0.13
11000 x 10^9 mtce	$73/mtce	342	0.24
11000 x 10^9 mtce	$103/mtce	612	0.13
21000 x 10^9 mtce	$43/mtce	145	0.06
21000 x 10^9 mtce	$73/mtce	687	0.13
21000 x 10^9 mtce	$103/mtce	1230	0.06

[a]Source: Tables 2.11 and 2.14, Equation (6b), and the text of this section.
[b]In 1975 U.S. dollars.

1981 would not have been consistent with the spontaneous flexibility of the production function. Since the relevant secondary price of fossil fuel in 1975 is $62/mtce and not $116/mtce, the operative fossil fuel price for the first 25 years is adjusted linearly according to

$$[P^c(t)]' = [(25 - t)/25]62 + [t/25]P^c(t).$$

Notice, as illustrated in Figure 2.11, that $[P^c(t)]'$ and $P^c(t)$ are therefore coincident only after the year 2000.

The second adjustment is necessary to preclude the possibility that the world would unexpectedly exhaust all of its fossil fuel reserves. The notion here is that there exists a "backstop" technology (such as solar or fusion) that should become economically feasible before exhaustion and that entrepreneurs would provide that technology before the economic effects, perhaps collapse, that unexpected exhaustion would create. For our purposes, we model the backstop as a gradual contraction of reliance on fossil fuel once its price climbed to levels in excess of four times the price of nonfossil fuel. The multiple is selected to match current estimates of the cost of generating hydrogen from conventional sources; the subsequent rate of decline of fossil fuel consumption is assumed to be roughly 6% per year and is estimated from preliminary runs in which the supply of fossil fuel was exhausted in the absence of the backstop.

Consideration of the airborne fraction is the final order of business. Estimates from a variety of experts are cited in Clark (1982), but we found that they were mostly the products of statistically inefficient estimation procedures and highly sensitive to assumptions made about the contribution of carbon dioxide to the atmosphere from the biosphere. The latter sensitivity reflected misspecification of

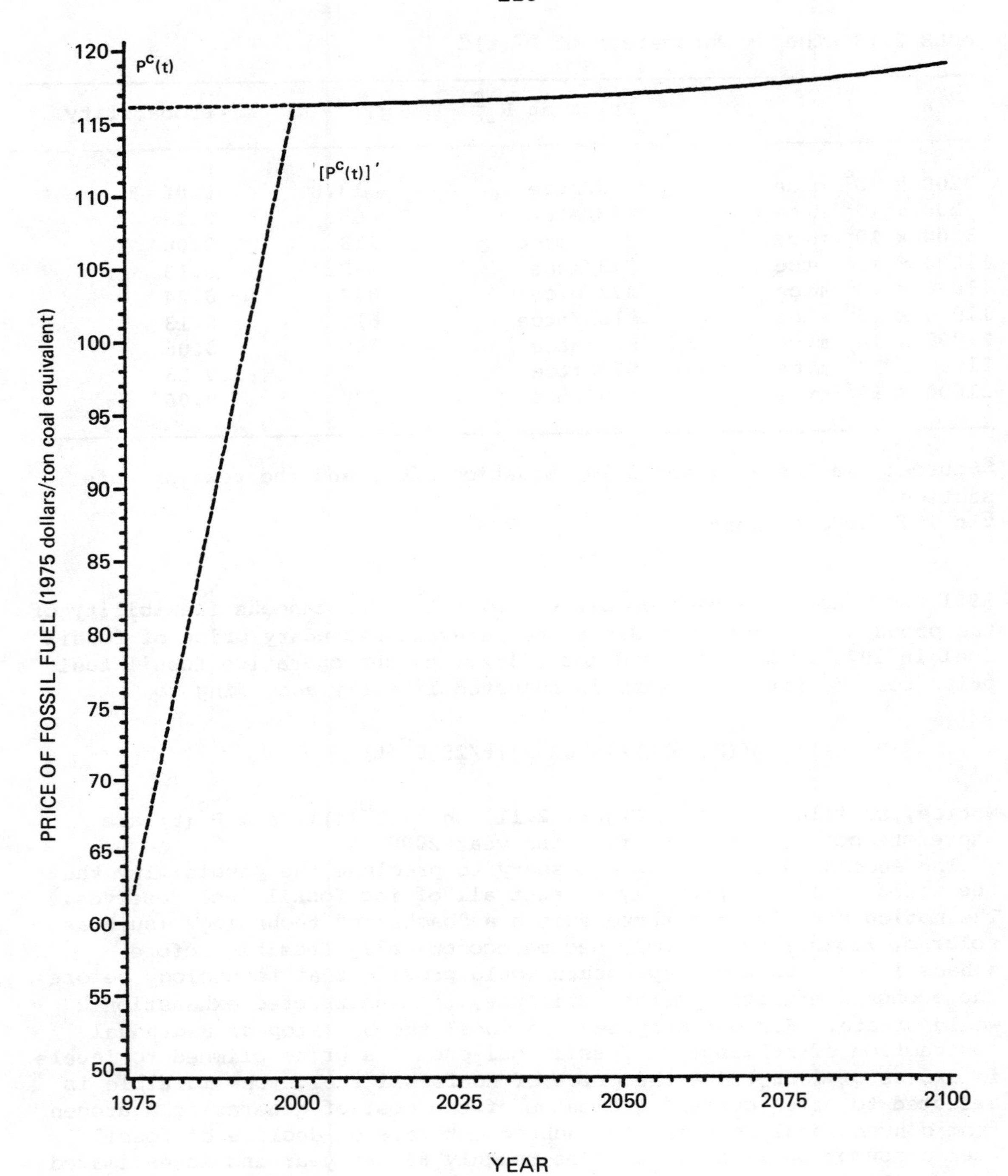

FIGURE 2.11 Price of fossil fuel (1975 dollars per ton of coal equivalent). A comparison of $P^c(t)$ and $[P^c(t)]'$--the adjustment required to accommodate the rapid increase in fossil fuel prices from 1975 through 1981.

the appropriate estimating equation and led us to our own study. At this point, we advance a maximum likelihood estimate of the marginal airborne fraction equal to 0.47, given an average seepage of 0.1% of ambient carbon dioxide into the oceans and an annual contribution from the biosphere of 1 Gt of C (the mean estimate). We differentiate between a marginal airborne fraction (the fraction of current emissions that remain in the atmosphere during their first year) and an average airborne fraction because of the seepage factor. The contribution of each year's emissions decays over time in our model, and that decay will have implications later when we turn to consider emission taxes as policies with which to address the carbon dioxide problem. The extreme cell estimates of the marginal fraction are also computed to discretize the usual normal distribution around a regression coefficient; 0.38 emerges as the low estimate, and 0.59 is the upper extreme.

2.1.2.2.1 Adjustment of Subjective Uncertainty

The inability of individuals, even those with statistical training, to deal efficiently with uncertainty about the future has come under increasing scrutiny. Studies in both the economic and psychological literature have argued, in particular, that individuals tend to underestimate the uncertainty about events.* For present purposes, two reasons for these systematic errors are relevant. First, when people look to previous studies in seeking guidance for their own views, they may place too much weight on the early studies. If one were then to view the range of resulting estimates as an indication of the true uncertainty, the computed variance would be too small.

A simple illustration of this phenomenon, sometimes known as the wine-tasting problem, can make this point. Suppose that a sample of two independent observations, x_1 and x_2, were taken from the same distribution, and let that distribution be normal with mean κ and variance σ^2. Each x_i would therefore be independently, normally distributed with mean κ and variance σ^2. The sample mean, $\overline{x} = (x_1 + x_2)/2$ would then be an unbiased estimate of κ, and $S^2 = (x_1 - \overline{x})^2 + (x_2 - x)^2$ would be an unbiased estimate of σ^2 (i.e., $\{E[S^2]\} = \sigma^2$). If, however, the second scientist looks over the shoulder of the first, he might allow his judgment to be influenced. Say that the reports of observation x_2 were weighted by x_1 so that reported values (y) are $y_1 = x_1$ and $y_2 = ax_2 + (1 - a)x_1$. The reported variance would decline. The reason is that the y_2 would display a variance

$$\sigma^2(y_2) = a^2\sigma^2 + (1 - a)^2\sigma^2 = [1 + 2(a^2 - a)]\sigma^2 < \sigma^2, \text{ for } 0 < a < 1.$$

*See Arrow (1982) for a summarizing review of both.

So, the variance of the reported values is biased downward from σ^2. The infusion of judgment allows the first researcher's result to influence the second, and the observed variance is smaller than the underlying variance.

Second, people seem reluctant to accept the true uncertainty inherent in small samples. Several studies report that individuals frequently base their expectations on one observation even when they are aware that historical experience has been widely varied.* The estimates for the more distant periods recorded in this section seem to suggest such a telescoping of uncertainty. In some cases, the range of estimates declined as the forecast period increased, even though the passage of time should have increased the uncertainties. It appears that, in the face of higher uncertainty, scientists may look to each other for guidance.

To correct for the resulting tendencies to underestimate the degree of uncertainty, estimate ranges are expanded around the computed means; i.e., the ranges are adjusted either to keep the ranges from contracting over time or to make them consistent with historical experience. The adjustments are based on our judgment but are undertaken only if they could be justified by one of these two rationales.

Table 2.14 presents the results of this procedure. The population growth ranges after the year 2025 are, for example, expanded to maintain the 0.5% deviation computed for the 2000-2025 period. The later productivity ranges are similarly expanded to match the uncertainty found in the first two periods. Energy price ranges are, finally, widened in response to the enormous political and economic uncertainties inherent in the world energy market. The survey of 23 projections for oil prices collected during the 1981 Stanford International Energy Workshop (ranging from 10% reductions to 100% increases in the price of imported crude oil) provide some very rough guidance for our energy price uncertainty (Manne, 1982).

2.1.2.3 Results

2.1.2.3.1 Levels and Uncertainties of Major Variables

Four types of experiments are conducted with the fully specified model. In the first, we investigate not only the most likely paths of emissions and concentrations but also the inherent uncertainty that surround those projections. This is accomplished by taking 1000 random samples from the 3^{10} different trajectories. The results of a sample of 1000 runs are recorded in Table 2.15. Figures 2.12 through 2.16 plot the first 100 of those runs for some of the more important variables. And Table 2.16 records the annual growth rates of the most likely path for those variables. Notice that these rates of growth, particularly those for energy consumption and GNP, conform well with

---

*See Arrow (1982) and the sources cited therein.

TABLE 2.14 Adjusted Ranges[a]

	1975-2000	2000-2025	2025 and Beyond
A. Population Growth			
High	2.0%	1.6%	0.8%
Middle	1.7%	1.1%	0.3%
Low	1.4%	0.6%	-0.2%
B. Productivity Growth			
High	3.4%	0.9%	0.1%(0.5%)
Middle	2.3%	1.6%	1.0%
Low	1.2%	2.3%	1.9%(1.5%)
C. Nonfossil Fuel Price Growth			
High	2.0%(1.3%)	1.0%(0.5%)	1.0%(n.a.)
Middle	0.5%	0.0%	0.0%
Low	-1.5%(-0.2%)	-1.0%(-0.5%)	-1.0%(n.a.)
D. Aggregate Carbon Emissions--no change			
E. Fossil Fuel Prices for $R_1 = 1100 \times 10^9$ mtce			
High	\$103/mtce (\$81/mtce)		
Middle	\$ 73/mtce		
Low	\$ 43/mtce (\$65/mtce)		
F. Airborne Fraction--no change[b]			

[a]Source: Previous tables and the present text. Unadjusted figures are indicated in parentheses when adjustments to widen the ranges have been made.
[b]The uncertainties cited were measurement problems and were biometrically evaluated from the carbon dioxide literature; they were not subject to the types of underestimation cited here.

the averages of the projections cited in Ausubel and Nordhaus (Section 2.2) through the year 2025. The 50-year averages predicted by the 1000 runs are, in fact, 2.1% and 3.3% for energy and GNP; the averages for the previous studies are 2.4% and 3.4%, respectively. The results presented here should not, therefore, be considered to be the products of a model that embodies radically different expectations about economic growth than the consensus of professional opinion.

The uncertainty surrounding the average path is, however, quite striking. The measured standard deviations of all variables expand over time, and that expansion is sometimes dramatic. For carbon emissions and concentrations, in particular, a fair amount of certainty through the year 2000 balloons to the point where, by 2100, standard deviations of their projections equal 60% and 23% of their means, respectively. Put another way, the extreme values for concentrations run from 377 ppm to 581 ppm in the year 2025, and from 465 ppm to 2212 ppm in the year 2100! Those interested in the actual distributions of

TABLE 2.15 Results of a Sample of 1000 Runs (Probability Weighted Means and Standard Deviations)

	1975	2000	2025	2050	2075	2100
Means						
Energy consumption	8.71	12.49	24.47	32.89	43.34	57.90
CO_2 emissions	4.59	5.37	10.21	13.94	17.55	19.39
Atmospheric concentration	340.	367.	425.	515.	634.	779.
Price of fossil fuel	0.067	0.133	0.144	0.163	0.207	0.274
Price of nonfossil fuel	0.257	0.295	0.299	0.303	0.309	0.315
Fossil consumption	7.90	9.14	17.09	21.2	24.2	26.
Nonfossil consumption	0.81	3.35	7.38	11.7	19.1	31.2
Output	6.93	17.8	36.8	53.49	77.94	113.
Alpha	0.894	0.881	0.880	0.877	0.874	0.870
Standard Deviations						
Energy consumption	0.18	3.70	8.45	13.5	21.3	36.5
CO_2 emissions	0.12	2.06	4.31	6.69	8.84	11.72
Atmospheric concentration	0.29	6.60	25.9	61.4	112.	181.
Price of fossil fuel	0.00011	0.0185	0.0253	0.0373	0.145	0.232
Price of nonfossil fuel	0.00300	0.0502	0.0633	0.078	0.094	0.113
Fossil consumption	0.216	3.50	7.22	9.81	11.73	15.81
Nonfossil consumption	0.055	1.92	4.96	9.15	17.6	32.0
Output	0.101	3.33	8.90	16.6	29.9	53.1

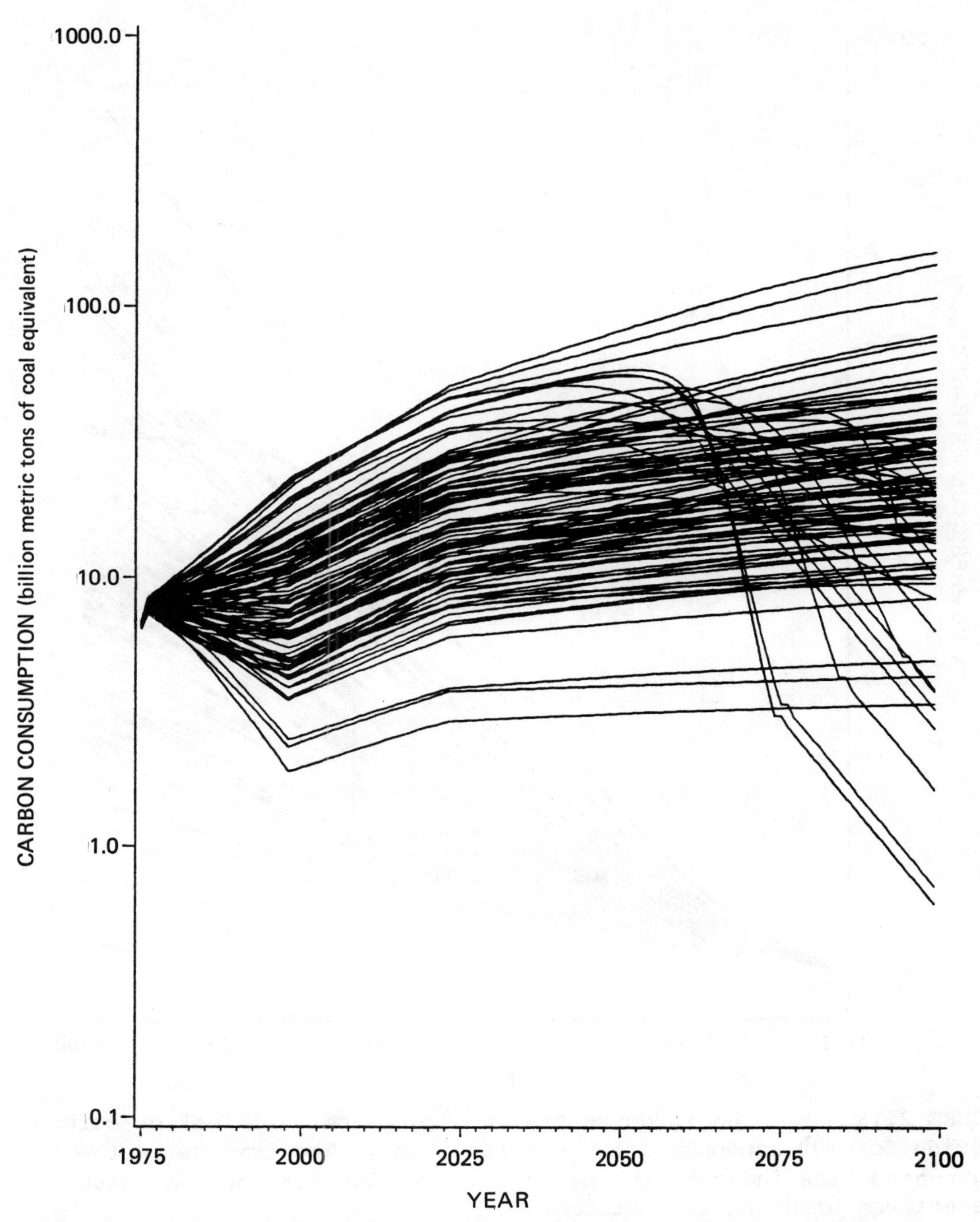

FIGURE 2.12 Fossil fuel consumption for 100 randomly drawn runs (billion metric tons of coal equivalent per year).

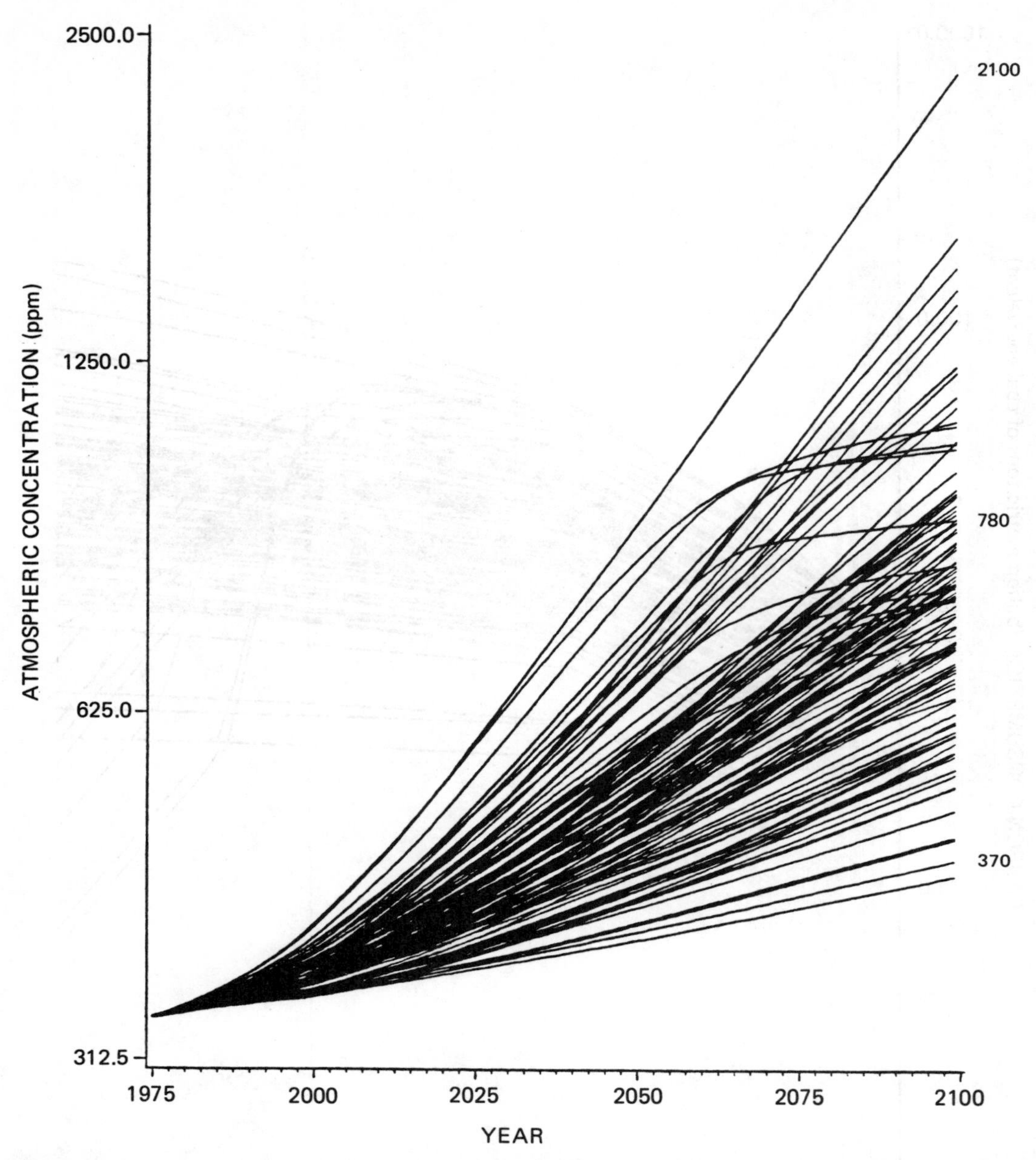

FIGURE 2.13 Atmospheric concentration (parts per million) of carbon dioxide for 100 randomly drawn emission runs. The numbers on the right-hand side indicate the mean concentration for the year 2100 and the extreme high and low outcomes.

emissions and concentrations for critical years are referred to Figures 2.17 and 2.18.

The model presented here finds that carbon dioxide emissions are likely to grow steadily over the next century or so, with an atmospheric concentration reaching 600 ppm, in our most likely case, shortly after 2065. If we call attainment of 600 ppm a "doubling," our

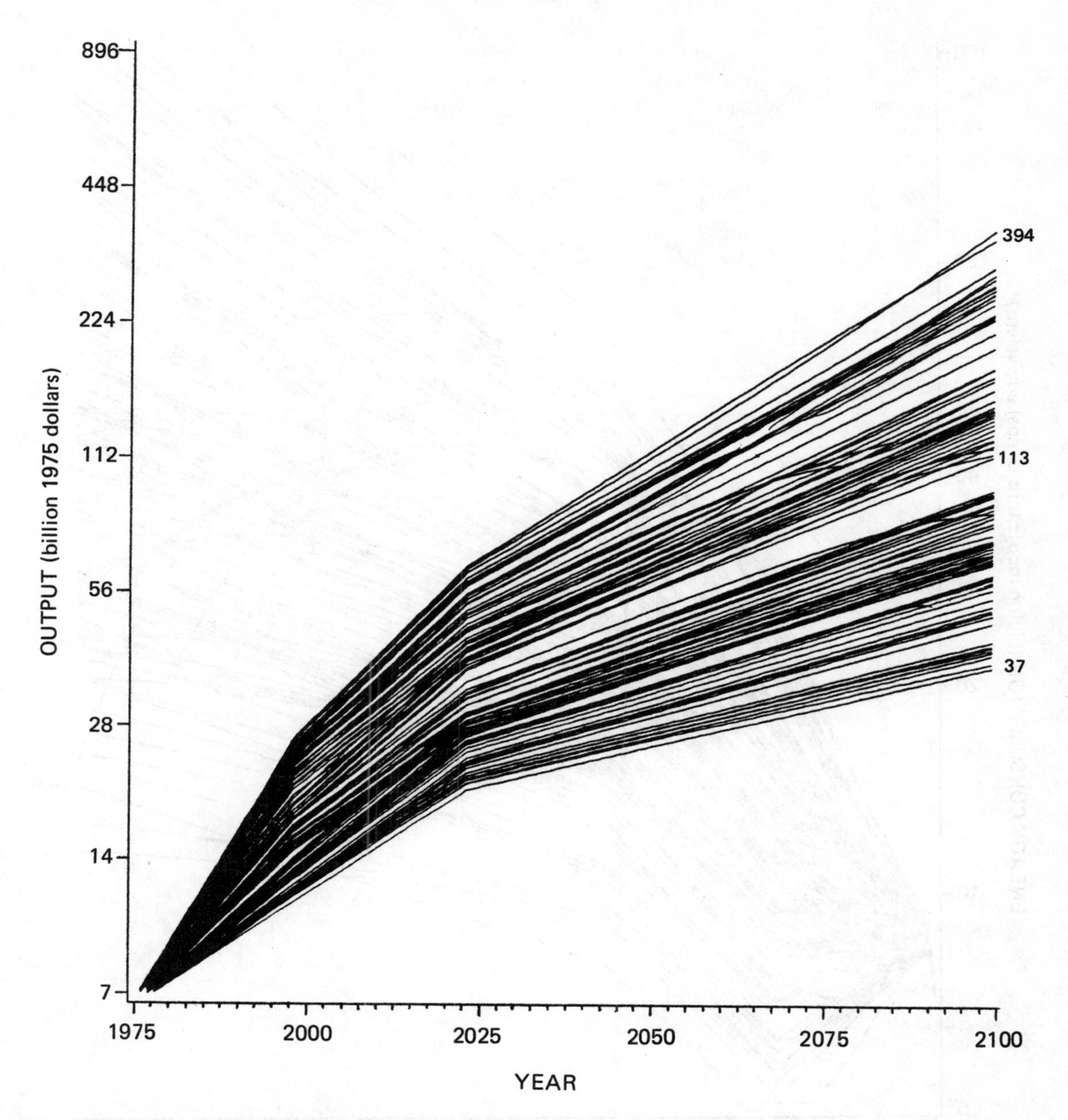

FIGURE 2.14 Gross world product for 100 randomly drawn runs (trillion 1975 dollars).

estimates indicate a doubling time longer than some earlier studies. This slower buildup arises primarily because we estimate a greater sensitivity of fossil fuel consumption to rising fossil fuel prices. But while this average result suggests a considerable time before a CO_2 doubling, our analysis also shows a substantial probability that doubling will occur much more quickly. Looking at the distribution for the year 2050, in fact, our results show a 27% chance that doubling will already have occurred. Unless this uncertainty can be reduced by

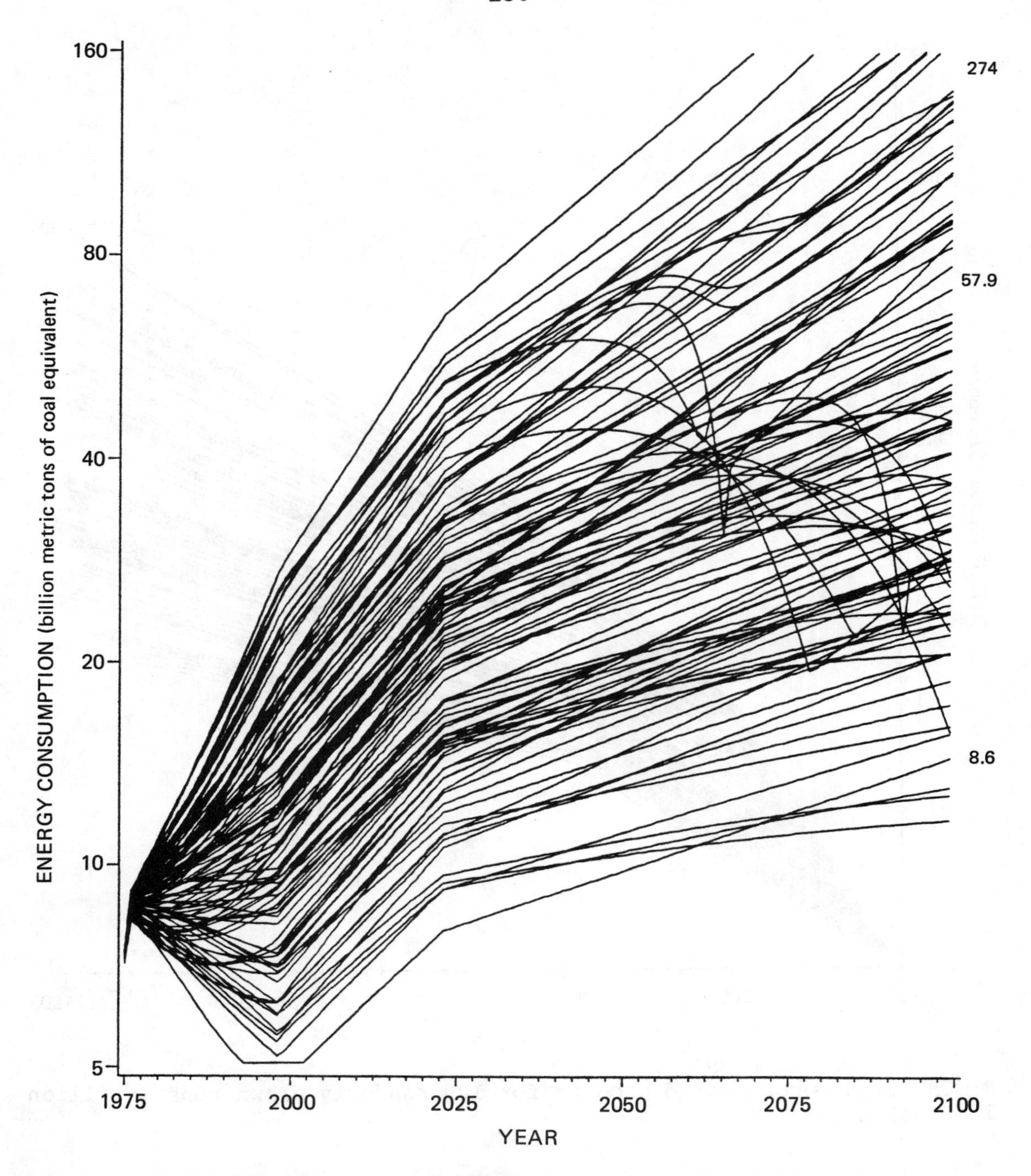

FIGURE 2.15 Energy consumption for 100 randomly drawn runs (billion metric tons of coal equivalent per year).

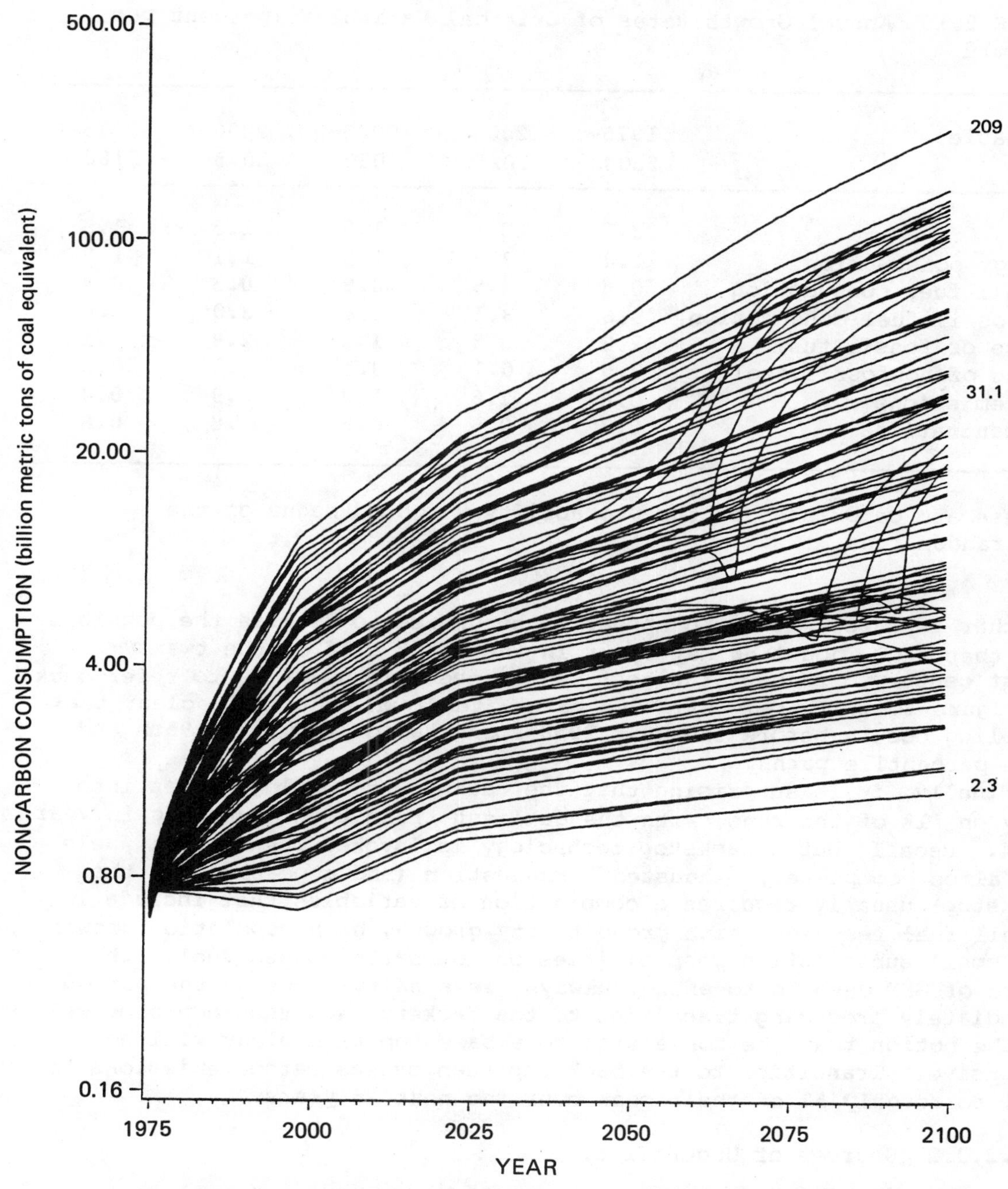

FIGURE 2.16 Nonfossil fuel consumption for 100 randomly drawn runs (billion metric tons of coal equivalent per year).

TABLE 2.16 Annual Growth Rates of Critical Variables (percent per annum)[a]

Variable	1975-2000	2000-2025	2025-2050	2050-2075	2075-2100
GNP	3.7	2.9	1.5	1.5	1.5
Energy consumption	1.4	2.7	1.2	1.1	1.2
Fossil fuel consumption	0.6	2.5	0.9	0.5	0.4
Nonfossil fuel consumption	5.6	3.1	1.8	2.0	2.0
Price of fossil fuel	2.8	0.3	1.2	2.9	1.1
Price of nonfossil fuel	0.5	0.1	0.1	0.1	0.1
CO_2 emissions	0.6	2.6	1.2	0.9	0.4
Concentrations	0.3	0.6	0.8	0.8	0.8

[a]These are calculated as the probability weighted means of the 100 random runs.

further research, it would appear to be unwise to dismiss the possibility that a CO_2 doubling may occur in the first half of the twenty-first century. Perhaps the best way to see this point is to refer back to Figure 2.4. There, only five paths are drawn, but it is clear that doubling occurs before the year 2050 along two of them, the 95th and 75th percentile paths.

Finally, it is surprising that the backstop technology comes into play on 11% of the runs, with the earliest transition occurring in year 2054. Recall that a backstop technology is invoked when fossil fuels are almost completely exhausted. Exhaustion (and appearance of the backstop) usually requires a combination of variables that include low fossil fuel reserves, high productivity growth, high population growth, and small substitution possibilities out of carbon-based fuel. The share of GNP devoted to energy always rises sharply during the period immediately preceding transition to the backstop and thus conforms well to the notion that the conversion to a backstop technology will be expensive. Transition to the backstop then causes carbon emissions to fall to roughly 5% of their peak over the next 25 years.

2.1.2.3.2 Sources of Uncertainty

A second experiment is designed to determine which of the 10 sources of uncertainty was most important in producing the ranges that have just been noted; it was conducted in two ways. In the first, each of the 10 random variables are, in turn, set equal to their two extreme values while all the others are fixed at their middle setting. In that way, the individual contributions of each source to the overall uncertainty of the projections is measured and compared. The second column of Table 2.1 records an index of the individual standard deviations that emerged. The second method is, in effect, the converse of the first. Each random variable is, in turn, held at its most likely (or "middle") value while random samples are taken across the other nine; the resulting reduction in the uncertainty is then taken to be a reflection of the marginal or incremental contribution of the fixed variable to overall uncertainty. The first column of Table 2.1 records an index of this measure.

Notice that both measures produce similar rankings. The ease of substitution between fossil fuels and nonfossil fuels rank first in both; this is an area that has thus far been almost ignored in prior study of the carbon dioxide problem. Second in both lists is the rate of productivity growth, but the importance of this variable is intuitively clear and has been apparent for some time. Below these two, a second echelon grouping of four variables appears in both columns: ease of substitution between labor and energy, extraction costs of fossil fuels, technological change in the energy sector, and the airborne fraction. The last member of this list has been heavily studied over the past few years, but even our wide range of uncertainty only pushed it into the middle of the ranking according to either scale. The fuel mix and the rate of growth in the population form a third grouping.

The bottom factors in terms of contribution to uncertainty are trends in relative energy costs and world fossil fuel resources. Holding these last two fixed actually produced <u>higher</u> variation in concentrations in 2100. The cost variable effect is small and probably cannot be statistically distinguished from zero, but the resources effect is pronounced. It is, however, easily explained. The backstop technology is invoked only when world resources are set at their lowest value. And when the backstop is imposed, emissions fall quickly to zero and concentrations tend toward roughly the same number. Removing the backstop as a possibility by removing the possibility that world fossil fuel resources might be quite small therefore removes circumstances that have a serious dampening effect on the range of possible atmospheric concentration. An expanded variance should therefore be expected.

2.1.2.3.3 Validation

Our third area of experimentation poses the problem of validating our results. Validation is, of course, a major issue arising in the estimation and use of very-long-run economic and energy models where the time period over which the data are available is typically much too short to permit testing and validating by usual statistical techniques. Moreover, economic systems evolve and mutate over time, so even models that use classical statistical time series tests would be suspect. Our time frame--125 years--clearly heightens concerns about these concerns. This raises the question of whether we have accurately estimated the uncertainty of future events by looking at each of the variables individually, imposing distributions on them and expanding those distributions to account for the likelihood that scientists systematically overestimate the confidence in their results. Two types of tests are run to attempt to answer this question.

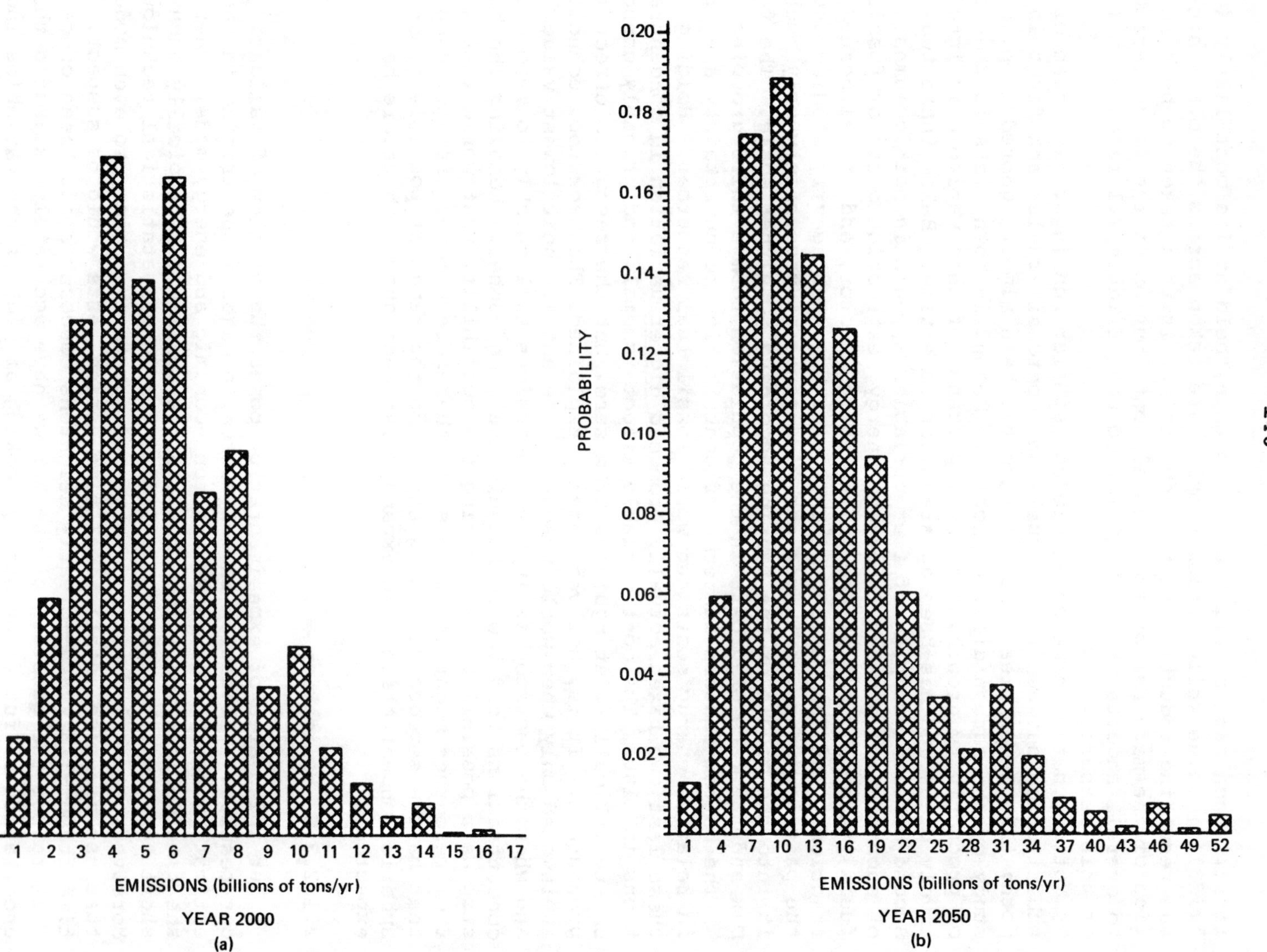

PROBABILITY
0.18
0.16
0.14
0.12
0.10
0.08
0.06
0.04
0.02
0 1 2 3 4 5 6 7 8 9 10 11 12 13 14 15 16 17
EMISSIONS (billions of tons/yr)
YEAR 2000
(a)
PROBABILITY
0.20
0.18
0.16
0.14
0.12
0.10
0.08
0.06
0.04
0.02
1 4 7 10 13 16 19 22 25 28 31 34 37 40 43 46 49 52
EMISSIONS (billions of tons/yr)
YEAR 2050
(b)

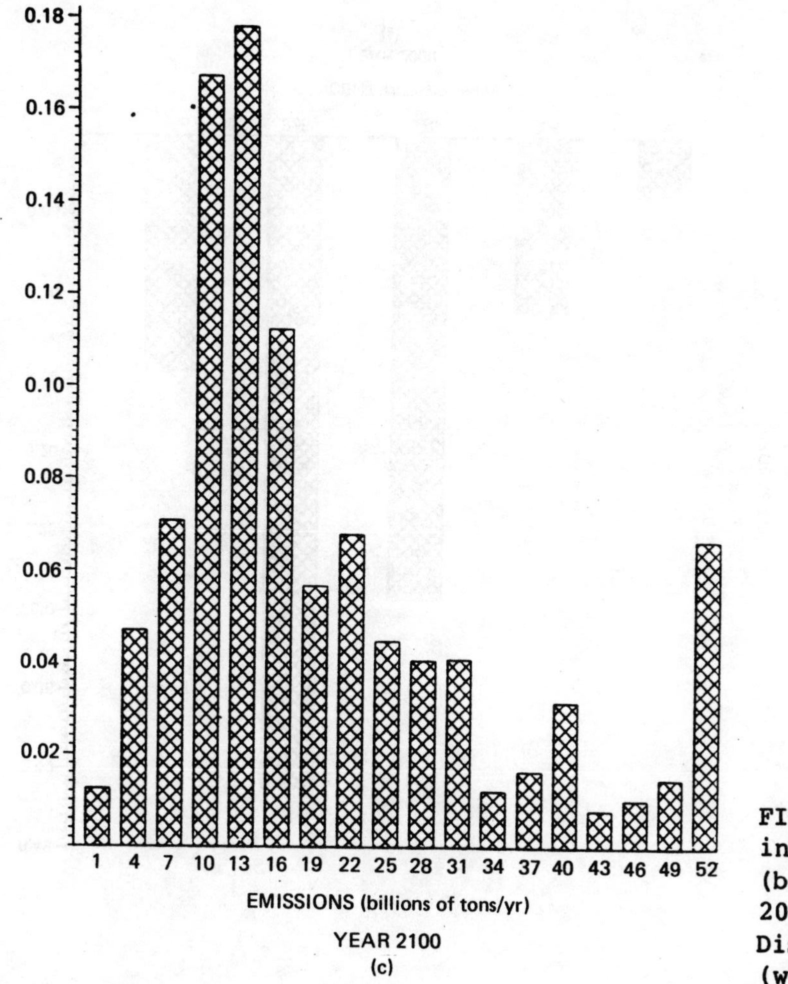

FIGURE 2.17 (a) Distribution of carbon emissions in the year 2000 (weighted sample of 1000 runs). (b) Distribution of carbon emissions in the year 2050 (weighted sample of 1000 runs). (c) Distribution of carbon emissions in the year 2100 (weighted sample of 1000 runs).

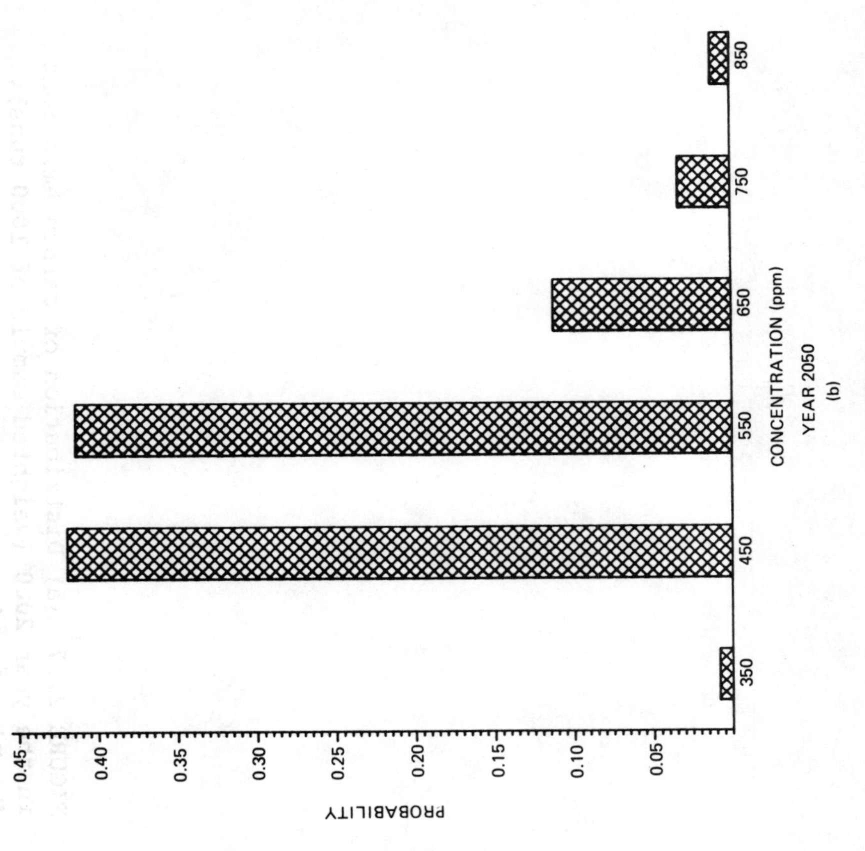
PROBABILITY
0.45
0.40
0.35
0.30
0.25
0.20
0.15
0.10
0.05
350
450
550
650
750
850
CONCENTRATION (ppm)
YEAR 2050
(b)

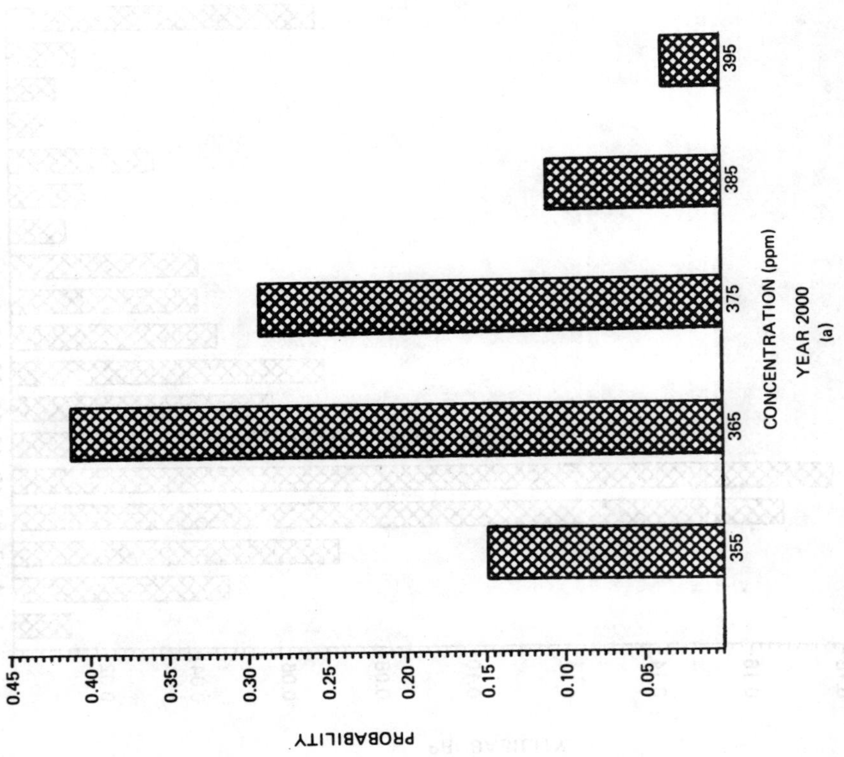
PROBABILITY
0.45
0.40
0.35
0.30
0.25
0.20
0.15
0.10
0.05
355
365
375
385
395
CONCENTRATION (ppm)
YEAR 2000
(a)

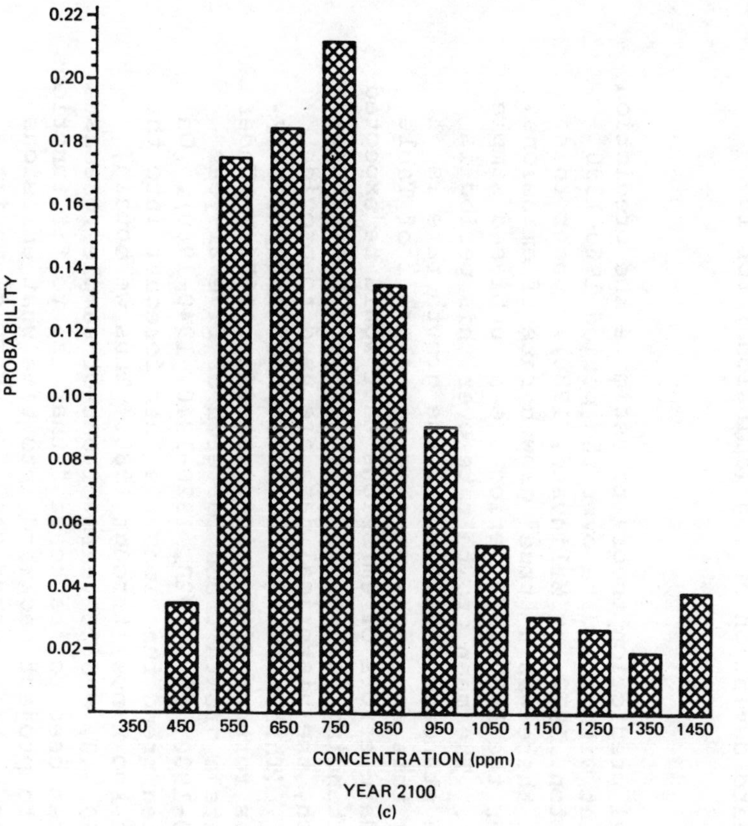

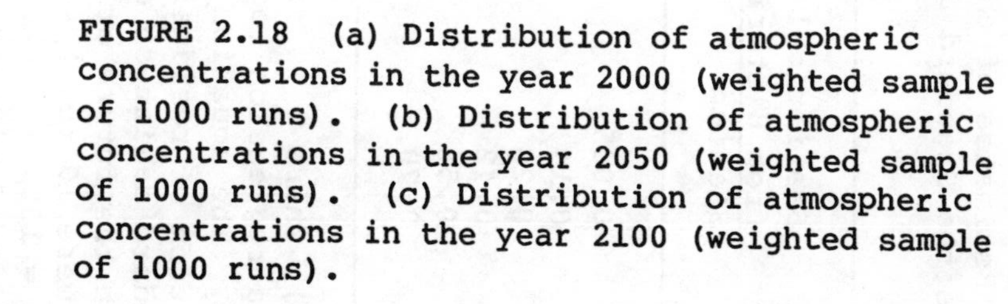

FIGURE 2.18 (a) Distribution of atmospheric concentrations in the year 2000 (weighted sample of 1000 runs). (b) Distribution of atmospheric concentrations in the year 2050 (weighted sample of 1000 runs). (c) Distribution of atmospheric concentrations in the year 2100 (weighted sample of 1000 runs).

TABLE 2.17 Alternative Estimates of the Uncertainty of Carbon Dioxide Emissions (Calculated as the Standard Deviation of the Logarithm of Emissions)

Time from Last Data (t)	Statistical (Calculated from Historical Data) Static[a]	Dynamic[b]	Model[c]
0	0.005	0.06	0.04
25	0.14	0.50	0.56
50	0.28	0.75	0.61
75	0.42	0.76	0.67
100	0.56	0.99	0.68
125	0.70	n.a.	0.85

[a]Calculated as t x se(g), where t is time in future and se(g) is the standard error of g in a regression log (emissions) = a + gt, over the sample period 1960-1980. The equation was estimated assuming first-order autocorrelation of residuals.
[b]Calculated as the standard deviation of a forecast of log (emissions) t periods in the future or past. The number of nonoverlapping samples were 10 for t = 0, 9 for t = 25, 6 for t = 50, 4 for t = 75, and 2 for t = 100.
[c]Calculated as the standard deviation of log (emissions) for 100 randomly selected runs.

We first use classical prediction theory to estimate the prediction errors that are consistent with the data over the period 1960-1980 (see, for example, Johnston, 1972; or Malinvaud, 1980). Under this approach, we assume that there was a "true" growth rate of emissions, g, and that the data over the 1960-1980 period are an unbiased sample of that true growth rate. The mean growth rate over this period is 3.67% per annum, and the standard deviation of the growth rate is 0.561% per annum. Using this approach, we show in column 2 of Table 2.17 the estimated standard errors of emissions that would be expected over forecast periods extending further and further into the future.

In the second approach, the historical data are used to provide out-of-sample forecasts. (This technique has been used infrequently. For an example as well as further discussion, see Fair, 1978.) Under this approach, we estimate a growth trend for each of five 21-year periods (1860-1880, 1880-1900, 1900-1920, 1920-1940, 1940-1960). On the basis of the estimated trend functions, we then forecast into the future (unknown then but known now) through 1980. Thus we obtain, respectively, 100, 80, 60, 40, and 20 years of out-of-sample forecasts. The same procedure is then used to "backcast," that is, to fit functions to recent data and then to project backward into time what emissions should have been. In the backcast exercises, for example, we fit a trend function to the data for 1960-1980, then use that estimated relation to calculate emissions over the period 1860-1960. Again, five 21-year trends are estimated and five different sets of backcasts are constructed.

From the 10 sets of forecasts and backcasts, we construct a set of out-of-sample errors, 0, 25, 50, 75, and 100 years away from the sample--future or past. The root-mean-squared errors are then calculated; they are labeled as "dynamic" statistical error forecasts and are shown in column 3 of Table 2.17.

The backcast procedure may at first appear bizarre. It is an implication of the model we are using, however, that estimates of the structure are equally valid forward and backward into time. This implication arises because the trend model has no explanatory variable but time, as well as because the trend model assumes that there is no change in the underlying economic structure. It should be emphasized, however, that use of this type of trend extrapolation model to forecast emissions in no way is an endorsement of such a technique for all purposes. We are employing it here only for validation purposes.

The results of this validation test indicate that the error bounds for emissions estimated by the model, recorded in column 4 of Table 2.17, are within the bounds generated by the two historical error estimation procedures. In general, the model produces error bounds greater than the classical statistical technique shown in column 2, but smaller than the dynamic estimates shown in column 3.

If we were to choose between procedures for estimating errors, we would be inclined toward the dynamic rather than the static as a realistic estimate of forecasting uncertainty. The reason for this inclination is that the static estimate assumes that there is no change in the underlying structure of the economy, so that future growth rates are drawn from the probability distribution generated by the in-sample growth rates. The dynamic model, on the other hand, recognizes that there is evolution in the structure of the economy, so that the distribution from which we draw observations is likely to drift around over time. Assuming that the pace of economic structural change over the next 100 years will be about as rapid as that over the last 100 years, the dynamic estimates give a better estimate of the realistic error bounds for a forecast of the future. To the extent that careful structural modeling allows us to improve on a naive extrapolation of trends--which is after all the major point of economic and energy models of the kind we introduce here--the error bounds of the model should be an improvement over the dynamic error bounds. And, finally, it should be noted that these exercises were conducted after the model had been constructed and estimated; no tuning of the model has been undertaken to bring it in line with either series recorded in Table 2.17.

2.1.2.3.4 Policy Experiments

Our final set of experiments considers the possibility that governments will intervene to curtail CO_2 emissions. There are clearly a wide variety of approaches to discouraging fossil fuel combustion and CO_2 emissions. Some might take the form of taxes on production or consumption of carbon-based fuels; nonfossil sources might be encouraged; countries that have large coal reserves might place export limitations or heavy taxes on coal exports. Government might agree to national CO_2 emissions quotas and then enforce these in a wide variety of ways. At this point, we do not pass on the likelihood or desirability of these different policies--rather, we attempt to investigate their impacts.

For purposes of analysis, it is convenient to convert all policies that discourage use of CO_2 into the carbon-equivalent taxes. As an example, say that a 2x% tax on carbon fuels would always produce an x% reduction in their use. We would then use this hypothetical formula as a way of representing any quantitative restriction on carbon-based fuels. Whatever the set of policies, we can derive the tax rate on fossil fuels that would produce the same restraint on CO_2 emissions. This equivalent tax is the carbon-equivalent tax investigated here. We consider only the most likely run--the case in which all the random variables are set equal to their middle values. Because of the crudeness of the policy, the results quoted below should be viewed as being extremely tentative.

Five different taxes are imposed on the supply equation for fossil fuel. Two are taxes increasing from zero to $20 per mtce over the course of 10 years and then declining back to zero over the next decade; one pulse begins in the year 1980, and the other begins in the year 2025. For a second set of runs, permanent taxes of $10 per ton are introduced linearly over the first 20 years. One begins in the year 1980, and the other begins in the year 2025. A final tax, modeled after the 100% control case presented in Nordhaus (1979) and called the "stringent tax," imposes a permanent tax that rose linearly beginning in the year 2000 from zero to $6 per mtce by the year 2020, then to $68 per mtce by the year 2040, and finally to $90 per mtce by the year 2060. These are all illustrated in Figure 2.19.

Table 2.18 and Figures 2.20 and 2.21 show the results of the tax runs. Notice there that while the pulse taxes accomplish very little, the permanent tax initiated in 2025 is the most effective among the first four alternatives. This paradox is explained as follows: burning more fossil fuel early and postponing CO_2 reductions lowers the eventual CO_2 concentration because it allows the atmosphere to cleanse itself slowly. To see this, recall that our model allows for a slow seepage (1 part per 1000 per year) of ambient carbon dioxide into the deep oceans. Earlier emissions, therefore, gradually disperse into the deep oceans and produce end-of-period concentrations that are somewhat lower.

The major conclusion concerns the extent to which concentrations appear to respond to taxes. As can be seen, a $10 per ton tax accomplishes only a modest reduction in CO_2 concentrations. Should emissions restraint become desirable, it will take extremely forceful policies to make a big dent in the problem. Even the "stringent"

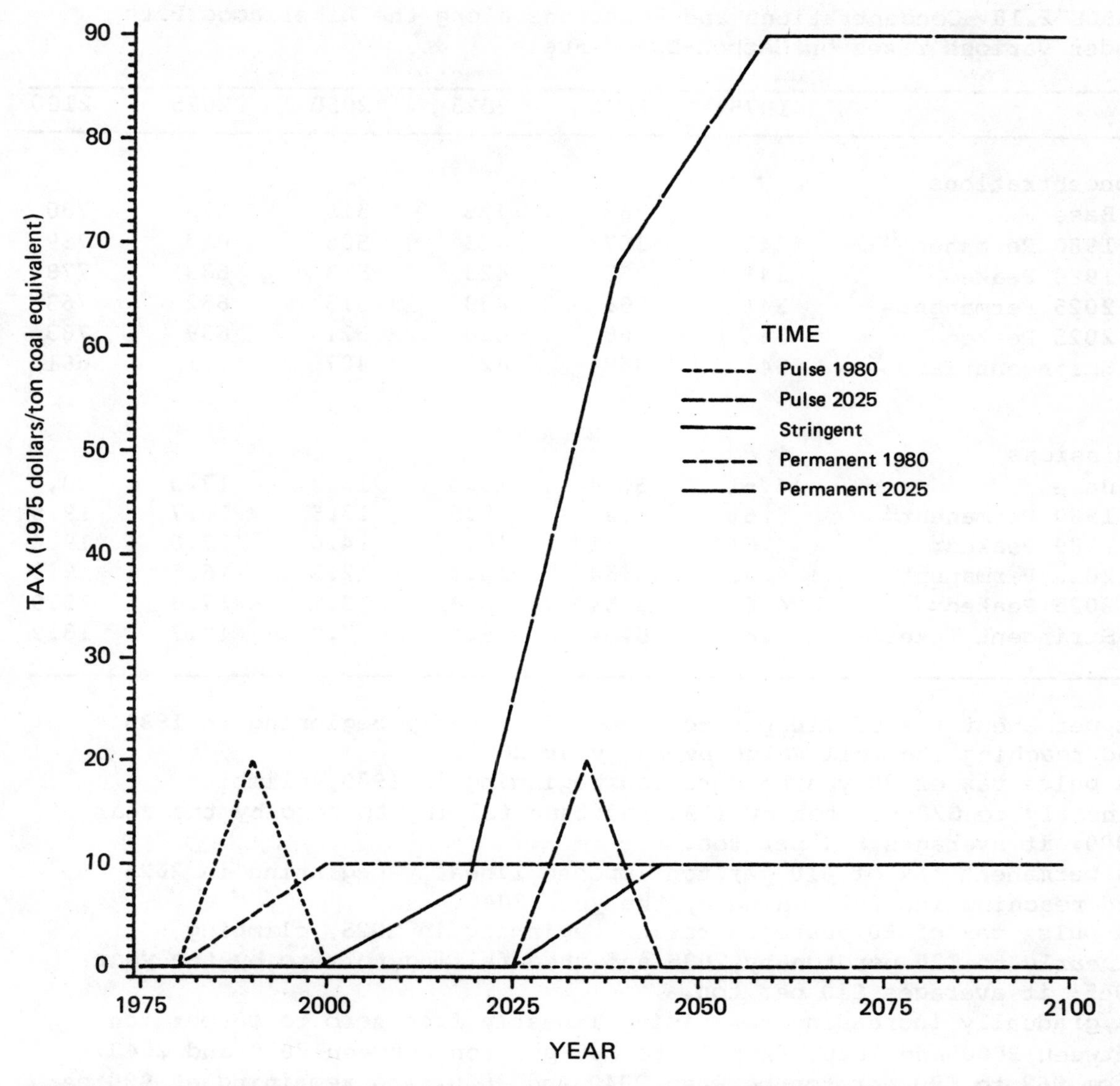

FIGURE 2.19 Taxation on carbon fuel price (1975 dollars per ton coal equivalent). The time tracks of a stringent tax and four alternative $10 per ton of coal equivalent taxes; the temporary taxes peak at $20 to accommodate the model.

TABLE 2.18 Concentrations and Emissions along the Likelihood Path under Various Taxes on Carbon-Based Fuels

	1975	2000	2025	2050	2075	2100
Concentrations						
Base	341	368	428	516	633	780
1980 Permanent[a]	341	367	423	506	617	759
1980 Peaked[b]	341	367	423	513	633	778
2025 Permanent[c]	341	368	428	513	632	763
2025 Peaked[d]	341	368	428	521	638	783
Stringent Taxes[e]	341	368	425	487	561	661
Emissions						
Base	4.61	5.54	10.3	13.3	17.5	20.0
1980 Permanent[a]	4.61	5.06	9.5	12.5	16.7	19.5
1980 Peaked[b]	4.61	5.31	10.3	14.0	17.6	19.4
2025 Permanent[c]	4.61	5.54	10.3	12.3	16.5	19.3
2025 Peaked[d]	4.61	5.54	10.3	13.1	17.3	19.9
Stringent Taxes[e]	4.61	5.54	8.4	7.9	10.7	13.9

[a]A permanent tax of $10 per ton imposed linearly beginning in 1980 and reaching its full value by the year 2000.
[b]A pulse tax of 20 years' duration beginning in 1980, climbing linearly to $20 per ton by 1990 and then falling to zero by the year 2000; it averages $10 per ton.
[c]A permanent tax of $10 per ton imposed linearly beginning in 2025 and reaching its full value by the year 2045.
[d]A pulse tax of 20 years' duration beginning in 2025, climbing linearly to $20 per ton by 2035 and then falling to zero by the year 2045; it averages $10 per ton.
[e]A gradually increasing tax rising linearly from zero to $8 per ton between 2000 and 2020, from $8 to $68 per ton between 2020 and 2040, from $68 to $90 per ton between 2040 and 2060, and remaining at $90 per ton thereafter. This tax was drawn from Nordhaus (1979, Chapter 8).

taxes, which would place 60% surcharges on the prices of fossil fuels, did not prevent doubling before 2100 in our most likely case.*

*Nordhaus (1979) presented an earlier estimate of the taxes needed to curtail the carbon dioxide buildup in an optimizing linear programming framework. In that calculation the potency of carbon taxes was very close to the estimates here. The estimate here is 0.46% reduction in 2100 carbon dioxide concentration per $1 of carbon tax, while in the earlier work the estimate was 0.36% reduction per $1. Note that comparison is not completely appropriate because these reactions are likely to be nonlinear.

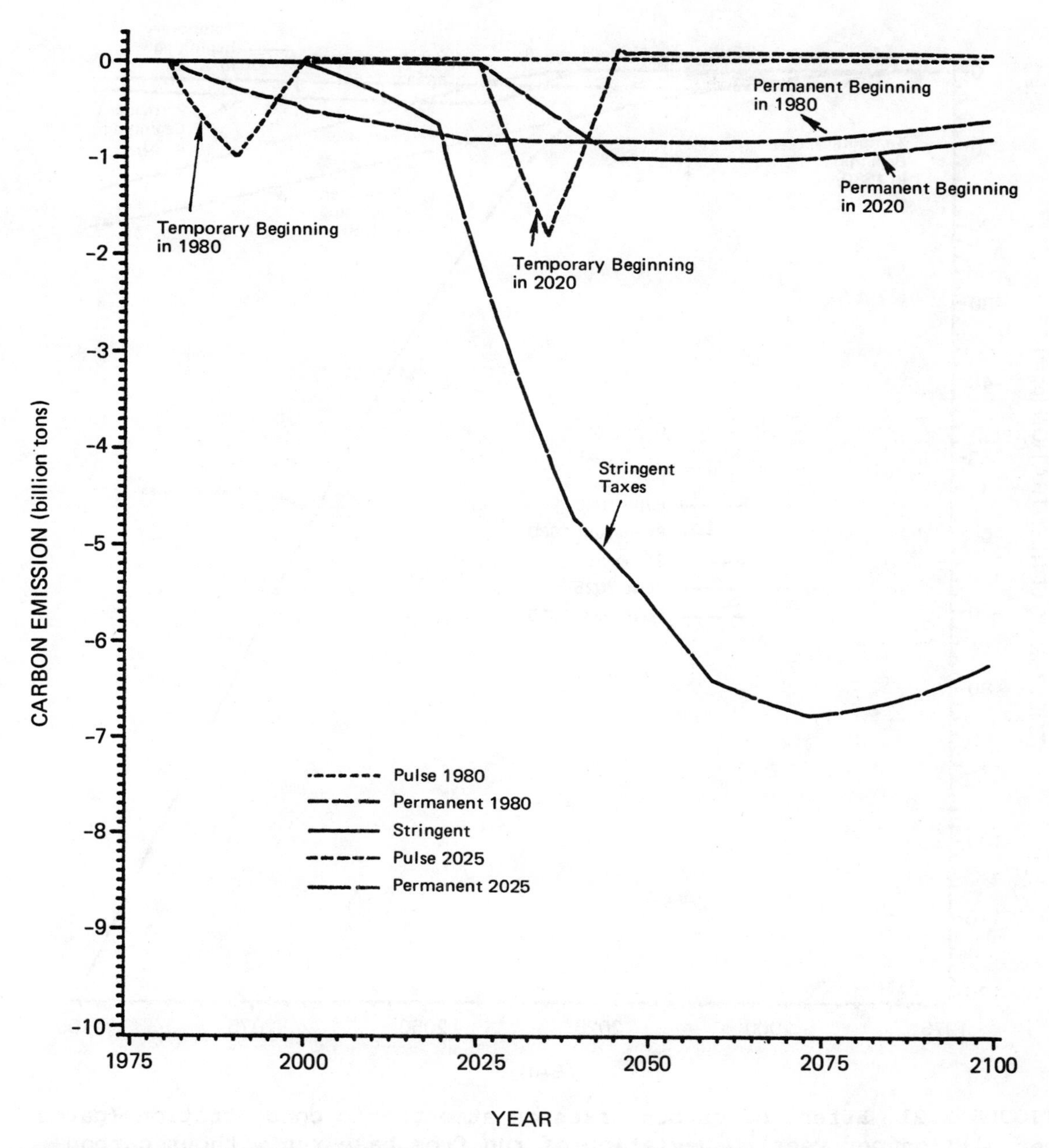

FIGURE 2.20 Plot of carbon emission versus time for taxed runs. Deviation in emissions from the base run for various taxes.

It should be emphasized that this conclusion about the potency of CO_2 taxes (or their regulatory equivalents) is extremely tentative. It is based on a model for which many of the parameters are known imperfectly. On the other hand, the model's conclusions appear to confirm results of a completely independent model, as reported in the last footnote and those of Edmonds and Reilly (1983).

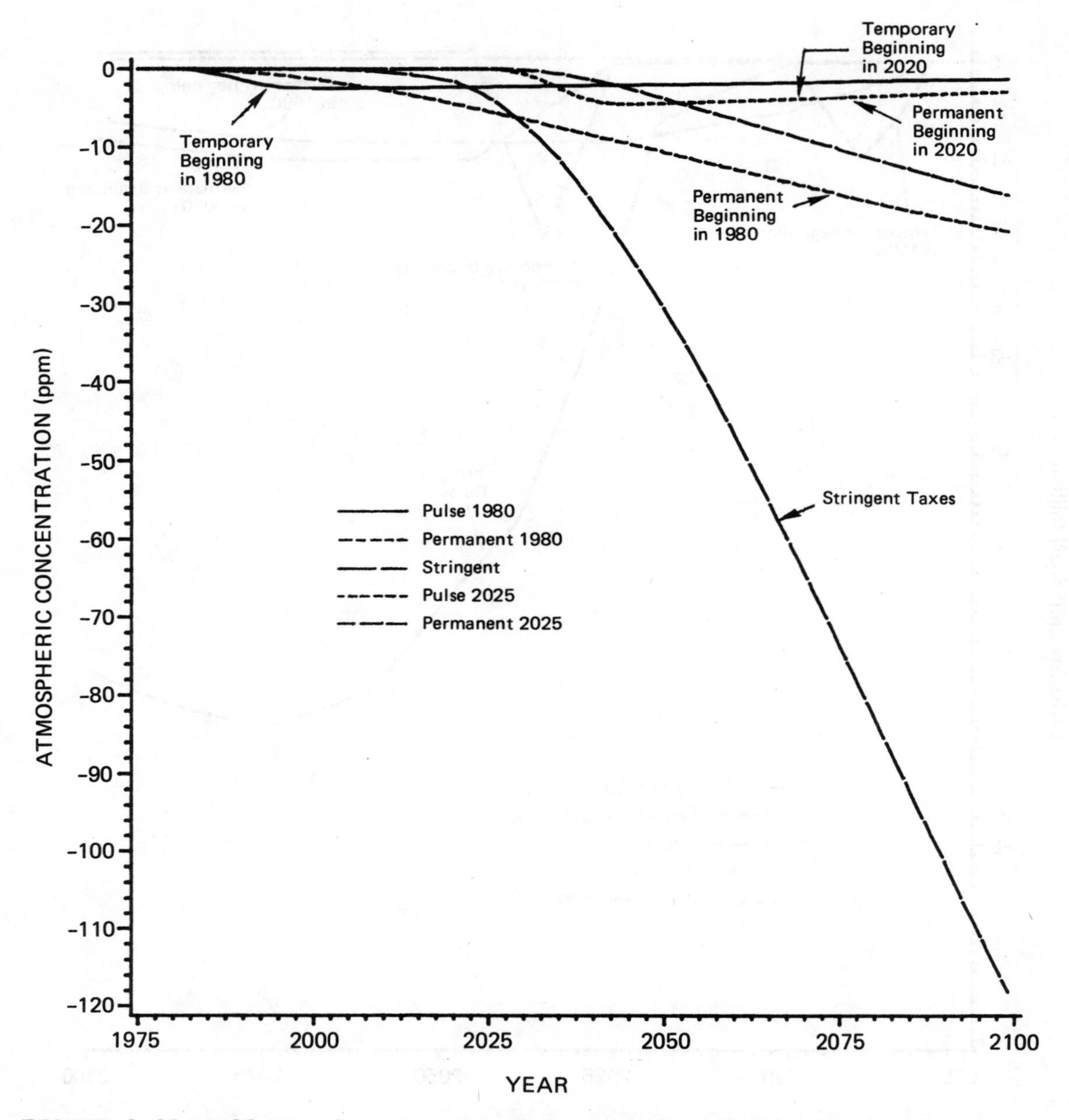

FIGURE 2.21 Effect of carbon taxes on atmospheric concentration (parts per million per year). Deviation of run from base run without carbon taxes.

Nevertheless, the conclusions about the potency of policy are sobering. They suggest that a significant reduction in the concentration of CO_2 will require very stringent policies, such as hefty taxes on fossil fuels. Global taxes of around $60 per ton of coal equivalent (approximately $10 per barrel of oil equivalent) reduce the concentrations of CO_2 at the end of our period by only 15% from the base run. Moreover, these taxes must be global; it is presumed that a tax imposed by only a fraction of the countries would have an effect roughly proportional to those countries' share of carbon emissions.

To the extent that such an approach can offer guidance, therefore, it suggests that there are unlikely to be easy ways to prevent the buildup of atmospheric CO_2. The strategies suggested later by Schelling (Chapter 9)--climate modification or simply adaptation to a high CO_2 and high temperature world--are likely to be more economical ways of adjusting to the potential for a large buildup of CO_2 and other greenhouse gases. Whether the imponderable side effects on society--on coastlines and agriculture, on life in high latitudes, on human health, and simply the unforeseen--will in the end prove more costly than a stringent abatement of greenhouse gases, we do not now know.

References

Arrow, K. J. (1982). Risk perceptions in psychology and economics. Economic Inquiry 20:1-9.

Bolin, B. ed. (1981). Carbon Cycle Modelling. SCOPE 16. Wiley, New York.

Brainard, W. C. (1967). Uncertainty and the effectiveness of policy. Am. Econ. Rev. 57(2).

Clark, W., ed. (1982). The Carbon Dioxide Review: 1982. Oxford U. Press, New York.

Council on Environmental Quality (1981). Global Energy Futures and the Carbon Dioxide Problem. Washington, D.C.

Duncan, D. C., and V. E. Swanson (1965). Organic Rich Shale of the United States and World Land Areas. U.S. Geological Survey Circ. 523.

Edmonds, J., and J. Reilly (1983). A long-term global energy economic model of carbon dioxide release from fossil use. Energy Econ. 5:74-88.

Fair, R. C. (1980). Estimating the expected predictive accuracy of econometric models. Int. Econ. Rev. 21(2).

Hogan, W. W. (1980). Dimensions of energy demand. In H. H. Landsberg, ed., Selected Studies on Energy; Background Papers for Energy: The Next Twenty Years. Ballinger, Cambridge, Mass., p. 14.

International Institute for Applied Systems Analysis (IIASA) (1981). Energy in a Finite World: Paths to a Sustainable Future. Report of the Energy Systems Group of IIASA, W. Häfele, Program Leader. Ballinger, Cambridge, Mass.

Johnston, J. (1972). Econometric Methods. McGraw-Hill, New York.

Kahn, H., W. Brown, and L. Martel (1976). The Next 200 Years. Hudson Institute Report, Morrow, New York.

Keeling, C. D. (1973). The carbon dioxide cycle. In Chemistry of the Lower Atmosphere, N. Rasool, ed. Plenum, New York.

Keyfitz, N. (1982). Population projections, 1975-2075. In Clark (1982), pp. 460-463.

Machta, L. (1972). The role of the oceans and biosphere in the carbon dioxide cycle. In O. Oyrssen and D. Jagnes, eds., The Changing Chemistry of the Oceans. Wiley-Interscience, New York, pp. 121-145.

Machta, L., and G. Telegades (1974). Climate forecasting. In Weather and Climate Modification, W. N. Hess, ed. Wiley, New York.

Malinvaud, E. (1980). Statistical Methods of Econometrics. North-Holland, Amsterdam.

Manne, A. S. (1974). Waiting for the Breeder. Research Report RR-74-5. International Institute for Applied Systems Analysis (IIASA), Laxenburg, Austria.

Manne, A. S., ed. (1982). Summary Report of the International Energy Workshop, 1981. Stanford University Institute for Energy Studies, Stanford, Calif.

Marland, G. (1982). The impact of synthetic fuels on global carbon dioxide emissions. In Clark (1982), pp. 406-410.

Meadows, D. H., D. L. Meadows, J. Randers, and W. W. Behrens (1972). The Limits to Growth: A Report for the Club of Rome's Project on the Predicament of Mankind. Universe Books, New York.

Modeling Resource Group (MRG) (1978). Energy Modeling for an Uncertain Future. Supporting Paper 2, Committee on Nuclear and Alternative Energy Systems (CONAES), chaired by T. C. Koopmans. National Research Council, Washington, D.C.

Moody, J. D., and R. E. Geiger (1975). Petroleum resources: how much oil and where? Technol. Rev. 77:38-45.

National Research Council (1979). Energy in Transition 1985-2010. Final Report of the Committee on Nuclear and Alternative Energy Systems (CONAES). National Academy of Sciences, Washington D.C.

Nelson, R. R., and S. G. Winter (1964). A case study in the economics of information and coordination: the weather forecasting system," Q. J. Econ. 78(3).

Nerlove, M. (1965). Estimation and Identification of Cobb-Douglas Production Functions. North-Holland, Amsterdam.

Nordhaus, W. D. (1977). The demand for energy: an international perspective. In W. D. Nordhaus, ed., International Studies of the Demand for Energy. North-Holland, Amsterdam, p. 273.

Nordhaus, W. D. (1979). Efficient Use of Energy Resources. Yale U. Press, New Haven, Conn.

Nordhaus, W. D. (1980a). Oil and economic performance in industrial countries. Brookings Papers on Economic Activity 2:341-388. Brookings, Washington, D.C.

Nordhaus, W. D. (1980b). Thinking about Carbon Dioxide: Theoretical and Empirical Aspects of Optimal Control Strategies. Discussion Paper No. 565. Cowles Foundation, Yale University, New Haven, Conn.

OECD Interfutures Project (1979). Facing the Future. Organization for Economic Cooperation and Development, Paris.

Raiffa, H. (1968). Decision Analysis: Introductory Lectures on Choices under Uncertainty. Addison-Wesley, Reading, Mass.

Reilly, J., R. Dougher, and J. Edmonds (1982). Determinants of Global Energy Supply to the Year 2050. Contribution 82-6 to the Carbon Dioxide Assessment Program. Institute for Energy Analysis, Oak Ridge Associated Universities, Oak Ridge, Tenn.

Ridker, R. G., and W. D. Watson (1980). To Choose A Future; Resource and Environmental Consequences of Alternative Growth Paths. Johns Hopkins U. Press, Baltimore, Md.

Tversky, A., and D. Kahneman (1974). Judgement under uncertainty. Science 185:1124-1131.

Tversky, A., and D. Kahneman (1981). The framing of decisions and the psychology of choice. Science 211:453-458.
Uzawa, H. (1962). Production functions with constant elasticities of substitution. Rev. Econ. Studies 29:291-299.
Varian, H. (1980). Microeconomic Analysis. Norton, New York.
Workshop on Alternative Energy Strategies (WAES) (1977). Energy: Global Prospects 1985-2000. Report of the Workshop on Alternative Energy Strategies, McGraw-Hill, New York.
Yohe, G. W. (1979). Comparisons of price and quantity controls--a survey. J. Comp. Econ. 3:213-234.

2.2 A REVIEW OF ESTIMATES OF FUTURE CARBON DIOXIDE EMISSIONS

Jesse H. Ausubel and William D. Nordhaus

2.2.1 Introduction

In analyzing prospects and policies concerning future carbon dioxide buildup, it is necessary to begin with projections of levels of CO_2 emissions. Because of the long residence time in the atmosphere of CO_2 emissions, along with the potential for large and durable societal impacts of higher CO_2 concentrations, there is great interest in long-term projections--those extending a half-century or more. While it is clearly necessary to make global long-term projections in this area, the projections are intrinsically uncertain, and the uncertainty compounds over time.

This section reviews methods involved in making projections of carbon dioxide emissions, describes the major projections, and offers some comparisons and comments. It is intended to serve three purposes. First, it should help to acquaint the reader with the state of the art in CO_2 forecasting and the range of previous forecasts. Second, this review may help to identify shortcomings of current efforts and point to directions for new research. Third, it should establish the context of the forecasts developed by Nordhaus and Yohe (Section 2.1) for this report.

Projections of future trajectories of CO_2 emissions can be roughly divided into three categories: (A) projections that are no more than extrapolations and that are primarily intended to be used to initiate studies of the carbon cycle or the climate system; (B) those based on relatively detailed examination of global energy supply and demand in which CO_2 emissions are largely incidental; (C) projections deriving from analysis of the energy system in which changing levels of CO_2 are themselves taken into account. Leading examples of category A, in which CO_2 emissions are projected with little more than passing reference to energy modeling, are Keeling and Bacastow (1977) and Siegenthaler and Oeschger (1978). These papers extrapolate emissions in order to predict future atmospheric CO_2 levels. Such efforts also appear in numerous reports and papers concentrating on calculating climatic change, for example, JASON (1979) and Hansen et al. (1981). The projections consist of little more than extrapolating rates of fossil fuel emissions growth from recent decades out a century and more into the future. These extrapolations can be regarded as simplifications or summarizations of more complete projections; they are useful for studies of the sensitivity of the carbon cycle and climate system but unpersuasive as elements of a comprehensive CO_2 assessment.

The projections based on relatively detailed analysis of an uncontrolled global energy-climate system, (B), which are the most important for purposes of this section, differ greatly in their design, in the extent to which formal models are employed, and in detail with respect to fuels, geography, and other factors. Leading examples include those made by H. Perry and H. H. Landsberg (1977) for the NRC Geophysics Study Committee, the several projections by Rotty and by Edmonds and Reilly of the Institute for Energy Analysis (IEA) of Oak Ridge Associated Universities (Rotty, 1977; Rotty and Marland, 1980; Edmonds and Reilly, 1983a), the projections of Nordhaus (1977) and (1979), and those made for the Energy Systems Program of the International Institute for Applied Systems Analysis (IIASA) (Niehaus and Williams, 1979; IIASA, 1981).

Category (C) projections, which require the basic analysis of category (B) as input, seek additionally to take into account the changing level of atmospheric CO_2 (or the costs of climatic change) in the calculations. That is, CO_2 is included as a possible eventual constraint on the energy system. Projections incorporating this perspective are found in Nordhaus (1979, 1980), Council on Environmental Quality (CEQ) (1980), A. M. Perry (1982), Perry et al. (1982), and Edmonds and Reilly (1983a).

Almost all of the scenarios applied to studies of CO_2 that are based on reasonably in-depth analysis of the energy situation project a continued growth of energy demand (or consumption) to between about 20 and 40 terawatt (TW) years per year (yr/yr) over the next 40 or 50 years, an increase of two and a half to five times the recent level.* These include scenarios developed for studies by the National Research Council (NRC), the International Institute for Applied Systems Analysis (IIASA), and the Institute for Energy Analysis (IEA) of Oak Ridge Associated Universities and by Nordhaus (1979). Several other energy scenarios, like those of the Interfutures Project (1979), the Hudson Institute (Kahn et al., 1976), the World Energy Conference (1978), and Stewart (1981) are in the same range. Whenever such scenarios do not project a large share of nonfossil energy, they lead to relatively serious concerns about climatic change in the next 50 to 100 years.†

*Estimated global primary energy supply in 1975 was roughly 8 TW yr/yr (IIASA, 1981).

†The market share of nonfossil energy sources (including noncommercial energy) is about 15% at present. The prominent scenarios

(continued on facing page)

Most of the estimates of CO_2 emissions from fossil fuels in the year 2030 lie in a range between about 10 and 30 gigatons of carbon (Gt of C). Thus, based on a review of past projections, it appears that the range of estimates of energy consumption 50 years hence is a factor of 2 or more, and consequent CO_2 emissions show a range of a factor of 3 or more.

It should become clear that the range of estimates is wider than the range of approaches. The large differences in the estimates are traceable in almost all cases to the sensitivity of the models to differences in estimates of the variables or parameters. Most prominent are assumptions about rates of population growth, economic growth, the ratio of energy demand to economic activity, and the mix of supply sources that will meet energy demand. Brief descriptions of the major projections in the three categories follow.

2.2.2 Projections Based on Extrapolations

The extrapolative (A) models are essentially one-equation global models. There are no nations, no economic sectors, no GNP or population projections. In these models, an idealized resource depletion function is customarily used to project the evolution of annual releases through future centuries. There are usually three key variables in the function. One is the total resource of carbon-based fuels. The second is the initial growth rate. The third is a parameter that embodies judgments about the future pattern of exploitation of the resource. It can be set such that peak exploitation occurs when the resource is, for example, 20% depleted, with the possible intention of reflecting a consumer response to rising prices. Or, it can be set to draw different patterns of exploitation, for example, short and intensive, or gradual.

In the very long run (past 2100), the key variable determining CO_2 buildup in this approach is the total carbon resource. The studies have typically taken a number in the vicinity of 5000 Gt of C for the total carbon resource. Such a figure is not out of line with estimates of ultimately available resources, although it is a factor of 10 larger than today's proved recoverable reserves (see World Energy Conference, 1980).

In the medium run (up to 2100), the central variable determining the CO_2 buildup in simple extrapolation models is the initial growth rate. It has been common in the literature to base this variable on work of Rotty (1977), who estimated from historical data that CO_2 emissions from fossil fuel burning (with a trivial addition for cement manufacture) increased 4.3% per year if one excludes the periods of the two world wars and the global economic depression of the early 1930s. This figure of 4.3% has been extremely influential and has been widely used to project future levels of atmospheric CO_2. Many papers and

---

mentioned above generally foresee either an unchanged share for nonfossil sources or a moderate expansion of nonfossil sources, to about 20-35% over the next 50 years.

reports on CO_2-induced climatic change written in the past few years mention it prominently. For example, a JASON report (1979) opens with the statement, "If the current growth rate in the use of fossil fuels continues at 4.3% per year, then the CO_2 concentration in the atmosphere can be expected to double by about 2035. . . ."

While the 4.3% figure is the one most mentioned in the climate literature, increasing debate has grown around it (e.g., World Climate Programme, 1981). One reason for the recent skepticism is that energy growth has slowed considerably--to an average of little more than 2% annually--since 1973. Projections reviewed below range from that of Lovins (1980; Lovins et al., 1982), who suggests there might be a global decrease in use of energy and fossil fuels, to the 50 TW yr/yr case proposed by Niehaus and Williams (1979), in which energy demand grows at an average of about 4% in coming decades and all of this high projected energy demand is covered by fossil fuels.

It is worth noting that the highest projections of CO_2 emissions have generally come from the simple extrapolative models and rarely from studies that incorporate explicit supply and demand models for energy. To illustrate, the Keeling and Bacastow (1977) "preferred scenario" projects emissions somewhat larger than the high coal scenario developed by Niehaus (1979) as an upper limit scenario from the energy perspective, and the Siegenthaler and Oeschger (1978) "upper-limit" scenario generates emissions at about twice the rate of the Niehaus high scenario.

The extrapolations might be best characterized as "gedanken experiments" devised for study of the carbon cycle: "suppose x thousand tons of carbon exist as fossil fuels and all will be used at a certain rate. . . ." While extrapolative models have been successful in drawing attention to the CO_2 issue, they are of limited interest in projecting likely outcomes for CO_2 emissions and concentrations. The main virtue of the approach is simplicity, for a constant growth or logistic curve has great transparency, particularly relative to the enormously complicated energy models. On the other hand, these models do not respect fundamental aspects of economic and energy sector behavior, such as conservation based on rising energy prices. It is not surprising that these models will, therefore, lack realism during periods (after 1973, for example) when changes in economic and political structures have been profound.

2.2.3 Energy System Projections

The major class of forecasts of CO_2 emissions arises from formal or informal energy modeling. Most of this work dates from the 1973 "energy crisis" and is only recently published. In general, it forms the most reliable basis on which to draw for projections. Note that only global studies are sufficient for projecting CO_2 emissions; the numerous national and regional energy studies may provide a consistency check on global studies, but they cannot be used independently to project CO_2 emissions.

2.2.3.1 Perry-Landsberg (NAS)

Perry and Landsberg (1977) assembled projections of world energy consumption and emissions to the year 2025 for the NAS report, Energy and Climate. The projections are for 11 geographic regions, which are sometimes large nations and sometimes groups of nations. Regional demand for energy is derived from projections of population, GNP, and the relationship of GNP per capita and energy consumption. A "high-population/low-economic-growth" situation is postulated for developing countries and a "low-population/low-economic-growth" situation for developed countries. Global population in 2025 is at 9.3 billion, about 20% higher than IIASA and Rotty. The net result is a total energy demand forecasted to reach about 39 TW yr/yr in 2025.

Emissions are calculated for two situations chosen to stress the contrast between a strategy based on "renewables" (i.e., noncarbon-based, abundant energy sources) and one based on coal. In the first case, if regional demand exceeds regional production, an estimate is made assuming the new noncarbon-based energy resource is available to meet the deficiency of nonrenewable resources. In the second case, an estimate is made for the situation for which regional deficiency would be met by coal. Based on these assumptions, annual world CO_2 emissions in 2025 would be between 13 and 14 Gt of C in the first case and about 27 Gt of C in the second, or about 2.5 to 5 times current levels.

The Perry-Landsberg study forms a careful baseline for comparison. It is comprehensible and plausible. A major shortcoming is that it omits any explicit role for prices to play in driving demand toward or away from energy in general or individual fuels in particular. In addition, while the total demand for energy grows out of a well-specified model, the fuel mix is based on arbitrary assumptions.

2.2.3.2 IIASA

2.2.3.2.1 Niehaus and Williams

The IIASA Energy Systems Program analyzed several hypothetical energy strategies for the period up to the year 2100 for their implications for atmospheric CO_2 (Niehaus and Williams, 1979; IIASA, 1981). As such, it could not directly employ the so-called "IIASA energy models," which were run only to the year 2030. Rather, distribution of energy supply among coal, oil, gas, solar, and nuclear is derived from a very-long-term energy model developed by Voss (1977). The Voss model employs principles similar to that of the Forrester-Meadows (system dynamics) school and is structured into six sectors: population, energy, resources, industrial production, capital, and the environment. It is global; there is no geographic disaggregation.

Among the strategies explored (Niehaus and Williams, 1979) are four in which global demand levels out to either 30 TW yr/yr or 50 TW yr/yr in the mid-twenty-first century and remains at that level to 2100. In both the lower- and higher-demand cases there is an analysis in which nuclear and solar energy play an important role and in which they do not. Table 2.19 shows the reserves of fossil fuels used in each strategy. The relation between total coal use and CO_2 emissions is characteristic of projections leading to high or low CO_2 emissions.

The scenarios with reliance on nuclear and solar energy lead to peak annual CO_2 emissions of about 8 to 10 Gt of C around the year 2000, while the scenarios with reliance on fossil fuels lead to emissions of about 22 and 30 Gt of C in 2030, increasing somewhat thereafter. While consideration is given to available fossil resources at the global level, there is no study of regional or national implications.

The Niehaus-Williams projections are based fundamentally on judgments external to economic analysis or modeling. Once the energy growth path and fuel mix are set, the outcome for CO_2 emissions is determined--the use of the system model plays but a small role in the outcome.

The most important issue is whether the _ex cathedra_ judgments as to the ultimate levels of global energy demand (the 30 and 50 TW yr/yr levels discussed above) are reasonable. While such figures are conventional, and indeed so often used that they become comfortable assumptions, they have no grounding in a physical or economic constraint or in the outcome of an energy model. The notion of "saturation" at these levels is a popular idea that has no particular basis, other than the hope that human society will pass through a transition to a stable plateau over the next couple of generations. Thus, while the critical assumptions of energy demand and fuel mix in these studies do not appear implausible, their grounding is weak.

2.2.3.2.2 The IIASA Energy Models

IIASA used a set of extremely detailed models to delineate two scenarios, a "high" and a "low" case culminating in 2030 with world energy consumption at 35 and 22 TW yr/yr, respectively. The models are oriented toward engineering and technical considerations for specific demand sectors and global consistency of supply among the seven regions

TABLE 2.19 Reserves of Fossil Fuels Used in Different Outcomes (1975-2100)[a]

Strategy	Coal (Gt of C)	Oil (Gt of C)	Gas (Gt of C)
30 TW with solar and nuclear	170	170	110
50 TW with solar and nuclear	230	210	130
30-TW fossil fuel	1980	190	120
50-TW fossil fuel	3020	230	140

[a]After Niehaus and Williams (1979).

into which the world is disaggregated. The distribution of supply sources in the actual IIASA scenarios is quite different from Niehaus and Williams, even though the Niehaus and Williams runs were originally chosen to be broadly consistent with the global energy demand pattern. Both the high and low IIASA scenarios are hybrids, with expanded use of many supply sources, so that in 2030 11 TW yr/yr are coming from non-fossil fuel sources in the high case and 7 TW yr/yr in the low case. In the high case emissions are above 16 Gt of C in 2030, and in the low case they are nearing 10 Gt of C.

The IIASA models have probably been the closest existing approach to an appropriate disaggregated technique for forecasting CO_2 emissions. In principle, they are grounded in engineering and economic relations, with attention to feasibility and response of supply and demand to price. In practice, because of the need to accommodate differing views, world energy consumption was adjusted judgmentally to be "reasonable," as well as on the basis of the formal methods. In this respect, the outcome shares the problems outlined in the last paragraph of the discussion of the Niehaus-Williams approach.

Another issue raised by the IIASA model is whether a high degree of disaggregation is appropriate. Such an approach allows considerations such as those involving trade and national policies; however, it also makes the models difficult to comprehend, manipulate, change, and verify independently.

2.2.3.3 Rotty et al.

For several years, Rotty and co-workers at the Institute for Energy Analysis of Oak Ridge Associated Universities emphasized extrapolation of the 4.3% estimate of historic annual increase in CO_2 emissions and figures tapering off from this (Rotty, 1977, 1978, 1979a,b; Marland and Rotty, 1979). Based on demand and fuel-share projections made for six world regions, an annual fossil fuel release of CO_2 containing 23-26 Gt of C from energy use of 36-40 TW yr/yr in the year 2025 is calculated. The work is partly based on a more formal analysis by Allen et al. (1981) developed for the year 2000. In extension of the projections to 2025 Rotty assumes supply will meet demand without examination of balancing economic factors such as prices. For projections of emissions beyond 2025, a different extrapolative technique involving application of arbitrary global fossil resource depletion rates is employed (see Section 2.2.3.2 above).

A more recent paper (Rotty and Marland, 1980) includes some discussion of constraints on fossil fuel use. Three kinds of constraints are examined: resource, environmental, and fuel demand. With respect to resource supplies, Rotty and Marland conclude that "the fraction of total resources used up to the present is so small that physical quantities cannot yet be perceived as presenting a real constraint." However, it is mentioned that unequal geographic distribution of the resources probably will continue to be a source of international stress. Climatic change as an environmental issue is dismissed as a constraint to fossil fuel use.

In contrast, Rotty and Marland (1980) discuss at some length the likelihood that slower growth in fuel demand dictated by social and economic factors will limit fossil fuel use. Reduced rates of economic growth are projected as a result of very recent trends and anticipated problems with capital and escalating costs and shifts toward conservation and less energy intensive industries. No formal modeling is offered to substantiate the position. Regardless of the precise causes, summing up estimates for about a dozen countries and half a dozen composite regions leads Rotty and Marland to project CO_2 emissions in 2025 of about 14 Gt of C in a 26 TW yr/yr global energy scenario--an annual growth rate of 2% per year from today. Thus, the range of Rotty emission projections are quite similar to those of IIASA and Perry and Landsberg.

The strengths and weaknesses of the Rotty approach are partly those of the extrapolative models and partly those of the Perry-Landsberg approach. The repeated adjustment of assumptions is evidence of the uncomfortably arbitrary nature of the endeavor.

2.2.3.4 Nordhaus

Nordhaus (1977, 1979) estimated the uncontrolled path of CO_2 emissions in a modification of a model developed for studies of efficient allocation of energy resources (or of a competitive market for energy). Nordhaus's approach was fundamentally based on economic modeling and assumptions--with interaction of forces of supply and demand leading to a path of prices and energy consumption over time. By comparison, prices play a lesser role in the IIASA and Perry and Landsberg approaches and virtually no role in Rotty's projections.

Nordhaus employs a medium-sized linear programming model, with basic components being an objective function (based on demand functions for energy) and a supply function centered on geological considerations and technology. The outcome is calculated by finding the lowest cost way of meeting the demand function, using a linear programming (LP) algorithm. The demand function (technically, the objective function) for the LP is drawn from data on market behavior. It is built up from four energy sectors (electricity, industry, residential, and transportation), with demand in each sector a function of population, per capita income, and relative prices. The technology or constraint set is derived from engineering and geological data on resource availability and costs of extraction, transportation, and conversion. The model incorporates constraints on new technologies, adaptation of demands, and upper bounds on rates of growth.

Running the model involves balancing supply and demand over time, with prices playing the central role of equilibration. Results are given in terms of both activity levels (for example, production of coal or oil in a given period) and prices. The calculation provides for six different fuels used in the four energy sectors, for two different regions (United States and rest of world), for ten time periods of 20 years each. The macroeconomic assumptions are that rapid growth in GNP per capita will continue in both regions, but at a diminishing rate after 2000, and that population will also slow to reach a world level of 10 billion in 2050.

Nordhaus calculates that the uncontrolled path leads to large changes in the level of atmospheric CO_2. In the uncontrolled case, annual emissions are at 18 Gt of C in 2020 and steeply increasing, so they reach 40 Gt of C in 2040. Global energy demand is about 40 TW yr/yr in 2030. A key in these high projections is high initial GNP growth rates; for 1975-1990 the assumed growth per year is 3.7% in the United States and 6.5% in the rest of the world.

The results of the Nordhaus analysis exhibit the strengths and weaknesses of pure (nonjudgment-based) economic modeling. On the one hand, the outcome is based on objective data (such as market prices and resource availability) and is thus reproducible and can be easily modified over time. On the other hand, results are very sensitive to assumptions about future price and growth trends. The actual model runs were based on an assumption of rapid future economic growth and low fuel prices--leading thereby to rapid estimated growth of CO_2 emissions and atmospheric concentrations.

2.2.3.5 Edmonds and Reilly

Rotty's work is now being followed at IEA by a more formal CO_2 emissions model developed by Edmonds and Reilly. The model takes as inputs key economic, resource availability, and demographic variables such as income, energy costs, resource constraints, labor force, and population. From these it calculates consistent energy-use paths. Consistency is defined as a balancing of supply and demand in the face of resource constraints, with energy prices adjusting to assure an equilibrium solution. Energy use is disaggregated into nine world regions and all major possible fuel types, including oil, gas, coal, coal liquefaction, and shale oil. The model is intended to be applicable out about 100 years, with calculations feasible at intervals that the user selects, for example, 10 or 20 years. The model is extensively documented (Edmonds et al., 1981; Reilly et al., 1981; Edmonds and Reilly, 1983c), and a base case is now being developed (Edmonds and Reilly, 1983a,b).

Initial results show a quite steady increase in energy demand of about 2.5% per year from now to 2050, so that demand has reached about 29 TW yr/yr in 2025 and 50 TW yr/yr in 2050. CO_2 emissions increase by 1.5% per year from now to 2000, and by 2.3% per year between 2000 and 2025, reaching 12 Gt. Because of increasing reliance on coal, oil shale, and synthetic fuels, emissions then rise quite steeply, by more than 3% per year, and reach an annual rate of 26 Gt of C by 2050.

The Edmonds-Reilly model has the potential of being an extremely useful follow-up to earlier detailed studies, such as that of IIASA. It contains sufficient regional and sectoral disaggregation that experts in individual areas (such as analysts specializing only in the U. S. economy or a particular fuel source) can evaluate the detailed forecasts and assumptions. It also appears to be flexibly designed, so that results of different assumptions can be examined easily.

At the same time, the current effort contains some of the problems that have plagued earlier large-scale energy models. Perhaps the most important is the decoupling of energy demand from output. The current model has energy demand sensitive to prices and incomes, but incomes and outputs are not directly related to energy, labor, and other inputs. (Technically, energy is not treated as a derived demand, that is, derived from a production function relating inputs to output.) A second problematic feature is the extensive use of logistic curves that are not sensitive to prices for determining supply.

It should also be noted that the Edmonds-Reilly model is quite large and somewhat forbidding for a casual user. The benefit from technological and regional detail is partially vitiated by the difficulty of understanding the structure and workings of the model. As in many large-scale models, the size makes identification of critical parameters or assumptions a formidable task.

Notwithstanding these reservations, the Edmonds-Reilly work stands out today as the only carefully documented long-run global energy model operating in the United States.

2.2.3.6 Other Projections

Marchetti (1980) has made a forecast of the amount of CO_2 that will be emitted to the year 2050 based on a logistic substitution model of energy systems (Marchetti and Nakicenovic, 1979). This model treats energy sources as technologies competing for a market and applies a form of market penetration analysis. A logistic function is used for describing the evolution of energy sources and is fitted to historical statistical data. The driving force for change in this model appears to be the geographical density of energy consumption, and the mechanisms leading to the switch from one source to another are the different technical characteristics associated with each energy source. For example, in the Marchetti view oil succeeded coal primarily because of the advantages achievable by a system operating on fluids.

With data on energy consumption back to 1860 and including both commercial and noncommercial (wood, farm waste, hay) energy sources, the slope of the fitted curve of energy demand implies an annual growth of 2.3%. [This contrasts strongly with Rotty (1979b), who emphasizes that commercial energy supply, excluding times of world conflicts and depression, has grown at a rate of about 5.3% since 1860.] Applying a future growth rate of 3% per year, Marchetti calculates energy consumption for the various sources for the period 1975-2050 based on the logistic equations. The model predicts a relatively rapid phaseout of coal, a dominant role for natural gas, rapid growth of nuclear power, and a negligible role for new sources other than nuclear over the next 50 years. The model implies an increase in annual CO_2 emissions to about 14 Gt of C in 2030, an amount close to the lower estimates of Perry and Landsberg, IIASA, and IEA, and a cumulative emission of carbon to the atmosphere between the years 1975 and 2050 of about 400 Gt of C [to somewhat less than 450 ppm(v)]. Perhaps more important, it predicts a gradual reduction in emissions and atmospheric CO_2 thereafter, rather than continuing increase.

While Marchetti's projection of fuel shares is singular, his analysis of the long-term pattern of energy demand is not. Stewart (1981) also uses an empirical approach leaning on application of logistic growth curves, chosen to fit historical data extending back to 1850. Stewart argues additionally that energy growth is likely to evolve in surges or cycles rather than monotonically. Stewart identifies historical "cycles" in energy use with periods of around 50 years (perhaps a manifestation of the frequently cited Kondratieff cycle of economic activity) and notes that deviations of plus or minus 20% around a long-term logistic growth curve were experienced.

On the basis of an assumed stable cyclical structure, Stewart projects world energy consumption to the year 2025. For the period 1975-2000 a 40% growth is indicated; this breaks down into zero energy growth in the United States and a 60% growth for the world outside the United States. This overall projection for 2000 is lower than most. However, Stewart's projection for 2025 is close to other high values. After the relatively depressed period between 1975 and 2000, world energy growth between 2000 and 2025 is projected at a rate of about 4%, increasing from about 13 TW yr/yr to almost 36 TW yr/yr.

Legasov and Kuz'min of the Atomic Energy Institute of the USSR have also made a projection employing a logistic approach. The key variable in their function is one that describes the level of stabilization of per capita energy consumption (Legasov and Kuz'min, 1981; Report of the US/USSR Workshop, 1982). Legasov and Kuz'min explore two cases, one in which global average annual per capita energy consumption by 2100 reaches 10 kW (roughly the level in the United States today) and one in which it reaches 20 kW. Population, meanwhile, stabilizes at a level of 12 billion people. Under these assumptions global energy use in 2020 is either 50 or 60 TW yr/yr, with a population of 8.8 billion. Legasov and Kuz'min project coal and nuclear power as the principal energy sources for the coming decades, with nuclear power gradually becoming dominant. Under these assumptions, CO_2 emissions in 2020 are about 15 Gt in the lower case and 18 Gt in the upper case and roughly stable for several subsequent decades.

The three approaches described above are more sophisticated than the extrapolation approach (see Section 2.2.3.2), but the underlying methodology is similar. All assume that there is a stable underlying dynamic (exponential, logistic, or logistic-cum-sinusoidal) and forecast off that base. These approaches allow for no structural relation between exogenous variables like population and resources and endogenous variables like energy consumption. Such autoregressive or inertial models do relatively well at prediction in the short run, but their level of aggregation is so high that for most purposes one must still turn to the more structural models.

A final source is Lovins (Lovins, 1980; Lovins et al., 1982), who projects very low CO_2 emissions because of a shift to conservation and renewable (nonfossil) sources. With a 4.6-fold increase in global economic activity during 1975-2080 and a doubling of world population, total energy needs will, according to Lovins, be below the 1975 level, indeed dropping over the next century to less than half the present level. A projected increase in energy efficiency in end uses along with renewable sources for energy production might, according to Lovins, largely or wholly eliminate the global use of fossil fuels. A case study of the Federal Republic of Germany, a diverse heavily industrialized economy in a rigorous climate, is used as an "existence proof" (basis for extrapolation) for the efficiency and renewables strategy.

Lovins's results appear to be wishful with respect both to rapid development and diffusion of solar technologies and to lifestyle changes involving energy conservation. He does not present a formal model or develop the implications of the analysis for capital and labor needs. In addition, some of the trends identified, such as increasing efficiency of end-use devices, may raise the demand for energy, an outcome not accounted for. While Lovins may turn out to be correct, the analytical basis for his views remains elusive and characterized by strong cultural bias.

2.2.4 Projections with CO_2 Feedback to the Energy System

The energy projections reviewed in Section 2.2.3 share a potential deficiency when used to generate long-term CO_2 emission trajectories. Calculation of the ejected CO_2 is largely incidental. An energy path is plotted for a variety of reasons, and CO_2 is merely the outcome of the chosen path.

There are two ways in which such an approach may be deficient. First, by focusing on CO_2 directly, it may be possible to get more accurate CO_2 forecasts, as secondary issues (such as coal versus oil) can be ignored. Second, if a CO_2 buildup takes place and leads to serious social consequences, there may be some impact on the economy directly (through output) or indirectly (through policy reactions). Put differently, models that allow very high CO_2 but do not allow feedback from environmental change to energy policy must be regarded with caution; they mask significant assumptions about the behavior of people and governments (Stahl and Ausubel, 1981).

Projections that include increased CO_2 levels as a possible eventual constraint on CO_2 emissions include Nordhaus (1979, 1980), Edmonds and Reilly (1983a,b), CEQ (1980), and Perry (1981; Perry et al., 1982). These projections generally require that some threshold concentration of CO_2 (or similar constraint) be set, presumably by political intervention. Trajectories are then calculated that keep ambient levels from exceeding this threshold. Thus, in these approaches, rather than begin from high- and low-energy scenarios, the approach is to work backward from a desired or specified terminal condition to defining energy demand and fuel mix patterns that satisfy it.

2.2.4.1 Nordhaus

Along with estimating the uncontrolled path described earlier, Nordhaus (1977, 1979) also estimates time paths of emissions given particular carbon dioxide constraints. Efficient allocation of energy resources is calculated using the model described earlier under the assumption that it would be necessary to prevent atmospheric CO_2 from exceeding either 1.5, 2, or 3 times the preindustrial level [about 450, 600, or 900 ppm(v)].

The optimal path does not differ from the uncontrolled path for the first period (up to 1990). Abatement measures become necessary only in the second period (1990-2010) for the stringent control (450 ppm) and in the third period (2010-2030) for the milder control programs. To illustrate, in 2020 emissions for the uncontrolled case and the tripling are identical at 18 Gt of C, and the doubling case is only marginally lower at 16 Gt of C, but the 50% increase limit requires emissions of only 4 Gt of C. In 2040 the stringent-case emissions have trailed off to barely more than 2 Gt of C, the doubling case leaves carbon emission steady at 16 Gt of C, while the tripling and uncontrolled case have both reached the vicinity of 40 Gt of C per year. This technique allows estimates of the costs of controlling CO_2 emissions as well as the "carbon taxes" necessary to induce such responses.

Nordhaus (1980, 1982) also develops an optimal control framework, which seeks to identify the most economical way to balance the exploitation of both carbon fuels and climatic resources. The analysis is at a highly aggregate, global level; implications for sectors or for regional or national policies are not explored. Nordhaus weighs CO_2 control strategies according to two criteria: their effects on the paths of consumption that are generated by the control strategy and maximization of the discounted value of consumption streams, where the discount rate combines both a temporal and a growth factor.

The framework consists of four simple equations. These are a description of the carbon cycle and climatic effects of CO_2 elevation, estimates of the costs of reducing or abating CO_2 emissions, an equation that incorporates estimates of economic impacts of CO_2 buildup, and an equation that represents intertemporal choice between consumption paths.

Since there is great uncertainty about the economic and social impact of elevation of CO_2 concentration, Nordhaus tests the sensitivity of the model to different sets of costs. These are described by a "loss parameter," which indicates the fractional loss of consumption per doubling of CO_2. By varying this and other parameters, a set of emissions trajectories is calculated. The outcome of the model was considered at best illustrative given the uncertainty surrounding key parameters (such as the economic impact of climate change). A major result, however, was that the best degree of CO_2 control was extremely sensitive to important uncertain parameters, that is, no obvious control strategy stood out.

2.2.4.2 Edmonds and Reilly

Edmonds and Reilly (1983a,b) have also begun to explore the effect of taxation policies of various kinds on CO_2 buildup. One question asked is what consequences a substantial CO_2 tax in the United States would have on the level of atmospheric CO_2. They find that global carbon emissions are reduced by much less than the U.S. reduction owing to the fact that decreased U.S. energy demand resulting from the CO_2 tax lowers world energy prices, which in turn spurs energy consumption in other regions. In contrast, when a global tax is combined with a U.S. embargo on coal exports, there are substantial reductions in U.S. and non-U.S. CO_2 emissions.

While all the studies on taxation of CO_2 are still quite tentative, the three sets of tax experiments that we have reviewed--the Nordhaus results discussed in Section 2.2.4.1, the Nordhaus-Yohe results in Section 2.1, and the Edmonds-Reilly results--appear broadly consistent. This finding is striking, given that the three approaches are very different.

2.2.4.3 CEQ

The CEQ study (1981) derives several curves to yield a buildup of atmospheric CO_2 equal to 1.5, 2.0, and 3.0 times the preindustrial level (slightly less than 450, 600, and 900 ppm, respectively). It employs a simple, two-equation global model consisting of a differential equation to explain buildup of atmospheric CO_2 and a logistic equation to forecast CO_2 emissions. The major unknown parameter is the initial growth rate of fossil fuel combustion. The model is run backward to calculate global fossil fuel releases that would produce the assumed buildups of carbon dioxide. Curves preferred for further analysis correspond in 2030 to fossil fuel energy production of about 8, 13, and 17 TW yr/yr and emissions of 6, 10, and 13 Gt of C, respectively.

The controlled curves are compared with two overall energy projections. The high global energy demand scenario is for a world whose population has leveled off at 10 billion by the year 2100 and an average per capita energy use equal to two thirds the present U.S. level. Energy use in 2030 is about 35 TW yr/yr (similar to the IIASA high scenario) and rises to about 75 TW yr/yr by 2100, a ninefold increase over current levels, with about one fourth accounted for by population growth. A lower world energy use scenario represents a world whose population has leveled off at about 8.5 billion by 2100 at an average per capita level of one third present U.S. consumption. In 2030 energy use is about 20 TW yr/yr (similar to the IIASA low scenario) and reaches a plateau well before 2100 of about 30 TW yr/yr, about a fourfold increase over current consumption in which one half the growth is attributable to population increase.

The CEQ study evaluates the significance of the gap between overall energy demand and the three assumed CO_2 limits. For example, to avoid exceeding a 50% increase in global CO_2 concentration and to meet the low-energy-demand scenario (a low growth, environmentally cautious world), nonfossil fuel sources would be required to increase from about 1 TW yr/yr today to more than 4 TW yr/yr by the year 2000, or about a 9% growth per year. By 2020 nonfossil fuel sources would have to contribute about 10 TW yr/yr, with their growth averaging 4%

between 2000-2020. The 10 TW yr/yr are more than the current total global annual energy use, and more than the nonfossil (solar and nuclear) supply estimated for 2020 in the IIASA high scenario. The CEQ study estimates that together hydropower and nuclear power could probably provide between 2 and 3 TW yr/yr (fuel equivalent) by the year 2000. Thus, with low-energy growth but also a low ceiling on CO_2 levels, a contribution of about 1 to 2 TW yr/yr would be needed by 2000 from other renewables, with rapid increases thereafter as hydropower potential is exhausted.

2.2.4.4 A. M. Perry et al.

Perry and colleagues (1982; Perry et al., 1982) begin by adopting global energy projections from IIASA, the World Coal Study (1980), and others as reference scenarios. The novel parameter in their analysis is the date when global fossil energy use must begin to deviate from the reference scenarios in order to meet various atmospheric CO_2 limits. This date is referred to as the action initiation time (AIT). All the analyses so far are at the global level; that is, they refer to when a global policy (national policies summing to a global policy) would need to begin to be followed.

The approach built around action initiation times stresses that the rate of change of energy strategies is extremely important. If some CO_2 limit were approached along the kinds of curves normally drawn, the limit would certainly be passed, because of the inertia or momentum of the energy system. If the ceiling were not to be exceeded, CO_2 production would have to fall abruptly to zero, a virtual impossibility. Thus, Perry (1982) proposes anticipatory scenarios, which involve a gradual slowing of growth of fossil fuel use, followed by an eventual slow decline. With a high initial growth rate or a late AIT, the transition required in order to remain below a given CO_2 target may be too rapid and the subsequent decline too steep--the required transition may be infeasible.

According to arbitrary feasibility criteria relating to historic evolution and behavior of energy systems, several scenarios are drawn that should allow sufficient time for the necessary changes in energy demand patterns and supply technologies. Table 2.20 lists some AITs thought to be of intermediate difficulty.

In the Perry study, as well as the CEQ study, it is apparent that by fixing only a few parameters, principally energy growth, CO_2 limits, and a few characteristic times like market penetration, the overall trends of fossil and nonfossil energy use become approximately determined. With further work it may be possible to judge more reliably whether the different patterns of energy use designed to limit CO_2 concentrations would be easy or difficult to attain. Without such information, it seems premature to employ these models for prescriptive purposes.

TABLE 2.20 Required Action Initiation Times for Various CO_2 Ceilings[a,b]

CO_2 Limit (ppm)	Initial Growth Rate of Annual Carbon Emissions 1.5%/yr	2.5%/yr	3%/yr
500	2005	1995	1990
600	2025	2010	2000
700	2040	2025	2010
800		2035	2020

[a]Source: Perry (1982).
[b]For example, if a global limit for CO_2 in the atmosphere of 500 ppm is to be met, and emissions are growing at the outset by 1.5%/year, actions to reduce the share of fossil fuels would need to begin in the year 2005. If emissions are growing by 3% in the coming decade and we wish to meet a limit of 500 ppm, policies to discourage use of fossil fuels might need to become effective as early as 1990. These action initiation times are for transitions away from fossil fuels judged by Perry to be of intermediate difficulty.

2.2.4.5 General Comments

Studies that attempt to include feedback from CO_2 concentrations to energy policy are in their infancy. A particular problem is the confusion and combination of "positive" and "normative" approaches. In a "positive" model, the attempt is to describe how a system will behave under given boundary conditions. In a normative approach, one sets up a policy goal or objective function and then asks how the system ought to behave in order to optimize the objective function. While the distinction is seldom clearly delineated in global energy models (see particularly the comment on Lovins above), potential confusion is most likely to arise concerning the class of models discussed in this section. For the most part, the best interpretation would seem to be the following: the energy systems are based on a positive description, and CO_2 constraints are viewed as alternative normative policy constraints. However, assumptions of inaction (or absence of feedback) at very high levels of CO_2 emissions are also in a sense normative.

A second issue concerns the actual limits imposed. While the limitation of a doubling of CO_2 is the policy most often analyzed, it does not arise from a well-developed line of reasoning. An ideal (or even a "good") set of CO_2 policies will depend on the costs and benefits of climate change and CO_2 controls; costs and benefits are so poorly understood that no clear line of policy stands out as appropriate (see Schelling, Chapter 9).

In terms of conclusions, the Nordhaus, CEQ, and Perry studies seem to be largely consistent in their projections of what emission trajectories would look like under particular CO_2-induced constraints. As long as fossil fuel growth rates continue at the low level of the past few years and concentrations of 400-450 ppm are judged acceptable, there is little urgency for significant reductions in CO_2 emissions below an uncontrolled path before 1990. Emissions would need to be reduced below an uncontrolled path around 2000 if a limit in the vicinity of 450-500 ppm is desirable. To limit concentrations to 600 ppm (a doubling from preindustrial levels) would require that serious reductions be initiated in the 2010 to 2030 period. These long lead times before CO_2 reductions are necessary may be misleading, however. To effect a significant reduction of CO_2 emissions in an orderly and efficient way probably requires planning and policy measures decades in advance, for the infrastructure and capital stock associated with fossil fuels cannot quickly be scrapped and replaced without high economic cost. Also, it is probably necessary to consider policies with regard to climatic change on the basis of possible combined effects of CO_2 _and_ other greenhouse gases.

2.2.5 A Note on the Biosphere

In the past the biosphere may have been a cumulative source of CO_2 as a result of human activities within a factor of 2 as great as burning of fossil fuels (Clark et al., 1982; Woodwell, this volume, Chapter 3, Section 3.3). However, it appears that in projecting future CO_2 emissions resulting from human activities the role of the biosphere is swamped by the potential contribution of fossil fuel combustion.*

An estimate for the maximum possible future addition from all biospheric sources is 240 Gt of C (Revelle and Munk, 1977). Baumgartner (1979) estimates that clearing of all tropical forests might contribute about 140 Gt of C. The total carbon content of the Amazon forest is estimated at about 120 Gt of C (Sioli, 1973). Chan et al. (1980) develop a high deforestation scenario in which total additional transfer of carbon from the biosphere to the atmosphere by the year 2100 is about 100 Gt of C. The World Climate Programme (1981) group of experts adopted a range of 50 to 150 Gt of C for biospheric emissions in the 1980-2025 period. Machta (this volume, Chapter 3, Section 3.5) estimates that massive oxidation of the biota might increase atmospheric CO_2 by 75 ppm by A.D. 2100.

Projections of future atmospheric CO_2 concentrations embracing both burning of fossil fuels and terrestrial sources have all been dominated by growth rates in fossil fuel emissions, except in cases where fossil fuel emissions are extremely low or in cases like that described by Woodwell (Chapter 3, Section 3.6), where all the world's forests are entirely destroyed in a few decades. While annual bio-

---

*While the role of the biosphere may be marginal in projecting future _emissions_, it is, of course, important in calculating how the emissions are distributed among ultimate reservoirs.

spheric emissions from human activities may average as high as 1 to 3 Gt of C per year in future decades, fossil fuel emissions are typically projected to be an order of magnitude larger. From a different perspective, Schelling (Section 9.2.4) estimates that massive destruction or plantation of forests might accelerate or retard the growth of atmospheric CO_2 to a particular level by a decade or so during the second half of the next century.

2.2.6 Projections of Non-CO_2 Trace Gases

Changes in atmospheric concentrations of several infrared absorbing gases besides CO_2 may result from human activities (see Machta, Chapter 4, Section 4.3). These activities include the following:

(a) Stratospheric flight. Increasing supersonic air traffic may lead to changes in the O_3 and H_2O content of the stratosphere.

(b) Use of nitrogen fertilizers. Denitrification of fertilizers in the soil releases nitrous oxide (N_2O) to the atmosphere. Less significant increases in NH_3 and HNO_3 may also result.

(c) Use of chlorofluorocarbons (CFCs)--CCl_2F_2 and CCl_3F--as refrigerants and propellants in aerosol spray cans, for example.

(d) Extraction and burning of fossil fuels. Methane (CH_4) may be released as a result of mining of coal and extraction of oil and gas. CH_4 is also a conversion product of CO, and its presence is thus correlated with burning of fossil fuels.

(e) Agricultural and livestock production. Increasing methane emissions may be associated with large livestock herds and expansion and intensification of rice production.

Projections of future emissions of these non-CO_2 trace gases are generally at a more primitive stage than are CO_2 projections.* Researchers studying biogeochemical cycles and the atmosphere typically have used simple assumptions of linear increase or exponential growth based on a short segment of recent years (see Flohn, 1980, pp. 22-23). Wang et al. (1976), in a widely cited article, assumed that by 2020 stratospheric H_2O, N_2O, and CH_4 would all double and that the CFCs would increase by a factor of 20. Alternatively, projections may be extrapolated from more detailed studies, like the Climatic Impact Assessment Program (CIAP, 1975), which developed scenarios of stratospheric flight. The time horizons of the studies of stratospheric flight, agricultural production, industrial use of chemicals, and other activities vary, and the macroeconomic assumptions employed vary as well. There is a lack of studies of the combined greenhouse effect that use assumptions consistently in generating both CO_2 emissions

---

*In addition to the references given in the text of this section, see Hameed et al. (1980), Lacis et al. (1981), Logan et al. (1978), Ramanathan (1980), and Rowland and Molina (1975).

and emissions of other infrared-absorbing trace gases. Given the very large inertia and modest rate of technological change in energy systems, projection of CO_2 emissions over periods of 50 years and longer has large but manageable error bounds. In human activities--like use of CFCs or stratospheric flight--where more rapid technological change is occurring, and where less inertia is imposed by a large and expensive capital stock, projections extending many decades are much more hazardous.

2.2.7 Findings

2.2.7.1 The State of the Art

2.2.7.1.1 Recent Progress

Few serious attempts at global long-range energy perspectives were undertaken before the 1970s. There has been rapid methodological progress in methods of making energy and CO_2 projections over the last decade. Much important work is a spinoff from energy analysis spurred by the 1973 oil shock. With some exceptions, methods developed independent of CO_2 studies in energy modeling, statistics, and econometrics should be adequate for the task of projecting future anthropogenic CO_2 emissions, when brought together with knowledge from geology, engineering, and other relevant fields.

2.2.7.1.2 Nature of Modeling Exercises

Modeling is a way of organizing thinking about a problem, one that should allow improved scrutiny of data, assumptions, and relationships. There is unlikely to be one "correct" approach to energy modeling for CO_2 applications. The systems involved are too complex, too uncertain; questions we ask may differ; methodological improvements occur frequently.

Moreover, there are cultural factors that influence forecasting. It is obvious to even a casual observer of the energy scene that there are deeply held and diverse views about energy futures, which are, after all, views of the character and relationship of man, nature, and society. Even if all could agree on a single model to use, we would certainly disagree on values for many variables. This is not merely a question of uncertainty; it is a question of coexisting contradictory certainties, points about which different groups and individuals hold highly assured but also highly different views.

Historical fashions in forecasting may be equally significant. It would be myopic to think that the current set of projections is free of today's implicit assumptions or biases. One cannot help but notice the tendency in energy forecasting to extrapolate the most recent past, whether one of relatively rapid or slow growth, far into the future. When the price of electricity was going down in the 1950s, people spoke of nuclear electricity becoming too cheap to meter; when the price of oil increased in the 1970s, people spoke of a barrel rising to a price of $100 or $200. The tendency of many forecasters to move in parallel (so that when one makes an upward or downward turn, all do) is also noteworthy.

How historical trends and trendlines in forecasts should affect both choice of method and our interpretation of contemporary forecasts and the spread of forecasts remains to be explored further. Probabilistic approaches, like that of Nordhaus and Yohe (Section 2.1), are one natural response. However, in view of the fickleness of forecasts, it is clearly useful to encourage a variety of approaches: large and small; formal and informal; stochastic and deterministic.

A pervasive question in research in CO_2 and energy is how much disaggregation is useful for accurate predictions of future CO_2 emissions. (A similar question arises, indeed, in climate modeling, where the question of the optimal level of refinement of spatial grid and time steps also occurs.) It is often assumed that more disaggregation is better. Careful investigation of the issue shows, however, that no general result holds. The potential improvement from disaggregation depends on the purposes of the study, the structure of microrelations, and the quality of the microdata. (See Grunfeld and Griliches, 1960.)

There are at least two possible reasons why disaggregation in CO_2 projections might not produce more reliable estimates. First, disaggregated data may be less reliable than aggregated data. This problem can lead to errors in variables and biased statistical estimates of microrelations. For example, we might have a good estimate of global energy production and, therefore, consumption but not of the distribution of global consumption. Second, there may be interdependence across regions that would be taken into account in aggregate models but not in disaggregated models. An example is the constraint that the balance of trade of the world be zero (or the net oil imports of the world be zero). A pasting together of studies of individual countries would generally not respect the constraint. In both cases, it is possible that aggregate models could provide superior prediction to disaggregated models.

2.2.7.1.3 Assessment of Current Efforts

The current modeling and knowledge of future CO_2 emissions appears marginally adequate today; we have a general idea of likely future trends and the range of uncertainty. It may be that further effort could increase the accuracy of our forecasts substantially. Given the large uncertainty that future energy growth and energy projections are contributing to the CO_2 issue, this area may well merit more research attention and support than it has received in the past. Future research efforts might be designed with four points in mind.

1. In general, the most detailed and theoretically based projections of CO_2 have been a spillover from work in other areas, particularly energy studies. This fact suggests that continued support of energy modeling efforts will be of importance in further pushing out the frontier of knowledge about future CO_2 emissions, as well as the interaction between possible CO_2 controls and the economy.

2. We have identified a serious deficiency in the support of long-run economic and energy models in the United States. There is not one U.S. long-range global energy or economic model that is being developed and constantly maintained, updated with documentation, and usable by a wide variety of groups. This shortcoming is in stark contrast to climate or carbon cycle models, where several models receive long-term support, are periodically updated, and can be used by outside groups. Another striking contrast is with short-run economic models, which are too plentiful to enumerate.

3. The bulk of CO_2 projections have been primitive from a methodological point of view. Work on projecting CO_2 emissions has not drawn sufficiently on existing work in statistics, econometrics, or decision theory. There has been little attention to uncertainties and probabilities. Also, considerable confusion of normative and positive approaches exists in modeling of CO_2 emissions.

4. Application of models for analysis of policies where there are, for example, feedbacks to the economy from climatic change or CO_2 control strategies is just beginning. Efforts to evaluate the effectiveness for CO_2 control of energy policies of particular nations or groups of nations in a globally consistent framework have been lacking.

2.2.7.2 Likely Future Outcomes

It is possible to synthesize past work to obtain a likely range of future CO_2 emissions. Before doing so, it is important to reiterate the inhomogeneous character of the projections surveyed.

Some studies, like those of Rotty, Perry and Landsberg, and Marchetti, seek to be best guesses or forecasts of future energy demand; others, like IIASA, posit scenarios, seek to fill out the descriptions, and avoid making claims about probability. Not only do the studies vary in intent, they are also of limited comparability in structure. The studies differ widely in levels of detail, time horizon, data base, and geographical aggregation. While projections of CO_2 emissions may extend to the year 2100, few energy studies offer detailed analysis beyond the year 2000, and fewer still offer detail past 2025 or 2030. The relative reliance on economic, engineering, and ecological logic varies. In addition, the studies are not independent. For example, researchers who participated in the IIASA work also participated in IEA and Interfutures research; both Lovins and IIASA rely on Keyfitz's population projections.

2.2.7.2.1 Energy Growth

Figure 2.22 summarizes the energy consumption forecasts to the year 2030.

Projections of CO_2 emissions are basically products of projections of energy demand and fuel mix. Projections of growth in energy use involve, more or less explicitly, assumptions or estimates concerning population growth, changes in per capita production of goods and services, and changes in the primary energy input required per unit of

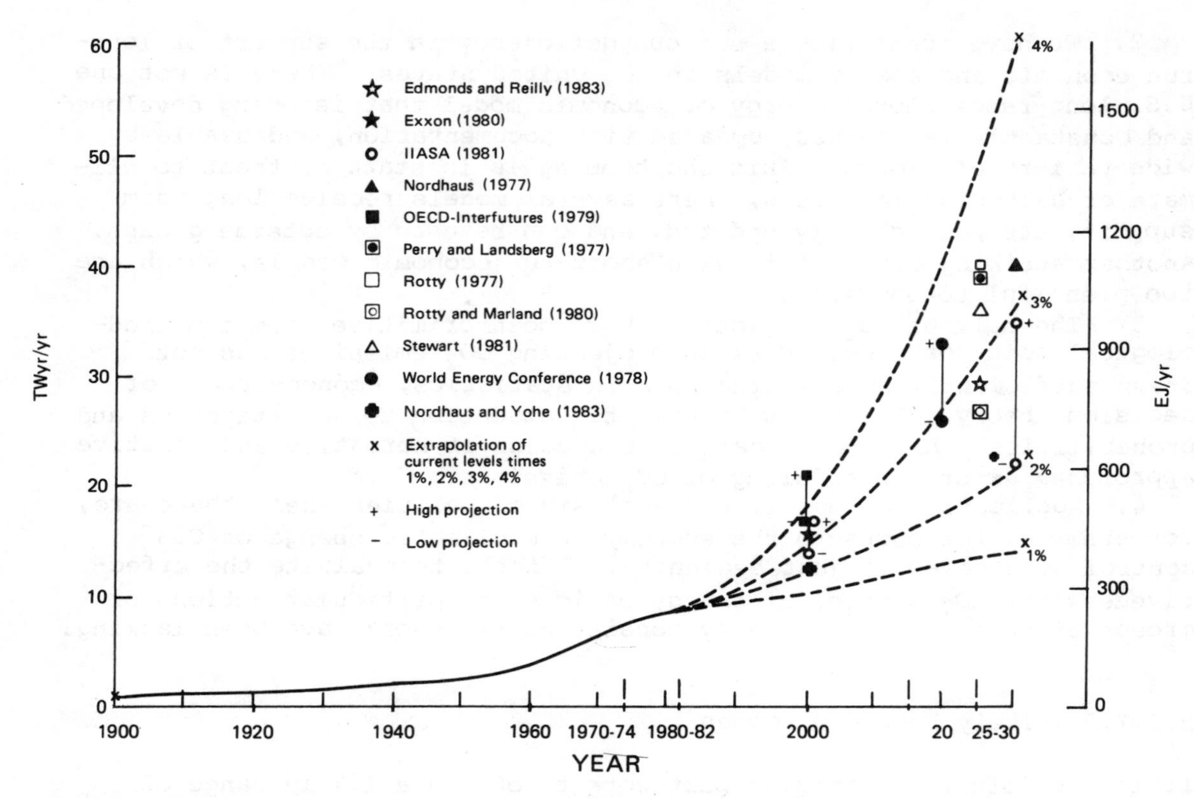

FIGURE 2.22 Past and projected energy consumption. Historical data are for primary energy consumption, including noncommercial (Nakicenovic, 1979; Schilling and Hildebrandt, 1977; Putnam, 1953). A similar figure with a different selection of estimates appears in Clark (1982).

output. Table 2.21 shows assumptions and estimates that major groups have offered. As described earlier, particular assumptions, for example, high population growth in Perry and Landsberg (1977) or high GDP growth in Nordhaus (1977), explain much of the resulting projection of world energy consumption.

While the sample of global, long-range energy projections is small and uneven and sporadically published, there is evidence of a reduction in projected rates of system growth over the past decade (Clark et al., in Clark, 1982; Lovins et al., 1982), perhaps spurred by the oil shock of 1973. A survey of energy demand projections for the United States shows the lower rates of growth expected in a major study from the late 1970s as opposed to studies in the mid-1970s (see Table 2.22). For example, the Rotty projections and even the perenially low projections of Lovins decline.

There are several reasons offered for lowering projected rates of growth to levels considerably below the past few decades: a lowering of projected rates of population increase, an assumption that economic development in developing countries will not imitate the pattern of the developed countries, and a reversal in the historical trend toward

TABLE 2.21 Long-Term World Energy Consumption: Estimates and Assumptions[a]

Reference	Scenario[b]	Estimate for Year	Population (10^9)	World Energy Consumption (TW yr/yr)	Average Annual Growth Rates (%/yr)					
					Interval	Pop.	GDP/ Cap.	GDP	Energy	CO_2 Emissions
Perry and Landsberg (1977)	Coal	2025	9.3	39	1980–2025	1.7	1.8	3.5	3.3	3.6
	Renewable	2025	9.3	39	1980–2025	1.7	1.8	3.5	3.3	2.0
World Energy Conference (1978)	High growth	2020	9	33	1972–2020	1.8	2.0	3.8	3.0	
	Low growth	2020	9	26	1972–2020	1.8	1.2	3.0	2.5	
OECD-Interfutures (1979)	High growth	2000	5.8	21	1975–2000	1.6	3.4	5	3.9	
	Low growth	2000	5.8	17	1975–2000	1.6	1.9	3.5	3.1	
Exxon Corporation (1980)		2000	--	16	1980–2000				2.4	
Rotty (IEA) (1977)	High growth	2025		40					3.5	3.3
Rotty and Marland (1980)	Low growth	2025	7.6	27	1975–2025	1.3	1.5	2.8	2.4	2
IIASA (1981)	High growth	2030	8	35	1980–2030	1.2	2.1	3.3	2.8	2.2
IIASA (1981)	Low growth	2030	8	22	1980–2030	1.2	1.1	2.3	1.9	1
Nordhaus (1977)		2030		40	1975–2030			4.4	3.4	3.3
CEQ (1981)	High growth	2100	10	75	1980–2100	1.1			2.3	
	Low growth	2100	8.5	30	1980–2100	1.08			1.3	
Stewart (1981)		2025		36	1975–2025				3.3	
Edmonds and Reilly (1982)		2025	7.4	29	1975–2025	1.2	1.7	2.9	2.5	1.9
Nordhaus and Yohe (this chapter, Section 2.1)		2025	7.8	24	1975–2025	1.4	1.9	3.3	2.0	1.6

[a]Sources: Perry (1982), which contains a similar table with a different selection of estimates; see references.
[b]"High" and "low" growth are not necessarily terms employed in studies cited.

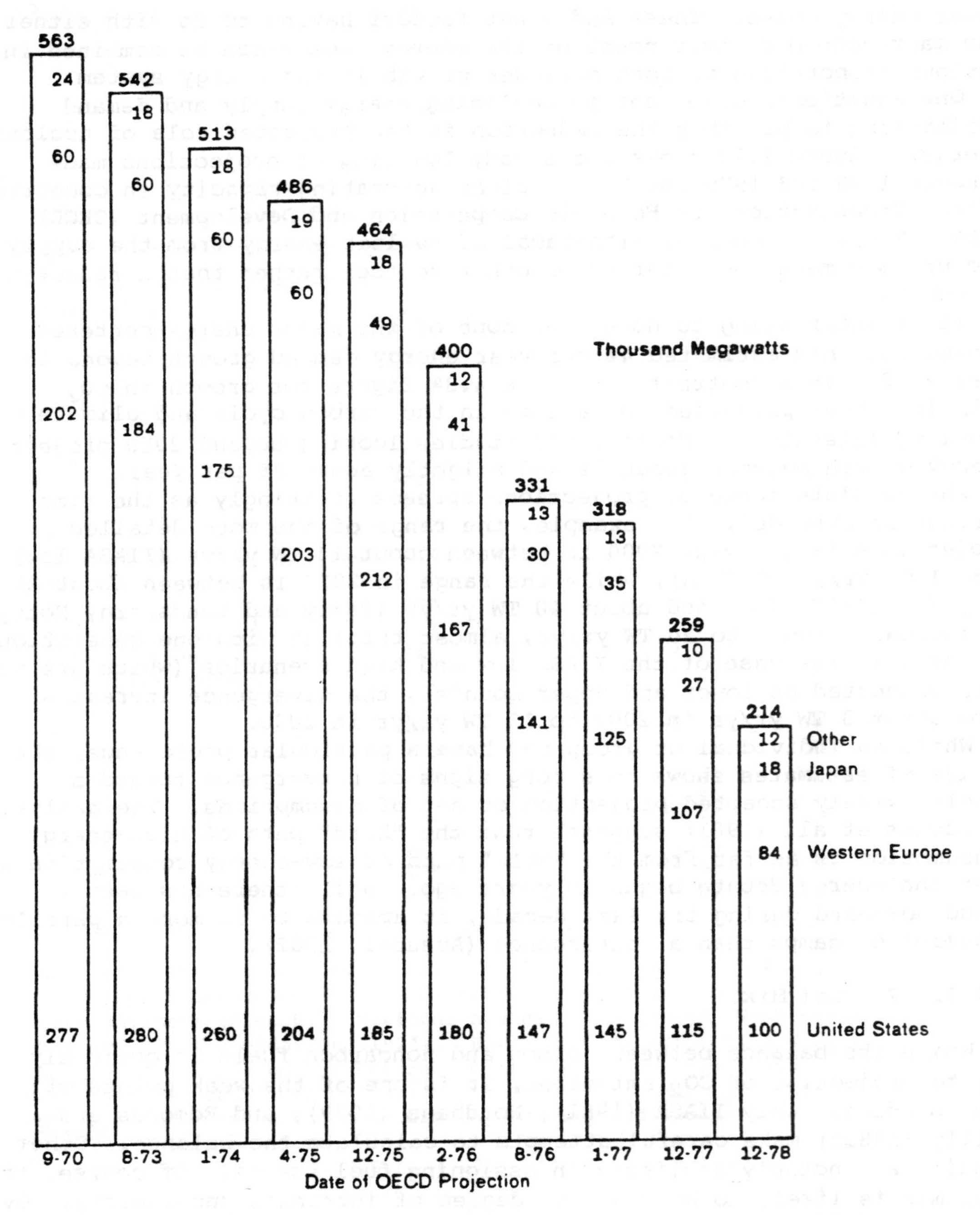

FIGURE 2.23 OECD: Past projections of year-end 1985 nuclear generating capacity. (Source: CIA, 1980.)

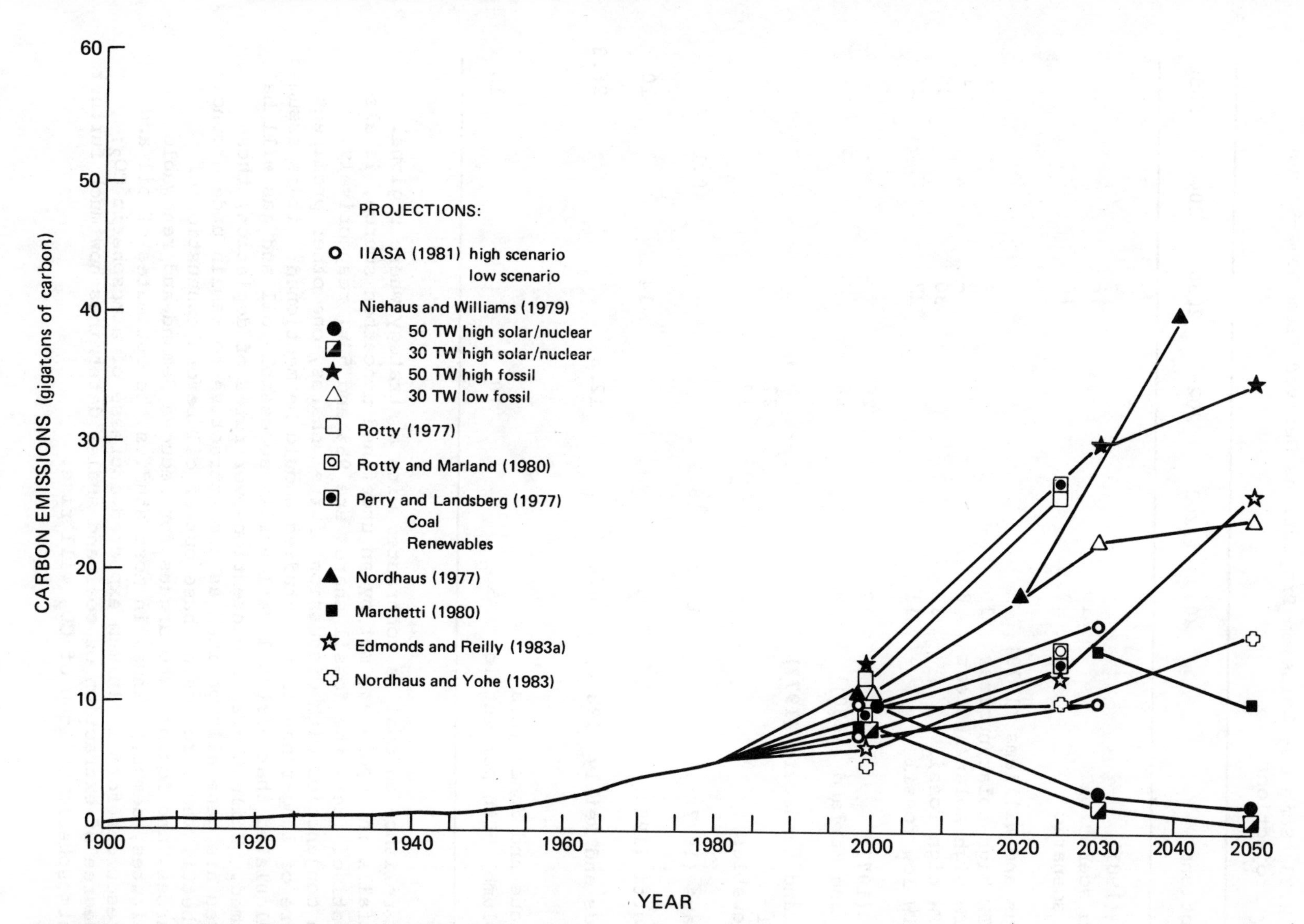

FIGURE 2.24 CO_2 emissions derived from long-range energy projections and historic production from fossil fuels. Sources: Historic data to 1949: Keeling (1973); 1950-1980: Rotty (1982). For data for projections see Table 2.23 and references.

TABLE 2.23 Some CO_2 Emission Projections Derived from Long-Range Energy Projections

Projections	2000	2020	2025	2030	2040	2050
IIASA (1981)						
High scenario	10			16		
Low scenario	7.5			10		
Niehaus and Williams (1979)						
50 TW high solar/nuclear	10			3		2
30 TW high solar/nuclear	8			2		1
50 TW high fossil	13			30		35
30 TW low fossil	11			22		24
Rotty (1977)	12		26			
Rotty and Marland (1980)	9		14			
Perry and Landsberg (1977)						
Coal			27			
Renewables			13			
Nordhaus (1977)	10.7	18.3			40.1	
Marchetti (1980)	8			14		10
Edmonds and Reilly (1983a)	6.9		12.3			26.3
Nordhaus and Yohe (this volume) 50th percentile	5		10			15

total reliance on coal is contrasted with a strategy where regional shortfalls in supply are met by an undefined noncarbon source, is also indicative. Here the fossil shares are 96% and 53%, respectively.

In conjunction with discussion of fuel shares, one other prominent feature of long-range energy studies should be mentioned: it is assumed or calculated that virtually all easily accessible oil and gas will be consumed. While there is contention over rates of depletion, these sources are generally posited as too attractive to remain underground. (Marchetti, who foresees a phase out of oil before exhaustion of resources, and Lovins, who argues for reduced demand and renewable substitutes, demur.) Thus, in most studies the estimates of oil and gas resources form a minimum expected increase of atmospheric CO_2; the degree of extraction of coal and shales determines how much further the atmospheric buildup of CO_2 will rise.

2.2.7.2.3 CO_2 Emissions

Combining estimates of energy and fuel mix leads to projections of CO_2 emissions. Figure 2.24 and Table 2.23 show CO_2 projections derived from long-range energy projections. Average annual rates of increase in CO_2 emissions to 2030 range from about 1% to 3.5%. Estimated annual emissions range from the past studies between about 7 and 13 Gt of C in the year 2000 and, with a couple of exceptions, between about 10 and 30 Gt of C in 2030.

2.2.8 Conclusion

Careful analysis of the economy, of energy, and of CO_2 emissions is vital. Such efforts are a key to better understanding of how the future atmosphere will evolve and what the likely costs and benefits of alternative CO_2 control or adaptation strategies will be. Considerable progress has been made over the last decade in developing more reliable and theoretically grounded models. As in other aspects of the issue of climate change, in economic and energy modeling a strong fundamental research program is a prerequisite for responding in an agile way to the concerns of today and images of the next century.

References

Allen, E. L., C. Davison, R. Dougher, J. A. Edmonds, and J. Reilly (1981). Global energy consumption and production in 2000. ORAU/IEA-81-2(M). Institute for Energy Analysis, Oak Ridge, Tenn.

Ausubel, J. H. (1982). Review of Least-Cost Energy: Solving the CO_2 Problem by Lovins et al. (1982). Climatic Change 4:313-317.

Baumgartner, A. (1979). Climatic variability and forestry. In Proceedings of the World Climate Conference. World Meteorological Organization, Geneva.

Bolin, B. (1979). Climate and global ecology. In Proceedings of the World Climate Conference. World Meteorological Organization, Geneva.

Central Intelligence Agency (CIA) (1980). OECD countries: prospects for nuclear power in the 1980s. NFAC/OER/M/IE. Washington, D.C., 11 July 1980.

Chan, Y.-H, J. Olson, and W. Emanuel (1980). Land use and energy scenarios affecting the global carbon cycle. Environ. Internat. 4:189-206.

Clark, W. C., ed. (1982). Carbon Dioxide Review: 1982. Oxford U. Press, New York.

Clark, W. C., K. H. Cook, G. Marland, A. M. Weinberg, R. M. Rotty, P. R. Bell, L. J. Allison, and C. L. Cooper (1982). The carbon dioxide question: a perspective for 1982. In W. C. Clark, ed., Carbon Dioxide Review: 1982. Oxford U. Press, New York.

Climatic Impact Assessment Program (CIAP) (1975). The Effects of Stratospheric Pollution by Aircraft. U.S. Department of Transportation, Washington, D.C.

Council on Environmental Quality (CEQ) (1981). Global energy futures and the carbon dioxide problem. CEQ, Washington, D.C.

Edmonds, J. A., and J. M. Reilly (1983a). Global energy and CO_2 to the year 2050. Institute for Energy Analysis, Oak Ridge, Tenn. Submitted to the Energy Journal.

Edmonds, J. A., and J. M. Reilly (1983b). Global energy production and use to the year 2050. Energy 8:419-432.

Edmonds, J. A., and J. M. Reilly (1983c). A long-term global energy-economic model of carbon dioxide release from fossil fuel use. Energy Econ. 5:74-88.

Edmonds, J. A., J. M. Reilly, and R. Dougher (1981). Determinants of Global Energy Demand to the Year 2050 (draft). Oak Ridge Associated Universities, Institute for Energy Analysis, Oak Ridge, Tenn.

Exxon Corporation (1980). World Energy Outlook. Exxon Corp., New York, December.

Flohn, H. (1980). Possible climatic consequences of a man-made global warming. RR-80-30. International Institute for Applied Systems Analysis, Laxenburg, Austria.

Grunfeld, Y., and Z. Griliches (1960). Is aggregation necessarily bad? Rev. Econ. Stat., pp. 1-13, February.

Hameed, S., R. D. Cess, and J. S. Hogan (1980). Response of the global climate to changes in atmospheric chemical composition due to fossil fuel burning. J. Geophys. Res. 85:7537.

Hansen, J., D. Johnson, A. Lacis, S. Lebedeff, P. Lee, D. Rind, and G. Russell (1981). Climatic impact of increasing atmospheric CO_2. Science 213: 957-966.

Interfutures Project (1979). Facing the Future. Organization for Economic Cooperation and Development (OECD), Paris.

International Institute for Applied Systems Analysis (IIASA) (1981). Energy in a Finite World: A Global Systems Analysis. Ballinger, Cambridge, Mass.

JASON (1979). The long term impact of atmospheric carbon dioxide on climate. Technical report JSR-78-07. SRI International, Arlington, Va.

Kahn, H., W. Brown, and L. Martel (1976). The Next 200 Years. William Morrow, New York.

Keeling, C. D. (1973). Industrial production of carbon dioxide from fossil fuels and limestone. Tellus 25:174.

Keeling, C. D., and R. B. Bacastow (1977). Impact of industrial gases on climate. In NRC Geophysics Study Committee, Energy and Climate. National Academy of Sciences, Washington, D.C.

Lacis, A., J. Hansen, P. Lee, T. Mitchell, and S. Lebedeff (1981). Greenhouse effect of trace gases, 1970-1980. Geophys. Res. Lett. 8:1035-1038.

Legasov, V. A., and I. I. Kuz'min (1981). The problem of energy production. Priroda(2).

Logan, J. A., M. J. Prather, S. C. Wofsy, and M. B. McElroy (1978). Atmospheric chemistry: response to human influence. Trans. R. Soc. 290:187.

Lovins, A. B. (1980). Economically efficient energy futures. In Interactions of Energy and Climate, W. Bach, J. Pankrath, and J. Williams, eds. Reidel, Dordrecht, pp. 1-31.

Lovins, A. B., L. H. Lovins, F. Krause, and W. Bach (1982). Least Cost Energy: Solving the CO_2 Problem. Brick House Publishing, Cambridge, Mass.

Marchetti, C. (1980). On energy systems in historical perspective. International Institute for Applied Systems Analysis, Laxenburg, Austria.

Marchetti, C., and N. Nakicenovic (1979). The dynamics of energy systems and the logistic substitution model. RR-79-13. International Institute for Applied Systems Analysis, Laxenburg, Austria.

Marland, G., and R. Rotty (1979). Atmospheric carbon dioxide: implications for world coal use. In Future Coal Supply for the World Energy Balance, M. Grenon, ed. Third IIASA Conference on Energy Resources, Nov. 28-Dec. 2, 1977. Pergamon, Oxford, pp. 700-713.

Nakicenovic, N. (1979). Software Package for the Logistic Substitution Model. Report RR-79-12, International Institute for Applied Systems Analysis, Laxenburg, Austria.

Niehaus, F. (1979). Carbon dioxide as a constraint for global energy scenarios. In Man's Impact on Climate, W. Bach, J. Pankrath, and W. Kellogg, eds. Elsevier, Amsterdam, pp. 285-297.

Niehaus, F., and J. Williams (1979). Studies of different energy strategies in terms of their effects on the atmospheric CO_2 concentration. J. Geophys. Res. 84:3123-3129.

Nordhaus, W. D. (1977). Strategies for the control of carbon dioxide. Cowles Foundation Discussion Paper No. 443. Yale U., New Haven, Conn.

Nordhaus, W. D. (1979). The Efficient Use of Energy Resources. Yale U. Press, New Haven, Conn.

Nordhaus, W. D. (1980). Thinking about carbon dioxide: theoretical and empirical aspects of optimal control strategies. Cowles Foundation Discussion Paper No. 565. Yale U., New Haven, Conn.

Nordhaus, W. D. (1982). How fast should we graze the global commons? Am. Econ. Rev. 72(2).

NRC Geophysics Study Committee (1977). Energy and Climate. National Academy of Sciences, Washington, D.C.

Perry, A. M. (1982). CO_2 production scenarios: an assessment of alternative futures. In The Carbon Dioxide Review: 1982, W. C. Clark, ed. Oxford U. Press, New York.

Perry, A. M., K. J. Araj, W. Fulkerson, D. J. Rose, M. M. Miller, and R. M. Rotty (1982). Energy supply and demand implications of CO_2. Energy 7:991-1004.

Perry, H., and H. H. Landsberg (1977). Projected world energy consumption. In NRC Geophysics Study Committee, Energy and Climate, National Academy of Sciences, Washington, D.C.

Putnam, P. (1953). Energy in the Future. Van Nostrand, New York.

Ramanathan, V. (1980). Climatic effects of anthropogenic trace gases. In Interactions of Energy and Climate, W. Bach, J. Pankrath, and J. Williams, eds. Reidel, Boston, Mass., pp. 269-280.

Reilly, J. M., R. Dougher, and J. A. Edmonds (1981). Determinants of Global Energy Supply to the Year 2050 (draft). Oak Ridge Associated Universities, Institute for Energy Analysis, Oak Ridge, Tenn.

Report of US/USSR Workshop on The Climatic Effects of Increased Atmospheric Carbon Dioxide, June 15-20, 1981. Published Leningrad, USSR, 1982. Available Division of Atmospheric Sciences, National Science Foundation, Washington, D.C.

Revelle, R., and W. Munk (1977). The carbon dioxide cycle and the biosphere. In NRC Geophysics Study Committee, Energy and Climate, National Academy of Sciences, Washington, D.C.

Rotty, R. (1977). Present and future production of CO_2 from fossil fuels. ORAU/IEA(O)-77-15. Institute for Energy Analysis, Oak Ridge, Tenn.

Rotty, R. (1978). The atmospheric CO_2 consequences of heavy dependence on coal. In Carbon Dioxide, Climate and Society, J. Williams, ed. Pergamon, Oxford, pp. 263-273.

Rotty, R. (1979a). Energy demand and global climate change. In Man's Impact on Climate, W. Bach, J. Pankrath, and W. Kellogg, eds. Elsevier, Amsterdam, pp. 269-283.

Rotty, R. (1979b). Growth in global energy demand and contribution of alternative supply systems. Energy 4:881-890.

Rotty, R. (1982). Distribution of and Changes in Industrial Carbon Dioxide Production. Institute for Energy Analysis, Oak Ridge, Tenn.

Rotty, R., and G. Marland (1980). Constraints on fossil fuel use. In Interactions of Energy and Climate, W. Bach, J. Pankrath, and J. Williams, eds. Reidel, Dordrecht, pp. 191-212.

Rowland, F. S., and M. J. Molina (1975). Chlorofluoromethanes in the environment. Rev. Geophys. Space Phys. 13(1).

Schilling, H. D., and R. Hildebrandt (1977). Primärenergie-Elektrische Energie. Glückauf, Essen, FRG.

Siegenthaler, V., and H. Oeschger (1978). Predicting future atmospheric carbon dioxide levels. Science 199:388.

Sioli, H. (1973). Recent human activities in the Brazilian Amazon region. In Tropical Forest Ecosystems in Africa and South America, B. J. Meggers et al., eds. Smithsonian Institution, Washington, D.C.

Stahl, I., and J. Ausubel (1981). Estimating the future input of fossil fuel CO_2 into the atmosphere by simulation gaming. In Beyond the Energy Crisis--Opportunity and Challenge, R. A. Fazzolare and C. B. Smith, eds. Pergamon, Oxford.

Stewart, H. (1981). Transitional Energy Policy 1980-2030. Pergamon, Oxford.

Voss, A. (1977). Ansätze for Gesamtanalyse des Systems Mensch-Energie-Umwelt. Birkhauser, Basel.

Wang, W. C., Y. L. Yung, A. A. Lacis, T. Mo, and J. E. Hansen (1976). Greenhouse effects due to man-made perturbations of trace gases. Science 194:685-690.

World Climate Programme (1981). On the assessment of the role of CO_2 on climate variations and their impact. Report of a WMO/UNEP/ICSU meeting of experts in Villach, Austria, November 1980. World Meteorological Organization, Geneva.

World Coal Study (1980). Coal--Bridge to the Future. Ballinger, Cambridge, Mass.

World Energy Conference (1978). World Energy Resources 1985-2020, An Appraisal of World Coal Resources and Their Future Availability; World Energy Demand (Report to the Conservation Commission). IPC Science and Technology Press, Guildford, U.K.

World Energy Conference (1980). Survey of Energy Resources 1980. IPC Science and Technology Press, Guildford, U.K.

3 Past and Future Atmospheric Concentrations of Carbon Dioxide

3.1 INTRODUCTION

Peter G. Brewer

This chapter on how the carbon content of the atmosphere and other reservoirs may change over time has been written in several sections by individual authors. In reviewing the material it is clear that all controversy in this area has not been resolved. There are three principal goals that we seek in our evaluation of the carbon cycle among atmosphere, oceans, and biota. First, the climate change that we anticipate is based on the atmospheric CO_2 rise. We only have good measurements of atmospheric CO_2 from the time of the International Geophysical Year in 1958 to the present. We need to know the preindustrial value, the time course of its change in the decades prior to 1958, and the factors causing this change. Second, we need to know as accurately as possible the fluxes of CO_2 among atmosphere, ocean, and biota today so as to be able to evaluate contemporary measurements; to separate natural, anthropogenic, and climatically modified effects; and to identify the role of other greenhouse gases. Third, if our projections of the future are to have credibility, we must understand the sensitivity of our carbon reservoirs to change and the linkages and feedbacks that exist between them.

We do have reasonable knowledge of the consumption of fossil fuels in the early part of this century. If we assume that the airborne CO_2 fraction has remained constant over this time, then simple backextrapolation yields an atmospheric CO_2 level of about 290 ppm at the turn of the century. Direct measurements of CO_2 in air at that time yield equivocal values, having a mean of about 290 ppm and a low value of about 270 ppm. Recent measurements of the CO_2 content of air trapped in glacial ice indicate a value of about 265 ppm for the middle of the last century. If a preindustrial value of 265 ppm is accepted, then two things are apparent. First, the discrepancy between the extrapolated 290 ppm and the observed 265 ppm must be accounted for; the difference of 25 ppm added to the atmosphere most likely would come from the terrestrial biosphere. There is little evidence for an oceanic source, although modest degassing of the ocean would occur on warming.

Second, the discrepancy is not an insignificant fraction of the atmospheric CO_2 rise but accounts for some 30% of the CO_2 signal that we see today. The warming due to CO_2 is complex but may be approximated by the logarithmic relationship (see Chapter 5)

$$T = \frac{3.0}{\ln 2} \ln\left(\frac{[CO_2]}{[CO_2]_0}\right),$$

where $[CO_2]$ is the present CO_2 value, $[CO_2]_0$ is the preindustrial value, and 3.0° represents the mean increase in temperature estimated for a doubling of atmospheric CO_2. The overall warming today then could be as high as 1.1°C, and the initial CO_2 difference between the low and high estimates is 0.4°C. In the long run the difference will be insignificant; for the present it is uncertain whether we have observed a CO_2-induced warming, and it is of great interest to know the theoretical size and shape of our signal.

In evaluating the fluxes of CO_2 among our atmospheric, oceanic and biotic reservoirs today we note that we have direct measurements only of the atmospheric reservoir. Measurements of carbon stocks in the global biosphere are complex and are inferred from local measurements, patterns of land use, soil changes, and deforestation. In the ocean we feel keenly the lack of a high-quality time series of measurements. Measurements of the contemporary ocean reveal the large natural CO_2 cycle and only hint at the anthropogenic signal. Our principal information comes from the observed fractionation of ^{14}C between air and sea from which we calculate oceanic CO_2 uptake. The result is an averaged signal, and resolution is poor on time scales less than a decade. The principal causes of the annual fluctuations in atmospheric CO_2 are the seasonal growth and decay of the terrestrial biota. It is as yet difficult to ascribe annual atmospheric fluctuations to oceanic changes. One exception to this has been the correlation by Bacastow and Keeling of atmospheric CO_2 trends correlated with the Southern Oscillation Index. The amplitude of the annual CO_2 fluctuations revealed at the Mauna Loa observatory shows a tendency to increase with time, suggesting increasing terrestrial photosynthetic and respiratory activity.

Ocean-atmosphere carbon models calculate the partioning of fossil fuel CO_2 released over the last two decades to be about 40% oceanic uptake and 60% atmospheric fraction, with minor net transfers from the terrestrial biota. If we express the global carbon balance as

$A = F - S \pm B$ (Woodwell, this volume, Section 3.3),

where A is the increase in the carbon content of the atmosphere over any period, F is the release of carbon to the atmosphere from combustion of fossil fuels in the same period, S is the net transfer to the oceans in the same period, and B is the absorption or release of carbon by the biota in the same period, then the terms S and B are the least well determined. The "airborne fraction" is not directly determined from either oceanic or biosphere experiments or models. Estimates of carbon changes in the terrestrial biota vary widely from small net increases to large net decreases. In this chapter the current net release of carbon from the biosphere is estimated as about 2 gigatons of carbon (Gt of C) per year. This is at the upper limit that can be accommodated by atmosphere-ocean models. Two things could confound our current thinking: If the rate of release of CO_2 from the biota was growing at an identical rate to the CO_2 release from fossil fuels, we would find it hard to detect; and if the biotic release was matched by some unknown CO_2 sink, the releases could indeed be large. So far we cannot unequivocally support either of these scenarios.

In summary, the recent estimates of an atmospheric CO_2 concentration of about 265 ppm around 1850 lead to a predicted warming greater than that yet observed today if we use the upper range of climate model results, and point to a net CO_2 source from the terrestrial biosphere contributing about 25 ppm to the atmospheric levels. This flux from the biota most likely occurred in the late nineteenth century and early decades of this century. The net release from the biosphere today could be about 2 Gt of C per year, although lesser or negative fluxes would not be inconsistent with oceanic and atmospheric models. Finally, we must keep in mind, as Revelle's analysis (Section 3.5) of methane hydrates in continental slope sediments suggests, that climate change may bring about surprising changes in fluxes of carbon.

3.2 CARBON DIOXIDE AND THE OCEANS

Peter G. Brewer

> The effect of solution of the gas by the sea water was next considered, because the sea acts as a giant regulator of carbon dioxide and holds some sixty times as much as the atmosphere. The rate at which the sea water could correct an excess of atmospheric carbon dioxide depends mainly upon the fresh volume of water exposed to the air each year, because equilibrium with the atmospheric gases is only established to a depth of about 200 m during such a period. The vertical circulation of the oceans is not well understood, but several factors point to an equilibrium time, in which the whole sea volume is exposed to the atmosphere of between two and five thousand years.
>
> --Callendar (1938)

3.2.1 Introduction

The quotation above is still, after more than 40 years of progress, as succinct a statement of the problem as one could desire. The rising atmospheric CO_2 level has been carefully measured since 1958. Calculations made on the amount of fossil fuel CO_2 released during this time, based on good records of oil, coal, and gas combustion, show that the observed increase in atmospheric CO_2 is a little more than half the amount of fossil fuel CO_2 input.

The CO_2 not present in the atmosphere must have been transferred to some other reservoir, and all investigators who have examined the problem over the last four decades have concluded that the ocean is, and will remain, the primary sink for fossil fuel CO_2. The ocean holds about 53 times the total atmospheric carbon dioxide content, or about 3.7 gigatons of carbon (Gt of C) as CO_2. The depth of the ocean mixed layer that establishes annual contact with the atmosphere is about 75 m, and the mean circulation time for the deep oceans is about 500 years (Stuiver et al., 1983). The ocean acts as a "giant regulator" not only of CO_2 but also of climate and thus occupies a central role in the debate over the effects of increasing atmospheric CO_2 levels on our society. The capacity of the ocean for CO_2 uptake is a function of its chemistry; the rate at which this capacity can be brought into play is a function of ocean physics. In addition to these direct and present contributions, the deep ocean carbonate sediments provide, on a longer time scale, a vast buffer against chemical change. The natural vertical gradient of CO_2 with depth in the oceans is driven by the biological flux of particulate matter.

There have been many recent papers and reviews on these topics (e.g., Broecker et al., 1979; Takahashi and Azevedo, 1982), and models designed to reproduce their effects (e.g., Oeschger et al., 1975; Killough and Emanuel, 1981). The Scientific Committee on Problems of the Environment (SCOPE) reports by Bolin et al. (1979) and Bolin (1981) provide excellent assessments of the carbon cycle. The scene is one of constant research and evaluation; some basic facts, however, hold constant, and some uncertainties are widely recognized. These form the basis of this review. In attempting to model future atmospheric CO_2 levels, the largest uncertainty of course surrounds the economic and energy resource decisions facing mankind. The CO_2 content of the future atmosphere will largely reflect how much CO_2 we choose to put in. The key word is choose, for however hard those decisions may be, they represent choices distinct from the natural laws that will inevitably be obeyed as the CO_2 level rises.

3.2.2 The Cycle of Carbon Dioxide within the Oceans

Measurements of the carbon dioxide system in seawater plainly reveal the natural cycle. If we wish to detect changes in the ocean resulting from anthropogenic CO_2, then a prerequisite is that the natural cycle be well understood. This cycle is intimately linked to that of oxygen, and the nutrient elements nitrogen and phosphorus, and to ocean circulation.

3.2.3 The Deep Circulation

The oceans are stably stratified, capped by a warm, less dense surface layer and increasing in density with depth. The deep ocean waters have a salinity of about 34.9 parts per thousand and a temperature of about 2°C. Most of the deep waters of the world's oceans are formed in wintertime in the Norwegian and Greenland Seas and in the Weddell Sea.

Here winter cooling increases the density of surface waters until the stratification of the water column breaks down and, by a poorly understood process, the deep source regions are renewed. Once formed, the bottom waters of the basins exit via various sills and proceed on their grand tour. From the Norwegian and Greenland Seas the flow is to the south; the residence time of deep water in the Atlantic Ocean has recently been estimated as 275 years (Stuiver et al., 1983). In the southern ocean these deep waters become entrained in the great clockwise circulation of the Antarctic Circumpolar Ocean. Here they branch, after a residence time of some 40 years, into either the Pacific or the Indian Oceans. The residence time of deep water in the Pacific Ocean is about 600 years, and in the Indian Ocean about 335 years. Within these ocean basins the deep waters are gradually entrained into the shallower flows of the intermediate waters and are eventually returned to the surface. The flows are not simply advective, and large-scale turbulent processes along and across density surfaces predominate. Reid and Lynn (1971) have elegantly shown that the salinity maximum of the North Atlantic deep waters can be traced on their trajectory around the globe. Stuiver et al. (1983) have followed the decay of radiocarbon within these waters and calculated their ages.

3.2.4 Biological Activity

Photosynthetic activity by phytoplankton in ocean surface waters fixes CO_2 into organic tissues. The global oceanic annual primary productivity is uncertain but is approximately one half that on land (Peterson, 1980). The amount, then, is large. Oxygen is produced in this process, and ocean surface waters typically show slight (1-2%) supersaturation with respect to equilibrium with atmospheric O_2.

In contrast to primary productivity on land, where large standing stocks of carbon are formed in woody tissues, oceanic production by phytoplankton is very rapidly consumed by grazing organisms. Some 90% of the organic matter formed is consumed within the euphotic zone. The remainder falls through the ocean-water column, as excreted fecal pellets and discrete cells, and is subject to oxidative decomposition by microbes. Production at the ocean surface is limited largely by the availability of the nutrient elements nitrogen and phosphorus. These combine with carbon in proportions generally represented by the reaction

$$106\ CO_{2\ (g)} + 16\ NO_3^- + H_2PO_4^- + 17\ H^+ + 122\ H_2O \longrightarrow C_{106}H_{263}O_{110}N_{16}P + 138\ O_{2\ (g)}. \tag{1}$$

where the subscript g denotes the gaseous state. Although local variations in these ratios are found, the mean oceanic signal is remarkably constant. The removal of CO_2 from surface waters raises the pH; the removal of nitrate and phosphate raises the pH and the alkalinity (Brewer and Goldman, 1976).

The pattern of oceanic primary productivity is such that large areas in the center of the great oceanic gyres are nutrient-impoverished oligotrophic regions, where little production occurs. In upwelling regions, such as those at the eastern sides of oceanic basins where nutrients are brought to the surface, intense biological activity occurs.

3.2.5 Deep Decomposition of Organic Matter

Below the euphotic zone reaction (1) proceeds in the reverse direction. Oxygen is consumed, and carbon dioxide and the nutrient elements are released to the dissolved state. The rate of this reverse reaction is quite variable. It is generally found to decrease quasi-exponentially with depth, with oxygen consumption rates ranging from about 0.5 ml of O_2/L/yr at shallow depths to less than 0.01 ml of O_2/L/yr in the abyss (Jenkins, 1980). Carbon dioxide is released in proportion to this. The result is that the ocean water column is characterized everywhere by an oxygen minimum below the euphotic zone, where decomposition is rapid and the water column is poorly ventilated.

Combining these processes with the scheme of the deep circulation given earlier, we see a progressive change in the chemistry of the deep water during its 500-year deep ocean tour. In traveling from the North Atlantic to the Antarctic to the North Pacific, the water becomes systematically depleted in oxygen and enriched in CO_2 and the nutrients. The GEOSECS series of atlases beautifully reveal these trends as the rain of material from above inexorably shifts the oceanic CO_2 chemistry.

3.2.6 Calcium Carbonate

The increase in deep ocean CO_2 due to oxidation of organic matter lowers the pH of seawater, increasing its corrosiveness to calcium carbonate. Surface seawater is supersaturated with respect to calcium carbonate, which is secreted by many marine organisms to form their shells. These rain down through the ocean to form calcareous sediments. Calcium carbonate solubility increases with increasing pressure and decreasing temperature, and at some point a horizon occurs below which calcium carbonate dissolves. In the North Atlantic Ocean this horizon occurs at great depth, approximately 5000 m. Progressing along the path of the deep circulation, oxygen and pH are lowered with increasing CO_2 levels, and progressively more calcium carbonate is dissolved. The dissolution horizon shoals markedly along this trajectory. The calcium carbonate thus dissolved from the sediments raises the CO_2 concentration of the deep water further and increases the alkalinity. Of the increase in CO_2 experienced by deep ocean waters, some 70% is attributable to decomposition of organic matter and 30% to the dissolution of calcium carbonate.

Baes (1982) has carefully reviewed this topic. Takahashi et al. (1981) have compiled the result of the GEOSECS expedition to illustrate the progressive change in these properties. In Figure 3.1 is shown the depth distribution of CO_2 in seawater for the various ocean basins from the North Atlantic to the North Pacific. In Figure 3.2, the equivalent change in alkalinity is shown.

With this background in mind we now examine some details of the chemistry of seawater.

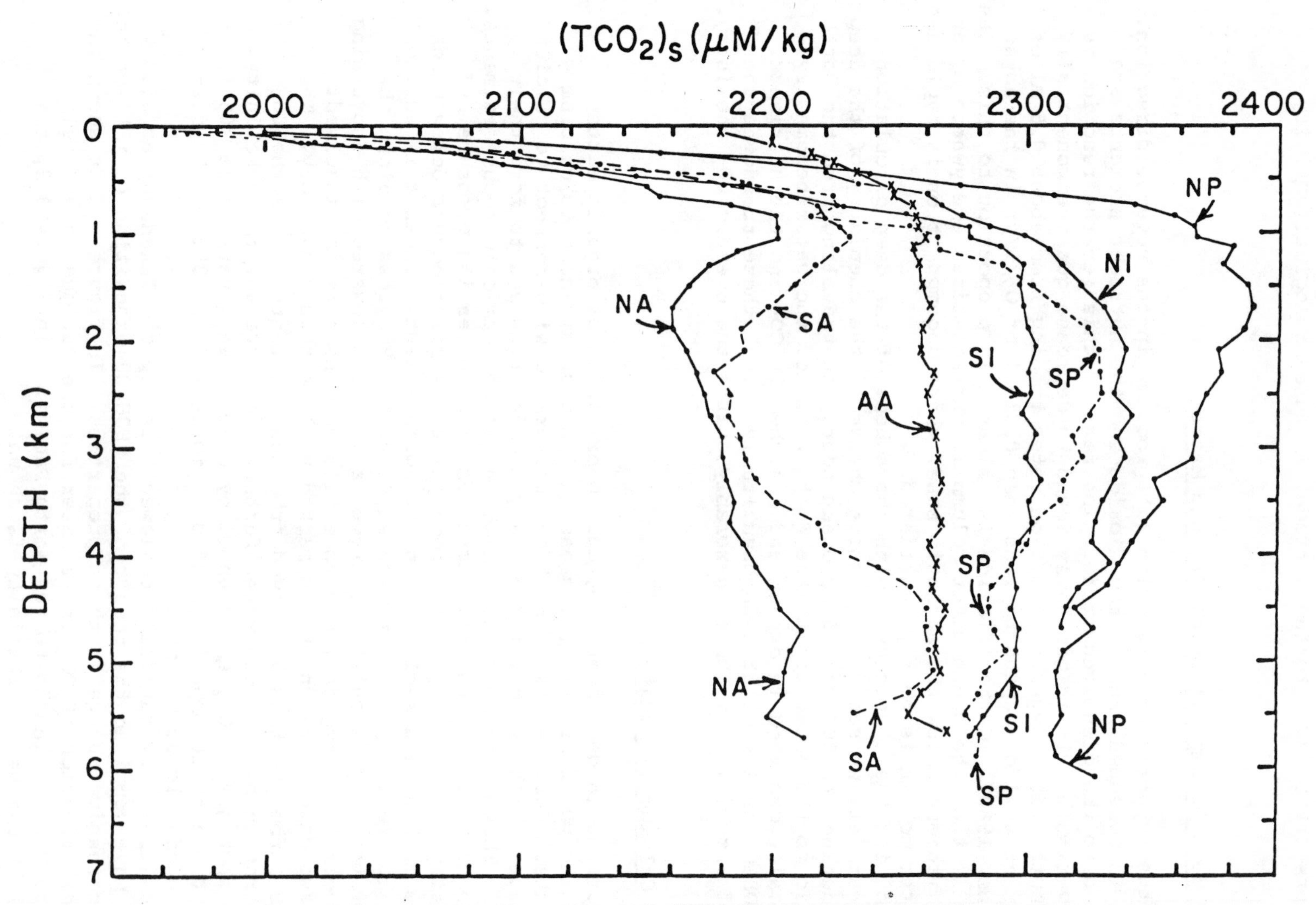

FIGURE 3.1 The distribution with depth of the total carbon dioxide concentration in seawater in various ocean basins. The concentrations are normalized to the mean world ocean salinity of 34.78‰. NA, North Atlantic; SA, South Atlantic; AA, Antarctic region south of 45° S; SI, South Indian; NI, North Indian; SP, South Pacific; and NP, North Pacific. (From Takahashi et al., 1980b.)

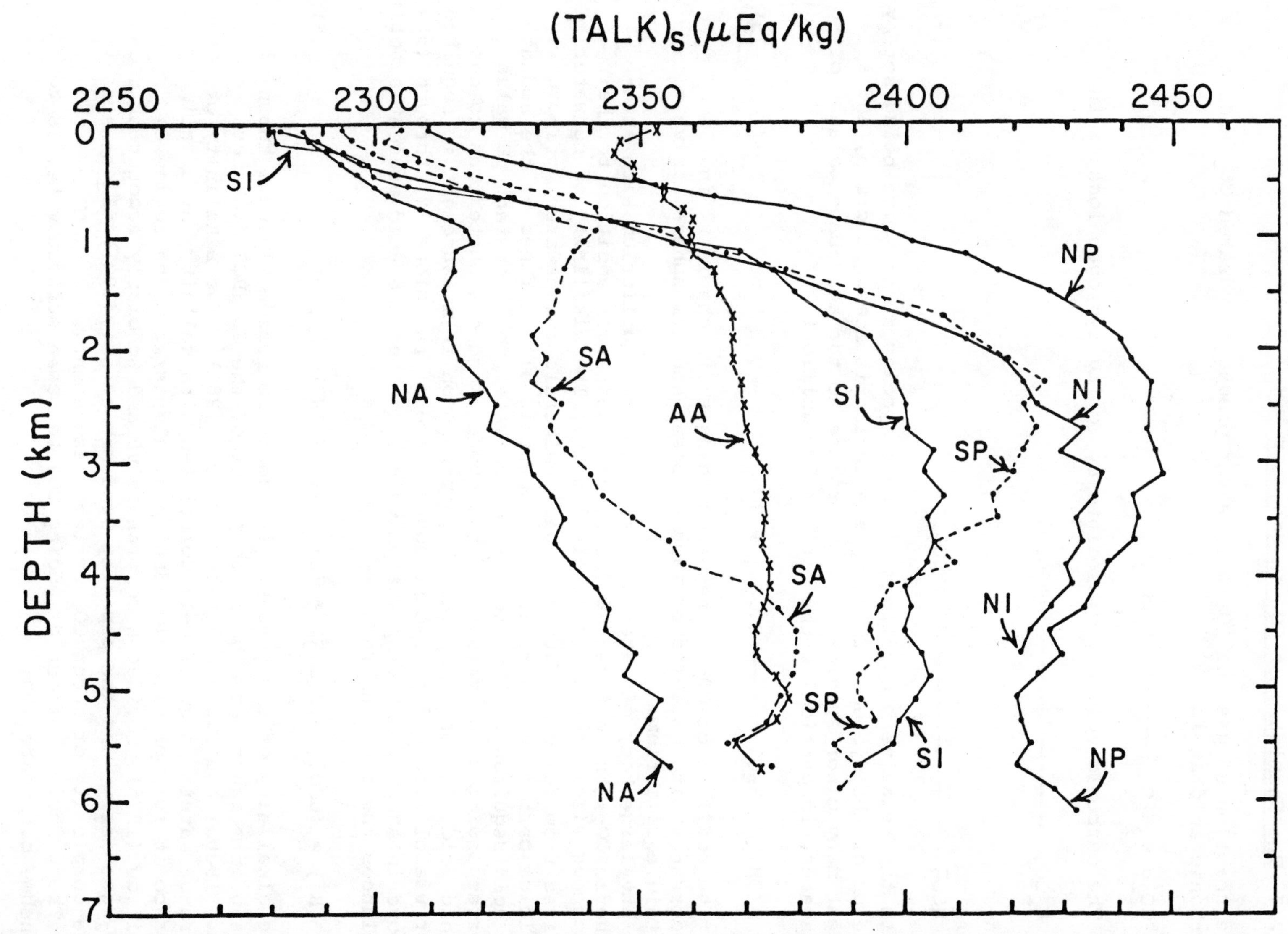

FIGURE 3.2 The distribution with depth of the total alkalinity of seawater in various ocean basins. The notation and normalization are as for Figure 3.1. (From Takahashi et al., 1981.)

3.2.7 The Chemistry of CO_2 in Seawater

The reaction of gaseous CO_2 with water produces hydrated CO_2 and carbonic acid as in

$$CO_2 g + H_2O \longrightarrow H_2CO_3. \qquad (2)$$

The carbonic acid may dissociate by losing hydrogen ions as in

$$H_2CO_3 \longrightarrow H^+ + HCO_3^- \qquad (3)$$

and

$$HCO_3^- \longrightarrow H^+ + CO_3^=. \qquad (4)$$

with the relative proportions of these species at any time being set by the pH of the system. This representation is somewhat crude, and a great many minor species also contribute to the acid-base balance of seawater, in particular the boric acid equilibrium

$$B(OH)_3 + H_2O \longrightarrow B(OH)_4^- + H^+. \qquad (5)$$

The addition of CO_2 to seawater changes its chemistry in accordance with established laws as ocean and atmosphere strive to attain equilibrium.

Any large body of water will tend toward equilibrium with atmospheric CO_2; the unique feature of the oceans, in addition to their enormous size, lies in their alkalinity. The alkalinity of seawater arises from the dissolution of basic minerals in seawater, principally calcium carbonate. Alkalinity is operationally defined as the amount of acid required to titrate 1 kg of seawater to a constant pH value corresponding to conversion of bicarbonate and carbonate ions to carbonic acid. In practice very high precision in measurement is required for useful work and a precise equivalence of reactions (2) through (5) above is sought, not simply a pH value, so that the acid present exactly balances the bases as in

$$[H^+] = [HCO_3^-] + 2[CO_3^=] + [B(OH)_4^-] + [OH^-]. \qquad (6)$$

The alkalinity of the present-day oceans is reasonably well known. It has been measured as a basic component of the GEOSECS (Takahashi et al., 1980a) and TTO expeditions (PCODF, 1981). The alkalinity of ocean surface waters is quite well correlated with salinity; at a salinity of 35°/oo it is approximately 2300 equivalents/kg. The review by Skirrow (1975) provides a comprehensive and scholarly account of ocean CO_2 chemistry, and the paper by Bradshaw et al. (1981) illuminates the complexity of ocean CO_2 system measurement.

The principal effect of adding CO_2 to ocean surface water is to consume carbonate ion:

$$CO_2 + CO_3^{2-} + H_2O \longrightarrow 2HCO_3^-. \qquad (7)$$

The reaction does not proceed entirely to the right, and considerable resistance to change occurs. This resistance is accurately reflected in the thermodynamics of the CO_2 system. The buffer factor, or Revelle factor as it is widely known, appropriate for this reaction may be presented by

$$\frac{(dpCO_2/pCO_2)\quad TA,T,S}{(dTCO_2/TCO_2)\quad TA,T,S} = R, \tag{8}$$

where TCO_2 is the total concentration of carbon dioxide in all its forms, pCO_2 is the partial pressure of carbon dioxide gas, TA is the total alkalinity, T is the temperature, and S is the salinity. Sundquist et al. (1979) have pointed out that this property is quite well known. It varies with temperature and has a numerical value of about 10. In essence a 10% change in pCO_2 produces only a 1% change in CO_2.

Takahashi et al. (1980b) have described the change in this factor that will inexorably occur as ocean CO_2 levels rise. Figure 3.3 shows the change as function of CO_2 for alkalinities of 2.2 and 2.4 milliequivalents/kg, taken from their paper. As the CO_2 content of the atmosphere and therefore the surface ocean increases, we move to the right on this figure encountering sharply rising values of R. The resistance to change increases, the ocean absorbs proportionately less CO_2, and the airborne fraction rises. This is a complex system, sensitive to the alkalinity/total CO_2 ratio, and hence pH. The sharp maximum that occurs in Figure 3.1 is readily understandable in terms of carbonate chemistry equilibria (Takahashi et al., 1980b) and occurs when the concentration of $CO_3^=$ becomes equal to that of H_2CO_3; thereafter a decrease in R will take place with R asymptotically approaching 1.

How accurately is the curve in Figure 3.3 defined, what processes are likely to alter it, and how will we know if the ocean does indeed proceed along the thermodynamic course that we have charted? The curve is a theoretical construct, based on sound principles of solution chemistry. We would like to have a series of field observations of the varying concentration of CO_2 in the ocean with time so as to follow these changes; however, there are no adequate measurements for this purpose. In practice, the buffer factor is not a measured variable but a calculated property.

The accuracy of these calculations depends on our knowledge of the solubility of CO_2 gas in seawater and on the thermodynamic constants describing the dissociation of carbonic and boric acids in seawater. Although these have long been investigated, it is only relatively recently that results of sufficient accuracy have been obtained.

The solubility of CO_2 gas has been determined by Murray and Riley (1971) and by Weiss (1974). The results of these experiments are in excellent agreement and have been fitted by Weiss (1974) to the equation

$$\ln \alpha' = -60.2409 + 93.4517\ (100/T) + 23.3585\ \ln\ (T/100) + [0.023517 - 0.023656\ (T/100) + 0.0047036\ (T/100)^2]S, \tag{9}$$

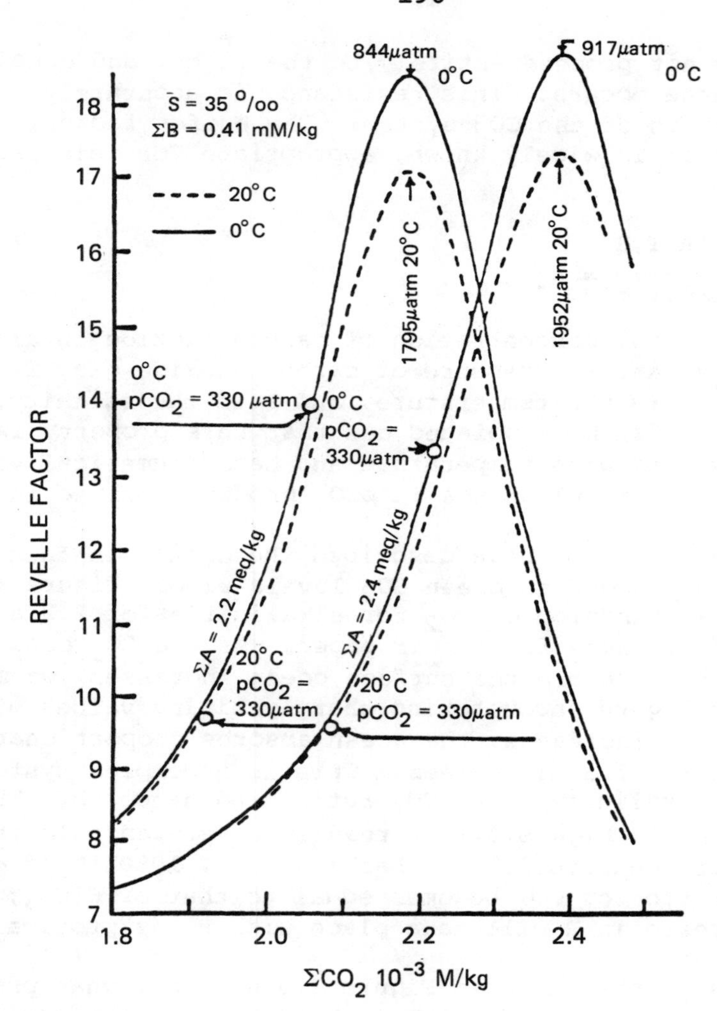

FIGURE 3.3 Variation of the buffer factor, or Revelle factor, (R) of seawater with changing total CO_2. The calculation is for seawater of 35% salinity and a total boron content of 0.41 mM/kg. Curves for waters of two different alkalinities are shown. Increasing CO_2 levels raise the buffer factor and diminish the oceans tendency to absorb CO_2. (From Takahashi et al., 1980b.)

where α' is the solubility in mol/kg of seawater/atm, T is the absolute temperature, and S is the salinity. This solubility equation has been used in virtually all recent models of ocean CO_2 uptake. The solubility of CO_2 is much greater than that of O_2 or N_2; the relative proportions of N_2:O_2:CO_2 in the atmosphere are about 2400:630:1, whereas in seawater the corresponding ratios are 28:19:1 depending on the salinity and temperature (Skirrow, 1975). There is little uncertainty in our knowledge of CO_2 solubility.

The dissociation constants (K_1 and K_2) of carbonic and boric acids in seawater have had a rich investigative history, and a complex literature attests to this. The dissociation constants are formally defined as

$$K_1 = \frac{H^+[HCO_3^-]}{[H_2CO_3]} \tag{10}$$

and

$$K_2 = \frac{H^+[CO_3^{2-}]}{[HCO_3^-]}, \tag{11}$$

where the square brackets denote the concentration of the species in seawater. These are apparent, not true, thermodynamic constants combining the activity of the hydrogen in with the concentrations of the CO_2 species. Much of the difficulty has surrounded the definition of the pH and ionic medium scales used by the various experimentalists who have determined these constants. Also, the lack of any convention for fitting the data obtained to mathematical functions has resulted in an arcane set of equations.

Millero (1979) has reviewed this situation. He finds that the data of Lyman (1956), Hansson (1973), and Mehrbach et al. (1973) all yield very similar results when the apparent constants (K_i) for the equilibria at various salinities (S) are fitted to equations of the form

$$\ln K_i = \ln K_{iw} + A_i S^{1/2} + B_i S, \tag{12}$$

where Ki is the constant for pure water, and A_i and B_i are temperature-dependent adjustable parameters. The discrepancies in calculated values of $[HCO_3^-]$ and $[CO_3^=]$ for water of fixed alkalinity and total CO_2 are about ± 10 μmol/kg among the various constants. These differences have almost no effect on our ability to model ocean CO_2 uptake but are a considerable irritant to researchers attempting to make and verify accurate CO_2 measurements under often trying field conditions.

In one area there is a degree of uncertainty: the dissolved organic matter in natural seawater is not represented in any way in these formulations. Typical dissolved organic matter concentrations in ocean water are 1 mg of C/kg (83 μmol/kg). The acid-base characteristics of this material, and thus its contribution to the alkalinity, are poorly known. Huizenga and Kester (1979) report about 11 μmol of sites per mg of C on this material with a dissociation constant of about $10^{3.5}$. The effect then is not likely to be large.

The preceding paragraphs show that for a seawater of constant alkalinity and chemical composition we can calculate quite well the effects of adding CO_2. However, changing the alkalinity of seawater will have a marked effect. Adding CO_2 gas to seawater [Equation (7)] does not change the alkalinity since charge balance is not altered; the dissolution or precipitation of $CaCO_3$ [Equation (6)] does.

The principal forms of $CaCO_3$ in the ocean are the polymorphs calcite and aragonite, and these are secreted by calcareous organisms to form their shells. Surface seawater is greatly supersaturated with respect to both calcite and aragonite, spontaneous precipitation being kinetically inhibited. The various chapters in the volume edited by

Andersen and Malahoff (1977) testify to the complexity surrounding $CaCO_3$ formation and dissolution in the oceans. The solubility of $CaCO_3$ in seawater increases with increasing pressure, decreasing temperature, and decreasing pH; thus, the deep oceans are undersaturated with respect to $CaCO_3$, and dissolution occurs.

As we add CO_2 to the surface ocean we decrease the pH and increase the tendency for $CaCO_3$ dissolution. If this dissolution occurs, then both the alkalinity and the total CO_2 increase; although this process generates an increase in total CO_2, the net effect of the alkalinity increase would be to enhance the ocean's capacity for CO_2 uptake by maintaining constant the factor R (Figure 3.3) and providing $CO_3^=$ ions.

Model calculations tend to assume a constant alkalinity scenario, and there is no evidence that the alkalinity of the ocean has increased in recent times. A skeptic could however point out that there is precious little evidence that it has not. Ambiguities in definition of alkalinity [e.g., the inclusion of minor species such as HPO_4^{2-} and $SiO(OH)_3^-$], imprecision in measurement, and lack of a historical time series leave us with a poor temporal record of the alkalinity and total CO_2 content of the ocean.

Several things make this problem complex. First, the concept of the solubility of pure $CaCO_3$ in seawater is moot; Morse et al. (1980) show that the surface undergoing dissolution rapidly becomes transformed into a Mg-Ca-CO_3 interfacial layer with complex kinetic and solubility controls. Second, biogenically produced magnesian calcites (such as in some algae and the spines of sea urchins), containing 15 mol % magnesium or more, commonly occur in ocean surface waters. The stability of this material is poorly understood, and its dissolution would change alkalinity. Garrels and Mackenzie (1981) recently reviewed the susceptibility of magnesian calcite phases to CO_2-induced dissolution. Their conclusion was that insufficient magnesian calcites existed globally to have major impact, if dissolved, on ocean CO_2 uptake. Dissolution of this material, however, would certainly be noticed on a local scale.

Finally, since the surface ocean is so strongly supersaturated with respect to calcite and aragonite and is likely to remain so, it is widely assumed that no dissolution of these minerals takes place there. Aller (1982) has pointed out that this is not so. Calcareous shells in nearshore muds are exposed to interstitial waters rich in respiratory CO_2 and low in pH. The shells are dissolved quite rapidly, resulting in a diffusive flux of Ca^{2+} and $CO_3^=$ ions to the overlying waters. As we change the CO_2 content of surface waters, we change the upper boundary condition controlling this flux. The effects of this process are still being explored.

It is quite possible to incorporate the effects of a postulated increase in ocean alkalinity into an atmosphere-ocean CO_2 model, such as has been done by Bacastow and Keeling (1979), although the accuracy of the conclusions is uncertain. In this calculation, the dissolution of deep calcium carbonate was found to have little immediate effect on rising atmospheric CO_2 levels, since the affected seawater would be sequestered in the deep ocean. As this water is brought into contact with the atmosphere, it will slowly draw down the atmospheric CO_2

levels some hundreds of years in the future. If we were to dissolve an average depth of 3 cm of pure calcium carbonate from the ocean floor, then in 1500 years the atmospheric CO_2 level would be some 30% lower than with no change in alkalinity. If shallow-water calcium carbonate dissolves, the effects are more dramatic. If we were to dissolve an average depth of 40 cm of pure calcium carbonate from shallow-water sediments, then the peak atmospheric CO_2 level would only be some 60% of that with no dissolution occurring, and the drawdown in the future would be more rapid.

This, as emphasized by Bacastow and Keeling (1979), is of course unrealistic. There are kinetic limits and controls on carbonate dissolution extrinsic to the model considered here (e.g., Sayles, 1981; Emerson and Bender, 1981), and the disappearance of such massive amounts of carbonate shells and corals and sediments from our shores and shallow seas would present a crisis for man arousing far greater concern than any incremental effect on CO_2 levels.

In the future it appears inevitable that fossil fuel CO_2-induced dissolution of calcium carbonate will take place in the ocean. The most sensitive site appears to be in the deep North Atlantic Ocean where waters enriched in industrial CO_2 begin their deep ocean tour (Broecker and Takahashi, 1977) and conceptually, since these waters have demonstrably "seen" fossil fuel CO_2, some dissolution has likely already occurred. However, there is no time series of measurements in the oceans adequate to confirm or deny these statements, and some thought must be given to this if progress is to occur.

3.2.8 Measurements of Ocean CO_2

Ocean surface water today contains about 2000 μmol/kg of CO_2. The amount varies with temperature, location, and season. We do not know the concentration in the past. The atmospheric CO_2 content in the last century appears to have been about 270 ppm, although uncertainty exists. Assuming the maintenance of overall equilibrium, we can calculate that modern-day surface ocean waters must contain about 35 μmol/kg more CO_2 than in the past, an approximate 1.8% increase. Although this appears to be so, there is no time series of ocean CO_2 measurements adequate to support this claim. The ocean surface pH is similarly variable in the range 8.0-8.3; we can calculate that ocean surface pH has been reduced by about 0.06 pH units. The carbonate ion content of surface seawater depends on the complex equilibria established between the species H_2CO_3, HCO_3^- and $CO_3^=$; it is typically 15% or so of the total CO_2 concentration. We calculate that ocean surface water today contains about 10% less carbonate ion than in the past.

How well can we measure these changes? The most problematic and least satisfactory measurement is pH; a measurement precision of ±0.01 pH unit is possible, but the long-term change would be hard to detect over seasonal fluctuations.

Alkalinity can be measured with a precision and accuracy of about 3 μequivalents/kg. The total CO_2 content can be determined to ±4 μmol/kg by potentiometric titration (Bradshaw et al, 1981); recently Keeling (1983) has presented ocean CO_2 data based on gas extraction and manometric measurement accurate to ± 0.5 μmol/kg. The current rate of increase of the total CO_2 concentration in ocean surface waters is calculated to be approximately 1 μmol/kg/yr. Plainly such an increase would be observable from time series measurements made today within the space of a few years.

The most sensitive measurement is pCO_2, which may be determined to within a few tenths of a part per million. By pCO_2 we mean the pressure of CO_2 gas that would be found in a small volume of air that had been allowed to reach gaseous equilibrium with a large volume of seawater. Surface seawater strives to maintain a pCO_2 globally in equilibrium with the atmosphere, lagging behind by some small value due to the finite time required for gas exchange to take place. There is a large natural variability in ocean surface pCO_2 (Keeling 1968; Miyake et al., 1974; Takahashi, 1979) as shown in Figure 3.4.

The data shown in Figure 3.4 are in fact the deviations of ocean surface pCO_2 from equilibrium with the atmosphere. It is not a synoptic data set but an ensemble of results from many expeditions, quasi-normalized in time and containing considerable seasonal noise. The essential features are strongly negative values at high latitudes, where rapid cooling and biological activity have markedly lowered pCO_2, and high values at the equator, where upwelling of CO_2-rich water and warming have raised pCO_2. Gaseous exchange of CO_2 is sufficiently slow that equilibrium with the atmosphere is not achieved locally, and these patterns persist. Negative values imply invasion of CO_2 from the atmosphere to the ocean; positive values indicate evasion of CO_2 from the ocean to the air.

This distribution results in a net flux of CO_2 between the equatorial and polar oceans (Bolin and Keeling, 1963). Pearman et al. (1983) have recently calculated that the net release of CO_2 by the equatorial oceans is currently about 1.3 Gt of C/yr; the net uptake by the high-latitude oceans is about 4.4 Gt of C/yr. The presence of fossil fuel CO_2 has enhanced the polar uptake and suppressed the equatorial source.

There is now evidence for a systematic change in ocean surface pCO_2 with time. Takahashi et al. (1983) have compiled their measurements of pCO_2 in Sargasso Sea surface waters from the IGY (1957), GEOSECS (1972), and TTO (1981) expeditions. The results are shown in Figure 3.5, together with the atmospheric trend. It is clear that ocean pCO_2 is rising; however, the interpretation of this signal requires caution. The problem is not with the precision of measurement but with compensating for the natural noise due to fluctuating ocean surface conditions. The atmospheric and oceanic lines are not parallel in Figure 3.5, and some explanation of this is needed. The Sargasso Sea exhibits low surface pCO_2 with respect to the atmosphere (Figure 3.4). This arises from the marked negative heat flux observed there (Bunker and Worthington, 1976), so that surface seawater is cooled during its residence time in the Sargasso Sea. The slope of the oceanic line in Figure 3.5 thus represents not only the anthropogenic

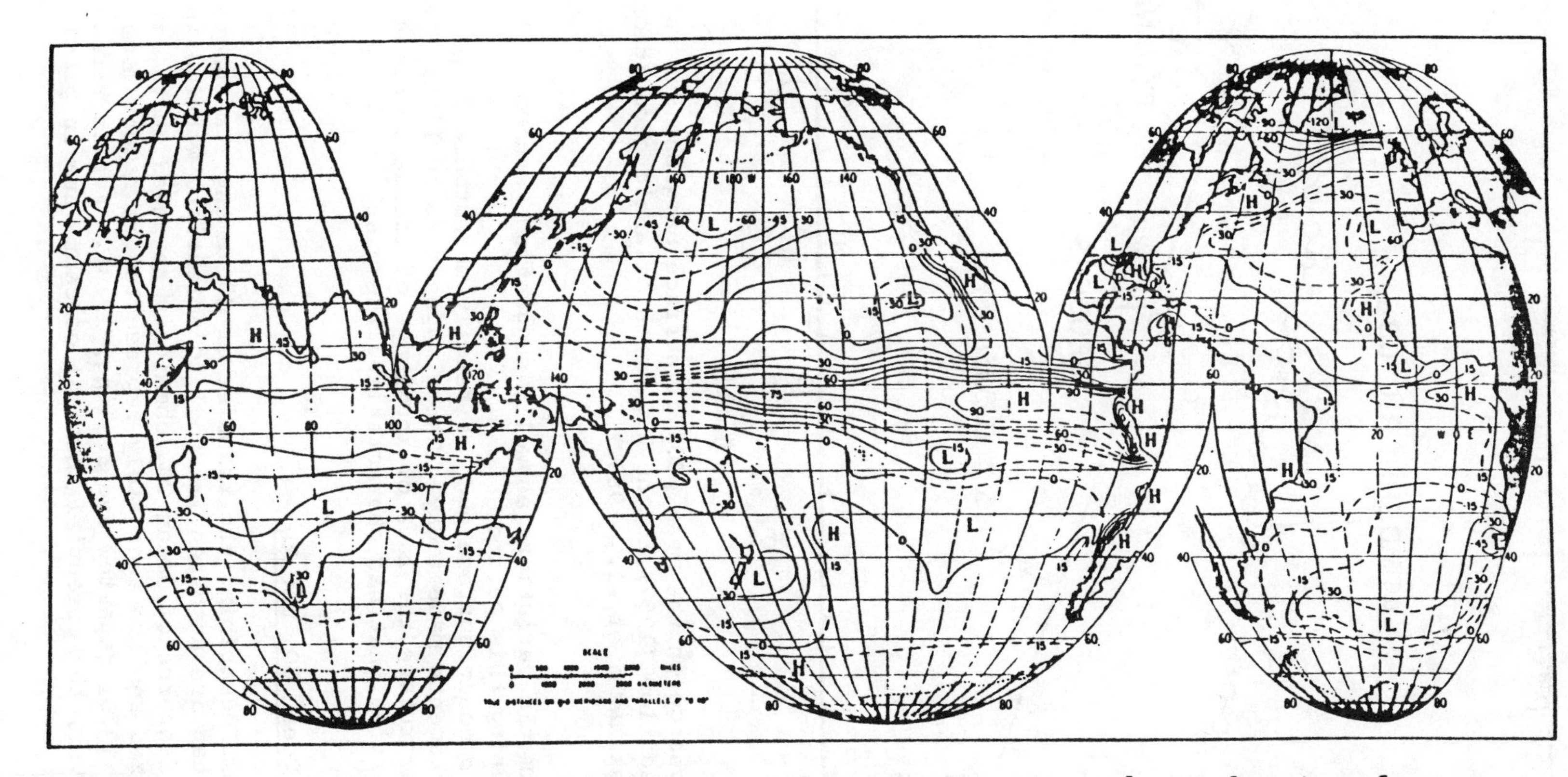

FIGURE 3.4 The partial pressure of CO_2 gas in surface seawater expressed as a departure from atmospheric equilibrium. The map is an ensemble of data from many sources. Negative values imply invasion of CO_2 from the atmosphere to the sea; positive values indicate evasion of CO_2 from sea to air. (From Keeling, 1968.)

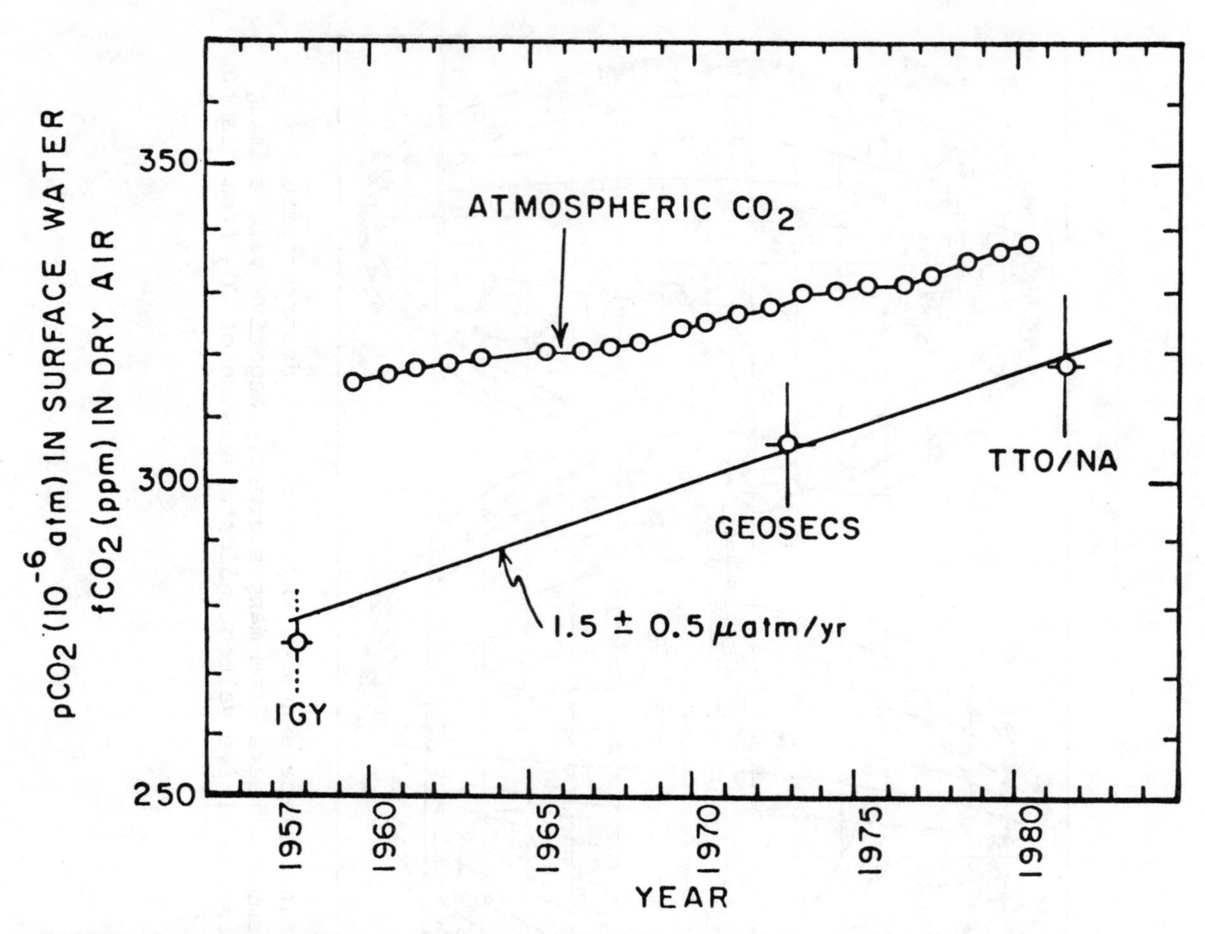

FIGURE 3.5 The change in the mean pCO_2 of surface Sargasso Sea water from the IGY expedition (1957) to the TTO-North Atlantic expedition (1981). (From Takahashi et al., 1983.)

CO_2 trend but climatic variations as well, suggesting perhaps a modest warming during this period.

The increase in ocean surface pCO_2 is readily measurable from time-series measurements made today. There appears to be little reason to maintain in the future the lack of knowledge of trends in ocean surface CO_2 properties that has characterized the past.

3.2.9 Models of Ocean CO_2 Uptake

Mathematical models describing the transfer of CO_2 among atmosphere, biosphere, and ocean are an essential tool to scientists working in this field. The features of ocean CO_2 chemistry reviewed above have long been recognized and are adequately represented in recent models (Oeschger et al., 1975; Bacastow and Keeling, 1979). Future refinement is probable; however, the essential chemical concepts are in place.

How do we model ocean CO_2 uptake, what data are available for model testing, and how do we know if the models are correct?

The physical basis for the modeling of ocean CO_2 uptake is reasonably well established. The first criterion is that the gas exchange rate be known. Broecker and Peng (1974) have examined the problem of determining gas exchange rates, and Broecker et al. (1979) have evaluated the effects of uncertainty in this property on calculations of ocean CO_2 uptake. Most field data have been obtained by the radon technique by which the deficiency of naturally produced radon-222, due to gas exchange at the sea surface, is measured. A simple stagnant film model is used to calculate the gas exchange parameter and to relate the observations to CO_2 exchange. The two-way CO_2 exchange rate obtained in this way is 16 mol/m^2/yr. A much larger-scale estimate is available from the natural balance of ^{14}C by radioactive decay in the interior of the ocean must be balanced by a fresh influx of $^{14}CO_2$. The result from this calculation yields a rate of 19 $\pm$ 6 mol/m^2/yr. CO_2 exchange in a wind-wave tunnel has been examined by Broecker et al. (1979), yielding results in substantial agreement with the above.

The calculations of gas exchange rates appear to rest on sound principles. The characteristic exchange time for CO_2 is about 10 times longer than for the nonreactive gases (N_2 and O_2, for example) and is about 1 year (Broecker and Peng, 1974). The time scale for $^{14}CO_2$ exchange is about 10 years, owing to the time required for complete isotopic equilibration with the carbon pool. Gas exchange for CO_2, though quite slow, does not appear to be the rate-limiting step in models of ocean CO_2 uptake.

The rate of vertical mixing in the sea is widely viewed as the critical parameter. The annual cycle over most of the ocean is such that in the summer the sea becomes capped by a warm surface layer, which undergoes continuous gaseous exchange with the atmosphere. Wintertime cooling and late winter storms increase the density of this upper layer so that it finally becomes unstable and undergoes turbulent exchange with waters of equivalent density below. This scheme of wintertime surface turnover, followed by lateral penetration along surfaces of constant density, is what is parameterized in various ways in ocean models. Most water mass formation takes place in a location and at a time when we cannot observe it. Some integrated measure of the effect is required, and this has been provided by the tracer approach.

The principal tracers used have been ^{14}C and ^3H; the transcendental virtue of ^{14}C is that it identically mimics CO_2, taking part in both the gaseous and biological exchange cycles. Revelle and Suess (1957) and Craig (1957) used ^{14}C box models to evaluate carbon dioxide exchange between atmosphere and ocean and within the ocean. As these models grew in complexity (Keeling and Bolin, 1967, 1968), the necessity for multiple tracers became apparent, as did the difficulty of assigning realistic transfer rates between boxes based on objective experimental evidence. The injection into the atmosphere of massive amounts of radionuclides in the nuclear bomb tests of 1958-1962 prior to the signing of the nuclear test-ban treaty provided such a tracer suite, and a decade later ocean scientists organized the GEOSECS

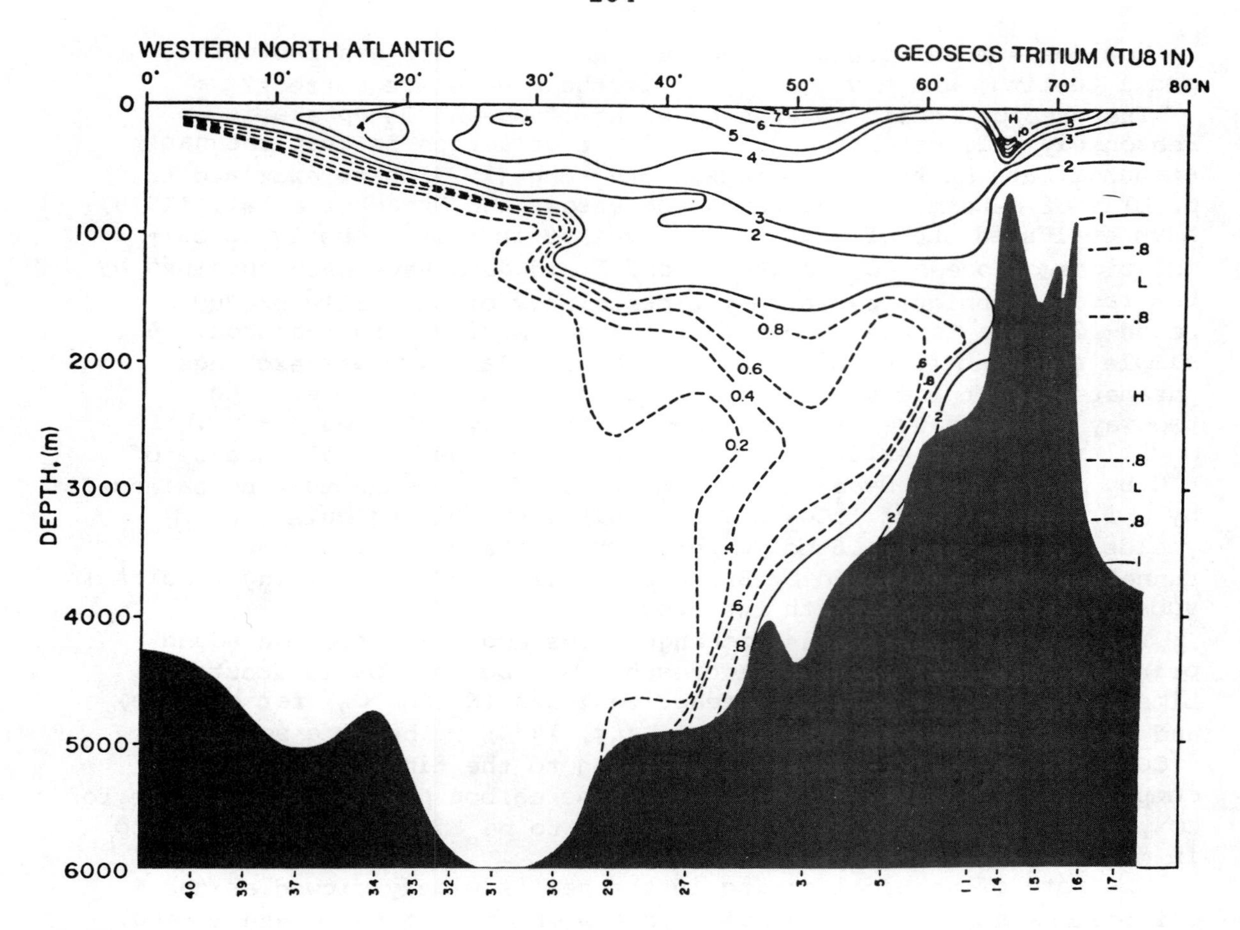

FIGURE 3.6 Penetration of tritium into the North Atlantic Ocean in 1972 along the GEOSECS cruise track. (From Ostlund et al., 1974.)

experiment to observe the oceanic distribution of these species. The tritium and radiocarbon results from this program (Ostlund et al., 1974) provided a snapshot of ocean tracer penetration on a 10-year time scale; Figure 3.6 shows tritium penetration into the Atlantic Ocean in 1972.

The results shown in Figure 3.6 represent a vertical slice through the western basin of the North Atlantic. They plainly reveal the formation of new deep waters; however, such a representation understates the great horizontal extent of the oceans. CO_2 is taken up by surface seawater globally, and between 50-70% of the ocean's anthropogenic CO_2 burden is stored in the surface and thermocline waters of the great oceanic gyres (Stuiver, 1978; Siegenthaler, 1983).

Although the penetration of tritium (Figure 3.6) provides a vivid pictorial representation of the extent of the ocean labeled by an invading tracer in the 10-year period 1962-1972, the conversion of this signal to that of CO_2 is complex. First, tritium was injected as a pulse early in the decade, and largely in the northern hemisphere. Second, the mean age of the fossil fuel CO_2 increase is about 28

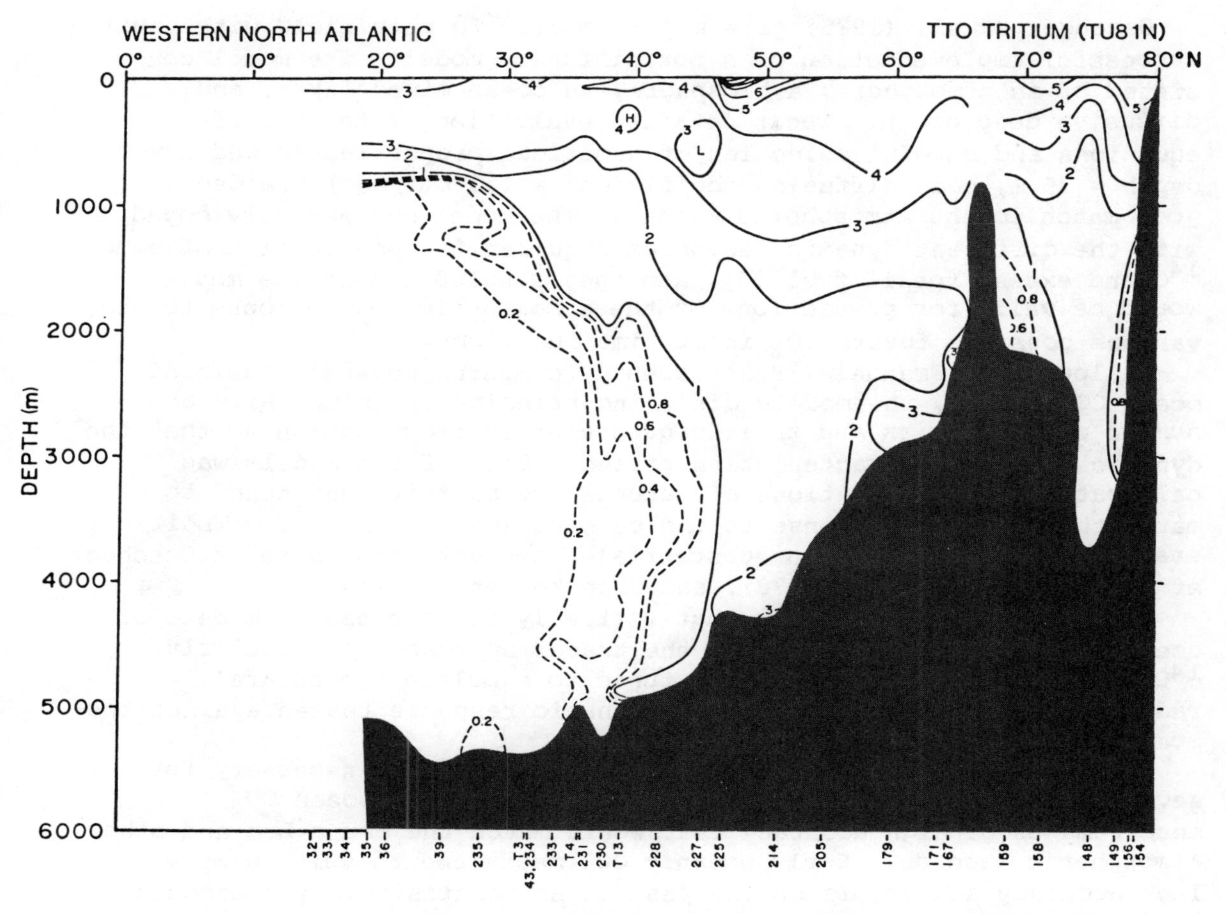

FIGURE 3.7 The re-occupation of the GEOSECS stations in 1981 on the TTO Expedition. Tritium was principally injected into the atmosphere of the northern hemisphere by nuclear bomb tests in 1962. The figure illustrates the labeling of the ocean by the invasion of a passive tracer in a 19-year period. The mean age of the fossil fuel CO_2 signal is about 28 years. (From Ostlund, 1983.)

years (Broecker et al., 1979). What fraction of the ocean will be labeled on the 28-year time scale appropriate for CO_2? Fortunately, the experiment shown in Figure 3.6 has been repeated. In Figure 3.7, results are shown for a reoccupation of this section in 1981, 19 years after the principal tritium pulse (Ostlund, 1983). The progression of the deep-water front is plainly to be seen.

These data provide a critical test for models of ocean CO_2 uptake, for unless the models can match these observations, they are unlikely to be correct. It should be noted, however, that the converse is not necessarily true. A model that simply matches data at one point in time and contains unrealistic physical principles or dynamic characteristics cannot be said to be "true" or "correct" in any satisfactory sense.

Oeschger et al. (1975) gave major impetus to this field with their successful implementation of a box diffusion model. The model consisted of an atmosphere, a biosphere, an ocean mixed layer, and a diffusive deep ocean. Their detailed evaluation of the transfer equations and careful selection of numerical properties (mixed layer depth = 75 m, eddy-diffusion coefficient = 1.3 cm^2/sec) yielded a good match to the atmospheric signal. The model successfully coped with the different dynamic responses required for penetration of bomb ^{14}C and excess fossil fuel CO_2, and they concluded that the model would be valid for predictions of the atmospheric CO_2 response to the various possible future CO_2 input time functions.

Killough and Emanuel (1981) recently compared several models of ocean CO_2 uptake, the models differing principally in the size and number of reservoirs and their sequence of interconnection so that the dynamic response characteristics varied. Each of the models was calibrated from observations of natural ^{14}C activity and tuned to match the observed response to the penetration of bomb ^{14}C. Their evaluation gave results in substantial agreement with those of Oeschger et al. (1975), Stuiver (1978), and Broecker et al. (1979).

At this point it is clear that virtually all successful models of ocean CO_2 uptake have relied on the tracer approach, particularly ^{14}C and tritium. The models are tuned to simulate the natural radiocarbon distribution and their dynamic response tested against the bomb transient.

The tracer, rather than direct, approach has been necessary for several reasons. First, there is no time series of ocean CO_2 measurements of high accuracy that would match the Mauna Loa and other atmospheric records. Early oceanic CO_2 measurements quite simply lack accuracy and precision and rest on an unsatisfactory thermodynamic basis. Second, models depending purely on physical principles and measurements (T,S) are frequently undetermined. Without the constraints supplied by independent tracers of differing source functions, satisfactory solutions are not easily achieved. Finally, measurements of ocean CO_2 made today are inadequate on their own to solve the fossil fuel CO_2 problem.

The models described above, which currently serve to calculate ocean CO_2 uptake, while highly ingenious, are nonetheless viewed with considerable skepticism by physical oceanographers. The concern is not that the overall estimate of ocean CO_2 uptake is substantially in error but that the parameterization of all of ocean physics into a single vertical eddy-diffusion coefficient is unacceptable. Garrett (1979) has carefully reviewed the evidence for vertical diffusion in the ocean and finds no physical basis for a vertical diffusion coefficient of ≥ 1 cm^2 sec^{-1} as required by Oeschger et al. (1975), and indeed by all one-dimensional vertical models. A value of one tenth that number appears realistic. Jenkins (1980) has followed the time history of the penetration of tritium, and its stable, gaseous daughter product helium-3, into the Sargasso Sea. He shows unequivocally that a one-dimensional model with high vertical diffusivity cannot explain the results and that a scheme of wintertime surface-water turnover followed by lateral penetration along density surfaces is required.

There are of course many models of ocean circulation that give a more realistic portrayal of ocean physics (Holland, 1971, 1978), and similar models are now being applied to study the time-dependent penetration of tracers into the interior of the ocean (Sarmiento, 1982). The application of these models to the CO_2 problem will be complex but seems to be possible. Rapid progress is to be expected in this field. However, the principal criticism of ocean CO_2 uptake models has come not from their representation of ocean physics but from their failure to include some components of the natural, or perturbed, CO_2 cycle.

The basic assumption for models such as those of Oeschger et al. (1975) or Killough and Emanuel (1981) is that the natural cycle of CO_2 within the ocean has been unchanged by the activities of man--i.e., primary production remains at past levels, and the vertical flux of particulate carbon has not been altered. The initial ocean CO_2 profile thus appears in the models as a constant value. The true ocean CO_2 cycle is enormously complex, so that the simplification introduced by this procedure is most attractive. The recent report by Bolin et al. (1982), who have incorporated realistic profiles of oceanic phosphate, oxygen, CO_2, alkalinity, and ^{14}C in a highly developed 12-box ocean model, illustrates this point well. The carbon dioxide concentration in the North Atlantic has been measured in parallel with the tritium section in Figure 3.7 (Brewer, 1983). The results (Figure 3.8) show the large natural variability of the oceanic CO_2 system. Hidden within these data lies the increase in CO_2 caused by the activities of man.

Kempe (1982) has reviewed the long-term trends in river fluxes to the ocean of carbon and nutrients in great detail and has shown that fertilization, land use, and industrial activities have altered many of these fluxes markedly. Walsh (1981) has suggested that overfishing off Peru and other perturbations (Walsh et al., 1981) have changed the storage of carbon on continental shelves. It is clear that the facets of carbon research are so many and varied that to attempt to satisfy all claims for omission would result in models of endless and labyrinthine complexity. We would like to know if these effects are offset by competing processes and if the net effect is of sufficient size to alter our conclusions regarding the fossil fuel signal.

Keeling (1983) has recently examined this problem. He has compared measurements of atmospheric CO_2 from a time series of samples obtained on board ships from north-south sections in the Pacific Ocean. Recent seasonal coverage has greatly increased the utility of these data. By assuming that the seasonal oscillations observed today hold true for the recent past, we can greatly improve the interpretation of previously obtained shipboard data. Figure 3.9 shows plots of such data grouped from the periods near 1962, 1968, and 1980 from Keeling (1983). The data are normalized relative to a constant value at the South Pole, shown as zero on this figure. The peak, centered at the equator, is attributed to degassing of CO_2 from the high pCO_2 zone in the equatorial Pacific Ocean (Keeling, 1968). This is a natural phenomenon. In the northern hemisphere a steady increase has occurred that is consistent with rising fossil fuel usage, over 90% of which

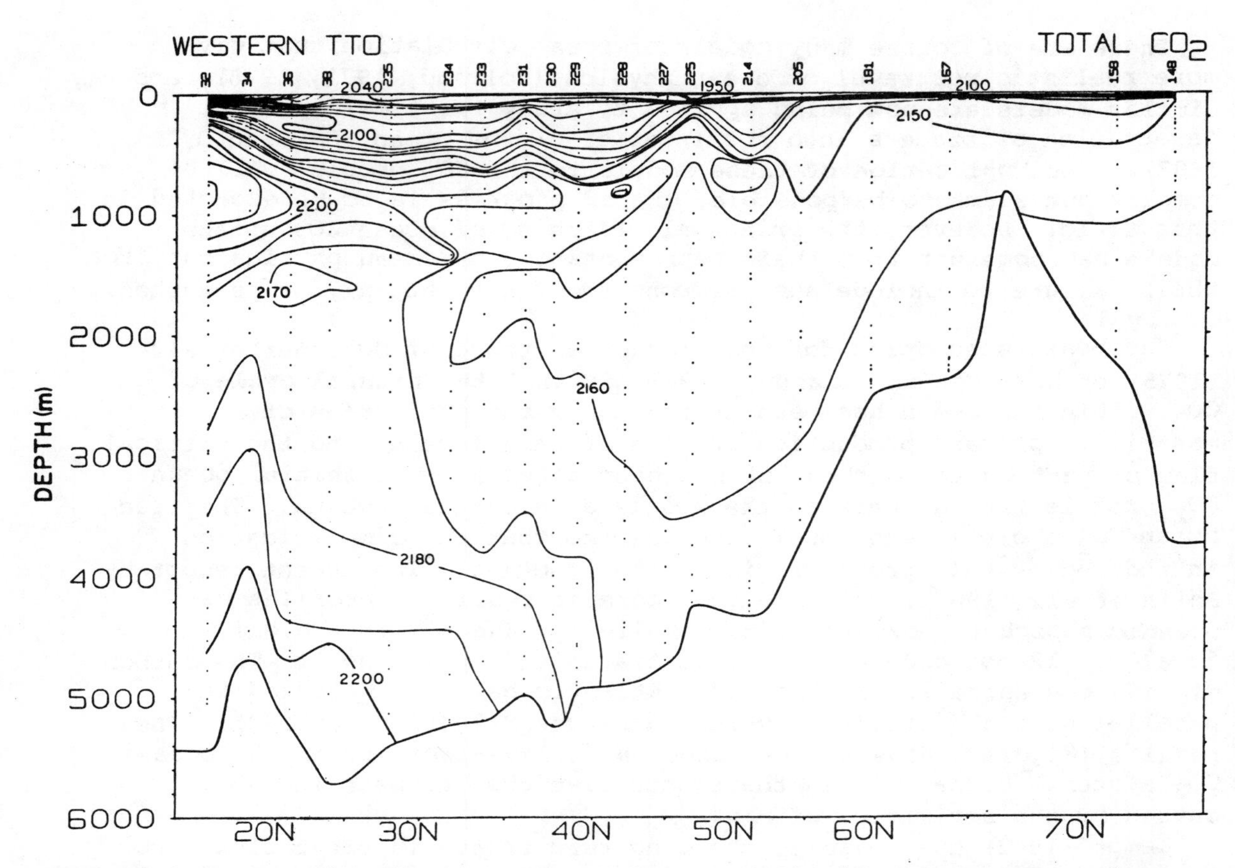

FIGURE 3.8 The distribution of total carbon dioxide in seawater along the section shown in Figure 3.7. The large natural variability seen here must have been perturbed by the invasion of fossil fuel CO_2, although we have no time series of measurements to document this change. (From Brewer, 1983.)

takes place in the northern hemisphere. By assuming that the change in atmospheric CO_2 is due solely to fossil fuel usage, with a constant oceanic uptake over this period, then model profiles can be developed that may be compared with the observed data. Figure 3.10, again from Keeling (1983), shows the result of subtraction of the model profiles from the observations. The residual signal reflects the expected constant equatorial degassing but does not indicate any significant perturbation of atmospheric CO_2 from other than the fossil fuel source. These data, obtained over approximately a 20-year period, appear to be the best available. They support, but do not prove, the idea of a constant, or very slowly changing, ocean and terrestrial biosphere. The atmosphere integrates the effects of changes in these systems, so that if a large change is postulated in one of them, then there appears to have been a most fortunate compensation in the other.

Pearman et al. (1983) have also examined this problem. They calculate that the dominantly northern hemisphere input of CO_2 to the atmosphere has changed the interhemispheric difference in atmospheric

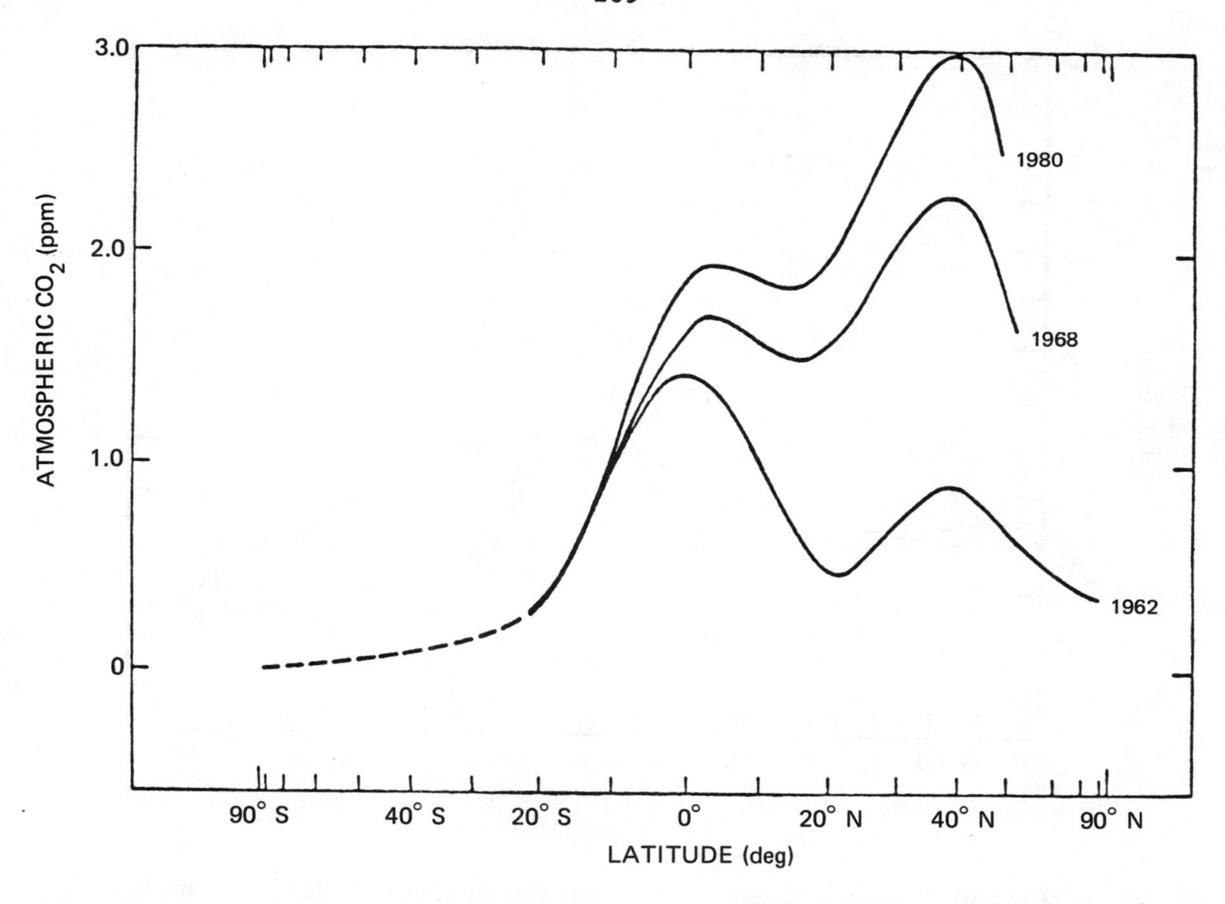

FIGURE 3.9 Atmospheric CO_2 levels along a Pacific Ocean transect, normalized to zero at the South Pole. (From Keeling, 1983.)

concentration (north-south) from -1 ppm in the last century to +4-5 ppm today. Their atmosphere-ocean model simulation points to a net upper limit of 2 Gt of C per year from a terrestrial carbon source.

The conclusion of Oeschger et al. (1975) was that the proportioning of fossil fuel CO_2 between atmosphere and ocean was 60 to 40%. Stuiver (1978) concluded that some 47% of fossil fuel CO_2 was stored in the ocean. Broecker et al. (1979) calculated that $37 \pm 4\%$ of the fossil fuel CO_2 generated between 1958 and the present has been taken up by the sea. Bolin et al. (1982) have recently concluded that the oceanic uptake falls in the range 30-38%. Since these estimates all fall in a narrow range, depend on accepted physical principles, and are verifiable by several different means, there does not appear to be the chance of significant error.

3.2.10 Future Studies and Problems

Although present-day models appear satisfactorily to answer the question of how much fossil fuel CO_2 the sea takes up, there are many things that we do not know.

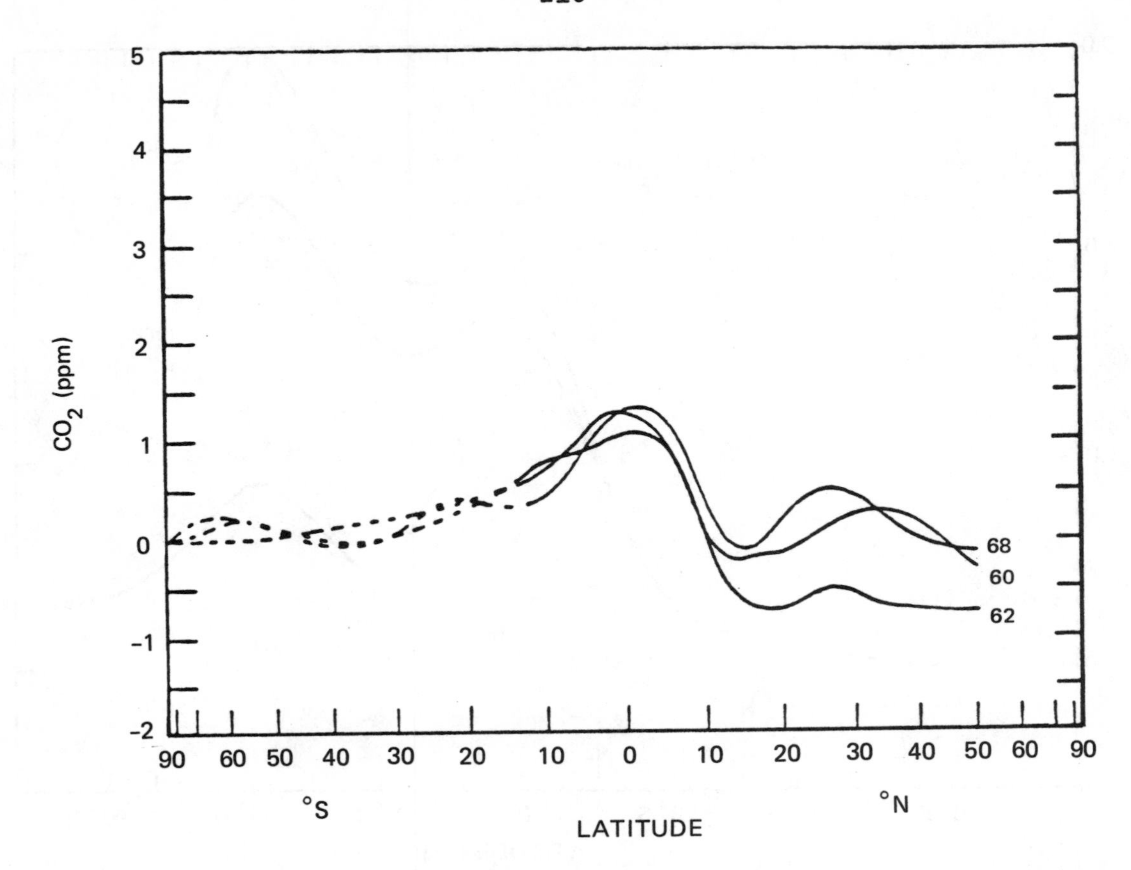

FIGURE 3.10 The results shown in Figure 3.9 corrected for seasonal and fossil fuel CO_2 effects alone. No perturbation of the signal from other than these effects is apparent. (From Keeling, 1983.)

We do not know the atmospheric or oceanic CO_2 levels in the last century. Recent ice core data (Neftel et al., 1982) promise to place new constraints on this. Calculations based on oceanic measurements (Brewer 1978, 1983) yield a preindustrial value of about 265 ppm of CO_2. However, this is likely a lower limit since oceanic deep waters are formed in areas of marked negative disequilibrium with the atmosphere. As our ocean data base grows, the one-dimensional models will become increasingly inadequate, and incorporation of the CO_2 and tracer data into new models will be required.

The warming accompanying the atmospheric CO_2 rise will affect the ocean. Storage of heat in the upper layers will mitigate, but not prevent, climate change (Bryan et al., 1982). Models of this heat storage currently treat it as passive uptake, not affecting water mass formation and vertical circulation. If such changes, however, did occur, it would affect ocean CO_2 uptake in unknown ways.

The warming of the ocean will reduce CO_2 solubility and expel further CO_2 to the atmosphere. This effect can be calculated reasonably well (Bacastow and Keeling, 1979) and is incorporated into current models (Killough and Emanuel, 1981) of ocean CO_2 uptake.

Whether we will detect this effect and separate it in our budgeting of CO_2 between atmosphere, ocean, and biosphere, or confuse it with some other signal, is problematic.

3.2.11 Summary

On average, each year the ocean currently takes up an amount of CO_2 approximately equal to 40% of the fossil fuel CO_2 added to the atmosphere by man. Calculations based on sound principles of solution chemistry and an estimated atmospheric CO_2 content of about 270 ppm in the last century show that ocean surface pH must have decreased by about 0.06 pH unit during this time; the total CO_2 concentration of surface seawater must have increased by about 35 μmol/kg (about 1.8%). The proportion of carbonate ion must have decreased by about 10%, increasing the tendency for calcium carbonate dissolution. The alkalinity is believed not to have changed. We do not know this on the basis of a direct time series of measurements, and until quite recently, oceanic CO_2 system measurements contained substantial inaccuracies.

Models of ocean CO_2 uptake have depended greatly on tracer data, particularly natural and bomb-produced ^{14}C. These models represent some features of ocean chemistry quite well, but they represent ocean physics by a simple vertical diffusion coefficient. The models treat only the CO_2 perturbation and do not yet attempt to mimic the natural and complex CO_2-oxygen-nutrient cycles within the ocean. Rapid progress in incorporating these features into more advanced ocean circulation models is anticipated.

It is quite possible to measure the changing CO_2 properties of the ocean with time using modern techniques, although no ongoing program yet exists to do so. There are some significant uncertainties awaiting. We do not know when and where calcium carbonate dissolution will occur. We do not know how future warming will affect ocean circulation and whether we will detect this warming in the CO_2 signal. We do not know adequately the time history of CO_2 releases from the biosphere so as firmly to close the atmosphere-ocean-biosphere triangle represented in our models.

Finally, we should not be too complacent. Nature has vast resources with which to fool us; the last glaciation was apparently accompanied by massive CO_2 transfers to and from the ocean, the causes, consequences, and explanations of which are poorly understood today (Broecker, 1982).

References

Aller, R. C. (1982). Carbonate dissolution in nearshore terrigenous muds: the role of physical and biological reworking. J. Geol. 90:79-95.

Andersen, N. R., and A. Malahoff, eds. (1977). The Fate of Fossil Fuel CO_2 in the Oceans. Plenum, New York, 749 pp.

Bacastow, R. B., and C. D. Keeling (1979). Models to predict future atmospheric CO_2 concentrations. In Workshop on the Global Effects of Carbon Dioxide from Fossil Fuels. U.S. Dept. of Energy Report CONF-770385, pp. 72-90.

Baes, C. F. (1982). Ocean Chemistry and Biology. In Carbon Dioxide Review: 1982, W. C. Clark, ed. Oxford U. Press, New York, pp. 187-211.

Bolin, B. (1981). Carbon Cycle Modelling. SCOPE Report 16. Wiley, New York.

Bolin, B., and C. D. Keeling (1963). Large-scale atmospheric mixing as deduced from the seasonal and meridional variations of carbon dioxide. J. Geophys. Res. 68:3899-3920.

Bolin, B., E. T. Degens, S. Kempe, and P. Ketner (1979). The Global Carbon Cycle. SCOPE Report 13. Wiley, New York, pp. 1-491.

Bolin, B., A. Bjorkstrom, K. Holmen, and B. Moore (1982). The simultaneous use of tracers for ocean circulation studies. Report CM-58, Dept. of Meteorology, U. of Stockholm, p. 70.

Bradshaw, A. L., P. G. Brewer, D. K. Shafer, and R. T. Williams (1981). Measurements of total carbon dioxide and alkalinity by potentiometric titration in the GEOSECS program. Earth Planet. Sci. Lett. 55:99-115.

Brewer, P. G. (1978). Direct observation of the oceanic CO_2 increase. Geophys. Res. Lett. 5:997-1000.

Brewer, P. G. (1983). The T.T.O. North Atlantic Study--A Progress Report. In Proceedings: Carbon Dioxide Research Conference: Carbon Dioxide, Science and Consensus. U.S. Dept. of Energy Report CONF-820970, Part II, pp. 91-122.

Brewer, P. G., and J. C. Goldman (1976). Alkalinity changes generated by phytoplankton growth. Limnol. Oceanog. 21:108-117.

Broecker, H. C., J. Peterman, and W. Siems (1978). The influence of wind on CO_2-exchange in a wind-wave tunnel, including the effects of monolayers. J. Mar. Res. 36:595-610.

Broecker, W. S. (1982). Ocean chemistry during glacial time. Geochim. Cosmochim. Acta 46:1689-1705.

Broecker, W. S., and T. H. Peng (1974). Gas exchange rates between air and sea. Tellus 26:21-35.

Broecker, W. S., and T. Takahashi (1977). Neutralization of fossil fuel CO_2 by marine calcium carbonate. In The Fate of Fossil Fuel CO_2 in the Oceans, N. E. Andersen and A. Malahoff, eds. Plenum, New York, pp. 213-241.

Broecker, W. S., T Takahashi, H. J. Simpson, and T. H. Peng (1979). Fate of fossil fuel carbon dioxide and the global carbon budget. Science 206:409-418.

Bryan, K., F. G. Komro, S. Manabe, and M. J. Spelman (1982). Transient climate response to increasing atmospheric carbon dioxide. Science 215:56-58.

Bunker, A. F., and L. V. Worthington (1976). Energy exchange charts of the North Atlantic Ocean. Bull. Am. Meteorol. Soc. 57:670-678.

Callendar, G. S. (1938). The artificial production of carbon dioxide and its influence on temperature. Q. J. Roy. Meteorol. Soc. 64:223-240.

Craig, H. (1957). The natural distribution of radiocarbon and the exchange time of carbon dioxide between atmosphere and sea. Tellus 9:1-17.

Emerson, S. R., and M. L. Bender, 1981. Carbon fluxes at the sediment-water interface of the deep sea: calcium carbonate preservation. J. Mar. Res. 39:139-162.

Garrels, R. M., and F. T. Mackenzie, 1981. Some aspects of the role of the shallow ocean in global carbon dioxide uptake. U.S. Dept. of Energy Report CONF-8003115.

Garrett, C. (1979). Mixing in the ocean interior. Dynamics Atmos. Oceans 3:239-265.

GEOSECS Atlases Vols. 1-6. U.S. Government Printing Office, Washington, D.C.

Hansson, I. (1973). A new set of acidity constants for carbonic acid and boric acid in sea water. Deep-Sea Res. 20:461-178.

Holland, W. R. (1971). Ocean tracer distributions. Tellus 23:371-392.

Holland, W. R. (1978). The role of mesoscale eddies in the general circulation of the ocean--numerical experiments using a wind-driven quasi-geostrophic model. J. Phys. Oceanog. 8:363-392.

Huizenga, D. L., and D. R. Kester (1979). Protonation equilibria of marine dissolved organic matter. Limnol. Oceanog. 24:145-150.

Jenkins, W. J. (1980). Tritium and ^{3}He in the Sargasso Sea, J. Mar. Res. 38:533-569.

Keeling, C. D. (1968). Carbon dioxide in surface oceans waters, 4. Global distribution. J. Geophys. Res. 73:4543-4553.

Keeling, C. D. (1983). The Global Carbon Cycle: What we know and could know from atmospheric, biospheric and oceanic observations. In Proceedings: Carbon Dioxide Research Conference: Carbon Dioxide, Science and Consensus. U.S. Dept. of Energy Report CONF-820970, Part II, pp. 1-62.

Keeling, C. D., and B. Bolin (1967). The simultaneous use of chemical tracers in oceanic studies, I. General theory of reservoir models. Tellus 19:566-581.

Keeling, C. D., and B. Bolin (1968). The simultaneous use of chemical tracers in oceanic studies, II. A three reservoir model of the North and South Pacific Oceans. Tellus 20:17-54.

Kempe, S. (1982). Long-term records of CO_2 pressure fluctuations in fresh waters. In Transport of Carbon and Minerals in Major World Rivers. Mitt. Geol.-Palaont. Inst. Univ. Hamburg, Federal Republic of Germany. SCOPE Report 52, pp. 91-332.

Killough, G. G., and W. R. Emanuel (1981). A comparison of several models of carbon turnover in the ocean with respect to their distributions of transit time and age responses to atmospheric CO_2 and ^{14}C. Tellus 33:274-290.

Lyman, J. (1956). Buffer mechanism of sea water. Ph.D. Thesis, University of California, Los Angeles, 196 pp.

Mehrbach, C., C. H. Culberson, J. E. Hawley, and R. M. Pytkowicz (1973). Measurement of the apparent dissociation constants of carbonic acid in sea water at atmospheric pressure. Limnol. Oceanog. 18:897-907.

Millero, F. J. (1979). The thermodynamics of the carbonate system in sea water. Geochim. Cosmochim. Acta 43:1651-1661.

Miyake, Y., Y. Sugimura, and K. Saruhashi (1974). The carbon dioxide content in the surface waters of the Pacific Ocean. Rec. Oceanog. Works in Jpn. 12:45-52.

Morse, J. W., A. Mucci, and F. J. Millero (1980). The solubility of aragonite and calcite in sea water of 35°/oo salinity at 25°C and atmospheric pressure. Geochim. Cosmochim. Acta 44:85-94.

Murray, C. N., and J. P. Riley (1971). The solubility of gases in distilled water and sea water, IV. Carbon dioxide. Deep-Sea Res. 18:533-541.

Neftel, A., H. Oeschger, J. Schwander, B. Stauffer, and R. Zumbrunn (1982). New measurements on ice core samples to determine the CO_2 content of the atmosphere during the last 40,000 years. Nature 295:220-223.

Oeschger, H., U. Siegenthaler, U. Schotterer, and A. Gugelmann (1975). A box diffusion model to study the carbon dioxide exchange in nature. Tellus 27:168-192.

Ostlund, H. G. (1983). Tritium and Radiocarbon: TTO Western North Atlantic Section GEOSECS Re-occupation. Tritium Laboratory Data Release 83-07, Rosenstiel School of Marine and Atmospheric Sciences, Miami, Fla., unpublished data.

Ostlund, H. G., H. G. Dorsey, and C. G. Rooth (1974). GEOSECS North Atlantic radiocarbon and tritium results. Earth Planet. Sci. Lett. 23:69-86.

PCODF (1981). TTO Preliminary Hydrographic Data Reports, Vol. I-IV. Scripps Institution of Oceanography Reports, La Jolla, Calif.

Pearman, G. I., P. Hyson, and P. J. Fraser (1983). The global distribution of atmospheric carbon dioxide: 1. Aspects of observations and modelling. J. Geophys. Res. 88:3581-3590.

Peterson, B. J. (1980). Aquatic primary productivity and the ^{14}C-CO_2 method: a history of the productivity problem. Ann. Rev. Ecol. Syst. 11:359-385.

Reid, J. L., and R. J. Lynn (1971). On the influence of the Norwegian-Greenland and Weddell seas upon the bottom waters of the Indian and Pacific Oceans. Deep-Sea Res. 18:1063-1088.

Revelle, R., and H. E. Suess (1957). Carbon dioxide exchange between atmosphere and ocean and the question of an increase of atmospheric CO_2 during past decades. Tellus 9:18-27.

Sarmiento, J. L. (1982). A simulation of bomb tritium entry into the Atlantic Ocean. J. Phys. Oceanog., in press.

Sayles, F. L. (1981). The composition and diagenesis of interstitial solutions, II. Fluxes and diagenesis at the water-sediment interface in the high latitude North and South Atlantic. Geochim. Cosmochim. Acta 45:1061-1086.

Siegenthaler, U. (1983). Uptake of excess CO_2 by an outcrop-diffusion model of the ocean. J. Geophys. Res. 88:3599-3608.

Skirrow, G. (1975). The dissolved gases-carbon dioxde. In Chemical Oceanography, Vol. 2, 2nd ed., J. P. Riley and G. Skirrow, eds. Academic, New York, pp. 1-192.

Stuiver, M. (1978). Atmospheric carbon dioxide and carbon reservoir changes. Science 199:253-258.

Stuiver, M., P. D. Quay, and H. G. Ostlund (1983). Abyssal water carbon-14 distribution and the age of the world oceans. Science 219:849-851.

Sundquist, E. T., L. N. Plummer, and T. M. L. Wigley (1979). Carbon dioxide in the ocean surface: the homogeneous buffer factor. Science 204:1203-1205.

Takahashi, T. (1979). Carbon dioxide chemistry in ocean water. In Workshop on the Global Effects of Carbon Dioxide from Fossil Fuels. U.S. Dept. of Energy Report CONF-770385, pp. 63-71.

Takahashi, T., and A. G. E. Azevedo (1982). The oceans as a CO_2 reservoir. In Interpretation of Climate and Photochemical Models, Ozone and Temperature Measurements, R. A. Beck and J. R. Hummel, eds. American Institute of Physics Conf. Proc. No. 82, pp. 83-109.

Takahashi, T., W. S. Broecker, A. E. Bainbridge, and R. F. Weiss (1980a). Carbonate chemistry of the Atlantic, Pacific and Indian Oceans: The results of the GEOSECS Expeditions, 1972-1978. Report 1, cv-1-80. Lamont-Doherty Geological Observatory.

Takahashi, T., W. S. Broecker, A. E. Bainbridge, and R. F. Weiss (1980b). Carbonate chemistry of the surface waters of the world oceans. In Isotope Marine Chemistry, E. Goldberg, Y. Horibe, and K. Saruhashi, eds. Uchida Rokakuho, Tokyo, pp. 147-182.

Takahashi, T., W. S. Broecker, and A. E. Bainbridge (1981). The alkalinity and total carbon dioxide concentration in the world oceans. In Carbon Cycle Modelling, B. Bolin, ed. SCOPE Report 16. Wiley, New York, pp. 159-199.

Takahashi, T., D. Chipman, and T. Volk (1983). Geographical, seasonal, and secular variations of the partial pressure of CO_2 in surface waters of the North Atlantic Ocean: The results of the North Atlantic TTO Program. In Proceedings: Carbon Dioxide Research Conference: Carbon Dioxide, Science and Consensus. U.S. Dept. of Energy Report CONF-820970, Part II, pp. 123-145.

Walsh, J. J. (1981). A carbon budget for over-fishing off Peru. Nature 290:300-304.

Walsh, J. J., G. T. Rowe, R. L. Iverson, and C. P. McRoy (1981). Biological export of shelf carbon is a sink of the global CO_2 cycle. Nature 291:196-201.

Weiss, R. F. (1974). Carbon dioxide in water and sea water: the solubility of a non-ideal gas. Mar. Chem. 2:203-215.

3.3 BIOTIC EFFECTS ON THE CONCENTRATION OF ATMOSPHERIC CARBON DIOXIDE: A REVIEW AND PROJECTION

George M. Woodwell

3.3.1 Introduction

The composition of the atmosphere is changing. It has changed greatly, of course, throughout the period of the Earth's evolution. Few doubt the dominant role of the biota in the evolution of the atmosphere: the oxygen is residual from storage in the crust of reduced carbon compounds fixed by plants over hundreds of millions of years. The current changes are due in part to mobilization of these fossil reserves as fuel, as well as to changes in the amount of carbon retained in the biota and soils globally. The immediate question is how large the recent and future influence of the biota may be.

The primary evidence for current change is the record of observations of the CO_2 content of the atmosphere. Modern records show not only a year-by-year increase of between about 0.5 and 2 ppm in CO_2 annually but also a seasonal fluctuation. The amplitude of the oscillation varies with latitude, altitude, and probably with other factors (see Machta, Section 3.4). The amplitude approaches 20 ppm in central Long Island (Woodwell et al., 1973a) and in Barrow, Alaska (Kelley, 1969). It is about 5 ppm at Mauna Loa and 1 ppm at the South Pole (Keeling et al., 1976a,b). The peak CO_2 concentration occurs in late winter; the minimum occurs in early fall. The oscillation is reversed in the southern hemisphere to coincide with the southern seasons. These observations are evidence that biotic factors are large enough to influence the CO_2 content of the atmosphere in the short term over large regions, possibly the Earth as a whole. We know, in addition, that the terrestrial biota and soils contain two to three times as much carbon as the atmosphere. A small change in the size of these reservoirs globally has the potential for storing or releasing a quantity of CO_2 sufficient to affect the amount in the atmosphere appreciably.

Despite the strength of these observations, answers to specific questions about the role of the biota have proven elusive. Can we use the history of the terrestrial biota over the past century to help determine the preindustrial atmospheric CO_2 concentration (and thus the sensitivity of climate to increasing CO_2)? What fraction of the increase in CO_2 observed since 1958 at Mauna Loa is due to oxidation of carbon compounds in plants and soils as opposed to combustion of fossil fuels? What proportion of future CO_2 emissions will be absorbed into biotic reservoirs? If current trends in land use continue, what effect will they have on future atmospheric concentrations? To what extent could the future CO_2 content of the atmosphere be controlled by management of land and forests?

In the sections that follow we discuss the factors that affect the role of the biota in determining atmospheric CO_2 concentrations. These factors include the size and location of major reservoirs of carbon, the various transitions in metabolism that affect the reservoirs, and direct human effects, such as the clearing of forested land for agriculture.

3.3.2 <u>How Much Carbon is Held in the Biota and Soils?</u>

3.3.2.1 The Biota

Over the past decade, the most widely used appraisal of the global carbon content of biotic systems has been that of Whittaker and Likens (1973). Whittaker and Likens estimated that the world of 1950 held in the biota a total of 829 gigatons of carbon (Gt of C), almost entirely on land. By far the largest biotic reservoir, 743 Gt of C, was estimated to be in forests. Forests have been reduced in area in the more than 30 years since 1950. More recent estimates (e.g., Ajtay et al., 1979) of terrestrial biomass have suggested additionally that the average standing crop of organic matter per unit area is lower than estimated by Whittaker and Likens (1973). The most comprehensive recent analysis is that of Olson (1982), who has suggested a total biomass for 1980 of 560 Gt of C. The differences among these estimates are probably due largely to differences in interpretation and in assumptions; they may be due in part to reduction in the area of forests between 1950 and 1980. There is no easy resolution. To the extent that areas in forests are in question, satellite imagery will offer greatly improved estimates. Better data on biomass will require additional field studies.

3.3.2.2 The Soils

The total carbon retained in soils has been estimated globally as between about 1400 and 3000 Gt (Table 3.1). Schlesinger (1977, 1983) has reviewed the estimates and has suggested that there is a total of about 1500 Gt of readily mobilized carbon available. This conclusion has been supported recently by Post et al. (1982), who summarized data from more than 2000 samples of soils from around the world.

3.3.2.3 Total Carbon Pool under Biotic Influences

The total amount of carbon readily transformable into CO_2 by metabolic processes is probably in the range of 2000-2500 Gt, about 3 times the 725 Gt of C currently held in the atmosphere. Most of this inventory is associated with forests.

The fraction of the biotic pool that will actually be transformed to atmospheric CO_2 in future decades is speculative. In principle, a large portion could be. Evidence is that when forests are disturbed or replaced by agriculture there is a substantial loss of carbon. Loss from the original standing stock of plants usually exceeds 90%, and loss from soils can also be substantial. Plough horizons of soils long in agriculture commonly contain 1-3% carbon or less on a total dry weight basis. Original soils of the forests from which the agricultural soils were developed may have contained as much as 40-50% carbon in the surface horizon. Total loss over the entire profile is probably 20-50% of the original amount in the soil.

TABLE 3.1 Amount of Carbon Retained in Soils Globally According to Various Recent Estimates[a]

Total Carbon in Soils of the Earth (Gt)	Source
3000	Bohn, 1978
2070	Ajtay et al., 1979
1456	Schlesinger, 1977
1395	Post et al., 1982
1477	Buringh (in press, 1983)

[a]The data of Post et al. (1982) are the most comprehensive and recent estimates and are based on more than 2000 samples from around the world.

3.3.3 Metabolism and the Storage of Carbon in Terrestrial and Aquatic Ecosystems

3.3.3.1 The Production Equation

The net flux of carbon between the atmosphere and any ecosystem is determined by the balance between gross photosynthesis and total respiration. The relationship is shown by the production equation for an ecosystem (Woodwell and Whittaker, 1968):

$$NEP = GP - (R_A + R_H),$$

where NEP is the net ecosystem production, the net flux of carbon into or from an ecosystem; GP is the gross production, total photosynthesis of the ecosystem; R_A is the respiration of the autotrophs, the green plants; R_H is the respiration of the heterotrophs, including all animals and organisms of decay; and $R_A + R_H$ is the total respiration of the ecosystem.

The potential range of values for NEP is from a large negative number, indicating a loss of stored carbon, to a positive value that approaches the net amount of carbon available from green plants after their own needs for respiration (R_A) have been met. This excess above respiration is commonly called net primary production (NP). Its relationship to gross production is given by the equation

$$NP = GP - R_A,$$

where GP is the total photosynthesis of the ecosystem, as above; R_A is the respiration of the autotrophs; and NEP is $NP - R_H$.

The production equations have been especially useful in analyses of the metabolism and carbon flux of forests, where the terms can be

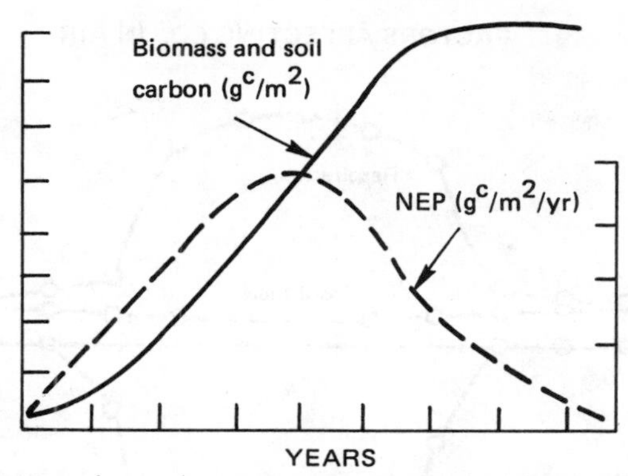

FIGURE 3.11 Relationship between net ecosystem production (NEP) and the total accumulation of carbon in a forest.

evaluated conveniently (for example, see Woodwell and Whittaker, 1968; Whittaker and Woodwell, 1969; Woodwell and Botkin, 1970; Reichle et al., 1973). The equations are applicable in aquatic systems as well (Woodwell et al., 1973b, 1979).

Net ecosystem production varies from zero at the start of the successional development of a forest to a maximum at midsuccession and back to zero at climax (Figure 3.11). The relationship emphasizes that undisturbed forests approach an equilibrium (climax) in which gross photosynthesis is equaled by total respiration. Any change in the relationship between gross production and either segment of the total respiration shifts the ratio and causes a positive or negative net ecosystem production. In general, photosynthesis is more vulnerable to disruption than total respiration. This generalization holds for both the individual plant and for the ecosystem as a whole. The reason is that photosynthesis is limited to certain plants and is dependent on many factors, all of which must be favorable; respiration is a general characteristic of all life and occurs in some form under a wide range of conditions. Virtually any disturbance (e.g., forest fire, land clearing, air pollution) favors respiration over photosynthesis, at least initially (see discussion below) and tends to result in transfer of carbon from the biota into the atmosphere.

3.3.3.2 A Basis in the Metabolism of Forests for the Oscillation in Atmospheric CO_2 Concentration

Our understanding of the metabolism of terrestrial ecosystems, including forests, agricultural systems, grasslands, and other communities, supports the hypothesis that the annual oscillation in CO_2 is due largely to the metabolism of temperate zone forests. The forests dominate because of their size, both in area and in magnitude of their metabolism. The annual course of metabolism of a temperate zone forest, taken in toto on a unit of land, expressed as gross photosynthesis and total respiration in separate curves, appears in Figure 3.12(a). Net ecosystem production at any time is the algebraic

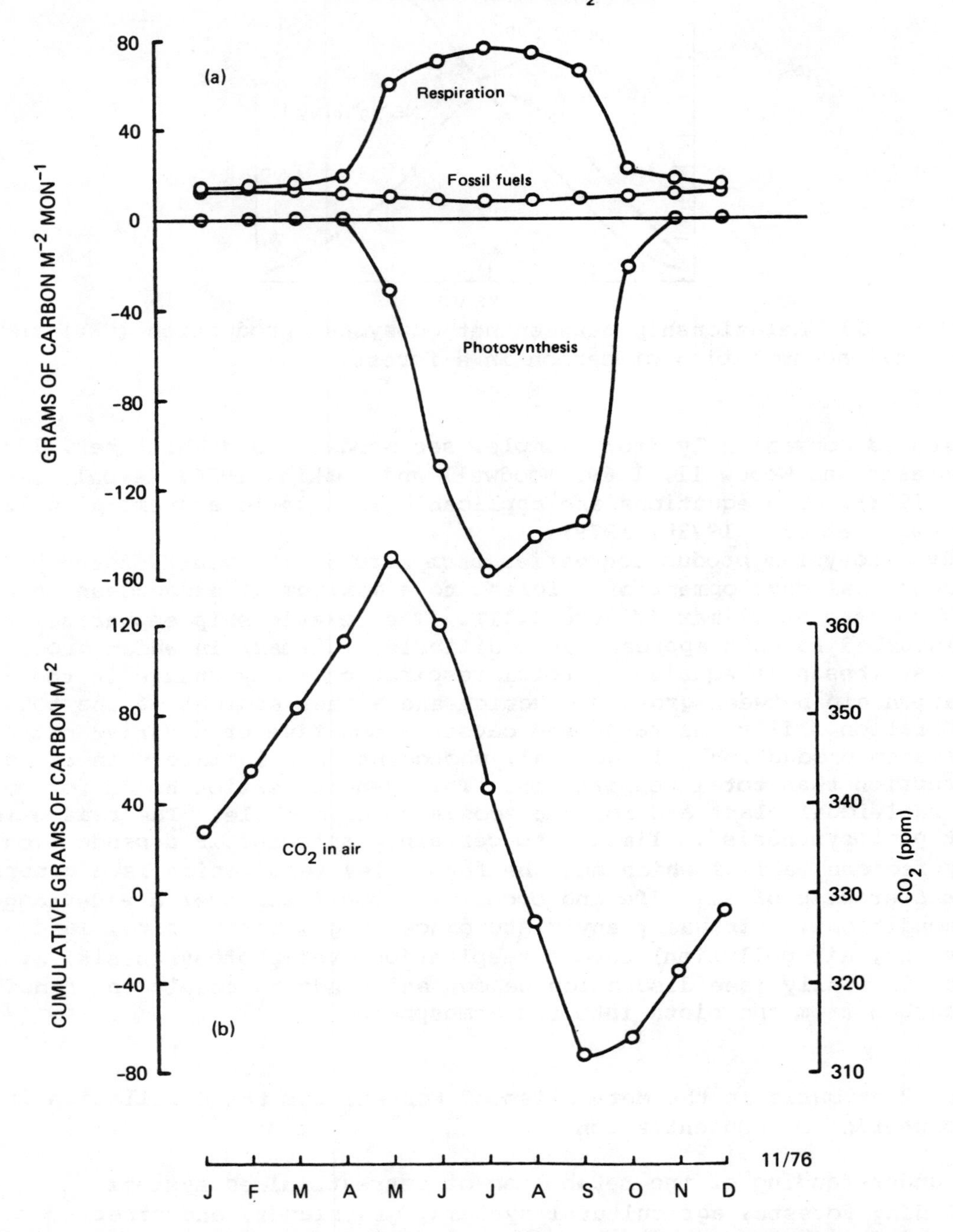

FIGURE 3.12 (a) The course of total respiration and gross photosynthesis of an oak-pine forest in central Long Island, New York. Integration of these two curves produced the prediction of the annual change in atmospheric CO_2 shown in the curve below. (b) The amplitude predicted in this way was considerably greater than observed (Woodwell et al., 1973a), apparently because of mixing with air from over the oceans.

sum of one point on each of the two curves. Integrating over the entire year produces a curve for net ecosystem production [Figure 3.12(b)]. This new curve follows closely the pattern of oscillation observed in the CO_2 concentration at Mauna Loa and elsewhere, although the amplitude observed around the world differs greatly from that calculated for the forest of central Long Island shown in Figure 3.12.

3.3.3.3 Factors Affecting Global Net Ecosystem Production

3.3.3.3.1 Succession and the Equilibrium Hypothesis

The equilibrium between gross photosynthesis and total respiration is achieved in forests after a period of successional development that may last decades to a century or so. It probably occurs in the pelagic segment of aquatic systems in days to weeks after disturbance. It is approximate: there is a residuum of the annual increment of fixed carbon that is stored in sediments both on land and in the sea, but the storage is a small fraction of annual production (Broecker, 1974). To illustrate, one of the largest accumulations of stored carbon on land is in the peat of the tundra and boreal forest, which we might estimate at 500 Gt of C. This mass is believed to have accumulated over the 10,000 years since the retreat of glacial ice, an annual rate of accumulation 0.05 Gt of C. The assumption of an equilibrium seems appropriate for this first approximation of the biotic flux.

Aquatic ecosystems move more rapidly than terrestrial systems toward equilibrium between gross photosynthesis and total respiration because the species are small-bodied and reproduce rapidly. Mechanical disturbance is common and has little effect on the ratio of gross production to total respiration under most circumstances. Chemical disturbance in the form of enrichment or pollution may simply speed the rate of turnover of carbon molecules, not increase the rate of sedimentation, at least through a wide range (Peterson, 1982), although Walsh et al. (1981) have suggested that enrichment of the coastal zone with nitrogen and phosphorus is causing increased accumulation of sediments on the shelf.

3.3.3.3.2 Gross Photosynthesis

Gross photosynthesis is awkward to measure directly in nature. It is normally measured as net photosynthesis of leaves plus total respiration of the ecosystem. Data are usually taken on a small scale in controlled environments or in carefully monitored natural ecosystems. Techniques are available now to provide more and better data on patterns and trends in the metabolism of ecosystems and should be exploited.

Several factors affect gross photosynthesis. The most important are light, moisture, and availability of nutrients, especially nitrogen, phosphorus, and CO_2. While it is common to think of one factor at a time as limiting the rate of any process such as photosynthesis, experience indicates that, throughout wide ranges, a change in the availability of any factor will produce a response. The dominant question for our consideration is whether the increase in atmospheric CO_2 is causing an increase in net ecosystem production globally. Waggoner (this volume, Chapter 6), Strain (1978), and a recent symposium (AAAS, in press) in Athens, Georgia, have reviewed studies in which efforts have been made to measure the effect of increased CO_2 on the growth of crop plants. While there is probably an important effect on growth of well-watered, fertilized plants, there is question as to whether these effects extend to natural communities. The following tentative generalizations may be offered (Kramer, 1981):

- Species differ greatly in response to enhanced CO_2.
- The response is greater in plants with indeterminate growth (cotton) than in plants with determinate growth (corn).
- The response is greater in C_3 plants such as soybeans than in C_4 plants such as corn.
- The largest response is in seedlings; in older plants the response decreases or ceases.

One of the most important factors is that any increase in growth observed is not always due to an increase in the rate of photosynthesis per unit of leaf area. Enhanced CO_2 concentrations cause changes in the morphology of growing plants, including an increase in branching of both woody and herbaceous plants, greater stem elongation, and an increase in the ratio between roots and shoots. One of the most persistent effects is an increase in the area of leaves, a result observed by Wong (1979) in a series of studies of effects of nitrogen nutrition and CO_2 on cotton and by others in several studies summarized by Kramer (1981) and Strain (1978).

While Waggoner (this volume, Chapter 6) reports generally beneficial effects on growth of crop plants, a positive response to increased CO_2 is not universal. Wong (1979) showed that corn plants assimilated less carbon under high CO_2 and speculated that contrary observations in other experiments may have been due to lower light intensities. Responses to enhanced CO_2 concentrations are particularly strong in younger plants, a factor that affects many of the experiments reported, such as that by Gifford (1979), in which a stimulation of growth was shown in wheat grown under moisture stress with enhanced CO_2 concentration. In longer-term experiments and in older plants response is diminished, disappears, or may involve a reduction in growth rate. These considerations should lead to caution in projecting effects of enhanced concentrations of CO_2 on photosynthesis in forests (Strain, 1978; Kramer, 1981).

Despite the expansion of agriculture, natural (unmanaged) forests still dominate the biotic segment of the global carbon cycle. Plants in natural forests live in conditions of extreme competition for light, water, nutrients, space, and, probably, CO_2 during daylight. None of the research reported applies directly to this circumstance, a fact that suggests great caution in predicting enhanced storage of carbon in natural systems due to increased atmospheric CO_2. In fact, Kramer (1981) concluded that in general, increase in CO_2 concentration will probably have the least effect on growth of plants in closed stands where light, water, and mineral nutrition, separately or collectively, already limit the rate of photosynthesis. Kramer's conclusion is supported by observations that the rate of photosynthesis per unit leaf area is not always increased by an increase in CO_2 concentration and that effects on plants often involve a change in the morphology in young plants. Forests are not modified rapidly in the latter respect.

Effects of temperature on gross photosynthesis are through effects on respiration, or other processes apart from the photochemical process, which is nearly independent of temperature (Larcher, 1980, p. 111).

3.3.3.3.3 Total Respiration

The data on respiration that are of significance in detecting a change in net ecosystem production are those that define the rate per unit of land area. They include the respiration of the community of higher plants, the community of animals, and the various communities of lichens, mosses, and organisms of decay. Such data have rarely been taken for terrestrial ecosystems; analysis must be based largely on inference from first principles or from data obtained for other purposes.

Rates of respiration are also affected by many factors, including availability of water, nutrients, especially nitrogen and phosphorus, and temperature. As in all chemical reactions, rates are affected by the availability of substrates and the accumulation of products. The greatest sensitivity is probably to temperature. A 10°C increase in temperature through the middle range of the response curve for a whole plant commonly produces a twofold to threefold increase in the rate of respiration (Table 3.2). Experimental evidence from tundra communities confirms the effect of warming (Billings et al., 1982). By comparison, the effects of other factors appear small. Direct effects of CO_2 concentrations in the range of 300-600 ppm on total respiration of an ecosystem are so small as to remain unmeasured and probably unmeasureable. The observation that rates of respiration in photosynthetic tissues differ in the light and in the dark (Zelitch, 1971) has little bearing on the total respiration, so large is this total relative to the fraction that occurs in photosynthetic tissues.

3.3.3.3.4 Net Ecosystem Production and the β-Factor

In an effort to resolve a discrepancy between the amount of carbon reportedly released through combustion of fossil fuels and the amount apparently transferred to the oceans, Bacastow and Keeling (1973) introduced a factor into their analysis that allowed an expansion of the biotic pool of carbon as a function of the increase in CO_2 in air. It was assumed that this so-called "β-factor" was the only important biotic consideration in the global carbon cycle. β was estimated to be a constant with a value of 0.26. The β-factor as formulated by Bacastow and Keeling was limited to the putatively positive effect of the increase in CO_2 on net ecosystem production.

TABLE 3.2 Respiratory Quotients (Q_{10}) for Plants and Plant Communities[a]

Zone/Species	Respiratory Q_{10}	Temperature Range (°C)	Reference
Pea seedlings	3.0	0-10	Giese, 1968
	2.4	10-20	
	1.8	20-30	
	1.4	30-40	
Plants	2.1-2.6	0-30	Fitter and Hay, 1981
Tundra/Taiga	2	Low temperature	Miller, 1981
	2	High temperature	
Tropics	3	10	Larcher, 1980
Clover	2.4	3-13	Woledge and Dennis, 1982
Greenland plants	2.0-2.7	0-25	Eckhardt et al., 1982

[a]The Q_{10} is the factor by which respiration is increased by a 10°C increase in temperature.

No consideration was given the possibility that processes other than CO_2 enrichment might affect the amount of carbon retained in the biota and soils or the possibility that the area of forests might be changing globally. The use of the β-factor was a pragmatic solution to a complex and puzzling issue that arose from attempts to analyze the global carbon cycle through a simple model. Its use should now be replaced by separate analyses of the effects of (a) changes in the area of forests and (b) potential changes in net ecosystem production caused by both increased atmospheric CO_2 and changes in climate. The latter will require modeling based on processes in terrestrial ecosystems.

The CO_2 content of the atmosphere has increased since 1860 to its current 340 ppm from a concentration now estimated at 260-280 ppm (see Machta, Section 3.4). The increase may be approaching 30%. Such a change in CO_2 content alone has probably had no effect on respiration. It may have affected gross photosynthesis, but if so, the change is detectable neither directly as a measurement of net or gross photosynthesis nor indirectly as a measurement of some segment of net ecosystem production such as the annual increment of wood in trees of seasonal forests. A widespread increase in annual tree growth of as little as 10% should be detectable as a universal or very common change in width of tree rings. No such stimulation is conspicuous. Rebello and Wagener (1976) found evidence in Europe of an increase in diameter growth, but Whittaker et al. (1974) found the opposite in North America. A smaller increase might remain undetected at present. Waggoner (this volume, Chapter 6) suggests a small (~5%) increase for crop plants in a 400-ppm atmosphere.

While some atmospheric changes may have favored growth in the total carbon held in the biota and soils globally, others may not. In particular, there is a need to consider the possible role of a long-term global warming of about 0.5°C since the lows of the late 1880s (see Weller et al., this volume, Chapter 5). If a 10°C increase in temperature increases rates of respiration twofold to threefold (Table 3.2), the 0.5°C warming observed may have increased total respiration of terrestrial ecosystems by 10-15%. Such a change would appear as a reduction in net ecosystem production; it might, of course, simply offset an increase in gross photosynthesis. The topic of changes in the biota as a result of enhanced CO_2 and climatic change requires detailed study through descriptive surveys and careful field experimentation under controlled circumstances. At the moment there is no direct evidence that net ecosystem production has changed per unit area of existing forests regionally or globally over the past century.

3.3.4 Changes in Area of Forests of the World

There have been many changes in the area of forests in postglacial time. The transitions have been caused by climatic changes such as those that accompanied the retreat of glacial ice, by shifts in patterns of distribution of rainfall, and by activities of man. The Levant, for instance, was probably largely deforested 5000 or more years ago through harvest of wood followed by intensive and prolonged grazing by goats. Other sections of the Mediterranean Basin were deforested more recently, but still 1000-5000 years ago. The British Isles were largely forested until the seventeenth century. Other areas, including much of Europe and northeastern North America, were cleared for agriculture and grazing two to three centuries ago. Agriculture was subsequently abandoned in some areas, and these have been partially reforested. Overall, the expansion of human population throughout history has been accompanied by an almost continuous decline in the area of forests globally.

The rate at which deforestation occurred in the past was slow by comparison with recent rates. The changes that affected areas as large as the Mediterranean Basin, the Levant, or the British Isles took place over centuries to millennia. Rates of CO_2 releases from biotic sources probably varied considerably and may have affected the CO_2 content of the atmosphere significantly, but there is no basis for measurements either of the biotic releases or of the CO_2 accumulation that followed them. During the past century, higher rates of deforestation may have been resulting in annual releases of CO_2 of as much as several billions of tons of carbon, in some years increasing the total atmospheric burden by as much as 0.5%.

The amount of the biotic contribution to the atmospheric increase is in question. The challenge has been measurement: how can information on rates of deforestation, and reforestation whenever it occurs, be summed for the Earth as a whole? The greatest uncertainty is in the rates of deforestation in the tropics, but there is uncertainty about the size of the contribution from loss of temperate zone and boreal forests as well. The largest areas of forest remaining in the world are in the tropics, especially the Amazon, and in the northern tem-

TABLE 3.3 Estimates of Annual Net Carbon Flux between Terrestrial Ecosystems and the Atmosphere in or about 1980[a]

Author	10^{15} g of C/yr
Adams et al., 1977	0.4 to 4
Bolin, 1977	0.4 to 1.6
Revelle and Munk, 1977	1.6
Wong, 1978	1.9
Woodwell et al., 1978	4 to 8
Hampicke, 1979	1.5 to 4.5
Seiler and Crutzen, 1980	-2.0 to 2.0
Brown and Lugo, 1981	-1.0 to 0.5
Moore et al., 1981	2.2 to 4.7
Olson, 1982	0.5 to 2.0
Houghton et al., 1983	1.8 to 4.7

[a]Positive values indicate net release to the atmosphere. Full citations in references.

perate and boreal zones. There are significant areas of forests remaining, however, in tropical Africa and in Southeast Asia.

There have been numerous attempts over the past decade to estimate the current annual net carbon flux between terrestrial ecosystems and the atmosphere. Estimates, summarized by Clark et al. (1982), range from a net absorption of 2.0 Gt of C to a net release of 20 Gt of C per year. Table 3.3 presents a selection of recent estimates. All the estimates suffer from a lack of persuasive detail as to rates of deforestation in key areas. The estimates differ in large part because they have treated different segments of the problem. When corrected to a common basis, the recent estimates converge considerably (Woodwell et al., 1982).

The most important advances in these analyses have come through recognition that sufficient information is available to allow prediction of the details of changes in forested areas if the time of (a) harvest or (b) transformation to agricultural or grazing land is known. A forest that is harvested by clearcutting and allowed to recover, for example, follows a predictable pattern of successional development. The observation that disturbance occurred is the critical point: the sequel is predictable. Similarly, a forest transformed to pasture or to row-crop agriculture loses its carbon stock predictably. If agriculture is abandoned, the forest recovers, again at predictable rates. Precision in the total inventory of carbon is less important than the evidence of change in the inventory. The evidence of change is abrupt and discontinuous, namely, the harvest of a forest or the abandonment of agriculture.

Evidence on changes in land use can be accumulated, tabulated, and summed in a model to offer an estimate at any time of the trends in carbon storage in forests, locally or globally. Such a model, constructed around the central principle of ecological succession, has been developed and used; results are reported below. Details of the construction of the model, including the data and assumptions used and tests of sensitivity, have been presented elsewhere (Moore et al., 1981; Houghton et al., 1983; Woodwell et al., 1983a). The model accommodates 12 geographic regions and 10 different types of vegetation in each region. Transitions in soils are also included. A few comments about quality and sources of data are necessary.

Data appraising rates of change in forested areas are surprisingly difficult to obtain and verify. The major factors to be measured are (1) change in the amount of carbon per unit area of forests and (2) change in the area of forests. Neither can be measured unequivocally for the globe at present.

Several factors contribute to the difficulties in obtaining sound data. One is that no nation is proud of the destruction of a valued resource, and national statistics are often unreliable. In addition, many nations lack equipment and personnel needed for gathering data on area of forests remaining and for evaluating such data. There is always question as to what is forest. Are the successional stands that replace moist tropical forests following harvest to be considered equivalent to the forests they replaced? Are impoverished, heavily grazed woodlands forests? Various economic considerations tend to bias reporting first one way, then another (Persson 1974, 1977). Recent studies of remote sensing using the LANDSAT and NOAA systems show that satellite imagery offers great promise for improving data on areas of forests globally (Woodwell, 1980; Woodwell et al., 1983b; Woodwell et al., in press).

Three sources for estimating rates of deforestation have been used extensively in recent analyses. One source is the series of *Production Yearbooks* of the UN Food and Agriculture Organization (FAO), published since 1949. They rely on data reported by governments. A second source is the work of Myers (1980), who has compiled detailed estimates of the rate of conversion of tropical forests using a variety of sources. The emphasis on the tropics is appropriate because of the extraordinary growth in the human population in the tropics and the surge of economic development that has affected the tropical regions since the Second World War. A third source of estimation is based on the assumption that there is a simple correlation between growth in human population and rate of deforestation. The basis of the assumption is that most of the loss of forests is due to conversion for agriculture. Revelle and Munk (1977) developed this approach initially.*

---

*Richards et al. (1983) have recently used historical records to determine land area converted from unmanaged ecosystems to regularly planted cropland. They estimate net conversion of 851 million hectares between 1860-1978; Revelle and Munk estimated clearing between 1860-1970 of 853 million hectares.

TABLE 3.4 Release of Carbon into the Atmosphere Using Three Sources of Data in the Marine Biological Laboratory (MBL) Terrestrial Carbon Model. Sensitivity Tests Show Factors of Greatest Importance in Affecting Releases[a]

	Total Release (Gt of C)		Release in 1980 (Gt of C)						
	1860-1980	1958-1980	Clearing for Agriculture	Harvest of Forests	Decay of Wood Removed	Abandonment of Agriculture	Afforestation	Clearing for Pasture	Total
A. Data Sources									
1. FAO Production Yearbooks	184	52	0.59	0.63	0.47	-0.066	-0.039	0.23	1.82
2. N. Myers	185	70	2.70	1.06	0.74	-0.066	-0.039	0.31	4.70
3. Data on population	180	57	1.30	0.63	0.56	-0.066	-0.039	0.23	2.61
B. Sensitivity Tests[b]									
4. Heavy clearing in 1860-1914	210	62	1.45	0.63	0.58	-0.062	-0.039	0.23	2.79
5. a. Forests cleared	228	76	1.97	0.63	0.76	-0.066	-0.039	0.23	3.49
b. Nonforests	135	38	0.62	0.63	0.33	-0.066	-0.039	0.23	1.72
6. Increased clearing and abandonment	186	60	1.49	0.60	0.60	-0.20	-0.039	0.23	2.70
7. Storage as charcoal	175	56	1.16	0.63	0.59	-0.066	-0.039	0.23	2.51
8. Reduced loss from soil	150	51	1.01	0.63	0.56	-0.041	-0.039	0.23	2.35
9. Doubled harvest of wood	175	55	1.30	0.29	0.78	-0.066	-0.039	0.23	2.49
10. Increased rates of recovery	152	47	1.30	0.055	0.55	-0.093	-0.050	0.23	1.98
11. Reduced C/area	152	45	1.03	0.43	0.49	-0.066	-0.039	0.23	1.98

[a]Adapted from Houghton et al. (1983). Biomass values in model are taken from Whittaker and Likens (1973).
[b]Using population-based approach (3) as baseline.

Analyses using the three sources have been reported in detail by Houghton et al. (1983), and comparisons have been made with other, less comprehensive analyses that appeared to reach different conclusions, such as those of Detwiler et al. (1981) and Brown and Lugo (1981). In each instance, when appropriate adjustments were made to include all major factors, such as decay of soil organic matter, the analyses converged to a narrower range. Olson (1982), on the other hand, estimated a net loss of slightly less than +1 Gt of C from the terrestrial sources in 1980; the estimate was based on lower appraisals of the mass of carbon per unit area than those used by Houghton et al. (1983) and interpretations of changes in the area of forests that are difficult to document.

The results of the application of the model using the three sources appear in Table 3.4. The range of the total release of carbon from the biota and soils into the atmosphere since 1860 varied little, between 180 and 185 Gt. Most of the difference occurred in recent decades: the range of estimates between 1958 and 1980 was 52-70 Gt of C. The annual net release in 1980 was estimated as between 1.8 and 4.7 Gt of C.

The model was used to test assumptions about the data and the significance of various biotic processes. These tests showed that lower total releases over 120 years could have occurred if agricultural land were taken only from nonforested areas (5b in Table 3.4), if there were reduced decay of soil carbon (8 in Table 3.4), if there were very much more rapid recovery of disturbed forests than experience dictated (10 in Table 3.4), and if there were substantially less carbon in the vegetation and soils than sources such as Whittaker and Likens (1973) had indicated (11 in Table 3.4). These changes were all applied to the population-based estimate. None of these modifications turned the biotic pools into a net sink for atmospheric carbon at any time in the 120-year period. The annual release in 1980 reached a minimum of 1.7 x 10 Gt of C when the assumption was made that all agricultural land was taken from nonforested areas. That assumption must be considered unrealistic; it was included in the analysis to provide a limit for comparison.

3.3.5 The Biota in the Context of the Global Carbon Balance

The effort to this point has been to summarize the most probable transitions in the biotic pools of carbon globally over the past century. The results are not consistent with current estimates of other segments of the global carbon cycle. The global carbon balance can be expressed as

$$A = F - S \pm B,$$

where A is the increase in the carbon content of the atmosphere over any period, F is the release of carbon to the atmosphere from combustion of fossil fuels in the same period, S is the net transfer to the oceans in the same period, and B is the absorption or release of carbon by the biota in the same period.

This equation can be evaluated to estimate the role of the biota. For example, for 1980 Houghton et al. (1983) reported (in Gt of C per year)

$$2.5\ (\pm 0.2) \neq 5.2\ (\pm 0.7) - 2.0\ (\pm 0.5) - 0.7(\pm 1.4).$$

The sources of the quantities in this case were A = 2.5 for 1980 from Bacastow and Keeling (1981), F = 5.2 for 1980 from Rotty (1981), and S = 2.0 calculated as 39% of the fossil fuel release following Broecker et al. (1979). If the estimates of A, F, and S are accepted, the biota absorbed 0.7 Gt of C in 1980. If A, F, and S are stretched to their proposed limits of uncertainty, the biota may have absorbed 2.1 Gt of C or released 0.7 Gt of C. The method is indirect, but it has the advantage of presenting simply the relationships among the various major fluxes of carbon.

Alternatively, methods based on isotopic dilution may be used to estimate total biotic effects over a long period of time. Results suggest a net release from the biota of between 70 and 195 Gt of C over a century or more (Table 3.5). These studies cannot provide precise estimates for a given year.

As we have seen, most direct analyses of recent changes in the biotic reservoirs of carbon, when they include releases of carbon from decay of organic residues from the plants, organic matter in soils, and the decay of wood and forest products removed from the site, show a net release of carbon from destruction of forests. The release is estimated currently here on the basis of extensive experience as between 1.8 and 4.7 Gt of C per year (Houghton et al., 1983; Woodwell et al., 1983a). This range is well outside the estimates above, including the ranges of uncertainty. Nonetheless, if the actual release is at the lower end of this range, there may be some basis for argument that the global carbon equation is balanced within the range of uncertainty associated with the other terms. If the actual value is near the upper end of the range, the equation is unreconcilable. A recent reinterpretation of

TABLE 3.5 Estimates of the Release of Carbon from the Terrestrial Biota and Soils to the Atmosphere during the Past Century Based on Studies of Isotopes of Carbon in Tree Rings[a]

Reference	Period	Gt of C
Stuiver, 1978	1850-1950	120
Wagener, 1978	1800-1935	170
Freyer, 1978	1860-1974	70
Siegenthaler and Oeschger, 1978	1860-1974	135-195
Tans, 1978	1850-1950	150

[a]Adapted from Houghton et al. (1982).

the data from Myers (1980) on forest conversion in the tropics suggests that the upper limit of this range may be as low as 3.0 Gt of C.

One hypothesis frequently proposed to balance the carbon equation is that net ecosystem production on land has increased in response to the increase in CO_2 in the atmosphere. Such an increase would have to have been substantial, as much as 100-200 Gt of C over the past 120 years, to overcome the biotic losses to the atmosphere described above. It would probably be detected as a universal increase in diameter growth of trees or storage of humus in soils. Any such stimulation of carbon storage would have been accelerating as deforestation has proceeded, reducing areas where storage could occur. Moreover, it would require an increased spread between gross production and total respiration globally. A warming trend would probably work counter to this by increasing rates of respiration of plants and soil organic matter without a corresponding increase in gross production. This relationship would persist unless other factors, too, were ameliorated, such as water supply and the availability of nutrient elements, especially N and P.

Additional insight on the role of the biota has been sought through exploration of the oscillation in CO_2 concentration observed at Mauna Loa (Hall et al., 1975; Bacastow et al., 1981b; Pearman and Hyson, 1981). The hypothesis is that a change in the area of forests or a change in net ecosystem production regionally should be reflected over the 25-year record in a systematic change in the amplitude of the oscillation. Analyses showed no identifiable trend over the first 15 years (Hall et al., 1975). There are now indications that the amplitude has been increasing in recent years (Bacastow et al., 1981a,b; Machta, this volume, Chapter 3.4).

How is such a change in amplitude to be interpreted? There appears to be little question that the oscillation itself is caused by the metabolism of forests. If the amplitude can be assured to be free of potentially confounding effects, it may offer an appraisal of the status of net ecosystem production at any moment for a segment of the northern hemisphere (Figure 3.12). Unfortunately, the coupling between the oscillation and metabolism of forests is too loose for the amplitude to be considered a unique measurement of metabolism. Many factors affect it. These include, for example, atmospheric mixing: the amplitude of the oscillation is reduced at higher elevations and at lower latitudes. Small changes in patterns of circulation of air can be expected to affect it, as well as variability in the temperature of seawater. The amplitude is also open to various direct effects of metabolism of forests. A warm winter in the northern hemisphere, for example, would increase the total respiration without affecting the photosynthetic withdrawal during summer appreciably; the excess CO_2 would appear as an increased late-winter peak in the Mauna Loa record. That CO_2 would be mixed into the rest of the atmosphere over the ensuing weeks and would have little or no effect on the late summer minimum. The amplitude of the oscillation would have been increased and the total biotic pool of carbon on land reduced. The converse, an increase in the storage of carbon during summer, whatever the cause, would also appear as an increased amplitude in the oscillation. Variations in fossil fuel use may also be a confusing factor. While the oscillation of the Mauna Loa record may be interpretable, reliance on the record to identify transitions in the biota must await considerably greater attention to detail and better data than have been available so far.

The possibility remains that aquatic systems have been stimulated in some way into accelerated storage of fixed carbon in sediments or in the deeper waters of the oceans. Freshwater systems can be ignored: they are very small in comparison with the oceans, which cover two thirds of the surface of the Earth. Baes (1982) and others (Smith, 1981; Walsh et al., 1981) have suggested various mechanisms by which biotic activity might result in sedimentation of carbon. To be significant in the global cycle these mechanisms would have to account for 1 Gt of C or more annually and would have to respond in some way roughly proportional to the increase in CO_2 in the atmosphere. While there is no question about the capacity of the oceanic biota, either in coastal areas (Walsh et al., 1981) or in the open oceans (Baes, 1982) to fix sufficient carbon to be significant in the global balance, there is question as to whether enough fixed carbon is sequestered in these waters to affect the global cycle in ways measurable now on a year-by-year basis. In pelagic aquatic systems gross production and total respiration tend to be closely coupled. An increase in photosynthesis is quickly followed by an increase in respiration; storage in sediments is small.

Peterson (1982) addressed the question raised by Baes (1982) as to the capacity of the marine biota for storing carbon in any form. His conclusion was that there is so much carbon in seawater as dissolved CO_2 in equilibrium with the oceanic carbonate-bicarbonate system that errors in estimates of oceanic absorption of CO_2 are most likely to involve rates of mixing of surface waters into intermediate or greater depths. The issue remains unresolved. If the terrestrial biota appear to be a substantial net source of CO_2 for the atmosphere beyond the fossil fuel source, the oceans must be absorbing substantially more CO_2 than has been measured.

The discrepancies are emphasized further by consideration of the fraction of the CO_2 released that remains in the atmosphere. The annual increase in CO_2 in the atmosphere is caused by the accumulation of some fraction of the total CO_2 released. Because the fossil fuel CO_2 has been thought to be the major, and sometimes the only, source of additional CO_2, a frequent practice has been to express the increase in atmospheric CO_2 as a fraction of the fossil fuel release. The fraction calculated in this way has the advantage of being based on two numbers that are measured with considerable accuracy. The fraction has the further advantage of being expected to approach a constant in the simplified models commonly used (Bacastow and Keeling 1979, 1981).

The airborne fraction, calculated solely on the basis of the Mauna Loa data and estimates of combustion of fossil fuels was 0.55 for the period 1959-1978, according to Bacastow and Keeling (1981). The range of airborne fractions consistent with current carbon-cycle models was explored by Oeschger and Heimann (1983). They suggest that a range from about 0.4 to 0.7 is possible. If the fraction is less than 0.4, there are deficiencies in current descriptions of the carbon cycle.

When the fossil fuel contribution alone is considered, the preindustrial atmospheric CO_2 concentration is usually thought to have been 290-300 ppm. Recognition that there has been a large additional release from the biota and soils is consistent with recent measurements and estimates of a lower preindustrial concentration (See Machta, Section 3.4). If the biotic release were larger during the latter part of the last century than now, the airborne fraction would have a curious time history. To illustrate, we can calculate the airborne fraction for two periods, 1860-1958 and 1959-1980, using the estimate of biotic releases based on population (see Table 3.4 above) and data from Rotty (1981, 1982) on releases from combustion of fossil fuels.

Time Period	Biotic Release	Fossil Release	Total Release	Atmospheric Increase	Airborne Fraction
1860-1958	123	76	199	106	0.53
1959-1980	57	86	143	49	0.34
Total	180	162	342	155	0.45

The estimates are made on the assumption, taken arbitrarily, that the preindustrial CO_2 concentration in the atmosphere was 265 ppm. During the earlier period, 1860-1958, 53% of the total CO_2 thought to have been released to the atmosphere appears to have remained there. During the latter period, about 34% of the total release seems to have accumulated. Over the entire 120 years about 45% of the total has remained in the atmosphere according to these estimates.

A circumstance in which the fraction of atmospheric carbon transferred into the oceans or other sinks increases as the concentration of CO_2 in the atmosphere rises is difficult to envision. No persuasive explanation is available. Mass balance considerations do not appear to support the hypothesis that entrophication of coastal waters, for instance, is causing accelerated sedimentation of carbon currently (Peterson, 1982). There do not appear to be mechanisms for sequestering sufficient carbon on land. The uncertainty of the basic numbers, especially the preindustrial CO_2 concentration and the magnitude and timing of biotic releases, both used to estimate the airborne fraction, re-emphasize that little should be inferred at present from calculations such as these.

There is, nonetheless, ample basis for arguing that the total release of carbon into the atmosphere has been and remains larger than the release from fossil fuels alone. Such a release means that the total accumulation in the atmosphere has been a lower fraction of the total release than estimated solely on the basis of combustion of fossil fuels. Presumably there has been a greater transfer to the oceans than commonly recognized.

3.3.6 A Projection of Further Releases from Biotic Pools

If the analyses above are correct, the most important biotic transition that affects the global carbon cycle is the destruction of forests. The rate of destruction and the rate of release of carbon to the atmosphere can be anticipated. If the rate is proportional to the increase in population as assumed in the intermediate analysis reported above from Houghton et al. (1983), and it is assumed further that population continues to grow through the year 2000 and that the rate of growth declines thenceforth to zero in the year 2100, the release from biotic sources might be expected to follow the solid line of Figure 3.13. Releases would range between the current estimate of approximately 2 Gt of C annually to a maximum of between 6 and 7 Gt of C in 2000. If the growth in population were to continue beyond 2000, the forests themselves would limit the release by the year 2030 to a maximum of about 10 Gt of C annually. Forests would then cease to exist as closed stands; the impoverished stands would decline progressively in carbon content and productivity. Rapid oxidation of 230 Gt of C, roughly the current total carbon content of tropical forests according to Olson (1982), would lead to an appreciable increase of atmospheric CO_2 (see Machta, Section 3.6).

3.3.7 Summary and Conclusions

The biota and soils of the Earth contain more than three times as much carbon as the atmosphere. The most powerful evidence for the importance of the biota in affecting the CO_2 content of the atmosphere is the annual oscillation in the CO_2 concentration observed at Mauna Loa and in virtually every other annual record of atmospheric CO_2. The extent of the biotic influence and the factors that govern it remain uncertain. The influence, however, is due primarily to global changes in forest biomass.

Forests have two types of effects on atmospheric CO_2, a shorter-term effect that is apparent in the oscillation in the concentration of CO_2 and a longer-term effect due to changes in the total amount of carbon stored in them. The largest change in the mass of carbon in forests appears to be a net global reduction due to deforestation to support the expansion of agriculture. The biotic release from all changes in area, expressed as a net for the world as a whole, is probably in the range of 1.8-4.7 Gt of C annually. This conclusion is derived from tabulations of data on rates of deforestation and forest harvest evaluated with a model. The model provides for the recovery of forests following harvest or abandonment of agriculture. Most studies, when expressed globally with adjustments to include soils and successional recovery, produce results that fall within the range stated.

The question remains as to whether the metabolism of forests is being affected to change the storage of carbon in otherwise untouched stands. Such a change would require an increase in the spread between gross photosynthesis and total respiration. The factors that are most likely to affect this spread are light, moisture, nutrients, and tem-

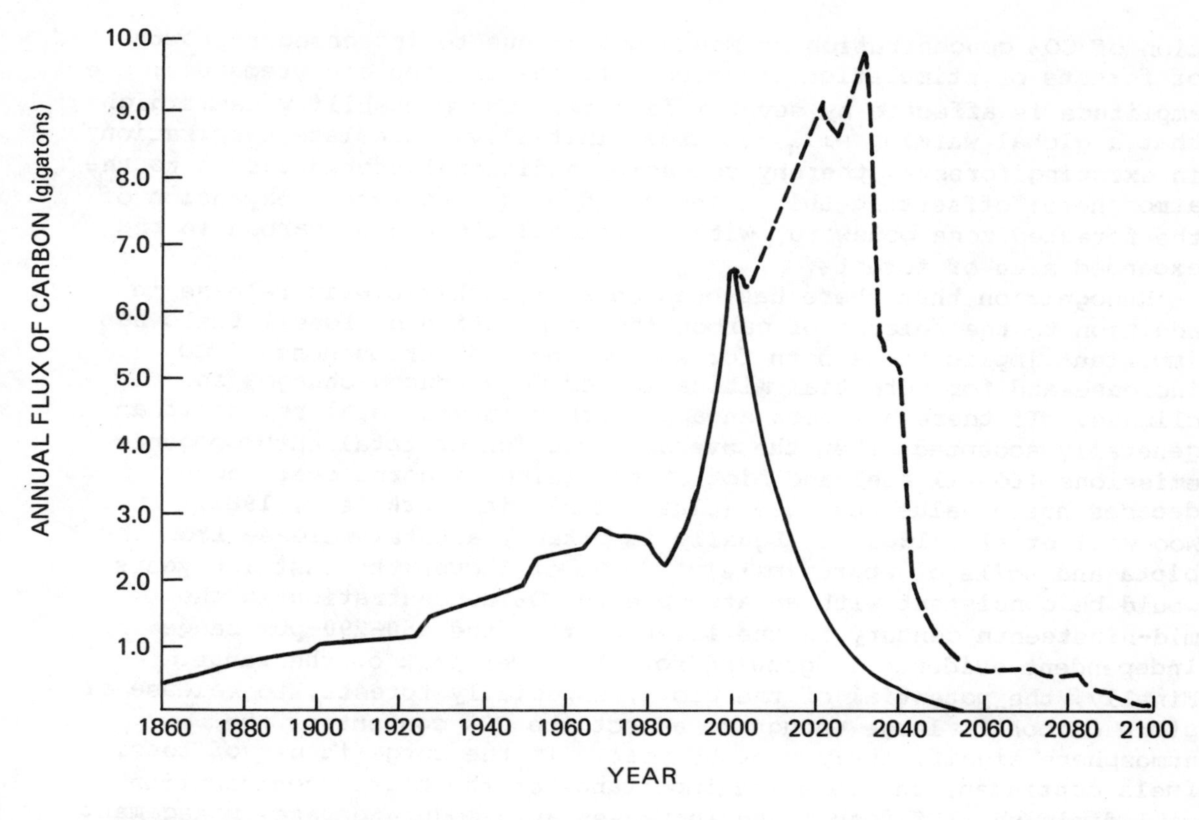

FIGURE 3.13 Predictions of the net release of carbon from the biota over the next several decades based on the assumption that deforestation is proportional to growth in population. The solid line is the net release on the assumption that the population continues to grow until the year 2000, when growth then declines exponentially to zero; the dashed line shows the release if the population continues to grow and forests simply disappear early in the next century. For comparison, Nordhaus and Yohe (this volume, Chapter 2, Section 2.1) offer a mean estimate of fossil fuel CO_2 emissions of about 5 Gt of C in the year 2000, 10 Gt of C in 2025, 15 Gt of C in 2050, and 20 Gt of C in 2100. Deforestation is not a simple function of growth in population; it is in part due to economic and industrial circumstances. Nonetheless, for this prediction the assumption as made here is adequate to show that releases of CO_2 from deforestation could increase well into the next century and then decline.

perature, in addition to the concentration of CO_2. Of these, temperature would seem to have the greatest potential: a 1°C change might be expected to produce a 20-30% change in rate of respiration. Any effect on total photosynthesis would be largely through an increase in the length of the growing season. An increase in the growing season with impact on photosynthesis comparable with that of temperature on respiration would not be expected from a 1°C increase in temperature. Interpretations of the apparent increase in amplitude of the oscillation of CO_2 concentration at Mauna Loa as due to increased regrowth of forests or stimulation of carbon storage on land are premature; the amplitude is affected by several factors. The probability seems high that a global warming will, at least initially, stimulate respiration in existing forests, thereby releasing additional stored carbon to the atmosphere; offsetting this trend could be a longer-term expansion of the forested zone poleward, with additional storage of carbon in the expanded area of forests.

Recognition that there has been an appreciable biotic release in addition to the release of carbon from combustion of fossil fuels has important implications both for estimating the seriousness of CO_2 increase and for potential mitigation of CO_2-induced changes in climate. If there has been an appreciably larger total release than generally accepted, then the average fraction of total anthropogenic emissions (fossil fuel and biotic) remaining airborne over recent decades has a value near 0.4 (Clark et al. in Clark, ed., 1982; Woodwell et al., 1983a). Equally important, a total release from the biota and soils of approximately 180 Gt of C over the past 120 years would be consistent with an atmospheric CO_2 concentration in the mid-nineteenth century in the lower part of the 250-290-ppm range. Independent evidence is growing for the lower part of the range. Finally, the potential of the biota, especially forests, to release or store carbon is large enough to affect the CO_2 content of the atmosphere significantly year by year. If the surge in use of fossil fuels continues, the relative importance of the biotic contribution will diminish. If fossil use increases at a moderate rate, management of the biotic pools of carbon can affect the time that any given atmospheric CO_2 concentration is reached by several decades. Better prediction, even control, of future CO_2 concentrations are possible but will require substantially strengthened research and understanding of the carbon cycle.

References

Adams, J. A. S., M. S. M. Mantovani, and L. L. Lundell (1977). Wood versus fossil fuel as a source of excess carbon dioxide in the atmosphere: a preliminary report. Science 196:54-56.

Ajtay, G. L., P. Ketner, and P. Duvigneaud (1979). Terrestrial primary production and phytomass. In The Global Carbon Cycle, B. Bolin et al., eds. SCOPE Report 13. Wiley, New York, pp. 129-181.

Bacastow, R., and C. D. Keeling (1973). Atmospheric carbon dioxide and radiocarbon in the natural carbon cycle. II: Changes from A.D. 1700 to 2070 as deduced from a geochemical model. In Carbon in the Biosphere, G. M. Woodwell and E. V. Pecan, eds. USAEC Symposium Series No. 30., U.S. Dept. of Commerce. NTIS, Springfield, Va., pp. 86-136.

Bacastow, R. B., and C. D. Keeling (1979). Models to predict future atmospheric CO_2 concentrations. In Workshop on the Global Effects of Carbon Dioxide from Fossil Fuels, W. P. Elliott and L.

Machta, eds. U.S. Dept. of Energy, CONF-770385, NTIS, Springfield, Va., pp. 72-90.

Bacastow, R., and C. D. Keeling (1981). Atmospheric carbon dioxide concentration and the observed airborne fraction. In Carbon Cycle Modelling, B. Bolin, ed. SCOPE Report 16. Wiley, New York, pp. 103-112.

Bacastow, R. B., C. D. Keeling, T. P. Whorf, and C. S. Wong (1981a). Seasonal amplitude in atmospheric CO_2 concentration at Canadian weather station P, 1970-1980. Papers presented at WMO/ICSU/UNEP Scientific Conference on Analysis and Interpretation of Atmospheric CO_2 Data, Bern, 1981. World Climate Programme.

Bacastow, R. B., C. D. Keeling, and T. P. Whorf (1981b). Seasonal amplitude in atmospheric CO_2 concentration at Mauna Loa, Hawaii, 1959-1980. Papers presented at WMO/ICSU/UNEP Scientific Conference on Analysis and Interpretation of Atmospheric CO_2 Data, Bern, 1981. World Climate Programme.

Baes, C. F. (1982). Effects of ocean chemistry and biology on atmospheric carbon dioxide. In Carbon Dioxide Review: 1982, W. C. Clark, ed. Oxford U. Press, New York, pp. 187-204

Billings, W. D., J. O. Luken, D. A. Mortensen, and K. M. Peterson (1982). Arctic tundra: A source sink for atmospheric carbon dioxide in a changing environment? Oecologia 53:7-11.

Bohn, H. L. (1976). Estimate of organic carbon in world soils. Soil. Sci. Soc. Am. J. 40:468-470.

Bolin, B. (1977). Changes of land biota and their importance for the carbon cycle. Science 196:613-615.

Broecker, W. S. (1974). Chemical Oceanography. Harcourt Brace Jovanovich, New York.

Broecker, W. S., T. Takahashi, H. J. Simpson, T.-H. Peng (1979). Fate of fossil fuel carbon dioxide and the global carbon budget. Science 206:409-418.

Brown, C. W., and C. D. Keeling (1965). The concentration of atmospheric carbon dioxide in Antarctica. J. Geophys. Res. 70:6077-6085.

Brown, S., and A. E. Lugo (1981). The role of the terrestrial biota in the global CO_2 cycle. Proceedings of a Symposium: A Review of the Carbon Dioxide Problem. Am. Chem. Soc., Div. Petrol. Chem. 26:1019-1025.

Buringh, P. (1983). Organic carbon soils of the world. In The Role of Terrestrial Vegetation in the Global Carbon Cycle: Measurement by Remote Sensing, G. M. Woodwell, ed. SCOPE. Wiley, New York (in press).

Clark, W. C., ed. (1982). Carbon Dioxide Review: 1982. Oxford U. Press, New York.

Detwiler, R. P., C. A. S. Hall, and P. Bogdonoff (1981). Simulating the impact of tropical land use changes on the exchange of carbon between vegetation and the atmosphere. In Global Dynamics of Atmospheric Carbon, S. Brown, ed. Proceedings of a symposium held at the annual meeting of the Ecological Society of America under the auspices of the American Institute of Biological Sciences, Bloomington, Indiana, 1981. U.S. Dept. of Energy, pp. 141-159

Eckhardt, F. E., L. Heerfordt, H. M. Jorgensen, and P. Vaag (1982). Photo-synthetic production in Greenland as related to climate, plant cover, and grazing pressure. Photosynthetics 16:71-100.

Giese, A. C. (1968). Cell Physiology. 3rd ed. Saunders, Philadelphia, Pa.

Gifford, R. M. (1979). CO_2 and plant growth under water and light stress: implications for balancing the global carbon budget. Search 10:316-318.

Hall, C. A. S., C. A. Ekdahl, and D. E. Wartenberg (1975). A fifteen year record of biotic metabolism in the northern hemisphere. Nature 255:136-138.

Hampicke, U. (1979). Sources and sinks of carbon dioxide in terrestrial ecosystems. Environ. Internat. 2:301-316.

Houghton, R. A., J. E. Hobbie, J. M. Melillo, B. Moore, B. J. Peterson, G. R. Shaver, and G. M. Woodwell (1983). Changes in the carbon content of terrestrial biota and soils between 1860 and 1980: A net release of CO_2 to the atmosphere. Ecolog. Monogr. (in press).

Keeling, C. D., J. A. Adams, Jr., C. A. Ekdahl, and P. R. Guenther (1976a). Atmospheric carbon dioxide variation at the South Pole. Tellus 28:552-564.

Keeling, C. D., R. B. Bacastow, A. E. Bainbridge, C. A. Ekdahl, Jr., P. R. Guenther, L. S. Waterman, and J. F. S. Chin (1976b). Atmospheric carbon dioxide variations at Mauna Loa Observatory, Hawaii. Tellus 28: 538-551.

Kelley, J. J. (1969). An analysis of carbon dioxide in the arctic atmosphere near Barrow, Alaska, 1961 to 1967. University of Washington, Dept. of Atmospheric Sciences, Scientific Report, Naval Research Contract No. 614-67-A-0103-0007. NR. 307-252, Seattle, Washington.

Kramer, P. J. (1981). Carbon dioxide concentration, photosynthesis and dry matter production. BioScience 31:29-33.

Larcher, W. (1980). Physiological Plant Ecology. Springer-Verlag, Berlin.

Moore, B., R. D. Boone, J. E. Hobbie, R. A. Houghton, J. M. Melillo, B. J. Peterson, G. R. Shaver, C. J. Vorosmarty, and G. M. Woodwell (1981). A simple model for analysis of the role of terrestrial ecosystems in the global carbon budget. In Carbon Cycle Modelling, B. Bolin, ed. SCOPE Report 16. Wiley, New York.

Myers, N. (1980). Conversion of Tropical Moist Forests. Prepared for Committee on Research Priorities in Tropical Biology, National Research Council, National Academy of Sciences, Washington, D.C.

Oeschger, H., and M. Heimann (1983). Uncertainties of predictions of future atmospheric CO_2 concentrations. J. Geophys. Res. 88:1258-1262.

Olson, J. S. (1982). Earth's vegetation and atmospheric carbon dioxide. In Carbon Dioxide Review: 1982, W. C. Clark, ed. Oxford U. Press, New York, pp. 388-398.

Pearman, G. I., and P. Hyson (1981). The annual variation of atmospheric CO_2 concentration observed in the Northern Hemisphere. J. Geophys. Res. 86: 9836-9843.

Persson, R. (1974). Review of the world's forest resources in the early 1970s. Dept. of Forest Survey Res. Notes No. 17. Royal College of Forestry, Stockholm. 265 pp.

Persson, R. (1977). Scope and approach to world forest resource appraisals. Dept. of Forestry Survey Res. Notes No. 23. Royal College of Forestry, Stockholm.

Peterson, B. J. (1982). In Carbon Dioxide Review: 1982, W. C. Clark, ed. Oxford U. Press, New York, pp. 205-206.

Post, W. M., W. R. Emanuel, P. J. Zinke, and A. G. Stangenberger (1982). Soil carbon pools and world life zones. Nature 298:156-159.

Rebello, A., and K. Wagener (1976). Evaluation of ^{12}C and ^{13}C data on atmospheric CO_2 on the basis of a diffusion model for oceanic mixing. In Environmental Biogeochemistry, Vol. 1: Carbon, Nitrogen, Phosphorus, Sulfur, and Selenium, J. O. Nriagu, ed. Ann Arbor Science Publishers, Ann Arbor, Mich., pp. 13-23.

Reichle, D. E., B. E. Dinger, N. T. Edwards, W. F. Harris, and P. Sollins (1973). Carbon flow and storage in a forest ecosystem. In Carbon and the Biosphere, G. M. Woodwell and E. V. Pecan, eds. U.S. Atomic Energy Commission, pp. 345-365.

Revelle, R., and W. Munk (1977). The carbon dioxide cycle and the biosphere. Energy and Climate. National Research Council, Geophysics Study Committee, National Academy Press, Washington, D.C., pp. 140-158.

Richards, J. F., J. S. Olson, and R. M. Rotty (1983). Development of a data base for carbon dioxide releases resulting from conversion of land to agricultural uses. ORAU/IEA-82-10(M). Institute for Energy Analysis, Oak Ridge, Tenn.

Rotty, R. M. (1981). Data for global CO_2 production from fossil fuels and cement. In Carbon Cycle Modelling, B. Bolin, ed. SCOPE Report 16. Wiley, New York, pp. 121-125.

Rotty, R. M. (1982). Distribution and changes in industrial carbon dioxide production. J. Geophys. Res. 88:1301-1308.

Schlesinger, W. H. (1977). Carbon balance in terrestrial detritus. Ann. Rev. Ecol. Syst. 8:51-81.

Schlesinger, W. H. (1983). The world carbon pool in soil organic matter: a source of atmospheric CO_2. In The Role of Terrestrial Vegetation in the Global Carbon Cycle: Measurement by Remote Sensing. SCOPE. Wiley, New York.

Seiler, W., and P. J. Crutzen (1980). Estimates of gross and net fluxes of carbon between the biosphere and the atmosphere from biomass burning. Climatic Change 2:207-247.

Siegenthaler, U., and H. Oeschger (1978). Predicting future atmospheric carbon dioxide levels. Science 199:388-394.

Smith, S. V. (1981). Marine macrophytes as a global carbon sink. Science 211:838-840.

Strain, B. R., ed. (1978). Report of the workshop on anticipated plant responses to the global carbon dioxide enrichment, held August 4-5, 1977. Dept. of Botany, Duke U., Durham, N.C. 91 pp.

Stuiver, M. (1978). Atmospheric carbon dioxide and carbon reservoir changes. Science 199:253-270.

Tans, P. P. (1978). Carbon 13 and Carbon 14 in trees and the atmospheric CO_2 increase. Thesis. Rijsuniversiteit te Groningen, The Netherlands.

Wagener, K. (1978). Total anthropogenic CO_2 production during the period 1800-1935 from carbon-13 measurements in tree rings. Radiat. Environ. Biophys. 15:101-111.

Walsh, J. J., G. T. Rowe, R. L. Iverson, and C. P. McRoy (1981). Biological export of shelf carbon is a sink of the global CO_2 cycle. Nature 291: 196-201.

Whittaker, R. H., F. H.Bovmann, G. E. Likens, and T. G. Siccama (1974). The Hubbard Brook ecosystem study: forest biomass and production. Ecolog. Manag. 44:233-254.

Whittaker, R. H., and G. E. Likens (1973). Carbon in the biota. In Carbon and the Biosphere, G. M. Woodwell and E. V. Pecan, eds. USAEC Symposium Series No. 30. NTIS, Springfield, Va., pp. 281-302.

Whittaker, R. H., and G. M. Woodwell (1969). Structure, production and diversity of the oak-pine forest at Brookhaven, New York. J. Ecol. 57: 155-174.

Woledge, J., and W. D. Dennis (1982). The effect of temperature on photosynthesis of ryegrass and whole clover leaves. Ann. Bot. 50:25-35.

Wong, C. S. (1978). Atmospheric input of carbon dioxide from burning wood. Science 200:197-199.

Wong, C. S. (1979). Elevated atmospheric partial pressure of CO_2 and plant growth. Oecologia 44:68-74.

Woodwell, G. M., ed. (1980). Measurement of changes in terrestrial carbon using remote sensing. U.S. Dept. of Energy CONF-7905176, UC-11. Available from NTIS, Springfield, Va.

Woodwell, G. M., ed. (in press). The Role of Terrestrial Vegetation in the Global Carbon Cycle: Measurement by Remote Sensing. SCOPE Report 23. Wiley, New York.

Woodwell, G. M., and D. B. Botkin (1970). Metabolism of terrestrial ecosystems by gas exchange techniques: the Brookhaven approach. In Ecological Studies, D. E. Reichle, ed. Analysis and Synthesis, Volume 1. Springer-Verlag, Berlin, pp. 73-85.

Woodwell, G. M., and R. A. Houghton (1977). Biotic influences on the world carbon budget. In Global Chemical Cycles and Their Alteration by Man, W. Stumm, ed. Report of the Dahlem Workshop, November 15-19, 1976. Dahlem Konferenzen, Berlin, pp. 61-72.

Woodwell, G. M., and R. H. Whittaker (1968). Primary production in terrestrial ecosystems. Am. Zool. 8:19-30.

Woodwell, G. M., R. A. Houghton, and N. R. Tempel (1973a). Atmospheric CO_2 at Brookhaven, Long Island, New York: patterns of variation up to 125 meters. J. Geophy. Res. 78:932-940.

Woodwell, G. M., P. H. Rich, and C. A. S. Hall (1973b). Carbon in estuaries. In Carbon in the Biosphere, G. M. Woodwell and E. V. Pecan, eds. Proceedings of the Twenty-fourth Brookhaven Symposium in Biology, Upton, New York. USAEC, Office of Information Sources. NTIS, Springfield, Va., pp. 221-240.

Woodwell, G. M., R. H. Whittaker, W. A. Reiners, G. E. Likens, C. C. Delwiche, and D. B. Botkin (1978). The biota and the world carbon budget. Science 199:141-146.

Woodwell, G. M., R. A. Houghton, C. A. S. Hall, D. E. Whitney, R. A. Moll, and D. W. Juers (1979). The Flax Pond ecosystem study: the annual metabolism and nutrient budgets of a salt marsh. In Ecological Processes in Coastal Environments, R. L. Jeffries and A. J. Davy, eds. The First European Ecological Symposium and the Nineteenth Symposium of the British Ecological Society, Norwich, September 1977. Blackwell Scientific Publications, Boston, Mass., pp. 491-511.

Woodwell, G. M., J. E. Hobbie, R. A. Houghton, J. M. Melillo, B. Moore, C. A. Palm, B. J. Peterson, and G. R. Shaver (1982). Report of the Woods Hole Conference on the Biotic Contributions to the Global Carbon Cycle at the Ecosystems Center. Ecosystems Center, Marine Biological Laboratory, Woods Hole, Mass., March 1982.

Woodwell, G. M., J. E. Hobbie, R. A. Houghton, J. M. Melillo, B. Moore, B. J. Peterson, and G. R. Shaver (1983a). The contributions of global deforestation to atmospheric carbon dioxide. Photocopies available from the Ecosystem Center, Marine Biological Laboratory, Woods Hole, Mass.

Woodwell, G. M., J. E. Hobbie, R. A. Houghton, J. M. Melillo, B. J. Peterson, G. R. Shaver, T. A. Stone, B. Moore, and A. B. Paru (1983b). Deforestation measured by LANDSAT: steps toward a method. DOE/EV/10468-1. NTIS, Springfield, Va.

Zelitch, I. (1971). Photosynthesis, Photorespiration, and Plant Productivity. Academic, New York.

3.4 THE ATMOSPHERE

Lester Machta

3.4.1 Introduction

It is the growing concentration of CO_2 in the atmospheric reservoir that has attracted most attention to the CO_2 issue. The pre-Industrial Revolution (e.g., 1850) concentration probably lay in the range 250 to 295 ppmv (parts per million by volume or mole fraction). Measurements by chemical analysis (Callendar, 1958; Keeling, 1978) and extrapolations backward based only on records of fossil fuel emissions suggest a late-nineteenth-century concentration at the upper end. Measurements from ice cores (Neftel et al., 1982; Oeschger, 1983) and reconstructed ocean measurements (see Brewer, Section 3.2) suggest preindustrial concentrations at the lower end. A WMO-sponsored Meeting of Experts in June 1983 concluded the most likely mid-nineteenth-century concentration was between 260 and 280 ppm, based on consideration of all the various estimates including carbon isotope data in tree rings (the meeting report will be issued at a later date). Concentrations significantly less than 290 ppm imply the existence of a large nonfossil fuel source of CO_2 and are thus consistent with a large early input from disturbances of the biosphere. By 1980 the atmospheric CO_2 concentration had risen to about 340 ppmv.

The behavior of CO_2 in air is simpler and better understood than in the other two major reservoirs--the land biosphere and the oceans. CO_2 is conservative in air, that is, it is not subject to chemical transformation at least up to an altitude of about 60 km. It moves with the other inert air molecules with which it is embedded. Most, if not all, of the known variations of CO_2 in time and space in the air appear to follow known meteorological principles. Since interest often focuses on time scales of years to decades, as a first approximation, it is usually acceptable to treat the whole atmosphere as a single well-mixed box and apply first-order kinetics to the CO_2 transfer to other reservoirs.

The growth of CO_2 in the air can be demonstrated at almost any location on Earth over a period of several years. Modern-day measurements were begun by C. D. Keeling of Scripps Institution of Oceanography during the 1957-1958 International Geophysical Year. Stations were established at the South Pole and at 11,150 feet aside Mauna Loa in Hawaii. The latter, the better record, is reproduced in Figure 3.14. Since the pioneering measurements of Keeling, other stations operated by many countries and organizations have been established. A map of the location of stations as of July 1982 as supplied by the World Meteorological Organization (WMO) appears as Figure 3.15. At most of the stations air is collected in containers for subsequent analysis in central laboratories. With few exceptions, both on-station and laboratory analyses are performed by nondispersive infrared analysis comparing the ambient samples with standard gases. Since the response of the analyzer depends on the carrier gas, it is now agreed that the carrier gas of the standard should duplicate air as closely as possible, e.g., nitrogen, oxygen, and argon instead of the former nitrogen gas. A transition to standards of CO_2 in air (or simulated air) now in progress around the world should proceed as quickly as possible and be consistent with ensuring long-term integrity of the standards. Carrier-gas corrections based on CO_2-in-N_2 standards are required to bring concentrations closer to their true values.

3.4.2 Changes in Atmospheric CO_2 Growth Rate with Time and Space

There appear to be two kinds of changes in the year-to-year growth rate at Mauna Loa (Figure 3.14): a shorter- and longer-term variation.

3.4.2.1 Shorter-Term Variation and Its Possible Cause

Here we follow the analysis of Machta et al. (1977); the same result is arrived at by the analysis of Bacastow (1976) and Newell and Weare (1977), although a different interpretation is offered by them. The monthly mean concentrations at a given station exhibit a seasonal oscillation and a long-term trend, both of which can be removed mathematically. The resulting monthly values of two stations, in this case

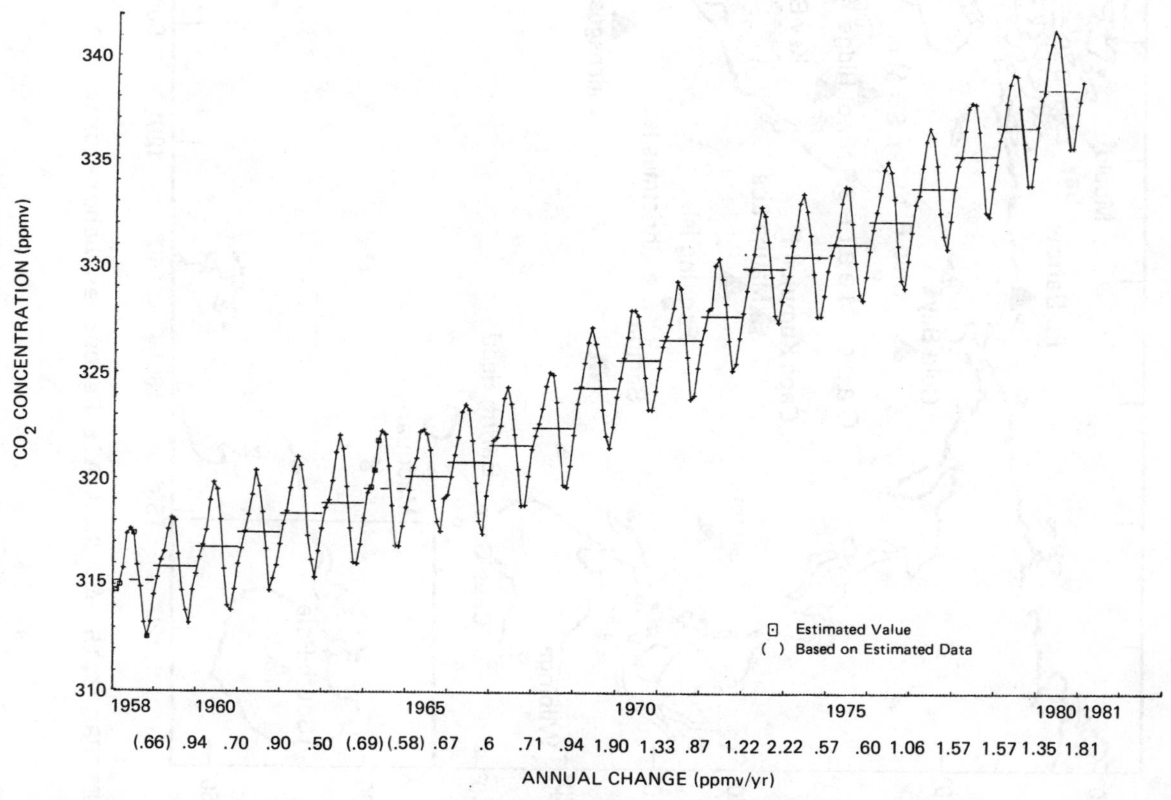

FIGURE 3.14 Mean monthly concentrations of atmospheric CO_2 at Mauna Loa.

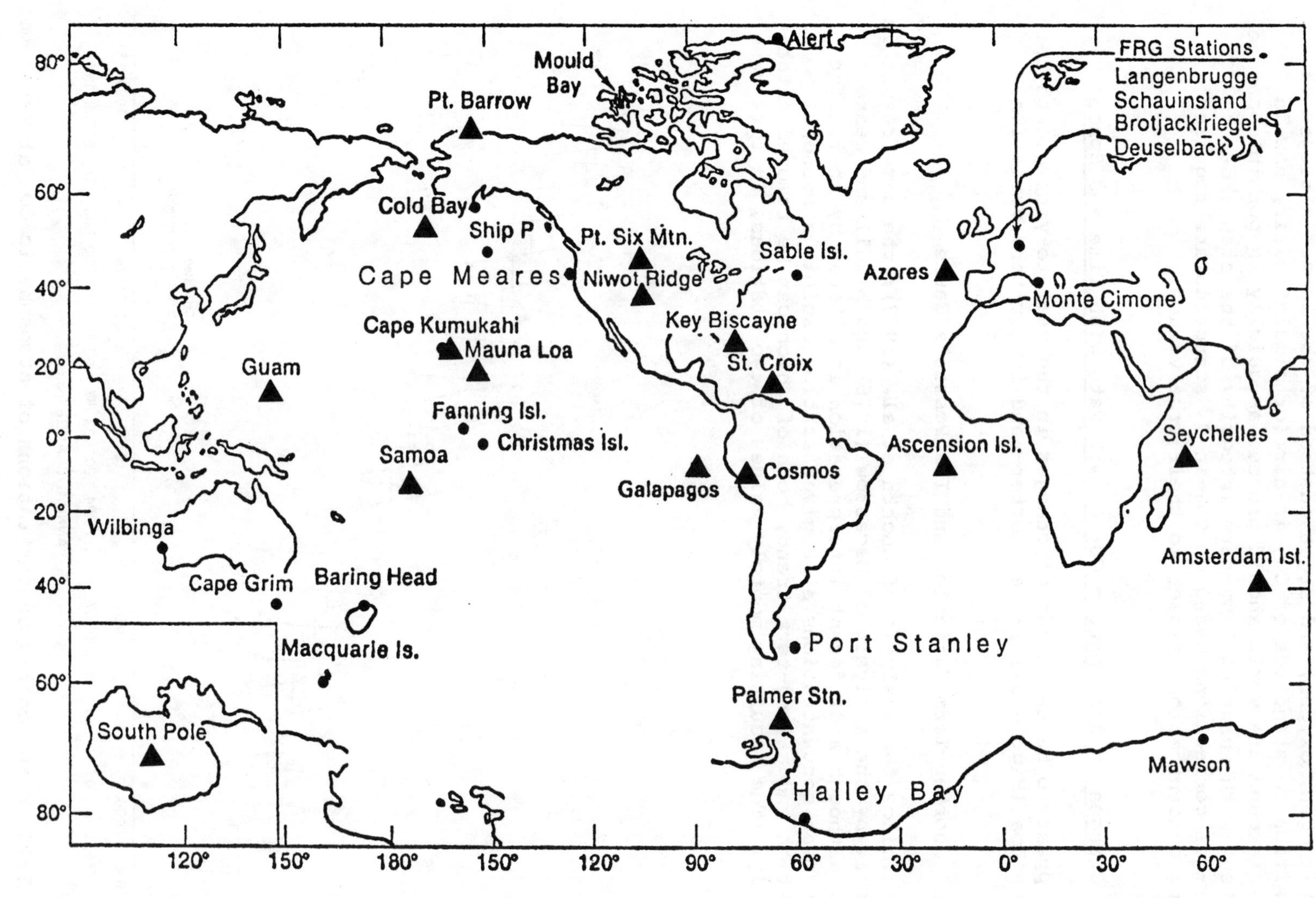

FIGURE 3.15 Δ, NOAA Air Resources Laboratories CO_2 sampling sites; O, other sampling sites.

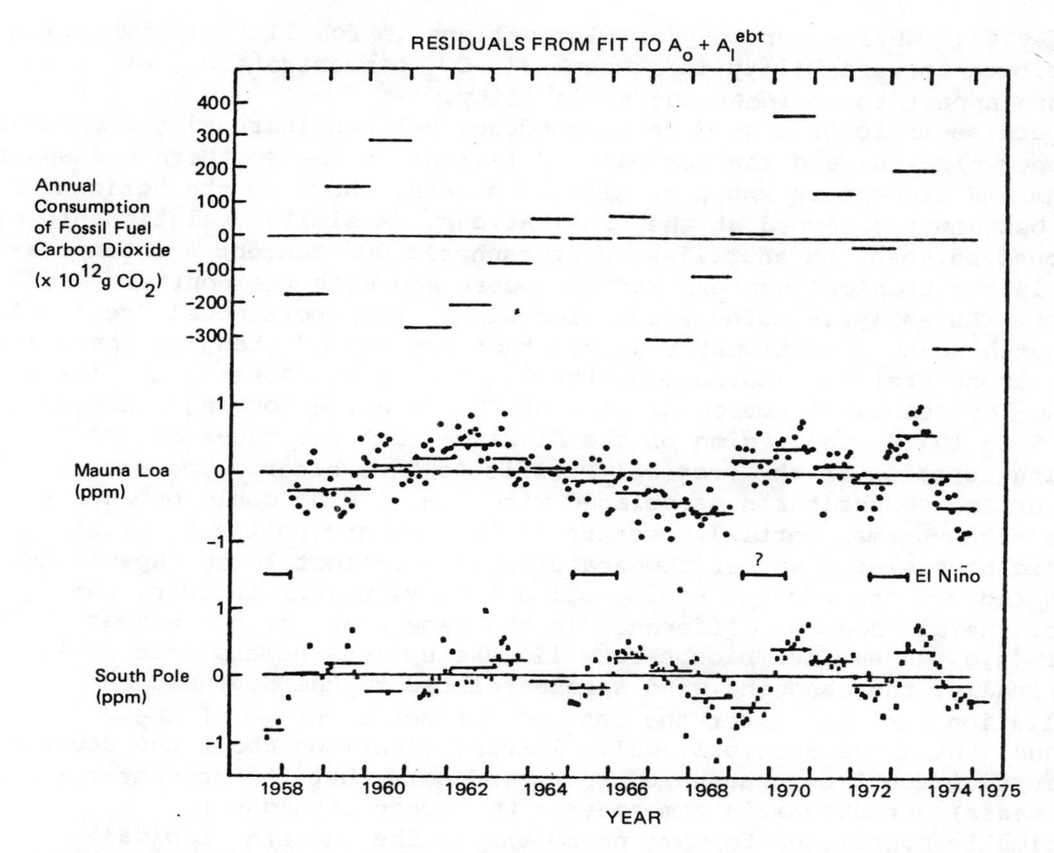

FIGURE 3.16 Time history of residuals of monthly carbon dioxide concentrations after removing the long-term upward trend and seasonal variability at Mauna Loa and the South Pole (lower section). The horizontal lines in the upper section represent the residuals of the annual consumption (actually production) of fossil fuels after removing the long-term upward trend. The periods of El Niño are also shown in the lower section. The horizontal lines in the lower section among the circles and crosses are the annual average residuals.

Mauna Loa (small circles) and the South Pole (crosses), appear in Figure 3.16. This figure shows a pattern of short-term variation with fluctuations reversing themselves every 2 to 6 years. The horizontal bars among the circles and crosses represent the annual values of the 12 monthly averages. The horizontal bars in the upper part of Figure 3.16 represent the departures from the best exponential fit of the annual fossil fuel emissions of CO_2 from the mean value for the 17 years.

Variation in annual emissions might be the reason for the year-to-year fluctuation in Figure 3.16 of the atmospheric CO_2 content; however, the correlation coefficient between concurrent anomalies of emissions and atmospheric content is only 0.42, not statistically significant. Allowing for a lag of up to 6 years between emissions and atmospheric CO_2 content simply reduces the correlation coefficient below 0.42. While year-to-year fluctuations in fossil fuel combustion contribute to variability in atmospheric CO_2 concentrations, other factors appear to dominate the variability.

There seems to be a good correspondence between increasing anomalies, the open circles, and the periodic variations in the southern hemisphere oceans and atmosphere known as El Niño events, shown as the horizontal bars between the record at the two stations. A similar relationship can be found between the anomalies in atmospheric CO_2 content and temperature in the tropical eastern Pacific Ocean and with the Southern Oscillation. An analysis using a two-dimensional transport model (vertical and north-south directions) suggests that the lag of changing concentrations among stations (Mauna Loa, the South Pole and Australia) fits a ground- or sea-level source or sink of CO_2 (i.e., a forcing function) near 5 to 10° S, the region of the El Niño. But the cause of the forcing function in the tropical Pacific is less clear. The warmer sea-surface temperatures associated with the El Niño could produce a higher-than-normal partial pressure of CO_2, enhancing the tropical oceanic source; the warmer temperatures also reflect lesser upwelling, which reduced the oceanic biological activity, which, in turn, can affect the air-sea CO_2 difference in the same sense as the warmer water (e.g., a smaller biosphere will take up less atmospheric CO_2); and finally, the changing wind speeds related to the Southern Oscillation can also alter the rate of air-sea exchange of CO_2.

Thus, the above analysis, while leaving questions about the cause of the forcing function unanswered, does indicate that the shorter-term (2 to 6 years) variations in atmospheric CO_2 concentrations are empirically correlated to some phenomena in the eastern tropical Pacific Ocean (El Niño, Southern Oscillation, biota change).

3.4.2.2 Longer-Term Variations

An inspection of year-to-year increases in concentration at Mauna Loa reveals that they are generally becoming larger with time. Figure 3.14 shows this trend in the values of annual changes in ppmv/yr. Through 1968 the annual increase was below 1 ppmv, while in recent years it has often been nearer to 1.5 ppmv. The emissions of man-made CO_2 (mainly fossil fuel combustion plus cement manufacture and flaring of natural gas*) are also becoming larger with time. Elliott (1983) estimates an overall annual growth rate of emissions from industrial activity of about 3.5% over the past 120 years, with wide variation due to economic fluctuations, and Nordhaus and Yohe (this volume, Chapter 2, Section 2.1) predict that emissions from fossil fuels will most likely grow at a rate of about 1% or 2% per year over the next hundred years. But there are other likely sources for atmospheric CO_2; in particular the CO_2 produced as a result of deforestation. Elliott and Machta (1981)

---

*Hereafter, the term "fossil fuel CO_2" is understood to include the other two minor contributions as well.

have sought to determine whether the Mauna Loa and South Pole records of CO_2 increase (the average of the two is taken as the average for the Earth in this analysis) are better fitted by the increasing fossil fuel combustion source of CO_2 or whether the addition of another significant source, such as from deforestation, results in a better fit to the measurements. In principle, if there were an accurate model of the carbon cycle, one might enter alternate amounts of CO_2 into it, and the best fit to the observed data would be the best size of the source to fit observed concentrations.

Elliott and Machta (1981) have tried to avoid the issue of defining the carbon cycle. They assume that each year roughly the same fraction of that year's CO_2 emissions remains airborne. This airborne fraction is determined from the ratio of the CO_2 increase in the atmosphere (as found from the average annual increases at Mauna Loa and the South Pole) to the amount added to the air from all sources that one wishes to assume. The interval of this analysis is from 1958 to 1981. The result indicates that the observed increases in atmospheric CO_2 are best fitted by only a fossil fuel source, without any additional constant or random CO_2 source. In fact, the analysis suggests that there could be a small loss of CO_2 from the atmosphere, possibly through CO_2 fertilization of photosynthesis because of the elevated atmospheric CO_2 concentration. There is one caveat, however. This analysis could not distinguish between fossil fuel and nonfossil fuel emissions, such as deforestation, that are increasing at the same rate. It does, however, provide estimates of how closely the growth of the nonfossil source had to match the fossil fuel growth to be undetectable. For sources averaging more than 2 Gt of C per year, the growth rate would have to have been quite close to the fossil fuel growth rate to be undetected.

3.4.2.3 Change in Annual Cycle

The annual cycle shown in Figure 3.14 for Mauna Loa CO_2 concentrations is almost certainly the result of the warm season uptake of CO_2 during land biosphere photosynthesis (see Woodwell, Section 3.3). There could be a contribution or diminution from the seasonality of fossil fuel combustion or the air-sea exchange of CO_2. The seasonality of the latter processes are believed to be small in determining the annual cycle of atmospheric CO_2 concentrations.

The long, high-quality record at Mauna Loa has been analyzed to determine whether the amplitude of the annual cycle is changing with time. The result is shown in Figure 3.17. The increase of amplitude suggested by a best-fit line, the dashed line, in large part depends on an increase that occurs only during the past 6 years. The finding of increasing amplitude at Mauna Loa is supported by analysis of a shorter and less convincing CO_2 record at the Canadian ship Papa located in the Gulf of Alaska, according to Keeling (1983). The most plausible explanation of the increasing amplitude is increased biological activity, such as, but not necessarily, a larger temperate and high-latitude biosphere (e.g., larger forests) (c.f., Woodwell, Section 3.3).

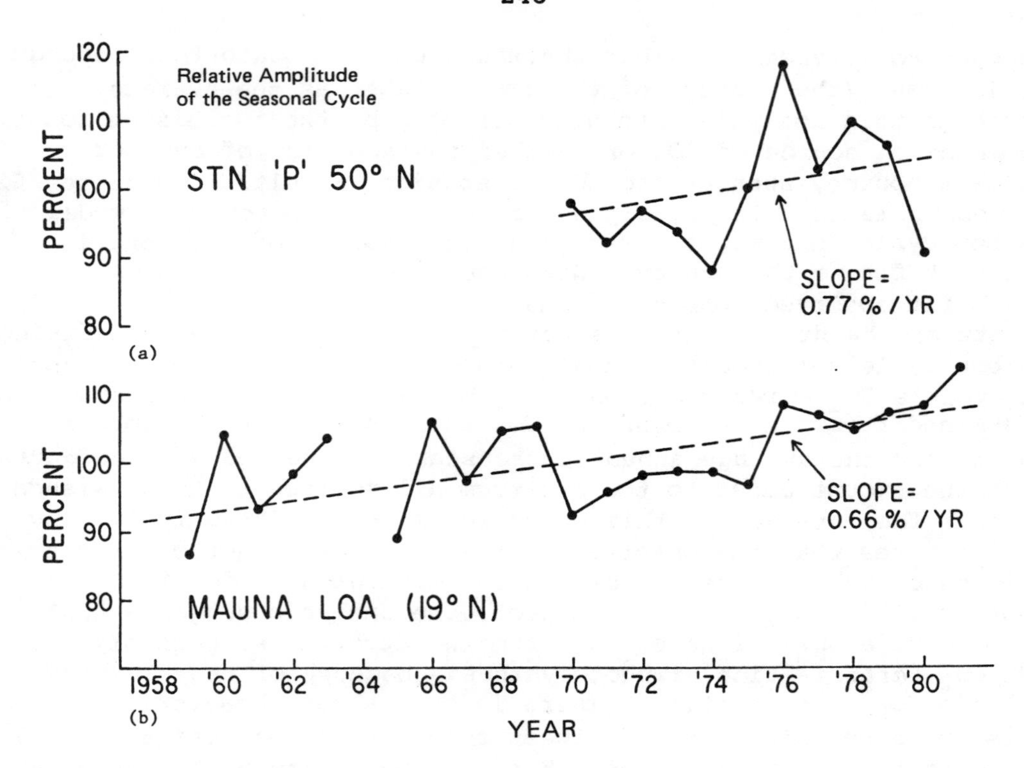

FIGURE 3.17 Seasonal amplitude in atmospheric CO_2 concentration at (a) Weather Ship P at 50.0° N and (b) Mauna Loa Observatory at 19.5° N. Dots connected by solid lines represent an estimation of the amplitude for individual years as determined by a best fit of a four-harmonic seasonal cycle as described by Bacastow et al. (1981). The dashed straight line is a least-squares fit of a linearly increasing amplitude over the entire period of record. (Source: Keeling, 1982.)

3.4.2.4 Spatial Distribution

Keeling (1982) has provided north-south profiles of ground-level air concentrations of CO_2 relative to that at the South Pole (see Brewer, Section 3.2). Profiles for 3 years, all adjusted to a common value at the South Pole are shown in Figure 3.9 in Section 3.2. It is evident that the secular growth in the northern hemisphere exceeds that in the southern hemisphere, as might be expected from the location of fossil fuel CO_2 sources mainly in the northern hemisphere. The concentration in each of the 3 years is also higher in the northern than southern hemisphere. Near the equator there is a secondary peak, which might be due to either the release of CO_2 from the tropical oceans or deforestation in the tropics. Keeling has estimated the transfer from the sea to air and compared this with the equatorial peak in Figure 3.18. His conclusion is that CO_2 from tropical deforestation during 1962-1980 is unlikely to exceed 1 or 2 Gt of C per year or no more than

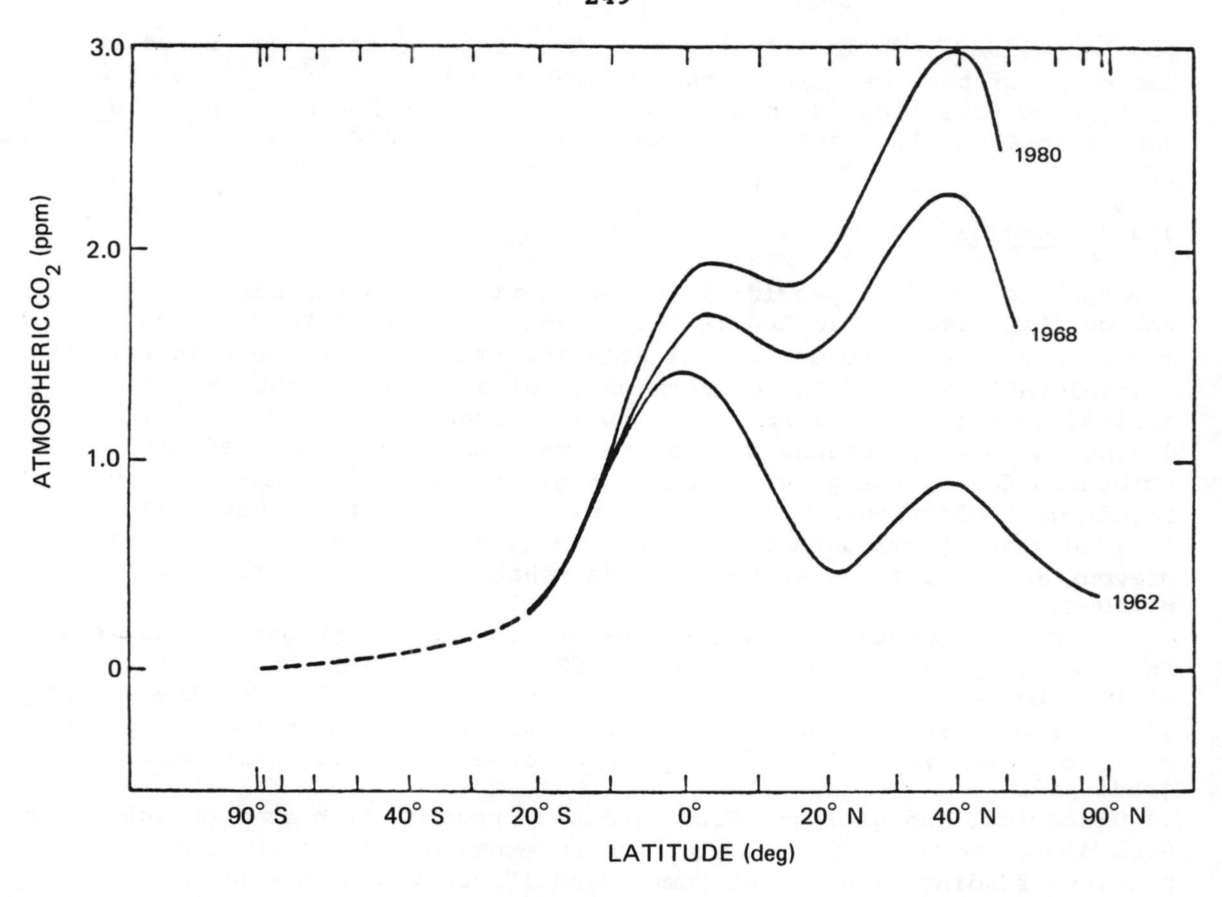

FIGURE 3.18 North-south profile of ground-level air concentration relative to that at the South Pole for 3 years.

40% of the current CO_2 released from fossil fuel combustion and likely much less.

3.4.2.5 Isotopic Content of Atmospheric CO_2

Isotopic (^{13}C) analyses of atmospheric CO_2 samples have been undertaken in a systematic fashion in only the past few years so that conclusions derived from these data must still be viewed with caution. Keeling (1983) contends that the seasonal cycle in ^{13}C measurements are consistent with land plants being the primary source of the annual cycle of CO_2 concentration for northern hemisphere stations. At the South Pole the isotopic data suggest an oceanic source for the cause of the much smaller annual cycle. Tentative results from Keeling from Fanning Island in the tropical Pacific Ocean and during oceanographic cruises in the tropics (the First Global Atmospheric Research Program Global Experiment expedition) support the contention that the peak CO_2 concentration in the equatorial region in Figure 3.18 is the result of air-sea exchange and not due to deforestation in the tropics. Leavitt and Long (1983), while allowing for other interpretations, report that the shape of the best-fit reconstruction of 50 yr of $^{13}C/^{12}C$ measurements from tree rings suggests that the biosphere has acted as a CO_2 source to about 1965 but has become a sink afterward.

3.4.3 Conclusions

Atmospheric CO_2 data provide information far beyond the single observation that atmospheric CO_2 is increasing. Through careful measurements, one is able to derive valuable information from the temporal and spatial variability. The pattern of results is highly suggestive of a minimal contribution of nonfossil fuel sources of CO_2. Globally, during the past 20 years, most of the variations are more readily accounted for by the growing fossil fuel source alone than from any significant additional source from, say, deforestation. Both the limited quantity of data and the possibility of alternate explanations prevent any definitive statement today that excludes nonfossil fuel sources.

The need to continue quality observations cannot be overemphasized. The spatial gradients of atmospheric CO_2 are so small that the total minimum to maximum concentration at a clean air location for the entire globe is no more than about 1% of the mean global concentration. Data collected with very high precision are needed to detect such small gradients.

Historical and geologic data from past records such as from ice cores have proven to be very valuable, and an expanded effort to confirm the previous findings (about 200 ppmv found 18,000 years ago) should be undertaken. The isotopic studies in tree rings and from current air samples offer potential to elucidate further the carbon cycle and the contribution of the nonfossil fuel CO_2. For example, following the fate with time of the nuclear weapons test $^{14}CO_2$ in the atmosphere can continue to provide new information on the atmospheric residence time of the fossil fuel CO_2.

References

Bacastow, R. B. (1976). Modulation of atmospheric carbon dioxide by the Southern Oscillation. Nature 261:116.

Bacastow, R. B., C. D. Keeling, and T. P. Whorf (1981). Seasonal amplitude in atmospheric CO_2 concentration at Mauna Loa, Hawaii 1959-1980. In papers presented at the WMO/ICSU/UNEP Scientific Conference on Analysis and Interpretation of Atmospheric CO_2 Data. World Meteorological Organization, Geneva, Switzerland, pp. 169-176.

Callendar, G. S. (1958). On the amount of carbon dioxide in the atmosphere. Tellus 10:243-248.

Elliott, W. P. (1983). A note on the historical industrial production of carbon dioxide. Climate Change 5:141-144.

Elliott, W. P., and L. Machta (1981). In papers presented at the WMO/ICSU/UNEP Scientific Conference on Analysis and Interpretation of Atmospheric CO_2 Data. World Meteorological Organization, Geneva, Switzerland, p. 191.

Keeling, C. D. (1978). Atmospheric carbon dioxide in the 19th century. Science 202:1109.

Keeling, C. D. (1983). The global carbon cycle: what we know and could know from atmospheric, biospheric, and oceanic observations. In Proceedings, CO_2 Research Conference: Carbon Dioxide, Science, and Consensus, Berkeley Springs, West Virginia. CONF-820970. NTIS, Springfield, Va. 22161.

Leavitt, S. W., and A. Long (1983). An atmospheric $^{13}C/^{12}C$ reconstruction generated through removal of climate effects from tree-ring $^{13}C/^{12}C$ measurements. Tellus 35B:92-102.

Machta, L., K. Hanson, and C. D. Keeling (1977). Atmospheric carbon dioxide and some interpretations. In The Fate of Fossil Fuel CO_2 in the Oceans, N. R. Andersen and A. Malahoff, eds. Plenum, New York, pp. 131-144.

Neftel, A., H. Oeschger, J. Schwander, B. Stauffer, and R. Zumbrunn (1982). New measurements on ice core samples to determine the CO_2 content of the atmosphere during the last 40,000 years. Nature 295:220-223.

Newell, R. E., and B. C. Weare (1977). A relation between atmospheric carbon dioxide and Pacific sea-surface temperature. Geophys. Res. Lett. 4:1-2.

Oeschger, H., and M. Heimann (1983). Uncertainties of predictions of future atmospheric CO_2 concentrations, J. Geophys. Res. 88:1258.

3.5 METHANE HYDRATES IN CONTINENTAL SLOPE SEDIMENTS AND INCREASING ATMOSPHERIC CARBON DIOXIDE

Roger R. Revelle

3.5.1 Methane in the Atmosphere

About 4.8 Gt of methane (CH_4) are present in the Earth's atmosphere, corresponding to 1.7 ppm by volume (see Machta, this volume, Chapter 4, Section 4.4). Methane is a strong absorber of infrared radiation in the part of the atmospheric "window" centered around a wavelength of 7.66 μm. According to Lacis et al. (1981), a doubling of the atmospheric methane concentration would cause an increase in global average surface temperature of 0.41°C. Chamberlain et al. (1982) estimate a larger value, 0.95°C for methane doubling and report lower and higher results by other groups as well. These calculations allow for positive feedbacks resulting from the increase in absolute humidity with rising temperatures and the consequent higher infrared absorption by water vapor, decreases of planetary albedo due to melting of snow and ice, and assumed cloud behavior. Chamberlain et al. estimate that methane is now being added to the atmosphere at a rate between 0.5 and 1.0 Gt per year, primarily from anaerobic fermentation of organic material in rice paddies, swamps, and tundras, plus enteric fermentation in the digestive tracts of ruminant animals. Anaerobic fermentation in the guts of termites, which contain cellulose-digesting symbiotic bacteria, is probably also a significant source. Some methane is being added to the ocean-atmosphere system from vents in the rift zones of the ocean floor (Welhan and Craig, 1979) and perhaps of East Africa (Deuser et al., 1973). As we shall see, ocean sediments on the continental slopes may be a relatively small source now but an important source in the future (MacDonald, 1982a).

Methane is removed from the lower atmosphere by a reaction with hydroxyl (HO) and is eventually oxidized to CO_2. With the above estimate of the rate of input of 0.5 to 1.0 Gt per year, the residence time in the air should be between 5 and 10 years. Measurements indicate that the methane content of the air is increasing perhaps by 0.07 Gt per year, or about 1.4% per year, doubling in 70 years (Rassmussen and Khalil, 1981; Craig and Chou, 1982). Part of this increase may be the direct result of such human actions as expansion of the area of rice paddies to meet the food needs of growing populations. A small part may represent release of methane from methane hydrates in continental slope sediments as the ocean responds to atmospheric warming.

MacDonald (1982a) defines methane hydrate as a "type of clathrate in which methane and smaller amounts of ethane and other higher hydrocarbons are trapped within a cage of water molecules in the form of ice." Though it is not a stoichiometric compound, about 6 mol of water are required for 1 mol of methane in the clathrate. Methane hydrates are stable at low temperatures and relatively high pressures (Claypool and Kaplan, 1974; Miller, 1974). They are found at depths between 200 and 1000 m below the ground surface in permafrost (Kvenvolden and

McMenamin, 1980; Chersky and Makogon, 1970) and should be present near the surface of marine bottom sediments below water depths of 290 to more than 800 m, depending on bottom-water temperature. Within the sediments the thickness of the clathrate zone will be limited by the geothermal gradient of about 30°C km^{-1}, which reflects heat conduction from the interior of the Earth.

As Bell (1982) has demonstrated, even with a CO_2-induced rise in surface air temperatures of around 10°C, virtually none of the clathrate in permafrost would become unstable during the next several hundred years because the surface heating of the "frozen ground" would first have to penetrate and melt the permafrost in a 200-m-thick clathrate-free zone. The enthalpy (latent heat of fusion) of the ice in permafrost would greatly slow the downward penetration of the heat wave.

But with a rise in ocean-bottom temperatures, the uppermost layers of sediments would also become warmer and methane hydrates would become unstable in the upper limit of their depth range, that is, about 300 m in the Arctic and about 600 m at low latitudes.

3.5.2 Formation of Methane Clathrate in Continental Slope Sediments

The quantity of clathrates that will be released from sediments under the seafloor as a result of ocean warming depends on the distribution of clathrates with depth and on their total abundance in the sediments. Estimates of total abundance by different authors differ by a factor of 500, from 10^3 to 5×10^5 Gt (MacDonald, 1982b). If the methane locked up in clathrates were produced by anaerobic fermentation of organic matter in the sediments, one would expect that most oceanic clathrates would be found in deep semienclosed basins and on continental slopes, particularly on passive continental margins such as those on both sides of the Atlantic.

The rate of sedimentation on continental slopes is relatively high (of the order of 10 to 20 cm/1000 years), and samples taken from near the surface of the deposits are high in organic matter--on the average about 2% of the dry weight (Trask, 1932). This organic matter is sometimes called "marine humus." It consists mainly of the partially decomposed tissues of marine plankton and nekton and to a lesser extent of the remains of terrestrial plants. On average, according to Trask, the carbon content of the organic matter is 56%; hence, organic carbon averages 1.12% of the dry weight of the uppermost layers of sediments.* The average density of the dry material is 2.6 g cm^{-3}.

---

*J. G. Erdman and colleagues of the Phillips Petroleum Company measured the organic carbon content of many samples collected by the Deep Sea Drilling Project from outer continental margins. The results were published in the Initial Reports of the Deep Sea Drilling Project (Volumes XXIV, XXXI, XXXVIII, XL, XLI, XLII, XLIII, XLIV, XLVII, XLVIII, L, and LXVI, published between 1975 and 1981, inclusive). Seventy-three samples of Quaternary to late Pliocene age from the

(continued overleaf)

These deposits are very porous; cores of freshly collected mud usually contain two thirds water by volume. Thus, an average liter of mud from near the surface of the deposits will contain 650 g of water and 870 g of solids, including silicate mineral grains, fragments of calcareous and siliceous skeletons, and shells, and about 17 g of organic matter containing close to 10 g of carbon.

Because of the relatively high rates of deposition and the abundance of decomposable organic material, free oxygen in the interstitial water of these sediments is rapidly depleted as they are buried, and "reducing" conditions prevail a short distance below the seafloor. The principal living organisms under these conditions are anaerobic bacteria, which are able to carry out their metabolic activities in the absence of free oxygen. Dissolved sulfate in the interstitial water will first be reduced to sulfide, and a small fraction (less than 0.5 g per liter) of organic carbon will be oxidized to CO_2. All the sulfate is usually depleted in the top meter of the sediments. Beneath this top layer, methane is produced (Claypool and Kaplan, 1974). Below water depths of 300 to 600 m, depending on bottom-water temperature, methane in excess of the quantity that can be dissolved in the interstitial water will be converted to methane hydrate as soon as it is formed.

There are no measurements of the actual methane concentration in deep-sea muds in situ. In several cores from areas of rapid deposition collected by the Deep Sea Drilling Project (DSDP), gas was observed escaping when the core liners were removed from the bore barrels on board _Challenger_. Some mud was ejected from the liners by the force of the escaping gas. Subsequent analysis showed this gas to be almost entirely methane (McIver, 1974). Presumably, most of the methane escaped from the samples while they were being raised from the seafloor and handled on deck, but the remaining amounts were surprisingly high--up to 15 mmol of methane per liter of interstitial water.

The organic carbon content of DSDP sample with high remaining gas content after shipboard handling (presumably those in which the methane was originally present as a clathrate) ranged from 0.28 to 1.14%--averaging 0.62% by dry weight (McIver, 1974). Assuming that this organic carbon represents the residue of organic matter after methanogenesis has been completed, and that the proportion of residual carbon to carbon in methane produced is roughly the same (1:0.51) as in the

---

(continued from overleaf)
northern Indian Ocean, eastern Pacific, north and south Atlantic, and the Japan, Mediterranean, and Black Seas have an average organic carbon content of 1.4% of the dry weight of the sediments. The depths beneath the sediment surface ranged from less than 1 to 1000 m, with most of the samples being from 30 to 300 m below the top of the sediment; the average depth of the overlying water was around 2000 m. Presumably the measured organic carbon represents the residue after sulfate reduction and methanogenesis. Erdman believes that Trask's analysis of surface sediment gave low results because the samples were poorly preserved.

"biogas" digesters described by Makhijani and Poole (1975), the calculated average methane content of these muds in situ is 0.42% by weight of dry sediment, or 3.6 g per liter of wet mud. The corresponding concentration of methane in the interstitial seawater would be about 330 mmol kg^{-1}.

The biochemical processes in the buried sediments are not well understood and may be quite different from those in methane-producing biogas digesters. J. G. Erdman (Bartlesville, Oklahoma, personal communication) has pointed out that the microbial population in marine sediments drops rapidly with depth in the sediments and that the organic matter, unlike terrestrial biomass, is relatively lean in the hydrolyzable constituents of plant materials that ferment easily. Erdman is convinced that the mass of methane hydrate in marine sediments is larger than the amount calculated above. He believes that most of this methane was formed from sedimentary organic matter by thermolytic processes under heat and pressure at substantial depth in the sediments. The methane then migrated upward until it became trapped in the zone of methane hydrate stability in the upper sedimentary layers. This hypothesis has the significant advantage that it does not require the methane in the upper part of the hydrate zone to have formed since the last interglacial approximately 125,000 years ago, when subsurface ocean warming may have been as great as that expected with a doubling of atmospheric carbon dioxide.

An estimate of the minimum concentration of methane can be made from the inferred existence of methane clathrate in muds from the Blake Plateau off the southeastern coast of the United States at a total depth (overlying water plus depth in the sediments) of about 4000 m (Stoll et al., 1971; Bryan, 1974). In order for a clathrate to form, the concentration of dissolved methane must have been close to 64-69 mmol kg^{-1} of interstitial water, which is the solubility of methane at a hydrostatic pressure of 400 atm (Claypool and Kaplan, 1974). This concentration is about 20% of that calculated above by comparison with observed methane production in biogas digesters.

For our present purposes, we may assume that the concentration of methane in continental slope muds is halfway between these two estimates, say 200 mmol kg^{-1} of interstitial water, or 2.2 g of methane per liter of mud. If all of this methane is present as clathrates, about 1200 mmol kg^{-1} of water will also be in the same state or 21.6 g kg^{-1} of interstitial water. This is 3.2% of the water in an average mud.

Miller (1974) shows that in seawater, the minimum hydrostatic pressure (P) at which methane hydrate is stable between temperatures (T) of 0° and 10°C is given by

$$\log_{10} P \text{ (atmospheres)} = 1.4613 + 0.0416T + 2.93 \times 10^{-4} (T)^2,$$

where P (atmospheres) is the partial pressure of methane, which is equal to hydrostatic pressure when the water is saturated with methane. The water depths below which methane hydrate in the uppermost layers of marine bottom sediments will be stable at different temperatures can be calculated from this equation, with the simplifying assumption that hydrostatic pressure in the ocean increases by approximately 1 atm for each 10 m of depth:

Bottom-Water Temperature (°C)	Minimum Depth of Clathrate Stability (m)
0	289
1	319
2	351
3	388
4	429
5	475
6	528
7	588
8	650
9	724
10	807

3.5.3 Effect of Carbon Dioxide-Induced Warming on Continental Slope Clathrates

With carbon dioxide-induced warming of the atmosphere, ocean surface temperatures will rise by a nearly equal amount, and heat will be carried downward by advection and eddy diffusion into the subsurface water layers. For a doubling of atmospheric CO_2 and the expected increase in other "greenhouse gases," with an assumed sensitivity of global average temperature of 3°C for a CO_2 doubling, the temperature increase in different latitudes at the water depths below which methane hydrate is stable at present can be estimated (see this volume, Chapter 8, Section 8.3). The corresponding increases in clathrate stability depths are shown in Table 3.6.

The depth of melting of the clathrate below the sediment-water interface at any time after the bottom-water temperature is raised will be much smaller than the depth at which warming will occur in the absence of clathrate. When bubbles of methane are formed, the latent heat (enthalpy) of vaporization of the methane must be added to that for melting of the water-ice in the clathrate. Miller (1974) has calculated that with 6 mol of water per mol of methane, the combined enthalpies are 120 cal g^{-1} of H_2O in the clathrate. Because most of the water in the sediment remains liquid even after clathrate has formed, we can compute a "virtual" latent heat, L, required for the depth of wetting to advance downward by 1 cm.

$$L = 0.032 \times 120 + 0.968 \times 1 = 4.8 \text{ cal } g^{-1} \text{ of } H_2O.$$

The problem of the rate of downward advance of the depth of melting has been solved by W. H. Munk (La Jolla, California, personal communication). He shows that the depth h (cm) of the melted zone at time t (sec) after an instantaneous rise in water temperature at the sediment-water interface is given approximately by,

$$h = \alpha (\kappa t)^{1/2},$$

where $\kappa(cm^2\ sec^{-1}) = K/\rho C$; K is thermal conductivity of wet sediment, in cal $sec^{-1}\ cm^{-1}\ °C^{-1}$; ρ is density of wet sediments, g cm^{-3}; C is specific heat, cal $g^{-1}\ °C^{-1}$,

and

$$\alpha(\text{dimensionless}) = \left[\frac{2C}{L}(T_1 - T_m)\right]^{1/2};$$

L is "virtual" enthalpy of melting for clathrate in sediments, cal g^{-1} of H_2O in both clathrate and liquid fractions of interstitial water; $T_1 - T_m$ is temperature at the sediment-water interface minus temperature of melting of clathrate.

With the following approximate numerical values: $K = 4 \times 10^{-3}$ (calculated from average geothermal heat flow through the seafloor and estimated temperature gradient in sediment: $K = 1.3 \times 10^{-6} \div 3 \times 10^{-4}$), $\rho = 1.54$, $T_1 - T_m = 1$, $C = 1$, $L = 4.8$, $\kappa = 2.6 \times 10^{-3}$, and $\alpha = 0.65$, then $h = 0.0331(t)^{1/2}$ cm.

After an instantaneous rise of bottom-water temperature of 1°C, the depth of melting would advance downward 18.6 m beneath the sediment-water interface during 100 years, or an average of 18.6 cm yr^{-1}. An average of 0.041 g of methane per cm^2 of seafloor should be released each year in the area where clathrates have become unstable.

The rate of exchange of dissolved materials between the sediments and the overlying water is very slow, and consequently methane will probably not diffuse significantly out of the sediments until its concentration becomes high enough to form bubbles of the gas. According to Miller (1974), bubbles of methane will form at a hydrostatic pressure of 400 atm when the concentration exceeds 350 mmol $liter^{-1}$. Provided the relation between the concentration required for bubble formation and hydrostatic pressure is approximately linear, the required concentration at about 500 m will be 44 mmol. With our assumed concentration of 200 mmol kg^{-1} of interstitial water, close to 80% of the methane released from clathrate should escape from the mud in bubbles and should rise rapidly to the sea surface before it can be oxidized in the water.

3.5.4 Future Rate of Methane Release from Sedimentary Clathrates

From Table 3.6 we see that the average depth interval on the continental slope over which the clathrates will become unstable with a CO_2 doubling is about 100 m. Using Kossina's 1921 estimate (see also Menard and Smith, 1966) that the area of seafloor between 200 and 1000 m is $15.5 \times 10^{16}\ cm^2$, and assuming that the depths of continental slopes increase linearly in this depth range, the area in which clathrates will become unstable is $1.94 \times 10^{16}\ cm^2$, and the total quantity of methane released annually as bubbles in the bottom muds should be $0.8 \times 0.041 \times 1.94 \times 10^{16}$ g, or, 0.64 Gt (10^9 metric tons). With an atmospheric residence time of 5 to 10 years, the quantity of methane in

TABLE 3.6 Depth Intervals over which Marine Sedimentary Clathrates Become Unstable with a Doubling of Atmospheric CO_2

Latitude	Present Depth of Clathrate Stability (m)	Present Bottom-Water Temperature at this Depth (°C)	Estimated Bottom-Water Temperature if Atmospheric CO_2 Doubles (°C)	Depth of Clathrate Stability for this Temperature (m)	Average Increase in Clathrate Stability Depth in Latitude Zones (m)
60° N	429	4.0	8.0	650	
					193
50° N	475	5.0	7.8	640	
					152
40° N	585	7.0	9.0	724	
					106
30° N	650	8.0	9.0	724	
					72
20° N	617	7.5	8.5	687	
					74
20° S	688	8.5	9.5	765	
					71
40° S	585	7.0	8.0	650	
					88
50° S	475	5.0	7.0	585	
					87
60° S	351	2.0	3.7	415	
				AVERAGE ~	100 m

the air could rise by 3.2 to 6.4 Gt, two thirds to four thirds of the present amount. This quantity would be in addition to most of the present rate of increase of about 7 Gt/century. The corresponding increase in global average surface temperature from the methane "greenhouse" effect toward the latter part of the twenty-first century could be 0.65 to 0.8°C if we accept the estimate of Lacis et al. (1981) for the equilibrium warming resulting from a doubling of methane, and 1.5 to 1.8°C, using the estimate of Chamberlain et al. (1982). As the CO_2 produced by oxidation of the methane accumulates in the air, there will be a slight further rise in temperature, of the order of 0.1 to 0.2°C in a hundred years.

Bell (1982) assumed that a CO_2-induced warming of the Arctic Ocean would release methane hydrates during 100 years from the top 40 m of the bottom sediments over half the area between water depths of 280 and 370 m. The estimated bottom area between these depths in the Arctic Ocean and Arctic Mediterranean is 0.240 million km^2 or about 1.5% of the area between 200 and 1000 m in the world ocean (Menard and Smith, 1966). Bell assumes that 10,000 Gt of C as methane hydrate are uniformly distributed in the top 250 m of the sediments over the world ocean area between 200 and 1000 m. The total release from the Arctic Ocean regions in 100 years would then be 12 Gt of C or 0.12 Gt/yr. (Bell's figure of 8 Gt of C/yr is an obvious computational error.)

Obviously, the uncertainties in calculations of release of methane from continental slope sediments are so great that the results cannot be thought of as a projection for the future. But it is equally obvious that an extensive sediment sampling program on continental

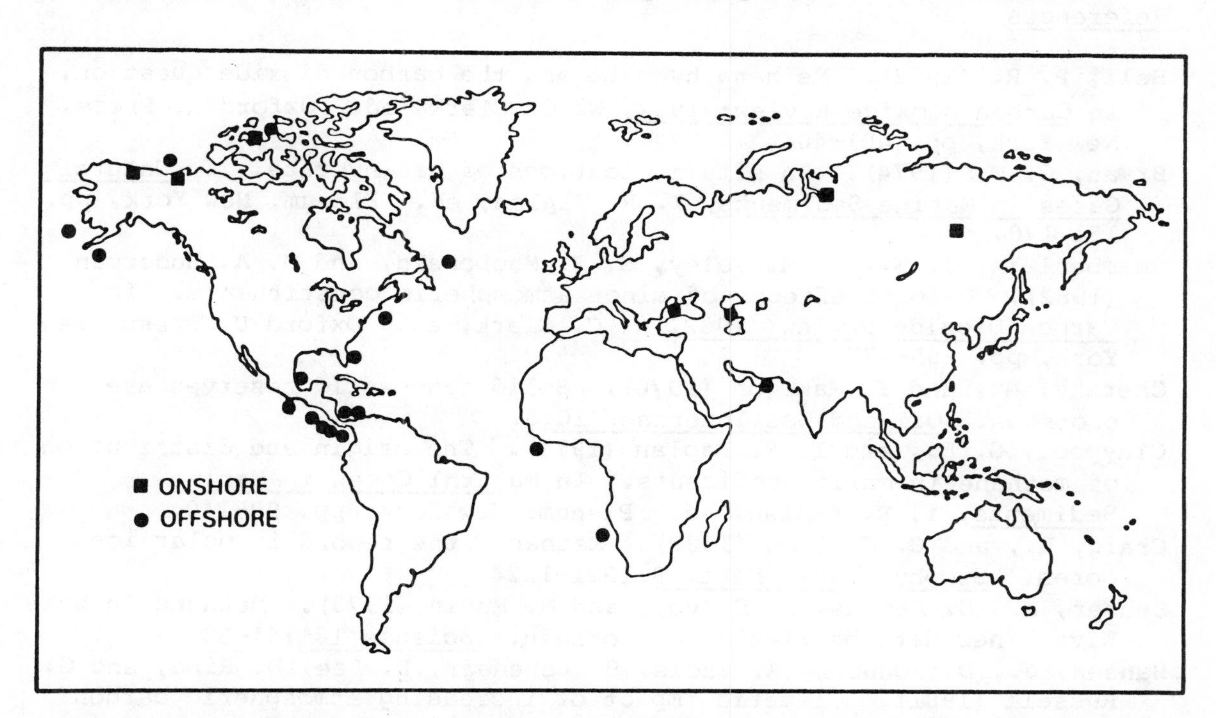

FIGURE 3.19 Reported occurrences of natural gas hydrates. (Updated from Kvenvolden and McMenamin, 1980.)

slopes throughout the world should be undertaken to determine the depth, thickness, and distribution of methane hydrate clathrates, especially where oceanfloor temperatures and depths are such that methane release is likely from ocean warming during the next century. A small release of methane clathrates may already be taking place as a consequence of the estimated 0.5°C increase of global air temperature (Hansen et al., 1982; Jones et al., 1982; Weller et al., this volume, Chapter 5, Section 5.2) during the last century, and the probable increase of ocean-bottom temperatures by 0.1-0.2°C from eddy diffusion and advection down to 500 m.

Some indications that clathrates may be widespread and abundant in ocean sediments are given in Figure 3.19 (from MacDonald, 1982b). This shows the locations in the continental slopes of the ocean floor and in the Black and Caspian Seas, in which the existence of methane clathrates has been inferred from high gas contents in cores of the Deep Sea Drilling Project or in which their presence is suspected from acoustic reflections from a layer below the sediment surface that parallels the ocean-bottom topography (Bryan, 1974; Stoll et al., 1971). It is believed that these are reflections from the bottom of the methane clathrate zone, although other explanations are possible. The likelihood of the widespread occurrence of clathrates in continental slope sediments gives force to our argument that a systematic survey should be made in an attempt to determine their abundance and distribution.

References

Bell, P. R. (1982). Methane hydrate and the carbon dioxide question. In Carbon Dioxide Review: 1982, W. C. Clark, ed. Oxford U. Press, New York, pp. 401-406.

Bryan, G. M. (1974). In situ indications of gas hydrate. In Natural Gases in Marine Sediments, I. R. Kaplan, ed. Plenum, New York, pp. 151-170.

Chamberlain, J. W., H. M. Foley, G. J. MacDonald, and M. A. Ruderman (1982). Climate effects of minor atmospheric constituents. In Carbon Dioxide Review: 1982, W. C. Clark, ed. Oxford U. Press, New York, pp. 255-277.

Chersky, N., and Y. Makogon (1970). Solid gas--world reserves are enormous. Oil and Gas Internat. 10:8.

Claypool, G. E., and I. R. Kaplan (1974). The origin and distribution of methane in marine sediments. In Natural Gases in Marine Sediments, I. R. Kaplan, ed. Plenum, New York, pp. 99-140.

Craig, H., and C. C. Chou (1982). Methane, the record in polar ice cores. Geophys. Res. Lett. 9:1221-1224.

Deuser, W., E. Degens, G. Harvey, and M. Rubin (1973). Methane in Lake Kiva: new data bearing on its origin. Science 181:51-53.

Hansen, J., D. Johnson, A. Lacis, S. Lebedeff, P. Lee, D. Rind, and G. Russell (1981). Climatic impact of increasing atmospheric carbon dioxide. Science 213:957-966.

Jones, P. D., R. M. L. Wigley, and P. M. Kelly (1982). Mon. Weather Rev. 110:59-70.

Kossina, E. (1921). Die Tiefen des Weltmeeres. Berlin U. Inst. für Meereskunde. Veroff. N.F., A Geogr. Naturwiss. Heft 9, 70 pp.

Kvenvolden, K. A., and M. A. McMenamin (1980). Hydrates of natural gas: a review of their geologic occurrence. U.S. Geological Survey Circ. 825. U.S. Dept. of the Interior, Washington, D.C.

Lacis, A., J. Hansen, P. Lee, T. Mitchell, and S. Lebedeff (1981). Greenhouse effect of trace gases, 1970-1980. Geophys. Res. Lett. 8:1035-1038.

MacDonald, G. J., ed. (1982a). The Long-Term Impacts of Increasing Atmospheric Carbon Dioxide Levels. Ballinger, Cambridge, Mass., 273 pp.

MacDonald, G. J. (1982b). The many origins of natural gas. Paper presented May 1982 at Deep Source Gas Workshop sponsored by Morgantown Energy Technology Center. U.S. Dept. of Energy.

McIver, R. D. (1974). Hydrocarbon gas (methane) in canned Deep Sea Drilling Project core samples. In Natural Gases in Marine Sediments, I. R. Kaplan, ed. Plenum, New York, pp. 63-70.

Makhijani, A., and A. Poole (1975). Energy and Agriculture in the Third World, Appendix B, Biogasification. Ballinger, Cambridge, Mass., pp. 143-160.

Menard, H. W., and S. M. Smith (1966). Hypsometry of ocean basin provinces. J. Geophys. Res. 71:4305-4321.

Miller, S. R. (1974). The nature and occurrence of clathrate hydrates. In Natural Gases in Marine Sediments, I. R. Kaplan, ed. Plenum, New York, pp. 151-170.

Rasmussen, R. A., and M. A. K. Khalil (1981). Atmospheric methane (CH_4): trends and seasonal cycles. J. Geophys. Res. 86:9826-9832.

Stoll, R. D., J. Ewing, and G. Bryan (1971). Anomalous wave velocities in sediments containing gas hydrates. J. Geophys. Res. 76(8):2090.

Trask, P. D. (1932). Origin and environment of source sediments of petroleum. American Petroleum Institute, Gulf Publ. Co., Houston, Tex., reprinted 1982, 332 pp. Quoted in H. Sverdrup et al. (1942), The Oceans. Prentice-Hall, New York, pp. 1009-1015.

Welhan, J., and H. Craig (1979). Methane and hydrogen in East Pacific Rise hydrothermal fluids. Geophys. Res. Lett. 6:829-831.

3.6 SENSITIVITY STUDIES USING CARBON CYCLE MODELS
Lester Machta

Sensitivity studies using carbon cycle models provide a way of estimating uncertainties in model predictions and aid in distinguishing between those factors in the model that require improvement and those whose uncertainty makes little difference in a final answer. On the other hand, sensitivity studies cannot identify defects that are incorporated in the model and for which no sensitivity analysis is possible. Furthermore, a model based on the present physical and biological world usually assumes that future behavior of that world can be derived from its past and current status. One should remember that all models of the real world limit their treatment to one or at most a very few forcing functions. In reality many other factors may also produce relevant changes.

There are a wide variety of carbon cycle models currently available into which one may enter values for fossil fuel or other CO_2 sources. These models have in common three reservoirs: the atmosphere, the oceans, and the biosphere. They also limit themselves to exchange processes that act on a time scale less than several thousand years; that is, many geologic processes are omitted. Each model subdivides the reservoirs and transports carbon and its isotopes in different ways. All models inject the past releases of fossil fuel CO_2 and, in a few cases other sources of CO_2, into the atmospheric reservoir and try to reconstruct the growth of CO_2 concentration in air as found at, say, Mauna Loa Observatory after 1958. In some instances, other observations, not involved in model development, may also be used to validate the model.

Few, if any, models of the carbon cycle would likely be published unless there were some agreement with the Mauna Loa CO_2 record. Thus, using predictions from the preindustrialized period to the 1958-1982 interval may not be useful for sensitivity studies. Rather, predictions into the future, as long as they use the same projected CO_2 releases, are a better approach to sensitivity. Note that this in no way implies that the future emissions are known; merely that they represent convenient and realistic numbers to use for sensitivity studies.

3.6.1 Comparison among Different Models

Killough and Emanuel (1981) have compared the projections made from two simple ocean box models and three other more complex ocean layer models. In each case, plausible parameters for the reservoirs, which include an atmosphere and biosphere as well as the oceans, are used to determine the size and exchange between reservoirs. Figure 3.20 illustrates the projection to year 2275. The topmost curve labeled "cumulative release +290 ppmv" represents the assumed curve of the input of CO_2 into the atmosphere expressed in units of the atmospheric concentration as though all the CO_2 remained airborne and were uniformly mixed in the air. The input scenario and the models suggest a peak concentration in the air about six times the preindustrial concentration of 290 ppmv.

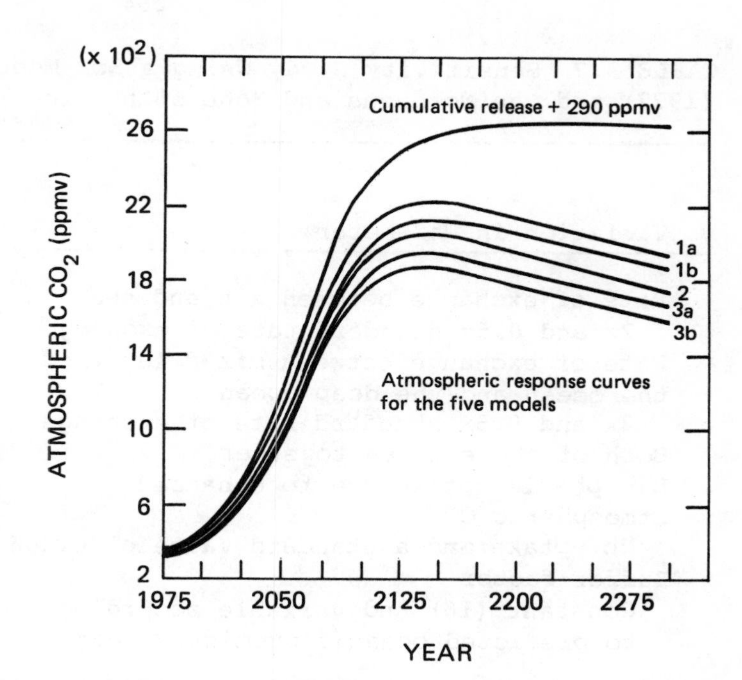

FIGURE 3.20 Atmospheric response levels of CO_2 based on cumulative release as shown. Net exchange with terrestrial biota was assumed to be zero after an asymptotic transition period within the decade after 1975. (From Killough and Emmanuel, 1981.)

At this peak, the range between the minimum and maximum concentrations from the five models was 350 ppmv or about 18% of the peak concentration. Were one to select the average among five models, the maximum deviation of the lowest and highest concentration from the five models would differ from the mean by less than 10%.

Laurmann and Spreiter (1983) have compared three simple box-type models (2, 3, and 4 boxes or carbon reservoirs). They conclude that not only are predictions of future concentrations from these three models sufficiently similar, but the use of a single airborne fraction (they use the term "retention fraction") yields similar results as the box models. Two exceptions are noted. First, the models do diverge if the growth rate of CO_2 emissions is small, an exponential growth rate of less than 1.5% per year. Second, if the transfer to the deep oceans is much faster than used in the models then all bets are off (i.e., the models do not represent nature). It is noted that problems will arise if there has been significant deforestation CO_2 during the past several decades; the three models assume none.

3.6.2 Comparison of Parameters within a Single Model

As part of the analysis of their carbon cycle model, Enting and Pearman (1982) have undertaken a sensitivity study. Although the model does not allow for geographical variation in any of the three major reservoirs (atmosphere, oceans, and biosphere), it contains many features beyond those of the other models described in this review of sensitivity studies. Virtually all of the parameters (or observations leading to the choice of a value for a parameter) were studied. The conclusions drawn by Enting and Pearman are: "The sensitivity analysis . . .

TABLE 3.7 Sensitivity Study Using a Box Model of Keeling and Bacastow (1977) and the Nordhaus and Yohe 50th Percentile CO_2 Emissions Scenario

Variation in Parameter	Range (ppmv)[a]	Range (%)[b]
Rate of exchange between air and sea		
2x and 0.5x standard rate of exchange	2	0.3
Rate of exchange between mixed layer of the ocean and the deep ocean		
2x and 0.5x standard rate of exchange	70	9
Both of above taken together	74	10
Biospheric uptake due to enhanced atmospheric CO_2		
No uptake and a standard value of 0.266	229	29
Buffer factor		
Constant (10) and variable according to predicted oceanic chemistry change	61	8

[a]The range is the higher minus the lower predicted by the changes in arithmetic number used for the parameter in the year 2100.
[b]Range divided by 784 ppmv, the predicted value for the year 2100, times 100.

indicated that for practical purposes the best-fit parameter set is not unique but that there is a range of values within which the model agrees with the data. . . . These uncertainties have only a limited effect on the model predictions for the year 2000, the main uncertainty being the future rate of fossil fuel release."

Using the Keeling and Bacastow (1977) box model, we have varied some of the parameters one at a time to determine the effect of such a change on the prediction of atmospheric CO_2 concentration in the year 2100. Table 3.7 shows some of the results based mainly on doubling or halving the most probable value of the parameter and expressing the uncertainty in the prediction as a range in predicted concentration. For all cases, the Nordhaus and Yohe (see this volume, Chapter 2, Section 2.1) 50 percentile scenario of CO_2 emissions was used to the year A.D. 2100. Again, if a departure from a mean value is preferred to the range, the extremes would be at most 15% from the mean CO_2 concentration, 784 ppmv.

3.6.3 Deforestation as a Source of CO_2

We consider deforestation CO_2 as a possible real and important source of atmospheric CO_2. If some reasonable amounts of future CO_2 from deforestation are added to CO_2 from future fossil fuel combustion, the error that would be introduced by the omission of the future deforestation CO_2 would still be small in the year 2100 assuming, say, a 2% per year growth rate in the emission of fossil fuel CO_2 after 1980. For example, oxidizing half of the living biosphere (or half of about 600 x 10^{15} g of C) would result in an increase of perhaps 75 ppmv in a predicted value of about 1,000 ppmv in the year 2100.

On the other hand, if significant deforestation is currently in progress, and has been for the past several decades, irrespective of future deforestation, the issue is different from that described in the above paragraph. If, for example, the deforestation CO_2 were about as large as that from fossil fuel sources, the current models would fail to reproduce the observed atmospheric CO_2 growth after 1958. The models would likely have to be modified since no reasonable adjustment of the parameters will allow a good fit of predictions to observations after 1958. The airborne fraction, the ratio of atmospheric increase in a year to the net amount added to the atmosphere, would be calculated to drop to about 0.3 from a value of almost 0.6. Instead of an increase of predicted concentration from 340 to 1000 ppmv, the increase might be only from 340 to 670 ppmv to the year 2100.

3.6.4 Conclusion

It may be worth noting what the limited survey of sensitivity studies does and does not reveal. The studies do suggest that if the carbon cycle models are accepted as valid representations of reality, reasonable variations in the numerical values of the parameters do not appear to affect significantly the predictions of future concentrations of atmospheric CO_2. Research that accepts the physics, chemistry, and biology of the existing models but tries simply to refine the parameters may not be so effective as research in other aspects of the carbon dioxide issue. On the other hand, the sensitivity studies reveal nothing about the ability of the models to represent nature either today or in the future except possibly indirectly.

The guidance provided by sensitivity studies suggests the need for research in those aspects of the carbon cycle likely to make a difference for predicting future atmospheric CO_2 concentrations.

References

Enting, I. G., and G. I. Pearman (1982). Description of a one-dimensional global carbon cycle model. Paper No. 42. Div. of Atmos. Phys., CSIRO, Australia.

Keeling, C. D., and R. B. Bacastow (1977). Impact of industrial gases on climate. Energy and Climate. Geophysics Research Board, National Research Council, National Academy Press, Washington, D.C.

Killough, G. G., and W. R. Emanuel (1981). A comparison of several models of carbon turnover in the ocean with respect to their distributions of transit time and age, and responses to atmospheric CO_2 and ^{14}C. Tellus 33:274-290.

Laurmann, J. A., and J. R. Spreiter (1983). The effects of carbon cycle model error in calculating future atmospheric carbon dioxide levels. Climatic Change 5:145-175.

4 Effects on Climate

4.1 EFFECTS OF CARBON DIOXIDE

Joseph Smagorinsky

4.1.1 Excerpts from "Charney" and "Smagorinsky" Reports

From the beginning of our Committee's work it was clear that at least one aspect of the CO_2 issue would require continuing attention: the linkage between increases in atmospheric CO_2 and changes in climate. This question had been addressed in 1979 by a panel chaired by the late Jule Charney (Climate Research Board, 1979), and I was asked to lead a similar group to take a second look in the light of subsequent research results. Our report, Carbon Dioxide and Climate: A Second Assessment (CO_2/Climate Review Panel, 1982), was to its authors both reassuring and disappointing. On one hand, we found no reasons to dispute the judgments of the Charney group: increased carbon dioxide can potentially produce climate changes sufficiently significant to merit concern. On the other, we continued to find large uncertainties in the timing, magnitude, character, and spatial distribution of these changes. At this writing, the results of our study and its predecessor still represent in my view a sober, balanced, and responsible consensus on the climatic implications of increased CO_2. Thus, the summarized conclusions of our study (and the earlier Charney report) are reproduced below. However, in the year that has elapsed since their formulation, the scientific issues have been further illuminated by research, and I will append in Section 4.1.2 some comments as an epilogue to the CO_2/Climate Review Panel report.

1
Summary and Conclusions

We have examined the principal attempts to simulate the effects of increased atmospheric CO_2 on climate. In doing so, we have limited our considerations to the direct climatic effects of steadily rising atmospheric concentrations of CO_2 and have assumed a rate of CO_2 increase that would lead to a doubling of airborne concentrations by some time in the first half of the twenty-first century. As indicated in Chapter 2 of this report, such a rate is consistent with observations of CO_2 increases in the recent past and with projections of its future sources and sinks. However, we have *not* examined anew the many uncertainties in these projections, such as their implicit assumptions with regard to the workings of the world economy and the role of the biosphere in the carbon cycle. These impose an uncertainty beyond that arising from our necessarily imperfect knowledge of the manifold and complex climatic system of the earth.

When it is assumed that the CO_2 content of the atmosphere is doubled and statistical thermal equilibrium is achieved, the more realistic of the modeling efforts predict a global surface warming of between 2°C and 3.5°C, with greater increases at high latitudes. This range reflects both uncertainties in physical understanding and inaccuracies arising from the need to reduce the mathematical problem to one that can be handled by even the fastest available electronic computers. It is significant, however, that none of the model calculations predicts negligible warming.

The primary effect of an increase of CO_2 is to cause more absorption of thermal radiation from the earth's surface and thus to increase the air temperature in the troposphere. A strong positive feedback mechanism is the accompanying increase of moisture, which is an even more powerful absorber

From Carbon Dioxide and Climate: A Scientific Assessment, Report of an Ad Hoc Study Group on Carbon Dioxide and Climate, National Research Council, 1979.

of terrestrial radiation. We have examined with care all known negative feedback mechanisms, such as increase in low or middle cloud amount, and have concluded that the oversimplifications and inaccuracies in the models are not likely to have vitiated the principal conclusion that there will be appreciable warming. The known negative feedback mechanisms can reduce the warming, but they do not appear to be so strong as the positive moisture feedback. We estimate the most probable global warming for a doubling of CO_2 to be near 3°C with a probable error of ± 1.5°C. Our estimate is based primarily on our review of a series of calculations with three-dimensional models of the global atmospheric circulation, which is summarized in Chapter 4. We have also reviewed simpler models that appear to contain the main physical factors. These give qualitatively similar results.

One of the major uncertainties has to do with the transfer of the increased heat into the oceans. It is well known that the oceans are a thermal regulator, warming the air in winter and cooling it in summer. The standard assumption has been that, while heat is transferred rapidly into a relatively thin, well-mixed surface layer of the ocean (averaging about 70 m in depth), the transfer into the deeper waters is so slow that the atmospheric temperature reaches effective equilibrium with the mixed layer in a decade or so. It seems to us quite possible that the capacity of the deeper oceans to absorb heat has been seriously underestimated, especially that of the intermediate waters of the subtropical gyres lying below the mixed layer and above the main thermocline. If this is so, warming will proceed at a slower rate until these intermediate waters are brought to a temperature at which they can no longer absorb heat.

Our estimates of the rates of vertical exchange of mass between the mixed and intermediate layers and the volumes of water involved give a delay of the order of decades in the time at which thermal equilibrium will be reached. This delay implies that the actual warming at any given time will be appreciably less than that calculated on the assumption that thermal equilibrium is reached quickly. One consequence may be that perceptible temperature changes may not become apparent nearly so soon as has been anticipated. We may not be given a warning until the CO_2 loading is such that an appreciable climate change is inevitable. The equilibrium warming will eventually occur; it will merely have been postponed.

The warming will be accompanied by shifts in the geographical distributions of the various climatic elements such as temperature, rainfall, evaporation, and soil moisture. The evidence is that the variations in these anomalies with latitude, longitude, and season will be at least as great as the globally averaged changes themselves, and it would be misleading to predict regional climatic changes on the basis of global or zonal averages alone. Unfortunately, only gross globally and zonally averaged features of the present climate can now be reasonably well simulated. At present, we cannot simulate accurately the details of regional climate and thus cannot predict the locations and intensities of regional climate changes with confidence. This situation may be expected to improve gradually as greater scientific understanding is acquired and faster computers are built.

To summarize, we have tried but have been unable to find any overlooked or underestimated physical effects that could reduce the currently estimated global warmings due to a doubling of atmospheric CO_2 to negligible proportions or reverse them altogether. However, we believe it quite possible that the capacity of the intermediate waters of the oceans to absorb heat could delay the estimated warming by several decades. It appears that the warming will eventually occur, and the associated regional climatic changes so important to the assessment of socioeconomic consequences may well be significant, but unfortunately the latter cannot yet be adequately projected.

Summary of Conclusions and Recommendations

For over a century, concern has been expressed that increases in atmospheric carbon dioxide (CO_2) concentration could affect global climate by changing the heat balance of the atmosphere and Earth. Observations reveal steadily increasing concentrations of CO_2, and experiments with numerical climate models indicate that continued increase would eventually produce significant climatic change. Comprehensive assessment of the issue will require projection of future CO_2 emissions and study of the disposition of this excess carbon in the atmosphere, ocean, and biota; the effect on climate; and the implications for human welfare. This study focuses on one aspect, estimation of the effect on climate of assumed future increases in atmospheric CO_2. Conclusions are drawn principally from present-day numerical models of the climate system. To address the significant role of the oceans, the study also makes use of observations of the distributions of anthropogenic tracers other than CO_2. The rapid scientific developments in these areas suggest that periodic reassessments will be warranted.

The starting point for the study was a similar 1979 review by a Climate Research Board panel chaired by the late Jule G. Charney. *The present study has not found any new results that necessitate substantial revision of the conclusions of the Charney report.*

SIMPLIFIED CLIMATE MODELS AND EMPIRICAL STUDIES

Numerical models of the climate system are the primary tools for investigating human impact on climate. Simplified models permit economically feasible

From Carbon Dioxide and Climate: A Second Assessment, Report of the CO_2/Climate Review Panel, National Research Council, National Academy Press, 1982.

analyses over a wide range of conditions. Although they can provide only limited information on local or regional effects, simplified models are valuable for focusing and interpreting studies performed with more complete and realistic models. *The sensitivity of global-mean temperature to increased atmospheric CO_2 estimated from simplified models is generally consistent with that estimated from more complete models.*

The effects of increased CO_2 are usually stated in terms of surface temperature, and models of the energy balance at the surface are often employed for their estimation. However, changes in atmospheric CO_2 actually affect the energy balance of the entire climate system. *Because of the strong coupling between the surface and the atmosphere, global-mean surface warming is driven by radiative heating of the entire surface–atmosphere system, not only by the direct radiative heating at the surface.*

Theoretical and empirical studies of the climatic effects of increased CO_2 must properly account for all significant processes involved, notably changes in the tropospheric energy budget and the effects of ocean storage and atmospheric and oceanic transport of heat. For example, studies of the isolated surface energy balance or local observational studies of the transient response to short-term radiative changes can result in misleading conclusions. Otherwise, such studies can grossly underestimate or, in some instances, overestimate the long-term equilibrium warming to be expected from increased CO_2. Surface energy balance approaches and empirical studies are fully consistent with comprehensive climate models employed for CO_2 sensitivity studies, provided that the globally connected energy storage and transport processes in the entire climate system are fully accounted for on the appropriate time scales. Indeed, *empirical approaches to estimating climatic sensitivity—particularly those employing satellite radiation budget measurements—should be encouraged.*

ROLE OF THE OCEANS

The heat capacity of the upper ocean is potentially great enough to slow down substantially the response of climate to increasing atmospheric CO_2. The upper ocean will affect both the detection of CO_2-induced climatic changes and the assessment of their likely social implications. The thermal time constant of the atmosphere coupled to the wind-mixed layer of the ocean is only 2–3 years. The thermal time constant of the atmosphere coupled to the upper 500 m of the ocean is roughly 10 times greater, or 20–30 years. On a time scale of a few decades, the deep water below 500 m can act as a sink of heat, slowing the rise of surface temperature. However, tracer data indicate that the globally averaged mixing rate into the deep ocean appears to be too slow for it to be of dominant importance on a global scale for time scales less than 100 years.

The lagging ocean thermal response may cause important regional differences in climatic response to increasing CO_2. The response in areas downwind from major oceans will certainly be different from that in the interior of major continents, and a significantly slower response to increasing CO_2 might be expected in the southern hemisphere. *The role of the ocean in time-dependent climatic response deserves special attention in future modeling studies, stressing the regional nature of oceanic thermal inertia and atmospheric energy-transfer mechanisms.*

Progress in understanding the ocean's role must be based on a broad program of research: continued observations of density distributions, tracers, heat fluxes, and ocean currents; quantitative elucidation of the mixing processes potentially involved; substantial theoretical effort; and development of models adequate to reproduce the relative magnitudes of a variety of competing effects. The problems are difficult, and complete success is unlikely to come quickly. Meanwhile, partially substantiated assumptions like those asserted here are likely to remain an integral part of any assessment. In planning the oceanographic field experiments in connection with the World Climate Research Program, *particular attention should be paid to improving estimates of mixing time scales in the main thermocline.*

Present knowledge of the interaction of sea-ice formation and deep-water formation is still rudimentary, and it will be difficult to say even qualitatively what role sea ice will play in high-latitude response and deep-water formation until the climatic factors that control the areal extent of polar pack ice in the northern and southern hemispheres are known. Field experiments are required to gain fundamental observational data concerning these processes.

CLOUD EFFECTS

Cloud amounts, heights, optical properties, and structure may be influenced by CO_2-induced climatic changes. In view of the uncertainties in our knowledge of cloud parameters and the crudeness of cloud-prediction schemes in existing climate models, *it is premature to draw conclusions regarding the influence of clouds on climate sensitivity to increased CO_2.* Empirical approaches, including satellite-observed radiation budget data, are an important means of studying the cloudiness–radiation problem, and they should be pursued.

Simplified climate models indicate that lowering of albedo owing to decreased areal extent of snow and ice contributes substantially to CO_2 warming at high latitudes. However, more complex models suggest that increases in low-level stratus cloud cover may at least partially offset this decrease in albedo. In view of the great oversimplification in the calculation of clouds in climate models, these inferences must be considered tentative.

OTHER TRACE GASES

Although the radiative effects of trace gases (nitrous oxide, methane, ozone, and chlorofluoromethanes) are in most instances additive, their concentrations can be chemically coupled. *The climatic effects of alterations in the concentrations of trace gases can be substantial.*

Since trace-gas abundances might change significantly in the future because of anthropogenic emissions or as a consequence of CO_2-initiated climatic changes, *it is important to monitor the most radiatively significant trace gases.*

ATMOSPHERIC AEROSOLS

Atmospheric aerosols are a potentially significant source of climate variability, but their effects depend on their composition, size, and vertical–global distributions. Stratospheric aerosols consisting mainly of aqueous sulfuric acid droplets, which persist for a few years following major volcanic eruptions, can produce a substantial, but temporary, reduction in global surface temperature and can explain much of the observed natural climatic variability. While stratospheric aerosols may contribute to the infrared greenhouse effect, their net influence appears to be surface cooling.

The climatic effect of tropospheric aerosols—sulfates, marine salts, and wind-blown dust—is much less certain, in part because of inadequate observations and understanding of the optical properties. Although anthropogenic aerosols are particularly noticeable in regions near and downwind of their sources, there does not appear to have been a significant long-term increase in the aerosol level in remote regions of the globe other than possibly the Arctic. *The climatic impact of changes in anthropogenic aerosols, if they occur, cannot currently be determined.* One cannot even conclude that possible future anthropogenic changes in aerosol loading would produce worldwide heating or cooling, although carbon-containing Arctic aerosol definitely causes local atmospheric heating. Increased tropospheric aerosols could also influence cloud optical properties and thus modify cloudiness–radiation feedback. This possibility requires further study.

THE LAND SURFACE

Land-surface processes also influence climate, and the treatment of surface albedo and evapotranspiration in climate models influences the behavior of climate models. Land-surface processes largely depend on vegetation coverage and may interact with climatic changes in ways that are as yet poorly understood.

VALIDATION OF CLIMATE MODELS

Mathematical-physical models, whether in a highly simplified form or as elaborate formulations of the behavior and interactions of the global atmosphere, ocean, cryosphere, and biomass, are generally considered to be the most powerful tools yet devised for the study of climate. Our confidence in them comes from tests of the correctness of the models' representation of the physical processes and from comparisons of the models' responses to known seasonal variations. Because decisions of immense social and economic importance may be made on the basis of model experiments, *it is important that a comprehensive climate-model validation effort be pursued, including the assembly of a wide variety of observational data specifically for model validation and the development of a validation methodology.*

Validation of climate models involves a hierarchy of tests, including checks on the internal behavior of subsystems of the model. The parameters used in comprehensive climate models are explicitly derived, as much as possible, from comparisons with observations and/or are derived from known physical principles. *Arbitrary adjustment or tuning of climate models is therefore greatly limited.*

The primary method for validating a climate model is to determine how well the model-simulated climate compares with observations. *Comparisons of simulated time means of a number of climatic variables with observations show that modern climate models provide a reasonably satisfactory simulation of the present large-scale global climate and its average seasonal changes.*

More complete validation of models depends on assembly of suitable data, comparison of higher-order statistics, confirmation of the models' representation of physical processes, and verification of ice models.

One test of climate theory can be obtained from empirical examination of other planets that in effect provide an ensemble of experiments over a variety of conditions. *Observed surface temperatures of Mars, Earth, and Venus confirm the existence, nature, and magnitude of the greenhouse effect.*

Laboratory experiments on the behavior of differentially heated rotating fluids have provided insights into the hydrodynamics of the atmosphere and ocean circulations and can contribute to our understanding of processes such as small-scale turbulence and mixing. However, they cannot simulate adequately the most important physical processes involved in climatic change.

Improvement of our confidence in the ability of climate models to assess the climatic impacts of increased CO_2 will require development of model validation methods, including determination of the models' statistical properties; assembly of standardized data for validation; development of observations to validate representations of physical processes; standardization of sensitivity tests; development of physical–dynamical and phenomenological diagnostic techniques focusing on changes specifically attributable to increased CO_2; and use of information from planetary atmospheres, laboratory experiments, and especially contemporary and past climates (see below).

PREDICTIONS AND SCENARIOS

A primary objective of climate-model development is to enable prediction of the response of the climate system to internal or external changes such as increases in atmospheric CO_2. Predictions consist of estimates of the probability of future climatic conditions and unavoidably involve many uncertainties. *Model-derived estimates of globally averaged temperature changes, and perhaps changes averaged along latitude circles, appear to have some predictive reliability for a prescribed CO_2 perturbation.* On the other hand, estimates with greater detail and including other important variables, e.g., windiness, soil moisture, cloudiness, solar insolation, are not yet sufficiently reliable. *Nevertheless, internally consistent and detailed specifications of hypothetical climatic conditions over space and time—"scenarios"—may be quite useful research tools for analysis of social responses and sensitivities to climatic changes.*

INFERENCES FROM CLIMATE MODELS

While present models are not sufficiently realistic to provide reliable predictions in the detail desired for assessment of most impacts, they can still suggest scales and ranges of temporal and spatial variations that can be incorporated into scenarios of possible climatic change.

Mathematical models of climate of a wide range of complexity have been used to estimate changes in the equilibrium climate that would result from an increase in atmospheric CO_2. The main statistically significant conclusions of these studies may be summarized as follows:

1. *The 1979 Charney report estimated the equilibrium global surface warming from a doubling of CO_2 to be "near 3°C with a probable error of*

±1.5°C." No substantial revision of this conclusion is warranted at this time.

2. Both radiative–convective and general-circulation models indicate a cooling of the stratosphere with relatively small latitudinal variation.

3. The global-mean rates of both evaporation and precipitation are projected to increase.

4. Increases in surface air temperature would vary significantly with latitude and over the seasons:

(a) Warming would be 2–3 times as great over the polar regions as over the tropics; warming would be significantly greater over the Arctic than over the Antarctic.

(b) Temperature increases would have large seasonal variations over the Arctic, with minimum warming in summer and maximum warming in winter. In lower latitudes (equatorward of 45° latitude) the warming has smaller seasonal variation.

5. Some qualitative inferences on hydrological changes averaged around latitude circles may be drawn from model simulations:

(a) Annual-mean runoff increases over polar and surrounding regions.

(b) Snowmelt arrives earlier and snowfall begins later.

(c) Summer soil moisture decreases in middle and high latitudes of the northern hemisphere.

(d) The coverage and thickness of sea ice over the Arctic and circum-Antarctic oceans decrease.

Improvement in the quality and resolution of geographical estimates of climatic change will require increased computational resolution in the mathematical models employed, improvement in the representation of the multitude of participating physical processes, better understanding of airflow over and around mountains, and extended time integration of climate models. It is clear, however, that local climate has a much larger temporal variability than climate averaged along latitude circles or over the globe.

OBSERVATIONAL STUDIES OF CONTEMPORARY AND PAST CLIMATES

Observational studies play an important role in three areas: (1) the formulation of ideas and models of how climate operates, (2) the general validation of theories and models, and (3) the construction of climate scenarios.

Studies based on contemporary climatic data have provided a useful starting point for diagnosis of climatic processes that may prove to be relevant to the CO_2 problem. The results of the Global Weather Experiment are now being analyzed and will provide a unique data base for model calibration and validation studies. *Further analyses and diagnostic studies based on contemporary climatic data sets, particularly the Global Weather Experiment data set, should be encouraged.* However, scenarios based on contemporary data sets do not yet provide a firm basis for climatic assessment of possible CO_2-induced climatic changes, nor should they be considered adequate at present for validation of CO_2 sensitivity studies with climate models.

Studies of past climatic data are leading to important advances in climate theory. For example, the large climatic changes between glacial and interglacial periods are being linked with relatively small changes in solar radiation due to variations in the Earth's orbit. If confirmed, these studies will improve our understanding of the sensitivity of climate to small changes in the Earth's radiation budget. A large multidisciplinary effort will be required to acquire the requisite data and carry out the analysis, and such work should be encouraged. Studies of past climate are also potentially valuable because they deal with large changes of the climate system, including the atmosphere, oceans, and cryosphere; because they can reveal regional patterns of climate change; and because there is knowledge of the changes in forcing (now including changes both in atmospheric CO_2 concentrations and in solar radiation) that are apparently driving the system.

4.1.2 Epilogue

Despite the relatively brief interval between the Charney report and our Panel's study, a considerable volume of additional work had been carried out. The virtually exponential growth in scientific research in the CO_2/climate area reflects increasing consciousness of the issue's social and scientific importance, burgeoning interest in the more general problems of climate, the development of an active community of scientists and institutions concerned with the problem, and--by no means least--continued strong support by funding agencies, notably the U.S. Department of Energy. The excellent reviews assembled by Clark (1982) and Reck and Hummel (1982) and conducted by the U.S. Department of Energy (1983) make a further detailed compendium of research unnecessary at this time. Their content, however, demonstrates clearly the health of this area of research and the need for periodic critical reviews. The consciously conservative assessments conducted by National Research Council groups revealed continued progress in basic understanding of the climate system. However, our understanding of the local and regional details of man-made climate changes is advancing only slowly. This frustratingly slow advance reflects no lack of talent, effort, or resources but rather the inherent difficulty of the task. Although our pace is slow, it is forward and provides us with increasingly clear views of future climate.

Our Panel considered a scenario of increasing CO_2 concentrations generally consistent with that postulated by an international assessment group in 1980, i.e., a slow growth leading to a concentration of about 450 ppm by about 2025. This scenario lies within the range of uncertainty suggested by the later studies of Nordhaus and Yohe (this volume, Chapter 2, Section 2.1) and Machta (this volume, Chapter 3, Section 3.5) presented in this report and is consistent with their estimates of a doubling of airborne CO_2 toward the end of the twenty-first century. As Machta describes below in Section 4.2, the growing concentrations of other radiatively active trace gases considerably complicate the problem. Tropospheric ozone, methane, nitrous oxide, and the chlorofluoromethanes also interact with thermal radiation and can produce significant additional perturbations to the Earth's heat budget. Thorough discussions have been presented by Chamberlain et al. (1982), Ramanathan (1982), and Hansen et al. (1982). While projection of the future concentrations of these gases is even more problematical than for CO_2, it does not seem improbable that the perturbation of the heat budget due to these trace gases could approach the magnitude of the perturbation due to CO_2 alone. If a doubling of atmospheric concentration of CO_2 is attained in the latter part of the next century, a concomitant rise in concentrations of other greenhouse gases would imply that the climatic equivalent of a doubling would be reached much sooner.

As Revelle and Suess (1957) observed, by changing the atmosphere's composition mankind is conducting a great and unprecedented geophysical experiment. Since we have no laboratory analog of this experiment, we must attempt to predict its outcome by recourse to some model--natural or analytical. In view of the complexity of the global climate system, with its myriad possibilities for unexpected and counterintuitive feedbacks and responses, the most desirable model would be the Earth itself. Indeed, Budyko and Yefimova (1981) attempted to relate paleoclimatic reconstructions to estimated contemporary CO_2 concentrations and have reached conclusions consistent with those drawn from numerical models. Others (e.g., Kellogg, 1977) have cited previous warm periods in the Earth's history as possible guides to the regional pattern of climatic changes in a CO_2-warmed Earth. Both approaches are hampered by inadequacies in data--particularly in chronology--and problems in causality. For example, glacial-interglacial oscillations seem to be paced by orbital changes, and in these cases CO_2 changes may be responses to rather than instigators of the associated climate changes. Similarly, the warm climate of the early Holocene, often cited as a model of a warmer Earth, has been plausibly explained (Kutzbach and Otto-Bliesner, 1982) in terms of changes in the Earth's orbit, although close study shows a complex time sequence of climate changes. Thus the search for a historical analog to CO_2-induced climatic change is strewn with pitfalls.

In the absence of satisfactory natural models, we must turn to mathematical models based on the most reliable physical and empirical relationships that we can muster. For example, Idso (1980, 1982) employed an empirically calibrated linear model relating radiant energy absorbed at the Earth's surface and surface air temperature, but our Panel found his analysis incomplete and misleading. More complete models treat the entire atmospheric column and calculate numerically the exchanges of sensible, latent, and radiative energy between layers and with space. Extension to two and three dimensions permits the energy balance and climate of the globe to be simulated with considerable realism and allows for interaction of climate with the oceans and with changes in surface characteristics and snow/ice cover.

The results of a number of model simulations of contemporary climate and the climate corresponding to increased CO_2 were reviewed by our Panel. Further comprehensive reviews have been made by Schlesinger (1983a,b), Reck (1982), and Budyko et al. (1982). In common with our Panel, these generally take as a convenient index the calculated equilibrium change in globally averaged surface air temperature for doubled (or sometimes quadrupled) CO_2. The values from studies with comprehensive models and realistic boundary conditions lie within the range suggested by the Charney Panel in 1979 and which our Panel found no grounds for changing. In this connection, one must recall that this range represents the best judgments of the two panels based on a small sample of inhomogeneous but--because of common physical assumptions--not independent numerical experiments. Continued research will, I hope, result in a sharpening of these estimates. Two recent simulations with comprehensive general-circulation models at the National Center for Atmospheric Research (Washington and Meehl, 1983) and the NASA Goddard Institute for Space Studies (Hansen, 1983) have yielded results also within or very near this range.

The global mean temperature index permits us to compare in general terms the potential magnitude and rate of climatic changes due to CO_2 with natural changes of the past. As suggested above, global mean surface temperature in the later decades of the next century may be 1.5 to 4.5°C warmer than today. At the lower end of this range, the change is comparable in magnitude with the difference between the cold decades of "the Little Ice Age" or the early Holocene warm period and today. The higher end approaches in magnitude the difference between the last glacial maximum and today and enters a climatic range with which mankind has had little experience. Rates of change due to CO_2 are projected to be a few tenths of a degree Celsius per decade. As discussed by the

Ad Hoc Panel on the Present Interglacial (1974) and by Clark et al. (1982), such rates are for short periods comparable with the rapid changes observed in some regions in the earlier part of this century or at the onset of the Little Ice Age. If sustained for a century or more, the most recent parallel is with the retreat of the Wisconsin glaciation.

While convenient and reasonably unambiguous, estimates of projected changes in equilibrium global mean temperature are not very useful. Parameters of climate other than temperature--precipitation, storm tracks, cloud cover, for example--and the frequencies of extreme events are at least as important in determining the real impact of climatic change. For example, tropical storm formation seems to be related to sea-surface temperature. Although hurricanes are not explicitly treated at present in general-circulation models, one might infer that warmer ocean temperatures would increase their penetration into midlatitudes; such inferences and the topic of climatic extremes deserve careful investigation. In general, the impacts of changing climate would be felt most acutely in terms of local changes that can be expected to vary widely across the Earth's surface. Moreover, as our Panel noted, the pace of change will be slowed and its evolution complicated by the ocean's buffering effects. Here recent measurements of transient tracers in the ocean (cf. Brewer, this volume, Chapter 3, Section 3.2) have tended to confirm the notions of the ocean's circulation and thermal capacity expressed in our report. Thus, understanding of the transient and local responses of climate to increasing CO_2 is far more relevant to our concerns than the ultimate globally averaged change in equilibrium temperature.

Despite the general agreement on the overall magnitude of the CO_2 effect, our understanding of these regional and temporal effects is only rudimentary. The Panel identified a few changes in zonally averaged quantities that appeared to have some statistical reliability. However, as noted by Manabe et al. (1981) and exhibited in the comparisons of Schlesinger (1982), more detailed local and regional responses are smothered in a sea of noise. Assessment of the truly relevant aspects of man-made climatic change has only just begun.

Climate changes over the United States and over other major agricultural regions of the world are naturally of great interest. Unfortunately, there is at present little that can be responsibly asserted about the details of such changes. In the Panel's report, we pointed out a few relevant inferences that seemed to be emerging from some experiments, notably a tendency toward summer dryness in midlatitudes (e.g., Manabe et al., 1981; Hayashi, 1982). These suggest, for example, some expansion of steppe and desert climates in the latitudes of the United States with increased CO_2. This inference is consistent with paleoclimatic data on warmer periods, although--as noted above--the analogies are by no means precise. As we gain more confidence in the regional details of climate simulations, analyses of climate model results in terms of climate categorizations tuned to the needs of impact analysis will become useful.

Climate models must be markedly improved and much more analysis of the implications of their results must be done before we will be able to place useful confidence in their detailed results. The most challenging, and perhaps the most intractable, problem is cloudiness. It is easy to show that small changes in cloudiness can alter the Earth's heat budget as much or more than the expected changes in CO_2 concentrations. Models with different formulations for cloudiness show great differences in global and regional climate sensitivity, even if their simulations of contemporary climate and of globally averaged changes are comparable. Our Panel concluded that "One should not trust model prediction schemes until they produce meaningful simulations of observed seasonal cloud cover and the seasonal radiation components." I believe that this reservation still stands and presents the outstanding unsolved problem in climate modeling. The parameterizations of boundary-layer convective processes--particularly in the tropics--and radiation transport also embody significant uncertainties (Kandel, 1981; Luther, 1982).

Despite our reservations about climate models we have no choice but to use them if we wish to assess the possibilities for changed climate in a changed atmosphere. We can shore up our confidence by conscientiously validating models through comparison with nature. The three-dimensional general-circulation models, in particular, can be closely compared with the real world. Indeed, as summarized for example by Gates (1982), the models simulate reasonably well the principal features of today's mean climate, the annual march of the seasons, and even the markedly different climates of the distant past. Of course, a most reassuring validation would be the unequivocal detection of the CO_2-induced climate changes that the models predict to be currently taking place. This problem is discussed at length in Chapter 5 by Weller et al., who also conclude from empirical studies that the real sensitivity of the climate system to CO_2 increases is in the lower part of the range indicated by models.

Quantitative estimates of the uncertainties in model predictions would be extremely useful. Some crude notion may be gained from studying the range of results obtained by different investigators employing different models and methodologies. Indeed, such results are the primary source for the uncertainty estimates proposed in the Charney report and left unchanged by our Panel. Attempts at more rigorous analysis are also being made (Hall et al., 1982). However, it must be recognized that all modelers incorporate similar ensembles of variables and physical processes and employ fundamentally similar algorithms (Schneider and Dickinson, 1974). One must always admit the possibility of some overlooked or underestimated feedback, e.g., cloudiness, that would markedly change the results. Careful probabilistic analysis of model simulations (e.g., Hayashi, 1982) can more clearly separate statistically significant conclusions from meaningless noise; the conclusions are usually discouraging, but the sparse scraps of significance that remain become even more precious. Finally, one may attempt to delimit the uncertainties attributable to various possible sources of sensitivity through numerical experimentation. Unsuspected--and possibly larger--sources of error may lurk in the wings. Nevertheless, we can hardly expect policymakers to lend credence to our predictions of climatic changes until we can demonstrate that changes are actually taking place (see Chapter 5) and to some degree quantify the credibility of our forecasts. Thus, the development of objective confidence limits for climate sensitivity estimates (e.g., Katz, 1982) is an important task for climate modelers.

Our Panel discussed at length some dissenting inferences of the magnitude of CO_2's effect on climate. I believe that our report fairly assessed these assertions and put them in proper perspective with respect to other research.

In summary, the conclusions of our study appear to remain valid. Man-made increases of CO_2 and other trace gases in the atmosphere may be reasonably expected to change climate significantly within the lifetimes of a large fraction of the world's inhabitants who are alive today. (According to Ausubel and Stoto, 1981, 40% of the current population will still be alive 50 years hence.) The change will be large and rapid; it will be greater in global terms than any natural climate changes that civilized man has yet experienced, although, as Schelling observes in Chapter 9 of this report, far less than the climate changes mankind has voluntarily undertaken through migration. We have some general notions of how the climate change will be distributed across the face of the Earth. In particular, there are indications of dryer and hotter summers for some already overheated and underwatered regions of our country, but these are as yet a very uncertain basis for decision making. There is a good prospect that further research can slowly sharpen and validate our predictive tools to give us more useful answers (see CO_2/Climate Review Panel, 1982, pp. 48-50).

References

Ad Hoc Panel on the Present Interglacial (1974). Report. Interdepartmental Committee for Atmospheric Sciences, Federal Council for Science and Technology. ICAS 18B-FY75. Washington, D.C.

Ausubel, J. H., and M. A. Stoto (1981). A Note on the Population Fifty Years Hence. International Institute for Applied Systems Analysis, Laxenburg, Austria.

Budyko, M. I., and N. A. Yefimova (1981). Impact of carbon dioxide on climate. Meteorol. Hydrol. 2:5-17.

Budyko, M. I., K. Ya. Vinnikov, and N. A. Yefimova (1982). The dependence of the air temperature and precipitation on carbon dioxide concentration in the atmosphere. Meteorol. Hydrol. 4:5-13.

Chamberlain, J. W., M. M. Foley, G. J. MacDonald, and M. A. Ruderman (1982). Climatic effect of trace constituents. In Carbon Dioxide Review: 1982, W. C. Clark, ed. Oxford U. Press, New York.

Clark, W. C., ed. (1982). Carbon Dioxide Review: 1982. Oxford U. Press, New York, 469 pp.

Clark, W. C., K. H. Cook, G. Marland, A. M. Weinberg, R. M. Rotty, P. R. Bell, L. J. Allison, and C. L. Cooper (1982). The carbon dioxide question: a perspective for 1982. In W. C. Clark, ed. (1982), pp. 3-43.

Climate Research Board (1979). Carbon Dioxide and Climate: A Scientific Assessment. National Academy of Sciences, Washington, D.C.

CO_2/Climate Review Panel (1982). Carbon Dioxide and Climate: A Second Assessment. National Research Council, National Academy Press, 72 pp.

Gates, W. L. (1982). Paleoclimatic modeling--a review with reference to problems and prospects for the pre-Pleistocene. In Climate in Earth History. Geophysics Study Committee, National Academy of Sciences, Washington, D.C., pp. 26-41.

Geophysical Fluid Dynamics Laboratory (1982). Activities FY80-Plans FY81. Environmental Research Laboratories, National Oceanic and Atmospheric Administration.

Hall, M. C. G., D. G. Cauci, and M. E. Schlesinger (1982). Sensitivity analysis of a radiative-convective model by the adjoint method. J. Atmos. Sci. 39:2038-2050.

Hansen, J. E., A. Lacis, and S. A. Lebedeff (1982). Commentary. In W. C. Clark, ed. (1982), pp. 284-289.

Hansen, J. E. (1983). Climate model sensitivities to changed solar irradiance and CO_2. In Climate Processes and Climate Sensitivity, Maurice Ewing Series, Vol. 4. American Geophysical Union, Washington, D.C.

Hayashi, Y. (1982). Confidence intervals of a climatic signal. J. Atmos. Sci. 39:1895-1905.

Idso, S. B. (1980). The climatological significance of a doubling of the earth's atmospheric carbon dioxide concentration. Science 207:1462-1463.

Idso, S. B. (1982a). A surface air temperature response function for earth's atmosphere. Boundary Layer Meteorol. 22:227-232.

Idso, S. B. (1982b). Carbon Dioxide: Friend or Foe? IBR Press, Tempe, Ariz., 92 pp.

Kandel, R. S. (1981). Surface temperature sensitivity to increased atmospheric CO_2. Nature 293:634-636.

Katz, R. W. (1982). Statistical evaluation of climate experiments with general circulation models: a parametric time series modeling approach. J. Atmos. Sci. 39:1446-1455.

Kellogg, W. W. (1977). Effects of Human Activities on Global Climate. Tech. Note No. 156 (WMO No. 486). World Meteorological Organization, Geneva, 47 pp.

Kutzbach, J. E., and B. L. Otto-Bliesner (1982). The sensitivity of African-Asian Monsoon climate to orbital parameter changes for 9000 years BP in a low-resolution general circulation model. J. Atmos. Sci. 39:1177-1188.

Luther, F. M. (1982). Radiative effects of a CO_2 increase: results of a model comparison. In Proceedings: Carbon Dioxide Research Conference: Carbon Dioxide, Science and Consensus. U.S. Dept. of Energy, CONF-820970, III-177-III-193.

Manabe, S., R. T. Wetherald, and R. S. Stouffer (1981). Summer dryness due to an increase of atmospheric CO_2 concentration. Climatic Change 3:347-386.

Ramanathan, V. (1982). Commentary. In W. C. Clark, ed. (1982), pp. 278-283.

Reck, R. A. (1982). Introduction to the Proceedings of the Workshop. In Reck and Hummel (1982).

Reck, R. A., and V. R. Hummel (1982). Interpretation of Climate and Photochemical Models, Ozone and Temperature Measurements. AIP Conference Proceedings No. 82. American Institute of Physics, New York, 308 pp.

Revelle, R., and H. E. Suess (1957). Carbon dioxide exchange between atmosphere and ocean and the question of an increase of atmospheric CO_2 during the past decades. Tellus 9:18-27.

Schlesinger, M. E. (1983a). Simulating CO_2-induced climatic change with mathematical climate models: capabilities, limitations, and prospects. In Proceedings, Conference on Carbon Dioxide, Climate, and Consensus, Coolfont, Virginia. U.S. Dept. of Energy, Washington, D.C.

Schlesinger, M. E. (1983b). A review of climate models and their simulation of CO_2-induced warming. Intern. J. Environ. Studies 20:103-114.

Schneider, S. H., and R. E. Dickinson (1974). Climate modeling. Rev. Geophys. and Space Phys. 12:447-493.

U.S. Department of Energy (1983). Proceedings: Carbon Dioxide Research Conference: Carbon Dioxide, Science, and Consensus. CONF-820970, February 1983.

Washington, W. M., and G. A. Meehl (1982). A summary of recent NCAR general circulation experiments on climatic effects of doubled and quadrupled amounts of CO_2. In Proceedings, Conference on Carbon Dioxide, Climate, and Consensus, Coolfont, Virginia. U.S. Dept. of Energy, Washington, D.C.

Washington, W. M., and G. A. Meehl (1983). General circulation model experiments on the climatic effects due to a doubling and quadrupling of carbon dioxide concentrations. J. Geophys. Res., in press.

4.2 EFFECTS OF NON-CO_2 GREENHOUSE GASES
Lester Machta

CO_2 is not the only gas or geophysical property capable of modifying the future climate. While a discussion of every potential climate modifier lies beyond this report, certain of them offer sufficient similarities to CO_2 (e.g., they are also "greenhouse" gases) that a brief survey is justified. Non-CO_2 greenhouse gases (and other climate modifiers) bear on the CO_2 issue in several important ways: (1) they can enhance the climate changes expected from rising atmospheric CO_2 and hence confuse the expected CO_2 changes; (2) the past climate responses to other forcing functions can aid in the interpretation of CO_2 warming; but (3) perhaps most important, predictions of the future climate require that all potential factors be taken into account.

In the past few decades the remarkable increase in interest in atmospheric chemistry along with improved technology have made it possible to measure changes with time of the concentration of many constituents of air not formerly possible. Despite this new capability, there may still be other greenhouse gases beyond those noted below that are not yet measured.

Chlorofluorocarbons. This class of gases originates from industrial activities and has been emitted to the atmosphere during the past 50 years. They are increasing in the atmosphere approximately as expected from their growth in emissions. CFC-11, CFC-12, and CFC-22, the three most abundant ones, all have long residence times in the air (tens of years) so that they can accumulate. Figure 4.1 illustrates typical time histories of CFC-11 and CFC-12 at ground level. Note that, like CO_2, with its fairly long atmospheric residence time, the concentration in air will usually increase even if the rate of emissions decreases; this is the case for CFC-11 and CFC-12 during the past few years. Both the sources and sinks of the chlorofluorocarbons are believed to be known. The emissions from industrial production and produce uses (such as aerosol propellants), for which good estimates are published by the Chemical Manufacturers Association, represent the only source of any consequence. Photochemical destruction, mainly in the stratosphere, and very slow uptake by the oceans are the only known significant sinks. Theoretically, chlorofluorocarbons are implicated as potential destroyers of stratospheric ozone, which in turn could result in human health and ecological damage from increased ultraviolet radiation. Since some consideration has been given to restricting their emissions, an extrapolation of current or past growth rates to predict future atmospheric concentrations may be unwise at this time.

Nitrous oxide. it is likely that most nitrous oxide in the air has come from denitrification in the natural or cultivated biosphere. One would therefore expect to find the largest part of atmospheric nitrous oxide to be derived from nature, unrelated to human activity. Recent, careful measurements by a few investigators (Weiss, 1981; U.S. Government, 1982) have suggested a small growth rate of the concentration of nitrous oxide in ground level air at remote locations as illustrated in

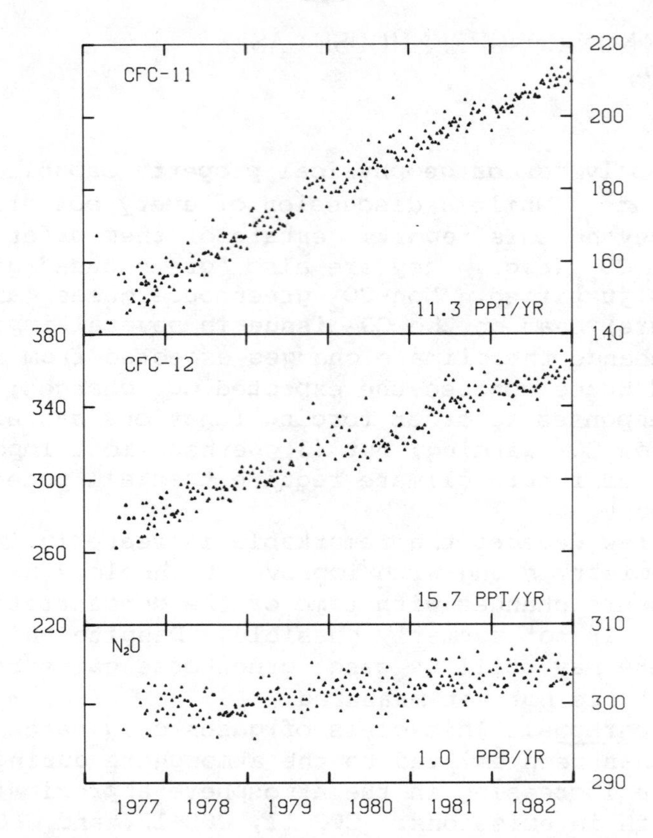

FIGURE 4.1 Upper panels: Measurements of chlorofluorocarbon (CFC-11, CFC-12) concentration in the atmosphere, Mauna Loa Observatory. Bottom panel: Measurements of atmospheric concentration of nitrous oxide in ground-level air at remote locations. [Source: Geophysical Monitoring for Climate Change (GMCC), Air Resources Laboratory, Rockville, Md.]

Figure 4.1. The source of the small increase is unknown, but a prime candidate is the continued expanded use of nitrogen fertilizers around the world to improve agricultural productivity. If so, the current slow increase is likely to continue into the foreseeable future since food demands will grow with population size. It has been suggested that since nitrous oxide is stable in the troposphere and is implicated in potential ozone destruction, there might be a move to try to restrict fertilizer usage. Figure 4.1 suggests, however, that the rate of increase of nitrous oxide in ground level air is so small, perhaps 0.25% per year, that many decades would pass before an increase of nitrous oxide would raise concern for ozone depletion.

Methane. The most abundant hydrocarbon, often called natural gas, is increasing in the atmosphere. It is thought to be a natural constituent of the air arising as it does from many biological processes and perhaps seeping out of the Earth. Measurements in the 1950s and 1960s had large error bars, and there were spatial differences so that the observed temporal variability was not viewed as an upward trend. However, in the late 1970s several investigators using gas chromatog-

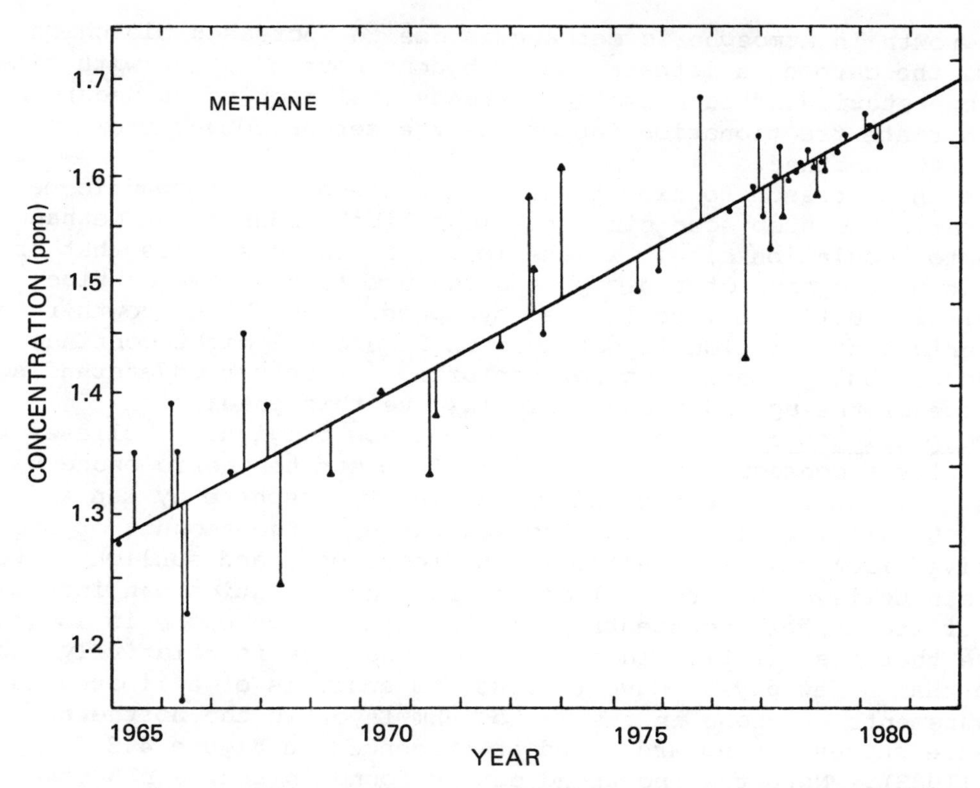

FIGURE 4.2 Northern hemisphere methane measurements, compiled from various sources, from 1965 to 1980 (log scale). The straight-line exponential best fit corresponds to an annual increase of about 1.7% per year (Rasmussen et al., 1981).

raphy have unequivocably demonstrated an upward trend. Rasmussen and Khalil (1981) have combined their recent measurements with earlier ones to suggest that the trend existed since at least 1965, as shown in Figure 4.2. Craig and Chow (1982) have found, from measurements of methane in ice cores, that concentrations prior to the sixteenth century were 0.7 ppmv.

Rasmussen and Khalil suggest that the expansion in the number of farm animals and rice production might well explain, at least qualitatively, the atmospheric methane growth. Other biological activities such as termite destruction of wood (Zimmerman et al., 1982) and possible leakage from man's mining and use of fossil methane might also contribute to methane in the air; their contribution to a growth of methane in air is less clear. The higher concentrations far north of the equatorial region suggest the termite source to be minor. The relatively rapid recent increase with time, about as fast as for CO_2, combined with the uncertainty as to its origin are both intriguing features of the methane growth in air. It is possible that the trend in the stable isotopic content of the carbon in methane might shed light on the source of the growth. Thus, atmospheric methane is now about -40%, while that derived from biological activity is about -60%.

If the growth in atmospheric methane is due to increased biological sources, the carbon in methane should become more negative with time. The interpretation of such isotopic trends will require an understanding of appropriate fractionation factors as the methane moves from one reservoir to another.

There is no reason to expect the upward trend in atmospheric methane concentration to stop soon since the most likely sources of methane are related to population size. In the long run, those sources that are dependent on the size of a biospheric feature (e.g., cows or rice paddies) will ultimately be limited by space. Thus, the growth rate in atmospheric concentration illustrated in Figure 4.2 might continue for many decades but probably not for centuries. A better understanding of the source of the upward trend would improve this prediction.

Tropospheric ozone. Tropospheric ozone was originally believed to be primarily a consequence of transport from stratospheric ozone by air motions. It can also be created within the troposphere by man and nature. Locally, as in the Los Angeles Basin, large amounts of ozone are derived from oxides of nitrogen, hydrocarbons, and sunlight. Few scientists believe that these local sources of pollution can increase the upper-tropospheric concentrations of ozone since ozone is so reactive that its lifetime in the lower atmosphere is relatively short, no more than a few days. Nevertheless, an analysis of a limited number of measurements of ozone in the 2- to 8-km layer in the northern hemisphere suggest an upward trend as evidenced in Figure 4.3 from Angell (1983). Note that no trend can be found in southern Australia. It has been suggested by Liu et al. (1980) that this increase of mid- and upper-troposphere ozone concentration of the northern hemisphere results from photochemical reactions of the oxides of nitrogen and hydrocarbons emitted by high-flying jet aircraft. Since the lifetime of an ozone molecule in the upper troposphere is also relatively short, little accumulation takes place. An increase in concentration must therefore reflect a continual increase in aircraft emissions, if that is the source. During most of the period illustrated in Figure 4.3, the number of jet aircraft as increasing in the northern hemisphere.

Some other gases. The ALE program of the Chemical Manufacturers Association and CSIRO have measured several other gases at the Australian Baseline station in Tasmania. At least two of three gases (in addition to some already noted above) show upward trends and may have absorption lines in the infrared window of the electromagnetic spectrum, making them potential greenhouse gases: carbon tetrachloride (CC_{14}) and methyl chloroform (CH_3CCl_3). The growth of carbon tetrachloride reported from Tasmania is about 1% per year since 1976, but the methyl chloroform is closer to 10% per year since 1979. Very likely both of these gases possess both natural and man-made sources. On the other hand, measurements at the Mauna Loa Observatory exhibit no or insignificant increases in carbon monoxide (CO). It is likely that the list of atmospheric gases studied for their trends and potential greenhouse effects will grow in years to come: the study of greenhouse gases other than CO_2 is still in its infancy.

The atmospheric concentrations of these trace gases are not all independent of one another. Complicated chemical reactions among these

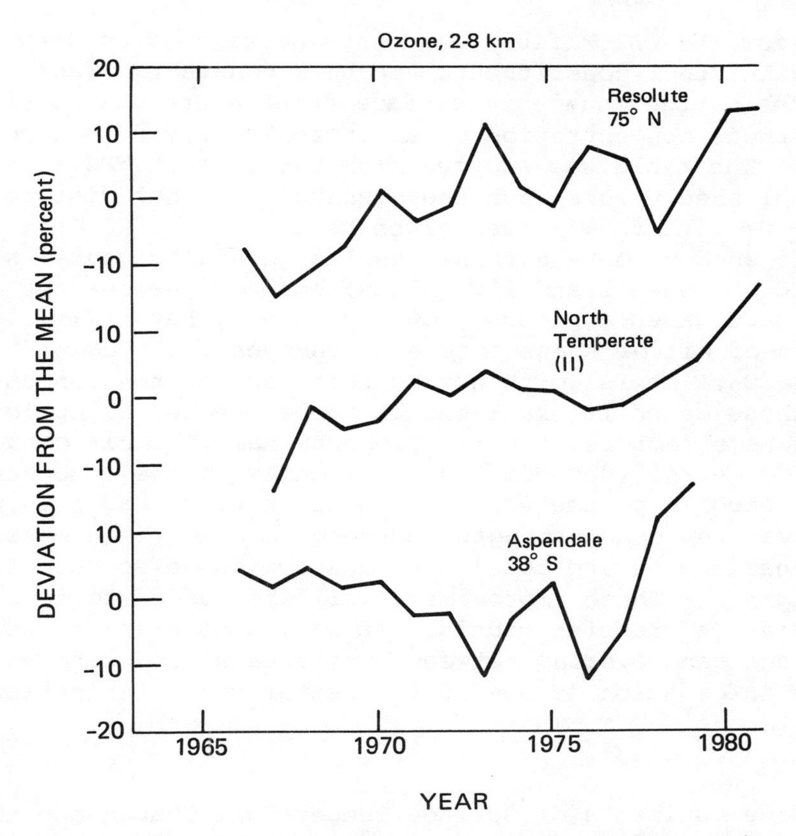

FIGURE 4.3 Ozone measurements in the 2- to 8-km layer at various latitudes. (Source: Angell, 1983.)

gases, as well as with other gases not particularly radiatively active, can affect their concentrations. For instance, it has been argued (WMO, 1983) that increases in carbon monoxide (CO) in the presence of NO can, by OH oxidation, produce an increase in O_3 and methane (CH_4).

In addition to chemical reactions with today's atmospheric composition, there would likely be new climate-chemistry interactions in the future. As the composition changes, the expected higher atmospheric water-vapor content will affect the atmospheric chemistry. An increase in OH accompanying an increase in H_2O could reduce the O_3 and CH_4 otherwise present.

Thus, unlike CO_2, which generally does not undergo chemical changes in the air, these trace gases frequently do. Not only can the mean concentration be affected by other chemicals and sunlight, but distribution particularly in the vertical can be influenced (ozone is a prime example). To estimate future concentrations will require more than estimates of natural and man-made emission rates, fundamental though those rates will be. Increased global coverage of the measurements of these gases will also help in separating natural from man-made sources of some of these gases.

Climate effect. Most of the estimates of the climate effects of the trace species have been based on 1-D radiative-convective models. Typically the calculation involves doubling a reference concentration of the gas (for the chlorofluorocarbons, increases from 0 to 1 or 2 ppb are used) while other constituents are held constant. Table 4.1 gives some estimates of the change in surface temperature due to either a doubling of their concentration or an increase from 0 to 1 ppb for the halocarbons. The table was adapted from Table 2a in WMO (1983). There are other published values, but they generally do not disagree by more than about ±30% with the figures given here.

The models used to obtain these results generally gave a sensitivity to doubled CO_2 between 2 and 3°C. Thus, none of the changes of individual trace gases approaches CO_2 by itself, but it is clear that the summation of all of these potential changes could be of the same magnitude as CO_2. It is worth noting that because the concentration of each of these gases is small enough to be treated as optically thin, the temperature effect is linearly proportional to their concentration, whereas the CO_2 effect depends logarithmically on the concentration.

The spectroscopic parameters of several of these gases is not well known, and even the band strengths of some have not been measured. The spectral transmittance and total band absorptions also need to be better determined. These improvements will be needed to develop more accurate radiative transfer models. It will also help in answering questions about band overlap between constituents and with water vapor. Such information is needed for better parameterization in climate models.

TABLE 4.1 Some Estimates of Surface Temperature Change due to Changes in Atmospheric Constituents Other Than CO_2

Constituent	Mixing Ratio Change (ppb) From	To	Surface Temperature Change (°C)	Source[a]
Nitrous oxide (N_2O)	300	600	0.3-0.4	1,3
Methane (CH_4)	1500	3000	0.3	3,4
CFC-11 ($CFCl_3$)	0	1	0.15	1,5
CFC-12 (CF_2Cl_2)	0	1	0.13	1,5
CFC-22 (CF_2HCl)	0	1	0.04	7
Carbon tetrachloride (CCl_4)	0	1	0.14	1,5
Carbon tetrafluoride (CF_4)	0	1	0.07	2
Methyl chloride (CH_3Cl_3)	0	1	0.013	1,5
Methylene chloride (CH_2Cl_2)	0	1	0.05	1,5
Chloroform ($CHCl_3$)	0	1	0.1	1,5
Methyl chloroform (CH_3CCl_3)	0	1	0.02	7
Ethylene (C_2H_4)	0.2	0.4	0.01	1
Sulfur dioxide (SO_2)	2	4	0.02	1
Ammonia (NH_3)	6	12	0.09	1
Tropospheric ozone (O_3)	F(Lat,ht)	2 F(Lat,ht)	0.9	4,6
Stratospheric water vapor (H_2O)	3000	6000	0.6	1

[a]Sources: 1, Wang et al. (1976); 2, Wang et al. (1980); 3, Donner and Ramanathan (1980); 4, Hameed et al. (1980); 5, Ramanathan (1975); 6, Fishman et al. (1979); 7, Hummel and Reck (1981).

References

Angell, J. K. (1983). Global variation in total ozone and layer-mean ozone: an update through 1981. Manuscript, Air Resources Laboratory, Silver Spring, Md.

Craig, H., and C. C. Chou (1982). Methane: the record in polar ice cores. Geophys. Res. Lett. 9:1221-1224.

Donner, L., and V. Ramanathan (1980). Methane and nitrous oxide: their effect on the terrestrial climate. J. Atmos. Sci. 37:119-124.

Fishman, J., V. Ramanathan, P. Crutzen, and S. Liu (1979). Tropospheric ozone and climate. Nature 282:818-820.

Hameed, S., R. Cess, and J. Hogan (1980). Response of the global climate to changes in atmospheric composition due to fossil fuel burning. J. Geophys. Res. 85:7537-7545.

Hummel, J. R., and R. A. Reck (1981). The direct thermal effect of $CHClF_2$, CH_3CCl_3 and CH_2Cl_2 on atmospheric surface temperatures. Atmos. Environ. 15:379-382.

Liu, S. C., D. Kley, M. McFarland, J. Mahlman, and H. Levy, II (1980). On the original of tropospheric ozone. J. Geophys. Res. 85:7546-7552.

Ramanathan, V. (1975). Greenhouse effect due to chlorofluorocarbons: climatic implications. Science 190:50-52.

Rasmussen, R. A., and M. A. K. Khalil (1981). Atmospheric methane (CH_4): trends and seasonal cycles. J. Geophys. Res. 86:9826-9832.

U.S. Government (1982). Summary report for 1981. Geophysical Monitoring for Climatic Change No. 10.

Wang, W. C., Y. Yung, A. Lacis, T. Mo, and J. Hansen (1976). Greenhouse effects due to man-made perturbation of trace gases. Science 194:685-690.

Wang, W. C., J. P. Pinto, and Y. Yung (1980). Climatic effects due to halogenated compounds in the Earth's atmosphere. J. Atmos. Sci. 37:333-338.

Weiss, R. F. (1981). The temporal and spatial distribution of tropospheric nitrous oxide. J. Geophys. Res. 86:7185-7196.

WMO (1983). Report of the Meeting of Experts on Potential Climatic Effects of Ozone and Other Minor Trace Gases. Report No. 14. WMO Global Ozone Research and Monitoring Project, Geneva, 38 pp.

Zimmerman, P. R., J. P. Greenberg, S. O. Wandiga, and P. J. Crutzen (1982). Termites: a potentially large source of atmospheric methane, carbon dioxide and molecular hydrogen. Science 218:563-565.

5 Detection and Monitoring of CO_2-Induced Climate Changes

Gunter Weller, D. James Baker, Jr., W. Lawrence Gates, Michael C. MacCracken, Syukuro Manabe, and Thomas H. Vonder Haar

5.1 SUMMARY

Two questions are addressed in this chapter: (1) have we already detected a climatic change attributable to increasing CO_2, and (2) what observations and analyses can most effectively enable us to detect such changes and to monitor their progress?

The most clearly defined change expected from increasing atmospheric CO_2 is a large-scale warming of the Earth's surface and lower atmosphere. A number of investigators have examined trends in globally or hemispherically averaged surface temperature for evidence of CO_2-induced changes. Although differing in detail because of varying data sources and analysis methods, the records of large-scale average temperatures reconstructed by a number of investigators are in general agreement for the period of instrumental records, i.e., about the last 100 years. Northern hemisphere temperatures increased from the late nineteenth century to the 1940s, decreased until the mid-1970s, and have apparently increased again in recent years. The mean temperature of the 1970s was about 0.5°C warmer than that of the 1880s. To the extent that one can judge from scanty data, southern hemisphere temperatures have increased more steadily by about the same amount. In view of the relatively large and inadequately explained fluctuations over the last century, we do not believe that the overall pattern of variations in hemispheric or global mean temperature or associated changes in other climatic variables yet confirms the occurrence of temperature changes attributable to increasing atmospheric CO_2 concentration.

This chapter was commissioned by and reviewed by the Climate Research Committee of the Board on Atmospheric Sciences and Climate at the request of the Carbon Dioxide Assessment Committee. The contributions of Hugh W. Ellsaesser, Frederick M. Luther, Robert A. Schiffer, David E. Thompson, and Donald J. Wuebbles are also gratefully acknowledged. John S. Perry and Jesse H. Ausubel provided logistical, administrative, and editorial support.

Other factors than CO_2--such as atmospheric turbidity and solar radiation--also influence climate. Attempts have been made to account for these influences on the temperature record, thereby making the sought-for CO_2 signal stand out more clearly. Unfortunately, only indirect sources of historical data are available for the time preceding the short period of instrumental records. Moreover, stratospheric turbidity has been inferred primarily from volcanic activity, and solar radiance from such phenomena such as sunspots. The quantitative reliability of these inferences is unknown.

Despite these difficulties, a number of investigators, employing various combinations of data and methodology, have related the global or hemispheric mean temperature record with indices of turbidity and solar radiance and with estimates of the effect of increasing CO_2. Although good agreement between modeled and observed variations has been obtained in some of these studies, it is clear that enormous uncertainties exist. When attempts are made to account for climatic influences of such other factors as volcanic and solar variations, an apparent temperature trend consistent with the trend in CO_2 concentrations and simulations with climate models becomes more evident. However, uncertainties preclude acceptance of such analyses as more than suggestive. Nevertheless, the studies done to date have been most helpful in raising questions, suggesting relationships, and identifying gaps in data and observations.

In essence, the problem of detection is to determine the existence and magnitude of a hypothesized CO_2 effect against the background of climatic variability, which may be in part due to internal processes in the atmosphere and ocean and in part explainable in terms of fluctuations in external factors. A reasonable approach is to attempt to decompose the record of some climatic parameter, e.g., temperature, into a hypothesized "natural" value, a perturbation due to CO_2, and a random component. The "natural" value may be taken as a constant long-run preindustrial mean or perhaps that mean corrected for variable factors such as volcanic and solar activity. The random component can probably be treated as "noise-like" but will have to differ substantially from both "white noise" and first-order autoregressive noise. It is clear that the magnitude of the derived CO_2 signal will depend markedly on the hypothesis chosen for the underlying climatic trend and the change in CO_2 assumed between the imperfectly known preindustrial value and the accurately measured current concentrations. The success achieved by several workers in explaining the temperature record in diverse ways demonstrates the availability of a number of sets of hypotheses that can fit the poorly defined historical data and estimated preindustrial concentrations.

The available data on trends in globally or hemispherically averaged temperatures over the last century, together with estimates of CO_2 changes over the period, do not preclude the possibility that slow climatic changes due to increasing atmospheric CO_2 projections might already be under way. If the preindustrial CO_2 concentration was near 300 ppm, the sensitivity of climate to CO_2 (expressed as projected temperature increase for a doubling of CO_2 concentration) might be as large as suggested by the upper half of the range of the study of the CO_2/Climate Review Panel (1982), i.e., up to perhaps 4.5°C; if the preindustrial CO2 concentration was well below 300 ppm and if other forcing factors did not intervene, however, the sensitivity must be below about 3°C to avoid inconsistency with the available record.

If, as expected, the CO_2 signal gradually increases into the future, then the likelihood of perceiving it with an appropriate degree of statistical significance will increase. Given the inertia created by the ocean thermal capacity and the level of natural fluctuations, we expect that achieving statistical confirmation of the CO_2-induced contribution to global temperature changes so as to narrow substantially the range of acceptable model estimates may require an extended period. Improvements in climatic monitoring and modeling and in our historic data bases for changes in CO_2, solar radiance, atmospheric turbidity, and other factors may, however, make it possible to account for climatic effects with less uncertainty and thus to detect a CO_2 signal at an earlier time and with greater confidence. Improved monitoring of appropriate variables can be of great importance here by allowing improvement and more effective validation of models. A complicating factor of increasing importance will be the role of rising concentrations of greenhouse gases other than CO_2. While the role of these gases in altering climate may have been negligible up to the present, their significance is likely to grow, and their effects will be difficult to distinguish from those due to CO_2.

A monitoring strategy should focus on parameters expected to respond strongly to changes in CO_2 (and other greenhouse gases)* and on other factors that may influence climate. Candidate parameters may be identified, their variability estimated, and their time evolution predicted through climate model simulations. Through analysis of past data, continued monitoring, and a combination of careful statistical analysis and physical reasoning, the effects of CO_2 may eventually be discerned.

Monitoring parameters should include not only data on the CO_2 forcing and the expected climate system responses but also data on other external factors that may influence climate and obscure CO_2 influences. Climate modeling and monitoring studies already accomplished provide considerable background for the selection of these parameters. Since fairly distinct climate changes are only expected to become evident over one or more decades, monitoring for both early detection and more rapid model improvement should be carried out for an

*It will be difficult to distinguish between the climatic effects of CO_2 and those of other radiatively active trace gases. Their expected relative contributions to climatic change will have to be inferred from model calculations and precise monitoring of radiation fluxes.

extended period. Parameters may be selected for early emphasis on the basis of the following criteria:

1. Sensitivity. How does the effect exerted on climate by the variables or the changes experienced by the variable on decadal time scales compare with that associated with corresponding changes in CO_2?
2. Response characteristics. Are changes likely to be rapid enough to be detectable in a few decades?
3. Signal-to-noise ratio. Are the relevant changes sufficiently greater than the statistical variability to be measured accurately?
4. Past data base. Are data on the past behavior of the variable adequate for determining its natural variability?
5. Spatial coverage and resolution of required measurements.
6. Required frequency of measurements.
7. Feasibility of technical systems. Can we make the required measurements?

Initial application of these criteria leads to the list of recommended variables for monitoring given below:

Priority	Monitoring Causal Factors by Measuring Changes in	Monitoring Climatic Effects by Measuring Changes in
First	CO_2 concentrations Volcanic aerosols Solar radiance	Troposphere/surface temperatures (including sea temperatures) Stratospheric temperatures Radiation fluxes at the top of the atmosphere Precipitable water content (and clouds)
Second	"Greenhouse" gases other than CO_2 Stratospheric and tropospheric ozone	Snow and sea-ice covers Polar ice-sheet mass balance Sea level

In the above list, emphasis has been given to parameters that may contribute, either directly or through model improvement, to detection of CO_2 effects at the earliest possible time. Over the long run it is important to build up a relatively complete data base of the possible causes and effects of climate change and the characteristics of climate variability, not simply for detection but also to assist in research on and calibration of models of the climate system. Once we become convinced that climate changes are indeed under way, we will seek to predict their future evolution with increasing urgency and with increasing emphasis on parameters of societal importance (e.g., sea level, rainfall). We should thus anticipate that a detection program will gradually evolve into a more comprehensive geophysical monitoring and prediction program.

It should be emphasized that the strategy proposed here is a single tentative step in what must be an iterative process of measurement and study. In subsequent steps, we urge more thorough evaluation with greater attention, in particular, to the following:

(a) Time: When could we have a record long enough to make a meaningful contribution to policy formulation?
(b) Long-run values for model calibration and verification.
(c) Cost.
(d) Societal importance.

Collection of the desired observations will require a healthy global observing system, of which satellites will be a major component. Satellites can provide or contribute to long-term global measurements of radiative fluxes, planetary albedo, snow/ice extent, ocean and atmospheric temperatures, atmospheric water content, polar ice sheet volume, aerosols, ozone, and trace atmospheric components; a well-designed and stable program of space-based environmental observation is essential if we are to monitor the state of our climate. Table 5.1 summarizes requirements and technical systems for monitoring high-priority variables.

We will also have to continue to improve our climate models in order to reduce the uncertainties in predictions of climate effects and to validate the models against observations (although we believe that climate models are at present sufficiently sound and detailed to enable us to identify a set of variables that could form the basis for an initial monitoring strategy). Also, statistical techniques for assessing the significance of observed changes will have to be improved so as to deal with the characteristics of the monitored variables. In the end, however, confidence that we have detected the effect of CO_2 will have to rest on a combination of both statistical testing and physical reasoning.

Finally, we must recognize that despite our best efforts there will always remain room for differing interpretations of data. Within our own country, different investigators have reached quite different conclusions from the same evidence. The detection issue is inherently global, and interest is growing throughout the world. It is only to be expected that investigators in different countries may also reach different conclusions. This diversity of judgments may lead to unnecessary confusion and division among nations. There is thus a clear need for an international focal point or clearinghouse for data, analyses, studies, and periodic assessments relevant to the climatic effects of increasing CO_2.

5.2 HAVE CO_2-INDUCED SURFACE TEMPERATURE CHANGES ALREADY OCCURRED?*

5.2.1 Introduction

In 1938, G. S. Callendar suggested that mankind's fossil fuel emissions were causing an increase in atmospheric CO_2 concentration and that this, in turn, was leading to the climatic warming that had been detected early in this century. With new radiative calculations performed by Plass (1953, 1956) and an improvement in understanding of oceanic chemistry due to Revelle and Seuss (1957), the quantitative basis for Callendar's suggestion became more acceptable. In January 1961, in one of his last papers, Callendar compiled global climatic data on temperature trends in order to assess the possible role of increasing CO_2 concentrations (Figure 5.1). He found "...that the observed distribution of recent climatic trends over the earth is not incompatible with the CO_2 hypothesis, and that in certain cases the latter [i.e., the CO_2 hypothesis] can supply a reasonable explanation." The carefully measured tone of this endorsement of his own hypothesis resulted from its failure, in his mind, to explain three troubling features of the observed climatic trends. These features included the following:

1. "...the rising temperature trend is very small in the south temperate zone as compared to that in the north."
2. "The tendency for precipitation to decrease in many warm regions, and remain sub-normal during the first three or four decades of this century (Kraus, 1955)."
3. "...the big rise of temperature in most parts of the sub-arctic zone during the 1920s and 1930s...as compared to the changes in the lower latitudes."

Callendar felt that some aspects of these features could be reconciled with the CO_2 theory, by, for example, considering the large thermal inertia of the oceans, greatly delayed transport of fossil fuel derived CO_2 to the southern hemisphere, radiatively induced stabilization of clouds, lower-latitude induced contraction of the polar vortex, and the consideration of other factors to explain short-term fluctuations.†

---

*Some of the material in this section also appeared in MacCracken (1983).

†The thermal inertia of the oceans, and not delayed cross-equatorial transport of CO_2, is now believed to be important in delaying the temperature response of the southern hemisphere (Climate Research Board, 1979; CO_2/Climate Review Panel, 1982). Radiatively induced stabilization of clouds is not now recognized as the cause of geographical variations in CO_2-induced rainfall changes; rather, the shifting circulation pattern, including contraction of the polar vortex, leads to these variations. Ice-albedo feedback and the presence of a polar temperature inversion are now believed to lead to the amplification of temperature changes in polar regions.

TABLE 5.1 High-Priority Measuring Networks Required for CO_2--Climate Monitoring

Parameter	Requirements	Technical Systems
a. Primary parameters for monitoring the causes of climate change		
Carbon dioxide	Present global network adequate	
Solar radiance	Continued satellite measurements	
Stratospheric aerosols	Lidar systems (2 new ones in S.H.) to calibrate satellites	Lidar/satellite measurements and in situ sampling
Other "greenhouse" gases and ozone		
Nitrous oxide	Present network adequate for monitoring	
Methane	Present network adequate for monitoring	
Chlorocarbons	Present network adequate for monitoring	
Stratospheric ozone	Satellites	
Tropospheric ozone	Total ozone (Dobson) network must be enlarged (especially in the southern hemisphere)	Dobson spectrometers/ satellite soundings
b. Primary parameters for monitoring the effects of climate change		
Atmosphere		
Global temperature		
Surface	Present global network adequate	
Tropospheric distribution	Present radiosonde/satellite soundings adequate	
Stratospheric distribution	Some key rocketsonde stations should be retained to calibrate satellite soundings	Rocketsonde/satellite soundings
Radiation		
Upward terrestrial radiation at the top of the atmosphere Reflected solar radiation at the top of the atmosphere	Need to continue ERBE measurements or similar observations beyond 1986	Satellite sensors

Cloud and Water Vapor		
Equivalent emission temperature	Satellites	
Precipitable water content of the atmosphere	Development and routine use of satellite techniques	Satellite imagers Satellite soundings
Cryosphere		
Sea Ice Extent, Albedo, and Thickness	Radar (SAR) systems on polar-orbiting satellites	Satellite SAR
Snow Cover Extent and Albedo	Present satellite/ground coverage is probably adequate	
Mass Balance of Polar Ice Sheets	Radar and laser altimetry from satellites	Radar/laser altimetry
Oceans		
Global Mean Sea Level	Expansion of present tide gauge network (Atlantic, Indian, and Antarctic Ocean)	Tide gauges
Sea Temperature	Improved satellite sensing of surface temperatures	Satellite sensors

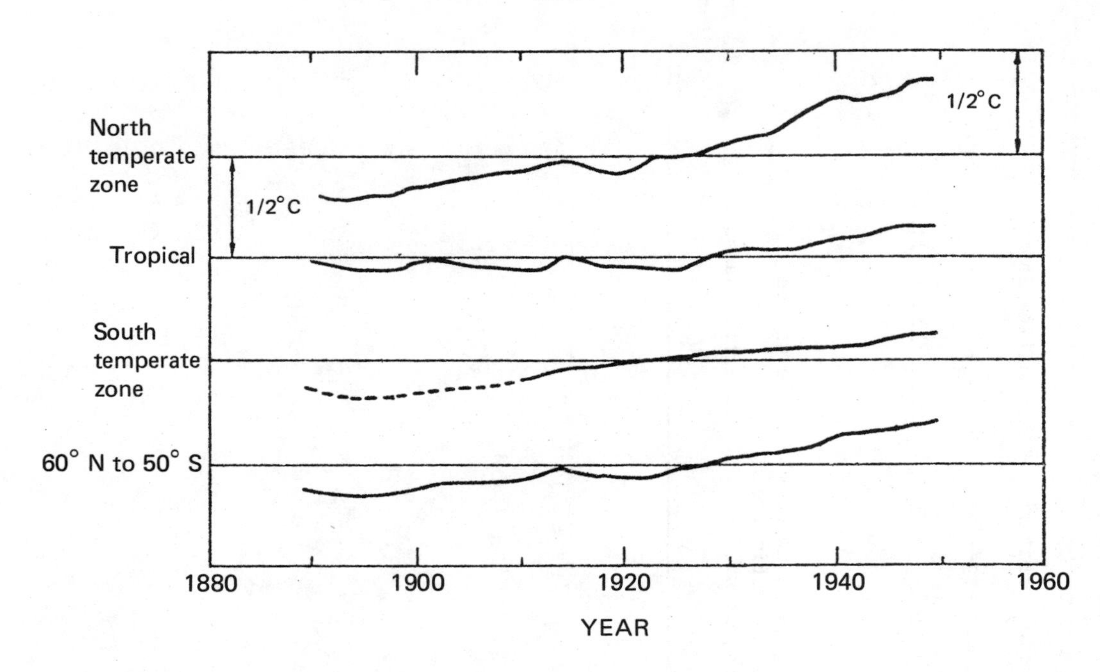

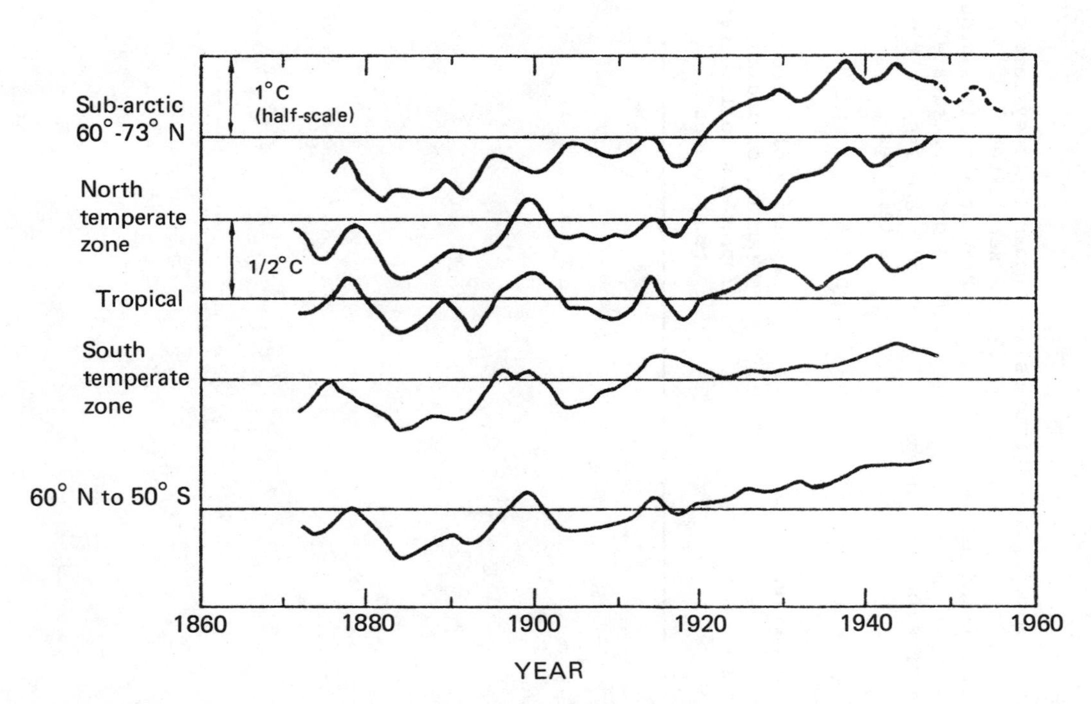

FIGURE 5.1 Reconstructions of changes in surface air temperature for different latitude regions by Callendar (1961) applying (top) a 20-year moving average and (bottom) a 5-year smoothing to the data.

He did not, however, believe that volcanic eruptions or solar variations could adequately explain the observed long-term climatic warming and the latitudinal pattern of the temperature change.

It is interesting to contemplate how much more forceful Callendar's endorsement of his CO_2 hypothesis would have been if he had had available the results of today's models. Although present climate model simulations indicate a sensitivity about half that found by Plass, they do project smaller temperature increases in the southern hemisphere (due presumably to ocean inertia and albedo effects), provide for regions of both increasing and decreasing precipitation, and indicate that there should indeed be a large polar amplification of the CO_2-induced global temperature increase. Predictions and observations would have seemed to be in agreement, and detection might have been claimed.

Perhaps quite fortunately, the absence of present-day model results and the lag in theoretical understanding prevented a consensus from developing, however, for later in 1961, J. Murray Mitchell, Jr., presented his now classic paper on global temperature trends. This paper reanalyzed the data on temperatures of the previous hundred years and found that the previously claimed warming, when properly weighted by area, was not really as large as had been thought by Callendar (1961) and Willett (1950). This finding actually would have improved Callendar's agreement with today's model results. More importantly, however, Mitchell found that the climate during the 1950s was, in fact, cooling (see Figure 5.2) at a time when the CO_2 trend should have exerted a warming influence.* The reversal in temperature trends found by Mitchell set the stage for intensive efforts during the last 20 years to untangle the web of factors that influence the climate.

The extensive monitoring, research, and analysis since then has taught us several lessons that should enable us to set a course for identifying the projected CO_2-induced climate response. This chapter will review some of the diagnostic studies that have been performed and the problems that have arisen in the attempt to detect CO_2-induced surface-air-temperature change. The chapter is not intended to serve as a comprehensive review of all studies analyzing the climate of the last 100 years but rather attempts to point out general problems and to use selected studies to illustrate how difficult these problems are to resolve and where we stand now with regard to identifying the climatic effects of increasing CO_2 concentrations.

5.2.2 Requirements for Identifying CO_2-Induced Climate Change

To achieve widespread confidence in the projected climatic effects of increasing CO_2 concentrations, both a consensus of climate model

---

*Concentrations were increasing relatively slowly, so very little warming should actually have been expected during this period. The cooling, however, is deserving of explanation.

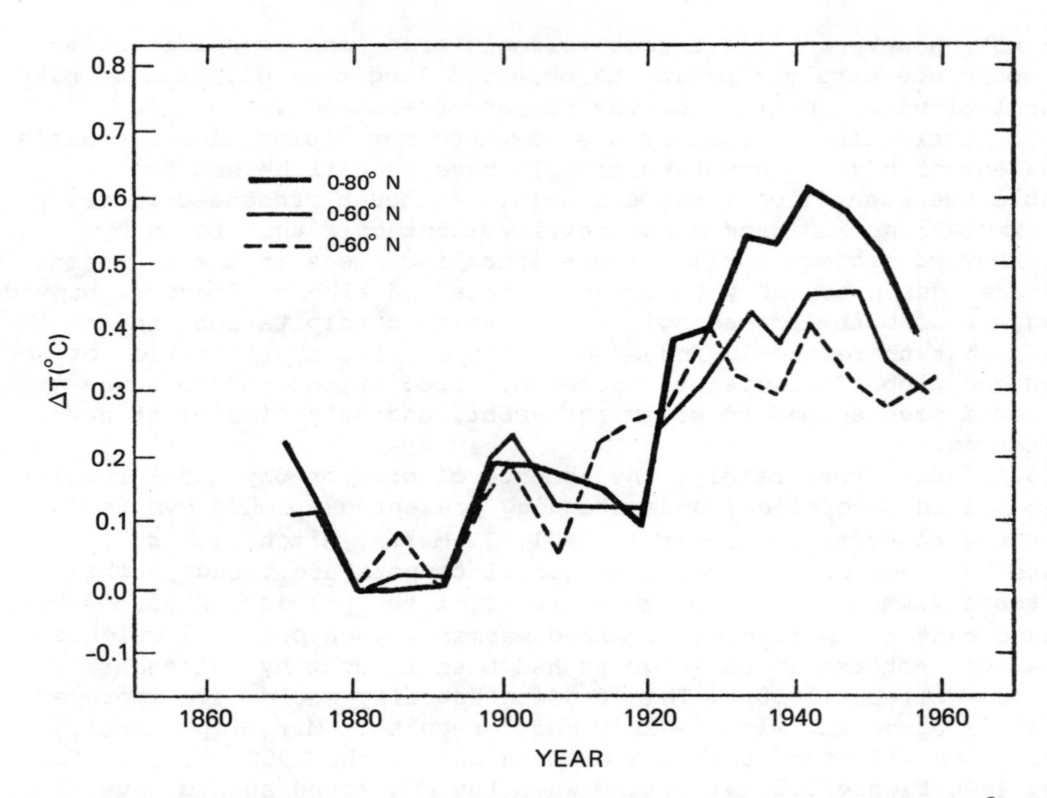

FIGURE 5.2 Construction of changes in surface air temperature for different latitude bands by Mitchell (1961). Average data are compiled for successive pentads starting with 1870-1874.

results and unambiguous identification in recent climatic records of the initial stages of the projected CO_2-induced warming are necessary. As expressed in recent reports of the National Research Council (Climate Research Board, 1979; CO_2/Climate Review Panel, 1982), the change in global average surface temperature (the most closely scrutinized model result) due to doubled CO_2 concentration is likely to be between 1.5 and 4.5°C, with more physically comprehensive models giving results in the range of 2 to 3.5°C. While increases in the global average temperature of even 1.5°C would lead to a historically unprecedented climatic situation, the range of estimated temperature (and other climate) changes remains wide and introduces considerable uncertainty into the detection debate. As the reasons for the spread in the estimates are becoming clear, we may expect that with further research the range will narrow.

If we can achieve a consensus of model results, convincing identification of the projected CO_2 warming will increasingly depend on observational data, inspiring confidence and certainty sufficient to allow comparison with model results. This will require (1) that the necessary climatic data bases be accurate, comprehensive, and of sufficient length to allow application of appropriate statistical analyses and, in particular, identification of whether a change is or is not occurring; (2) that data bases exist to evaluate the role of the factors that may have caused climate changes of comparable magnitude; (3) that the climatic effects of CO_2 and other factors be estimated with sufficient detail and confidence that the projected induced changes can be identified amid the fluctuations, perhaps similarly patterned, caused by other known causal factors, unrecognized causes, and natural fluctuations (see MacCracken and Moses, 1982). In the following subsections each of these points will be considered by example.

5.2.2.1 Climatic Data Bases

In the analysis of climatic change, and in particular in the search for evidence of CO_2-induced effects, one is faced with the limitations in the availability and accuracy of climatic data bases (see Table 5.2, for example).

As a result of these limitations, the primary indicator used in diagnostic studies of climate has been the change in surface air temperature during the last hundred years. (The problems illustrated and conclusions drawn in this section apply equally to analyses of variables other than temperature, such as sea ice extent and sea level rise.) This variable has been chosen because it is the only one for

TABLE 5.2 Causes of Differences in Temperature Anomaly Data Sets

(a) Differences Arising in Data Selection and Compilation
- Number of stations used in compiling average
- Methods for eliminating effects of unrepresentative stations (e.g., urban heat-island, station location changes)
- Relative distribution of stations over land and ocean
- Sources of data (ships, land sites, islands, for example)
- Treatment of stations with records starting in different years
- Differences or changes in observation times
- Unrepresentativeness of the sampling network
- Absolute temperature versus value of anomaly
- Methods of accounting for missing data
- Use of annual average versus monthly data
- Method of constructing daily average temperature.

(b) Differences in Averaging Methods
- Time periods (range from annual to multidecadal)
- Spatial domains
- Climatic baselines and periods of record
- Time-averaging methods (running averages versus finite period averages)
- Spatial interpolation techniques (sum of representative sites, interpolation onto grid, weighting of stations, for example)
- Means of selecting normals and identifying trends

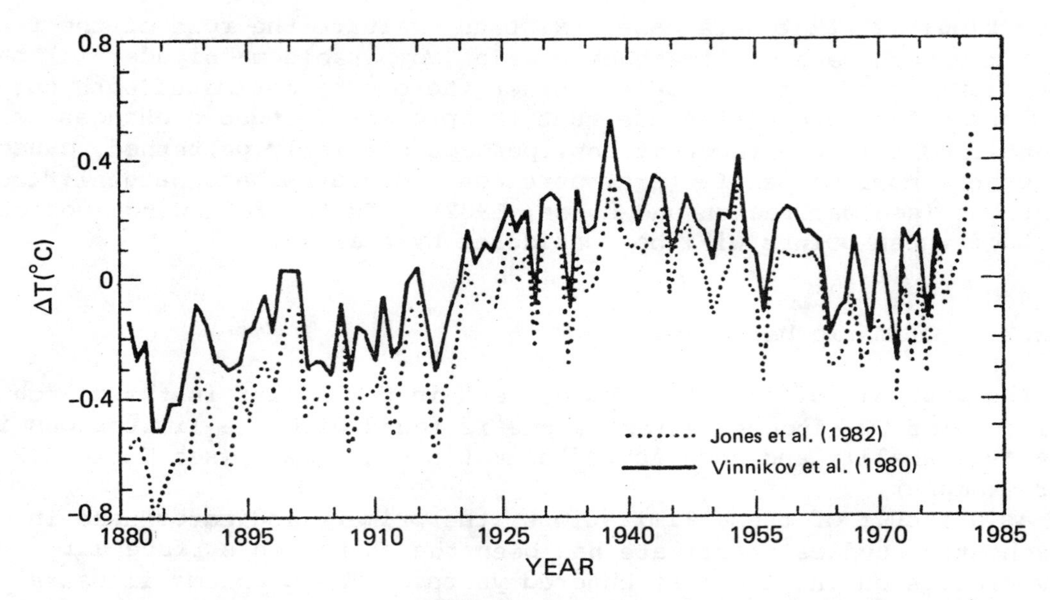

FIGURE 5.3 Comparison of the reconstructions of annual surface-air-temperature anomalies for the northern hemisphere from Jones et al. (1982) and Vinnikov et al. (1980). Figure from Clark (1982), but data for 1981 added to Jones et al. (P. D. Jones, Climatic Research Unit, University of East Anglia, Norwich NR4 7TJ, England, personal communication).

which a long record of measurements exists at many stations and because it is a convenient and straightforward, although not complete, measure of the climatic state. While some investigators have attempted to develop data bases of global and polar temperature change, the most used indicator has been the change in northern hemisphere average surface air temperature.

Comparison of the temperature records compiled by several investigators shows that not only has temperature fluctuated but also that the temporal pattern of the temperature anomalies is not uniquely established (Figure 5.3).

Estimates of surface-air-temperature anomalies using similar data sets and techniques agree well [e.g., Borzenkova et al. (1976), Vinnikov et al. (1980) and Jones et al. (1982)], although there remain differences in the details of the compilations that remain to be resolved (World Meteorological Organization, 1982a). Hansen et al. (1981) compiled data on the global temperature pattern and found a very different pattern of temperature change in the southern and northern hemispheres (Figure 5.4). Jones et al. (1982) show that the changes have different patterns by season.

The differences in estimates of actual climatic change arise primarily for two reasons: the choice of data selection/compilation methods and the choice of averaging methods. Table 5.2(a) lists some of the problems that arise in data selection and compilation; Table 5.2(b) lists problems related to averaging techniques. Consider just a

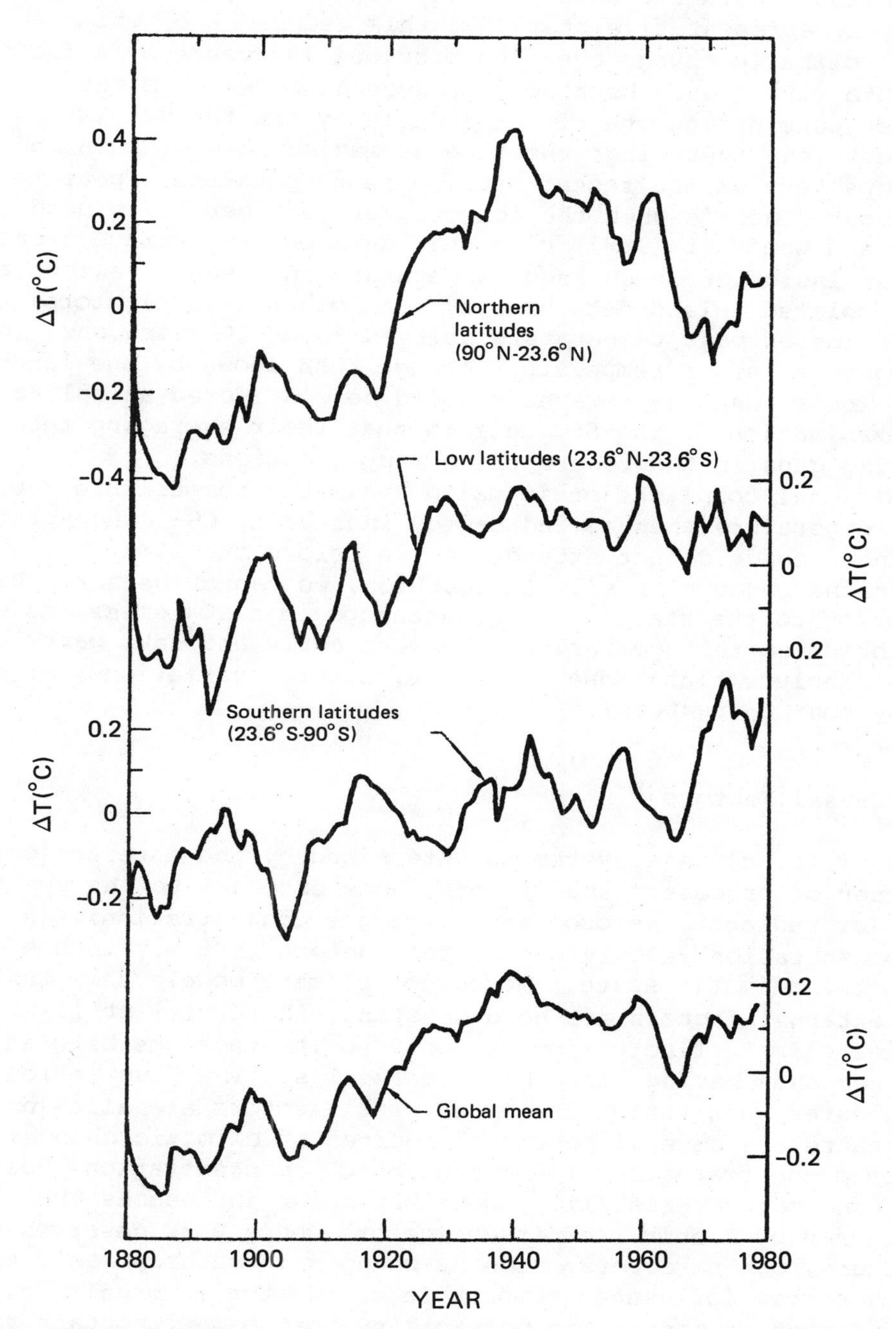

FIGURE 5.4 Reconstruction of surface-air-temperature anomalies for various latitude bands by Hansen et al. (1981).

few examples as illustrations of the problems. Mitchell (1961) used Willett's (1950) data and used an area weighting method rather than averaging of representative stations; this reduced the estimated wintertime climatic change over the previous 100 years by a factor of 2. Yamamoto (1980) used both 30-year averaging (which might be appropriate when considering the effect induced by the thermal lag of the ocean) and 5-year averaging; the time of maximum temperature and the start and pattern of the recent decline in temperature appear to be delayed about a decade when the longer averaging period is used. Paltridge and Woodruff (1981) used ship data on sea surface temperature (SST) as an indicator of surface temperature in oceanic regions rather than the isolated island data used by most other investigators; they found the time of peak temperature delayed about 20 years and, quite surprisingly, a larger temperature change than shown by the land-based records. Their results, however, should be considered as only an initial examination of the SST data in that their averaging techniques for handling gaps in the record raise many questions.

An additional complication in using available temperature records to estimate temperature changes induced by increasing CO_2 concentrations is that the records do not extend back to before the time that CO_2 concentrations began to rise. In addition, we cannot be sure that the climate prior to the start of large anthropogenic CO_2 emissions was in equilibrium so that comparisons between early and late parts of the record may include biases due to trends, natural variations, or changes induced by non-CO_2 factors.

5.2.2.2 Causal Factors

The state of the climate system is determined by the interactions of a large number of processes and factors, some external to the system (e.g., solar radiance, aerosol and trace-gas concentration). A change in CO_2 concentration is only one of the factors that may induce a change in the climatic state. Moreover, climate models show that, even when all external factors are held constant, there will still be substantial climate fluctuation due only to interactions between various processes having different time scales. The fluctuations will be even greater when variations in external factors are allowed.

While there are several potential sources of climatic change, our discussion concentrates for the moment on CO_2 concentration, volcanic aerosols, and solar variability, possible major influences that have been addressed by a number of investigators whose work deserves comment. In the future, other factors, such as changes in anthropogenic trace gases, may become influences comparable or greater in magnitude. Finally, we must recognize the possibility that some important factor may have been underestimated or remains unrecognized.

To identify the climatic signal caused by increasing CO_2 concentrations among the background of fluctuations requires either (a) that we treat the changes of all other factors as noise and wait until the CO_2-induced climatic change is large enough to cause a statistically significant change (assuming implicitly that no additional competing influences have come into play) or (b) that we account for the relative roles of the possible factors that may induce changes comparable in magnitude to increasing CO_2 concentration during the period of interest, thereby reducing the amount of unexplained variation and allowing easier identification of the CO_2 effect. In the record of the last 100 years, the standard deviation of monthly global mean temperature ranges from 0.23 to 0.65°C depending on month (Jones et al., 1982), while the change in annual global mean temperature is on the order of a half degree. Presumably the year-to-year variations were not due primarily to changes in CO_2 concentration. Since both observed variations and predicted changes are of similar magnitude, the latter approach of accounting for the role of as many causal factors as possible must be pursued if we are to have an early likelihood of identifying the projected CO_2 warming.

To attribute climatic changes to causal factors requires adequate data bases for the changes in each causal factor, as well as for the climate state itself, over a period sufficiently long that changes attributable to the causal factors are comparable with the level of natural variability. Moreover, for carbon dioxide the record is not yet adequate. The mid-nineteenth-century baseline concentration is thought to have been between 250 and 290 ppm, and the time history of CO_2 concentration between 1850 and 1950, when changes in the biosphere may have played a more important role than fossil fuel emissions, is not yet well defined (cf. Chapter 2). These uncertainties allow estimates of the change in CO_2 concentration from 1850 to 1980 to range from as little as 50 ppm to as much as 90 ppm, which in turn converts to a factor of about 2 in the estimated mean value of the warming attributable to changes in CO_2 concentration over this period if other factors remained constant. Combining the uncertainties caused by the range of possible change in CO_2 concentration, the range in model estimates of the temperature change for a CO_2 doubling, and the time constant of a thermal lag in the ocean, and using a logarithmic approximation to relate the CO_2 radiative effect to the temperature response, one finds that, assuming no net influence of other factors, <u>the expected CO_2-induced temperature change since 1850 may range from a few tenths of a degree to more than one-and-a-half degrees; estimates in the lower part of this range appear more consistent with the climatic record</u> (see Figure 5.5).

The data bases used to estimate the role of volcanoes--presumed to be an important factor in creating fluctuations over at least the last 100 years--are quite uncertain. Table 5.3 lists some of the factors contributing to the disagreements between the various data sets that have been used to account for changes in volcanically injected stratospheric aerosols. The different records show quite different relative magnitudes and temporal patterns of volcanic influences. Lamb's (1970) dust veil index, a measure of atmospheric aerosol content, is perhaps the most frequently used index, but various investigators modify it, often in ad hoc ways, thereby perhaps affecting later calculations (see Figure 5.6). Gilliland (1982) uses acidity as measured in a Greenland ice core as the basis for variations in stratospheric aerosol, suggesting that the volcanic chronology from this record should record the

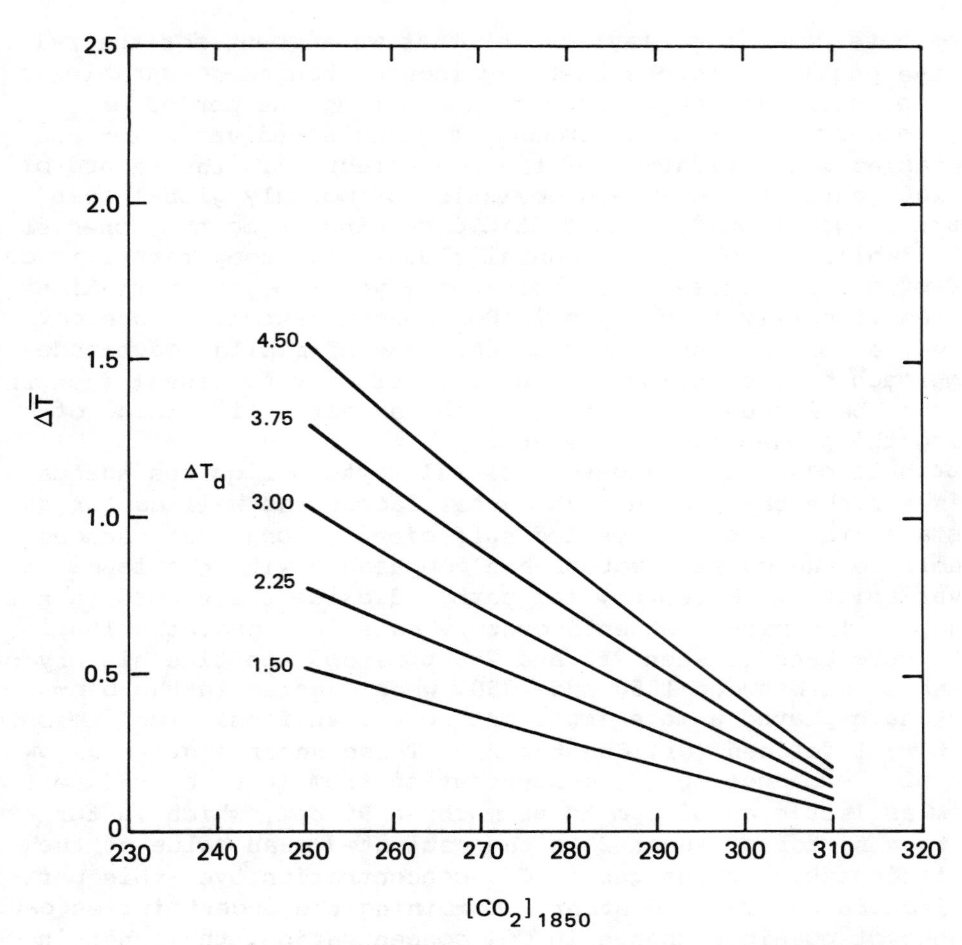

FIGURE 5.5 Relationship between CO_2 change, temperature change, and climate sensitivity assuming no other forcings. The abscissa represents a range of values for mid-nineteenth-century CO_2 concentration. The ordinate represents the increase ($\Delta\overline{T}$) in global mean equilibrium surface temperature between 1850 and the period 1961-1980. The response is calculated for a range of values of ΔT_d, the change of global mean equilibrium temperature for a doubling of CO_2 concentration (assumed independent of initial CO_2 concentration), and assumes that the temperature range is logarithmically related to the change in CO_2 concentration (Augustsson and Ramanathan, 1977). An ocean response time (mean thermal lag) of 15 years is used. The concentration of CO_2 was assumed in each case to increase linearly from the indicated value in 1850 to 310 ppm in 1950, and then linearly from 1950 to 340 ppm in 1980. Note that if the temperature increase from 1850 to the interval 1961-1980 is taken to be 0.5°C, then for consistency, ΔT_d may be as large as 4.5°C only if mid-nineteenth-century CO_2 concentrations were about 300 ppm, whereas ΔT_d may be as small as about 1.5°C if mid-nineteenth-century CO_2 concentrations were as low as 250 ppm. For ocean response times shorter than 15 years, the isolines slide upward. Varying the time of the start of the increase in CO_2 concentrations from 1850 to 1920 has little effect.

TABLE 5.3 Factors Contributing to Uncertainties in Creating a Volcanic Aerosol Data Base for Assessing the Climatic Effect of the Aerosol

Factors
Reliance on variables not directly related to the radiative effect of volcanic aerosol (e.g., ejecta volume, volcanic explosivity, ice-core-derived precipitation acidity)
Single or limited measurements extrapolated to global domain
Lack of latitudinal resolution or pattern of aerosol distribution
Assumptions about stratospheric lifetime of aerosols
Lack of seasonal resolution of aerosol distribution
Use of surface measurements to estimate changes in stratospheric aerosol (thereby introducing potential problems if trends exist in tropospheric aerosols)
Averaging period of volcanic aerosol loading
Definitions of major volcanoes and of the number of volcanoes considered
Assumptions about size distribution of volcanic particles and of chemical composition of gaseous emissions, leading to ad hoc adjustments
Assumptions regarding lifetime and distribution of aerosol due to season, latitude, and height of injection of volcanic dust and gases
Circularity caused by estimation of dust loading from observations of subsequent temperature change

high northern latitude volcanoes often cited as missing by other investigators--but strangely not evident in the core record. The recent Smithsonian compilation of volcanoes has also greatly expanded the number of volcanoes considered (Simkin et al., 1981). New listings, however, do not always reinforce confidence. For example, the recent compilation of an explosivity index by Newhall and Self (1982) attributed greater importance to the eruption of Mt. St. Helens than to Agung, whereas a sulfur-based stratospheric aerosol index would place Agung as being of much greater importance than Mt. St. Helens.

In addition to estimates of dust amount, a few indices also rely on actual radiative measurements (Figure 5.7). The general pattern of the two types of data show broad qualitative similarity, but details of timing and magnitude are quite different.

Even worse complications exist in generating data bases to evaluate the effect of changes in solar radiation, the third major factor often considered in these analyses. For this factor, the main problem is that there are several indicators that have been used as surrogates of the solar radiative flux reaching the Earth, but there is virtually no physical basis for deciding between these indicators, or even if any is relevant (i.e., is the solar constant constant?).

Quite clearly, without adequate historical data bases on the causal factors, identification of the CO_2 part of the climatic signal may not be possible. Therefore, examination and improvement of the data bases on potential causal factors must be an essential part of the early detection effort.

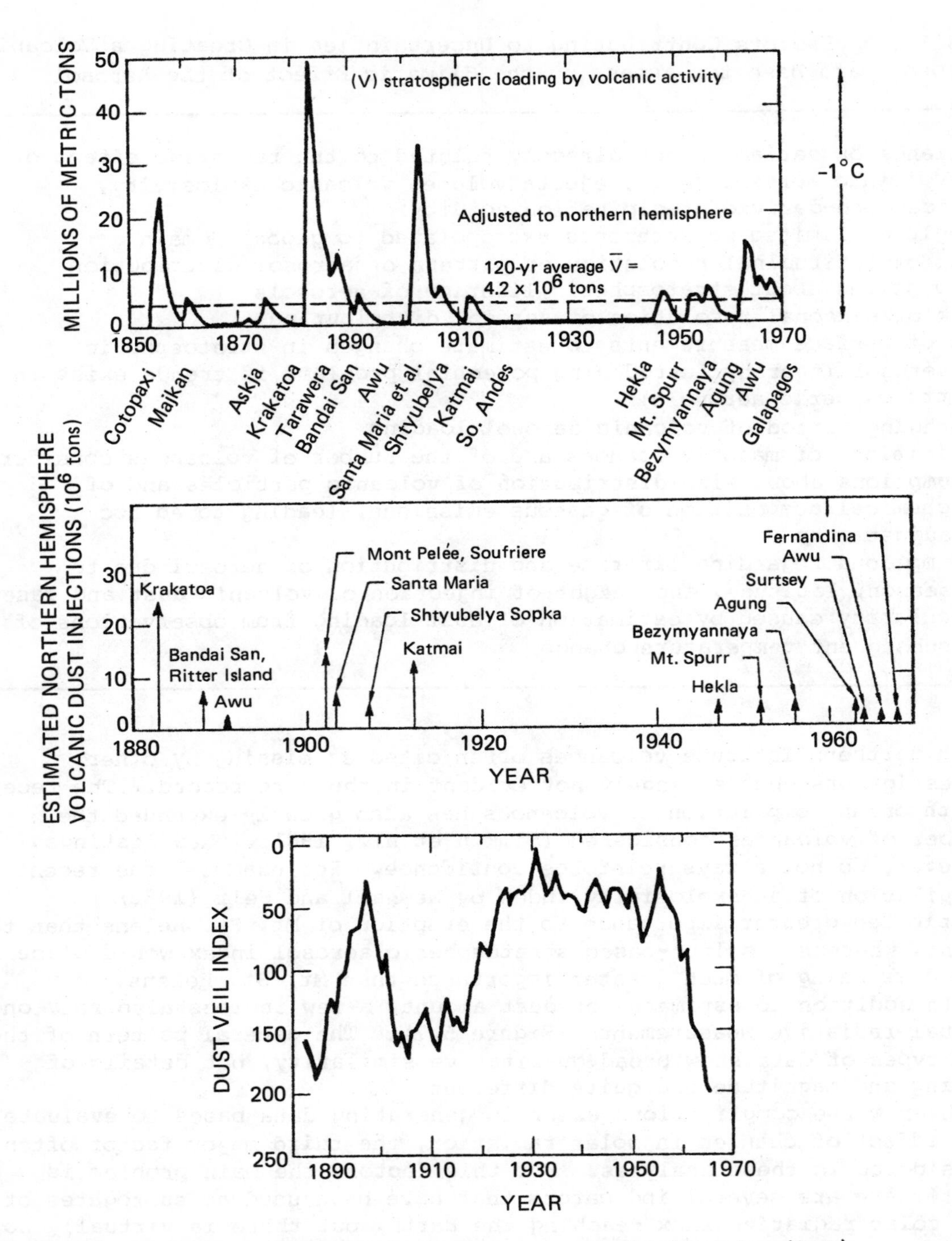

FIGURE 5.6 Estimates of stratospheric aerosol loading by (top) Mitchell (1970) and (middle) Oliver (1976) based on use of volcanic index of Lamb (1970) and by (bottom) Bryson and Dittberner (1976) using volcano index of Hirschboeck (1980). The model of Oliver (1976) uses a residence time approach to calculate aerosol loading after injection. The curve of Bryson and Dittberner (1976) represents a 10-year running mean in arbitrary units.

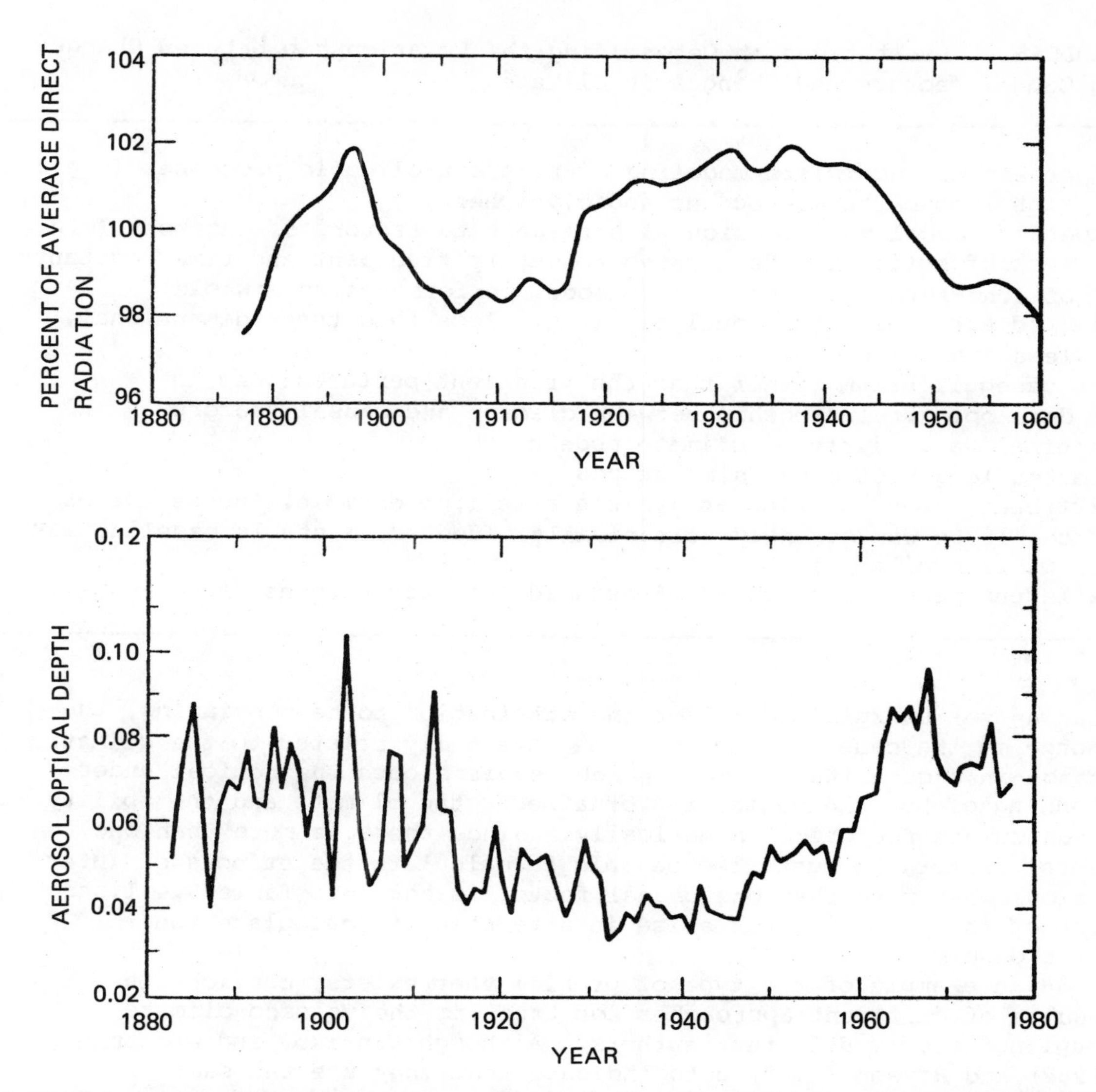

FIGURE 5.7 Estimates of stratospheric aerosol loading based on surface measurements of downward direct solar radiation (top) from Budyko (1969) using data from cloudless days of each month with 10-year smoothing, and (bottom) from Bryson and Goodman (1980) using data from 42 stations between 20° and 65° N. The apparent inverse relationship occurs because direct radiation measures the clarity of the stratosphere whereas aerosol optical depth measures the lack of clarity.

5.2.2.3 Relating Causal Factors and Climatic Effects

Even given precise records of how the climate has varied and the history of the important causal factors, a number of problems arise in attributing appropriate components of the climatic fluctuations of the last 100 years to the various causal factors, the remainder of the fluctuations being assumed to be natural (or, more properly, natural

TABLE 5.4 Limitations in Determining the Relationship between Changes in Causal Factors and Changes in Climate

Imperfect or incomplete modeling of relevant climatic processes (e.g., with respect to the oceans and cryosphere)
Imperfect model verification with respect to factors of interest (e.g., lack of sufficient test cases to verify treatment and time constants of long-term processes and of model performance as a whole)
Limited areal extent of analysis (e.g., less than three-dimensional, less than global)
Use of equilibrium rather than the transient perturbations in developing relationship between climate and causal factors
Internal variability of climate models
Limited length of model simulations
Different times at which analysis starts (for example, the decade of the 1880s was strongly volcanically affected; a stable baseline may not be available)
Different temporal patterns of assumed climatic response functions

plus as yet unexplained). For the attribution to be convincing, the change in the causal factor must be physically related to the climatic change in a quantitative way, which requires both theoretical understanding of how the causal factor affects the climate and the ability to calculate the effect numerically. Since there is more than one causal factor, we must also be able to calculate the effects of interactions when more than one causal factor is acting. Table 5.4 lists some of the problems that arise in attempting to calculate the relationship.

As an example of the type of problem that exists, consider the results of different approaches for treating the volcano-climate coupling used by different authors. Although Vinnikov and Groisman (1982) and Bryson (1980) both indicate that they use the same surface-air-temperature record and similar actinometric measures of stratospheric transmissivity, their different approaches to relating the causal factor to the induced climatic change lead them to very different conclusions about whether there has or has not been a temperature increase attributable to increasing CO_2 concentrations. Rather than specifying a response function, Oliver (1976) and Mitchell (1983) chose generalized volcanic response functions that were then optimized in order to fit the temperature data. They also come to differing conclusions about the role of CO_2 in the temperature record.

Rather than employing such empirical and statistical approaches, Hansen et al. (1981) used a one-dimensional radiative-convective model with a thermodynamically interactive ocean to simulate the climatic response to changes in the volcanic aerosol loading (and other factors). They then derived a quantitative relationship between CO_2 concentration and temperature to use in their analyses of the causes of climatic fluctuations. Such use of physically based relationships is an important requirement for studies of this type. Comparison of their one-dimensional calculations with the different time histories of the northern and southern hemisphere temperatures, however, makes clear that there is a need to analyze the observed fluctuations with models having, at least, greater spatial detail.

Such complications as arise in relating volcanic eruptions to the temperature response also arise in attempting to relate other causal factors to temperature change. In some cases, these complications can be resolved by just expanding the models to include omitted, but important, processes and domains or to treat the transient as opposed to the equilibrium response; in other cases, the complications can only be resolved by improving existing data bases (e.g., extending the temperature record back in time) or gathering new data (e.g., changes in Antarctic ice volume, if our climatic variable is sea level).

As illustrations of the difficulty in untangling the recent climatic record, the next section considers several recent attempts to acquire the necessary climatic and causal factor data bases and then to relate these data bases to climatic changes.

5.2.3 Attempts to Identify CO_2-Induced Climate Change

Mankind has always sought to relate climate fluctuations to causal factors. Certainly, the seasonal variations of solar position were widely recognized by early man as being related to the seasonal variations in temperature, even if the reasons for the change in solar position were not understood. Geologists once feared that the Little Ice Age was caused by the cooling of the sun as its fuel sources ran down, until they found evidence for very cold periods thousands and then millions of years earlier. Since discovery of these ice ages, there have been many suggestions about the causes of climate change. Current views of many scientists, for example, are that the sun's energy output has actually been increasing over geological time, that variations in the Earth's orbital parameters were major factors in the Pleistocene glacial cycles, and that the cold decades that characterized the Little Ice Age resulted from small fluctuations in solar output.

Although changes in CO_2 concentration were recognized as a possible factor in long-term climate change during the last century, the work of Callendar described earlier has served as the basis for numerous studies on that subject during the last 20 years (see Table 5.5). These studies have used different data sets, different analysis techniques, considered different causal factors and, perhaps not surprisingly, reached different conclusions. Because of these many differences, comparing the results is not straightforward.

As one measure of the differences in findings, Table 5.6 presents the results of a representative set of investigators in terms of a ratio measuring the relative importance of each causal factor in causing a climatic change to the maximum variation of the climatic variable in the record that was used. In making these comparisons, it should be noted that in many of the studies there has been some effort to achieve an optimum fit to observations by choice of data base or by adjusting

TABLE 5.5 Contribution of Change in CO_2 Concentration to Global Increase of Surface Air Temperature Based on Studies of Observed Climatic Change

Author	Estimate of Present Warming (K) over Period Indicated Attributable to and/or Consistent with Change in CO_2 Only[a]	Estimate of Equilibrium Warming (K) due to Doubling of CO_2 to 600 ppm (Excluding Other Effects)[a]	Comment
Callendar (1961)	0.4 (1891-1950)	3.6	Warming up to 1960; attribution to CO_2 based on consistency with modeling of Plass (1953)
Mitchell (1961)	0.6 (1840-1960)	3.0	Compared model estimate for CO_2 doubling interpolated to 1960 with observations, and found agreement only prior to 1950; rest of change attributed to sunspots
Bryson and Wendland (1970)	0.42 (1882-1960)	3.65	Values calculated from linear regression relationship given in paper
Budyko and Vinnikov (1973)	0.2 (1940-1970)	2.2	Opposed by a 0.5 K cooling due to man-made aerosol
Budyko (1974)	0.3-0.4 (1870-1970)	1.8-2.4	Opposed by a 0.5° cooling due to man-made aerosol
Willett (1974a,b)	None (1870-1970)	Little if any	Based on detailed study of temperature record and Humphrey's (1940) analysis of CO_2 absorption
Broecker (1975)	0.35 (1900-1975)	3.0	Did not consider volcanoes
Baldwin et al. (1976)	0.28 (1884-1975)	3.0	CO_2 effect based on Manabe and Wetherald (1975), not to optimize the fit to observations
Bryson and Dittberner (1976)	0.5 (1890-1960)	5.57[c]	CO_2 increase alone
	-0.44 (1890-1960)	-4.93	CO_2 increase plus associated dust release

Oliver (1976)	0. (1883-1968)	N/A	Volcanic effects studied in presence of an observed warming of unspecified cause
Miles and Gildersleeves (1978)	0.16-0.26 (1872-1972) 0.04-0.06 (1887-1922) 0.06-0.10 (1887-1937)	1.4-2.2	Including Lamb's (1977) DVI and Arctic ice feedback
Miles and Gildersleeves (1978)	0.10-0.16 (1872-1972) 0.03-0.04 (1887-1922 0.06-0.10 (1887-1937)	0.9-1.9	With Lamb's (1977) DVI only
Miles and Gildersleeves (1978)	0.30 (1665-1969) 0.28 (1715-1969)	1.6-2.6 or	Depending on linear or square-root dependence, including Lamb's (1977) DVI and Wolf sunspot No.
	0.18 (1870-1969) 0.06-0.08 (1750-1919)	1.3-2.1	CO_2 gives most consistent relationship with post-1870 warming
Robock (1978, 1979a)	0.22 (1880-1980)	1.68	Volcanic dust and CO_2 only external forces on climate identified as significant based on results of energy balance model
Robock (1979b)	0.1 (1620-1980)	0.53 (0.93 at equilibrium)	Improved snow and ice albedo feedback calculation
Hoyt (1979a)	0.3-0.4 (1880-1970)	2.0-3.1	Rest attributed to volcanoes and umbra/penumbra; statistical significance of CO_2 result is only 16%
Bryson (1980)	0 (1920-1973) ~0.1 (1887-1972)	~0 ~0.5	Observational analysis Model analysis

TABLE 5.5 Continued

Author	Estimate of Present Warming (K) over Period Indicated Attributable to and/or Consistent with Change in CO2 Only[a]	Estimate of Equilibrium Warming (K) due to Doubling of CO_2 to 600 ppm (Excluding Other Effects)[a]	Comment
Madden and Ramanathan (1980)	Below noise level	N/A	High-latitude analyses based on Manabe and Wetherald (1975) and Ramanathan et al. (1979) estimate of signal
Hansen et al. (1981)	0.25, 0.4 (1880-1980) 0.15-0.6[b] (1880-1980)	2.8 1.4-5.6	Based on one-dimensional model with and without thermocline below oceanic mixed layer
Vinnikov and Groisman (1981)	0.4-0.6 (1883-1977)	2.1-3.1	Empirical statistical model
Wigley and Jones (1981)	No statistically significant effect yet identified	Cannot yet be identified from observations	Based on Manabe and Stouffer (1980) estimate of signal; expect statistically significant detection around turn of century
Schonwiese (1981)	0.2 (1881-1979)	1.0	Based on 30-year filtered data
Gilliland (1982)	0.226 (1881-1975)	1.85	Rest attributed to volcanoes and 76-, 22-, and 12.4-yr solar cycles
Mitchell (1983)	0.11-0.38 (1883-1978)	0.8-2.8	Range for 12 cases that consider different lengths of record, CO_2 fossil fuel and biogenic emission scenarios, and temperature series; effect of oceanic lag is included in estimate of present warming

[a] Numbers that are underlined are derived via logarithmic interpolation/extrapolation in CO_2 concentration based on the predicted logarithmic variation in radiative forcing as a function of CO_2 concentration (Augustsson and Ramanathan, 1977). Extrapolations made for this table are dependent on past CO_2 concentratons that are not well known and assume, almost certainly incorrectly, that climatic equilibrium has been achieved, except as noted.

[b] Based on linear scaling from results for $T_{present}$ = 0.3 K and model with mixed layer and thermocline, assuming stated range of results for CO_2 doubling.

[c] Based on logarithmic extrapolation from 1% CO_2 change.

TABLE 5.6 Attribution of the Cause of Recent Changes in Surface Air Temperature (Values Given are a Measure of the Relative Significance[a] of Each Causal Factor Considered and Should be Viewed as Approximate)

	Causal Factor[b]				Solar Variations					
					Sunspots					
Author	CO_2 Increase	Volcanic Aerosol	Tropospheric Aerosol (dust)	Umbral/ Penumbral	Observed Wolf No.	12.4 yr	Magnetic Reversal, 22 yr	Solar Radius, 76 yr	Camp Century 80- and 180-yr Cycle ($\sigma^{18}O$)	Stochastic Forcing
Callendar (1961)	0.9	-	-	-	-	-	-	-	-	-
Bryson and Wendland (1970)	1.0	-	-1.1	-	0.8	-	-	-	-	-
Mitchell (1970)	0.3	-1.0	-0.1	-	-	-	-	-	-	-
Broecker (1975)	0.7	-	-	-	-	-	-	-	-0.8	-
Bryson and Dittberner (1976)	0.8	-0.9	-0.8	-	-	-	-	-	-	-
Oliver (1976)	-	-0.9	-	-	-	-	-	-	-	-
Robock (1978, 1979a)	6.4	-0.9	0.4	-	0.0	-	-	-	-	+1.0
Robock (1979b)	+0.1	-1.0	-	-	0.0	-	-	-	-	+0.5
Hoyt (1979a,b)	0.6	-	-	0.9	-	-	-	-	-	-
Hansen et al. (1981)	0.5	-0.5	-	0.3	-	-	-	-	-	-
Gilliland (1982)	0.26	-0.47	-	-	-	0.11	0.13	-0.32	-	-
Mitchell (1983)	0.3	-0.8(est.)	-	-	-	-	-	-	-	
Vinnikov and Groisman (1982)	0.7	-0.8	-	-	-	-	-	-	-	-

[a]Relative significance is defined as the ratio of the maximum temperature change caused by the particular causal factor alone divided by the maximum range of the temperature change in the record used, with both measures using common averaging times. A negative sign indicates that an increase in this causal factor causes a cooling.
[b]Additional causal factors, including non-CO_2 trace gases and surface albedo, are not included only because the studies listed do not consider them.

the strength of the relationship between variables, i.e., by model "tuning" or "curve fitting."

The factors considered by the studies included in Table 5.6 include CO_2, volcanic (stratospheric) and tropospheric aerosol loading, and various suggested measures of the variation in solar radiation reaching the top of the atmosphere. The results indicate that, depending on investigator, each of these variables can separately account for anywhere from nearly all to none of the observed climatic variations of the last 100 years. In addition, when the coupled effects of more than one factor are considered, there remains a significant range in the ratios.

Closer examination of a few of these studies shows more clearly the uncertainties and discrepancies that are involved, although identifying the exact causes of these disagreements is beyond the scope of this report.

In comparing the various studies, consideration should be given to how well their choice of causal factors is able to explain the major features of the climate of the last hundred years as indicated by almost all available data sets, including the warmth of the northern hemisphere from 1920 to 1950 and the relatively steady warming of the southern hemisphere.

5.2.3.1 Carbon Dioxide as a Causal Factor

If we consider every influence other than CO_2 that affects the climate to be part of the natural climatic variability, we can attempt to identify the CO_2 signal amid the natural fluctuations, or "noise," of the climate record. The level of noise that can, to first order, be estimated by examination of climatic records before CO_2 is thought to have had a substantial climatic effect. Past CO_2 concentrations have been frequently estimated by extrapolating back from current concentrations assuming that inputs from fossil fuel combustion have been the dominant source and that the fraction of each year's emission remaining airborne has been constant. If fossil fuels have been the primary factor perturbing atmospheric CO_2 concentrations, then the period prior to perhaps 1930 can be used as a baseline. In this case, a reasonably long record is available, and the variability of annual average temperatures may be estimated with some reliability to be a few tenths of a degree. If, however, the biosphere has been an important net source of atmospheric CO_2 over the last 150 years, or parts thereof, then the CO_2-induced trend in temperature may not now allow selection of an adequate baseline from which to estimate natural climatic variability.

Hansen et al. (1981) calculated the effect of a CO_2 increase on the global climate of the last 100 years, using a baseline CO_2 concentration of 293 ppm (which might be too high in 1880 if biospheric sources have been important) and assuming that a doubling of CO_2 concentrations would lead to an equilibrium warming of 2.8°C. As shown in Figure 5.8, their model-computed response to the CO_2 increase alone did not compare well with the global record. Interestingly,

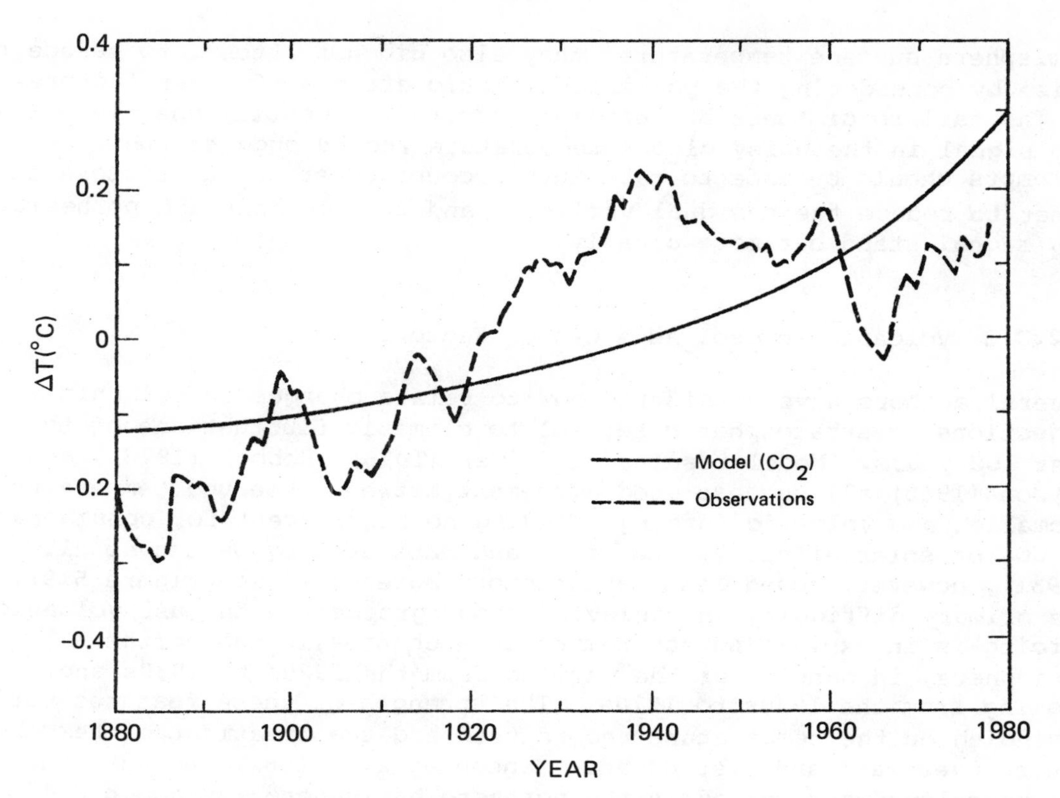

FIGURE 5.8 Comparison of observed global mean temperature anomalies of the last 100 years with anomalies predicted by a one-dimensional climate model assuming only that CO_2 concentrations are varying (Hansen, 1980).

however, although their result is in poor agreement with the northern hemisphere temperature record, it is in rather good agreement with the low-latitude and southern hemisphere temperature records (refer to Figure 5.4). These differences point to the need for conducting future analyses in more than one dimension; it may be, for example, that because of fewer volcanoes in the southern hemisphere and because of the thermal inertia of southern hemisphere oceans, the CO_2 warming may more easily be found in the southern rather than northern hemisphere.

Madden and Ramanathan (1980) also attempted to identify the CO_2 climatic effect, assuming that all other factors contributed to the climatic noise. They looked in the 50-60° N latitude band where the spatial coverage of the records is quite good and where equilibrium climate models predict that the temperature changes should be largest. Their results were negative (i.e., they found no statistically significant signal emerging from the noise), indicating, they suggested, that either the attempts to estimate present temperature changes from equilibrium climate models are inadequate or that the models are overestimating the CO_2 warming. Wigley and Jones (1981) also did not find evidence of a CO_2 signal when examining records of the northern hemisphere surface temperature; they also did not attempt to reduce the noise by considering the possible climatic effects of other factors.

The failure of these and earlier efforts to identify unequivocally a CO_2 signal in the noisy global temperature record suggests that attempts should be made to take into account other causal factors in order to reduce the residual variance, and thus to make a hypothesized CO_2 signal stand out more clearly.

5.2.3.2 Volcanic Aerosol as a Causal Factor

Several authors have considered how to relate changes in volcanic injections of stratospheric aerosol to climatic fluctuations of the last 100 years. Model results of Oliver (1976), Robock (1978), and Bryson (1980) all suggest good agreement between observed temperature anomalies and volcanic forcing, finding no requirement for consideration of CO_2 or solar effects. The model and data used by Hansen et al. (1981), however, found less satisfactory agreement (see Figure 5.9). The primary difficulty in achieving good agreement with just volcanic forcing is in explaining the temperature changes in the northern hemisphere, in particular the warming from the 1920s to 1930s and cooling from the 1940s to 1970s. The strength of these features varies depending on the temperature record used and the treatment of temperatures over land and over ocean. Hansen et al. (1981) indicate that the anomalous warm period in the northern hemisphere was about 0.4°C above a trend line through the rest of the record, whereas in Jones et al. (1982) and Budyko (1969) the anomaly is slightly less. Although most authors use the northern hemisphere temperature record as a basis for evaluation of the relationship between causal factors and climate changes, Hansen et al. (1981) used a global record that somewhat reduces the intensity of the 1920-1950 warming period that must be explained because of averaging in the rather steady warming trend in the southern hemisphere.

Bryson (1980) believes the problem that most investigators have in explaining the cooling that occurred prior to the Agung eruption of 1963 arises for two reasons. The first is the incompleteness of Lamb's (1970) volcanic record; in response to this criticism several volcanic eruptions in the 1950s have been added to most volcanic chronologies. The second reason is the inadequacy of estimates of the radiative effect of volcanic injections by the traditional methods of estimating the amount of dust injected. The recent El Chichon volcano and analyses of the effect of the Agung eruption have emphasized that the injection of sulfur-bearing gases rather than total dust is likely the more appropriate measure of the ultimate radiative effects. Although probably also affected by tropospheric aerosols, Bryson and Dittberner (1976) and Budyko (1969) urge use of actinometric data as a better measure of stratospheric turbidity.

It is also difficult to reconcile the relative effects on average temperature in the northern and southern hemisphere based on volcanic records, especially during the 1900-1910 period when (according to the temperature reconstruction of Hansen et al., 1981) the supposedly less

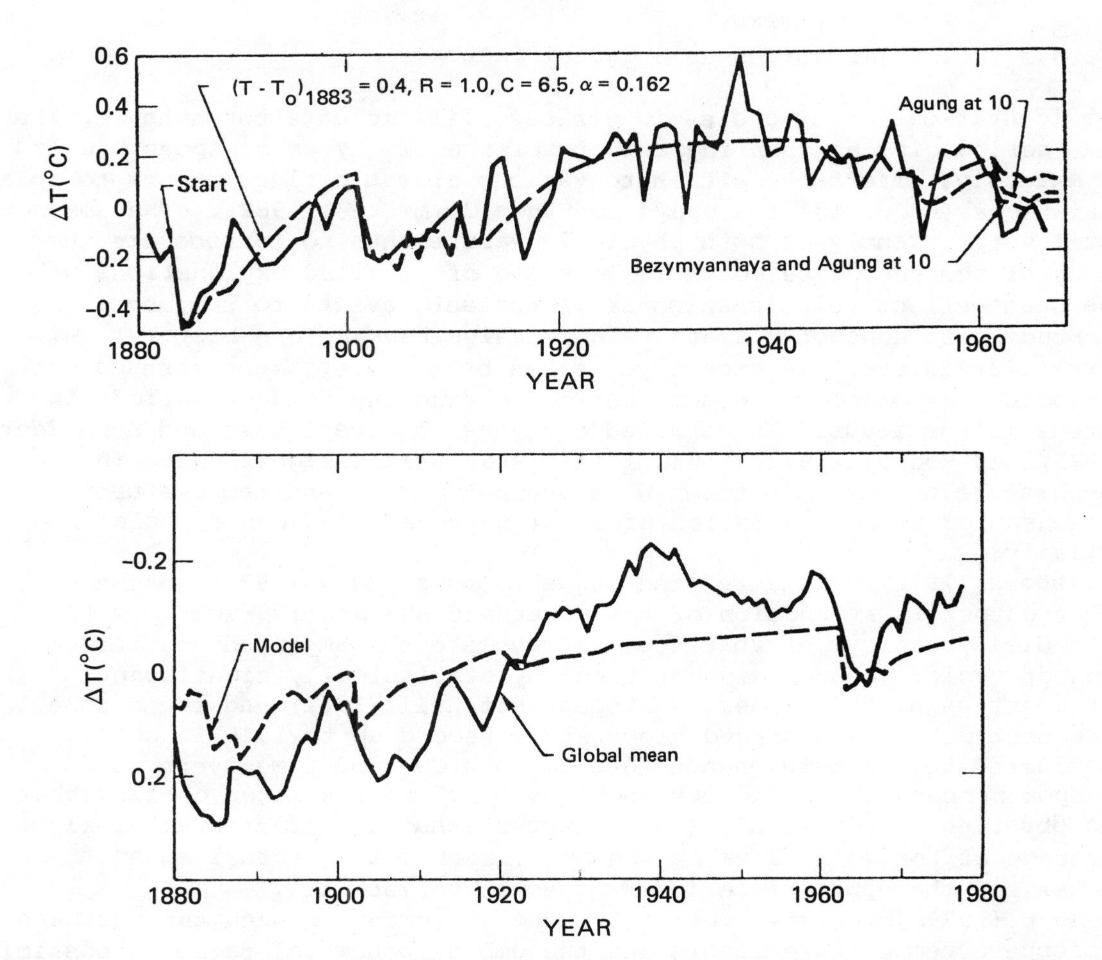

FIGURE 5.9 Comparison of observed surface-air-temperature anomalies of the last 100 years with anomalies predicted by models that included only changes in stratospheric aerosol loading as calculated by (top) Oliver (1976) and (bottom) Hansen (1980). Oliver compares his results with the northern hemisphere temperature record and Hansen to the global temperature record.

responsive southern hemisphere cooled more and prior to the northern hemisphere, even though listings of major volcanic activity (e.g., Mitchell, 1970) indicate that all important eruptions were in the northern hemisphere.

While the case for coupling between volcanic activity and climate is suggestive, the data bases and analysis methods still need much work in order to be able to account accurately for the climatic effects of volcanic injections with confidence. Analysis of the climatic fluctuations subsequent to the El Chichon eruption will be an important aspect of this effort.

5.2.3.3 Solar Variations as a Causal Factor

The identification of cycles in various climatic data bases having the same periodicity as such indices of solar activity as sunspots has led to extensive efforts to attribute various climatic fluctuations to solar variations (Eddy, 1977; Geophysics Study Committee, 1982). The absence until very recently of both physical measurements to corroborate the value of the surrogate solar indices and of detailed explanations of the cause-effect relationships are important caveats to remember.

Studies of sunspot-climate relationships have a long history. Of particular interest in recent years has been the apparent absence of sunspots (the Maunder minimum) noticed during the coldest periods in Europe in the 1600s. In detailed analyses, however, Mass and Schneider (1977) and Robock (1978) found little statistical significance in the in-phase relationship between Wolf sunspot number and temperature. This has led to consideration of other surrogate indices of solar activity.

Robock (1978) considered the suggestions of Eddy (1976) that the solar output is a function of the alternate 80- and 100-year cycles (the Gleissberg cycle) that seem to modulate the amplitude of the sunspot cycle. Robock did not find any statistically significant agreement using this index. Although Mitchell (1961) had found some agreement with the observed temperature record up to 1940 if he considered temperature changes due to both CO_2 and time-averaged sunspot number, these factors could not explain the later cooling that was observed. Eddy et al. (1982) suggest that the fractional areal coverage of the solar disk by sunspots, rather than actual sunspot number, is the appropriate index of solar variability.

Hoyt (1979a,b) finds apparently good agreement between the northern hemisphere temperature record and the umbral/penumbral ratio, a possible measure of the convective energy transport in the sun's photosphere, and therefore of its time-varying radiant flux. The range of his results was calibrated to agree with the temperature record; note that he found that umbral/penumbral ratio and CO_2 forcing, and not volcanic forcing, were sufficient to reproduce the variations in the observed temperature record. It would seem that, if valid, the solar effects should be similar in temperature records of both hemispheres except for the influence of differing amounts of ocean; the differences in the temperature records that exist would, it would seem, cast some doubt on this explanation.

Hansen et al. (1981) find that including the umbral/penumbral ratio improves their fit when also considering CO_2 effects (compare Figure 5.10, top, with Figure 5.8).

Gilliland (1981) uses the variation in solar radius as an alternative measure of solar irradiance. This cycle is about 76 years in length and is negatively correlated with the Gleissberg cycle. There are indications that the Greenland ice core exhibits cycles with a length about the same as the solar radius and Gleissberg sunspot cycles. Broecker (1975) suggested that the cooling associated with the 80 year Greenland ice core cycle was counteracting the warming expected from increasing CO_2, leading to only slight change in present temperatures.

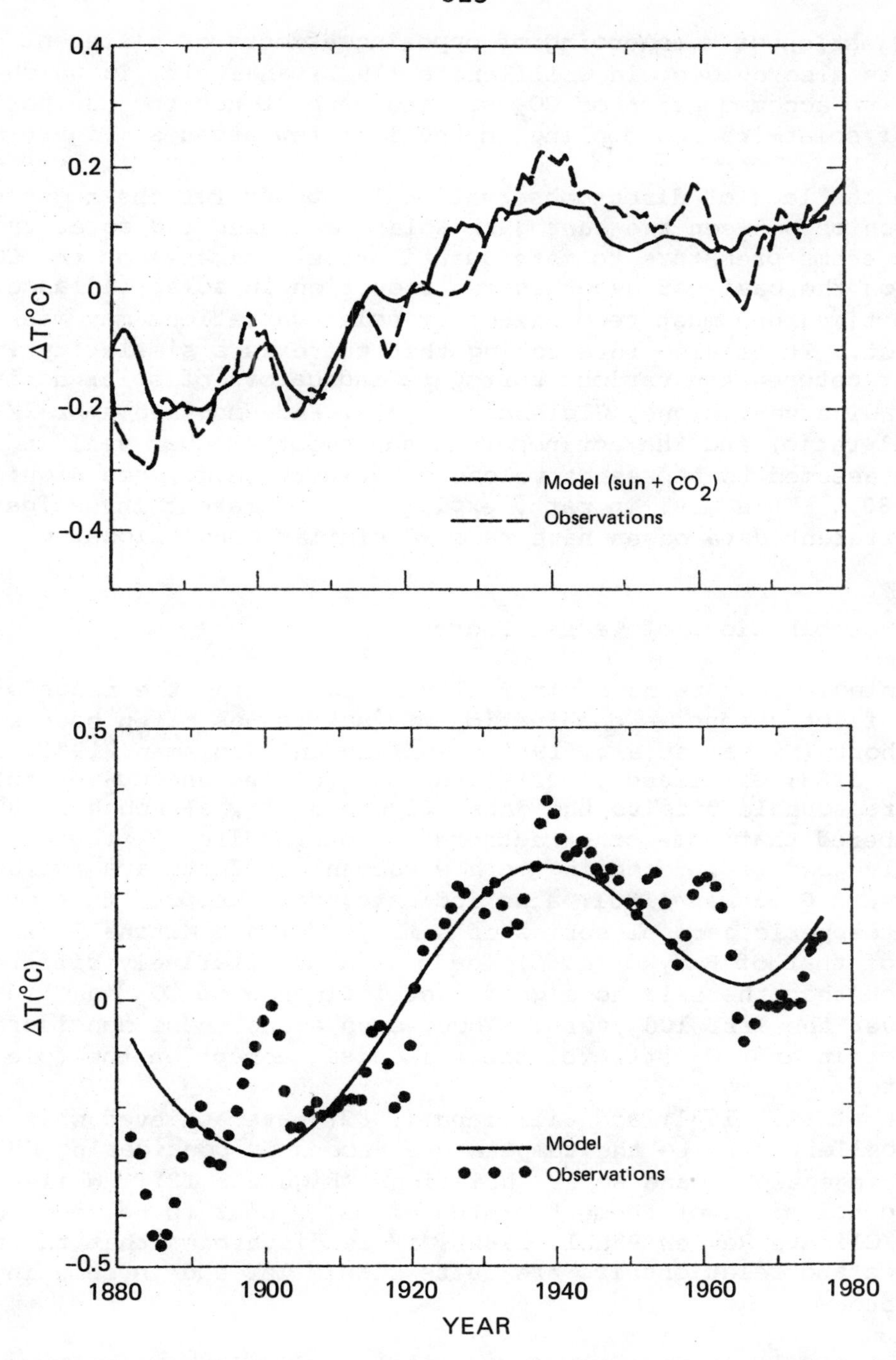

FIGURE 5.10 Comparison of the observed change in surface air temperature and model predictions of the change in temperature when considering the increase of CO_2 concentrations and changing solar radiance. (Top) Hansen (1980) uses umbral/penumbral ratio as an indicator of solar radiance and compares with global temperature change. (Bottom) Gilliland (1982) uses changes in solar radius as an indicator of solar radiance and compares with northern hemisphere temperature change.

This suggestion of a balancing of opposing effects of different causal factors is also evident in Gilliland's (1982) analysis, in which there could be no accommodation of CO_2 warming were it not for the postulation of solar-induced cooling in the last few decades (Figure 5.10, bottom).

Given the lack of direct observational support for the suggested relationships between the surrogate solar variables and solar radiation, it seems premature to make quantitative estimates of the CO_2 effects on the basis of hypothesized reduction in solar radiance. At the same time, one must recognize that solar variations may mask the CO_2 effect. It is also interesting that there is a similarity in character between the various surrogate indicators of solar activity (solar radius variations, Gleissberg cycle, and smoothed umbral/penumbral ratio) and the actinometric and smoothed dust veil indices that are assumed to represent volcanic activity (e.g., see Siquig and Hoyt, 1980). This may, in part, explain why different investigators using different data bases have reached similar conclusions.

5.2.3.4 Combinations of Causal Factors

Because single factors have difficulty in explaining the records of climatic fluctuations, a combination of factors has often been used. Some authors (Hansen et al., 1981; Vinnikov and Groisman, 1981, 1982; Mitchell, 1983; Gilliland, 1982) find that volcano and CO_2 effects lead to reasonable fits to the data (Figure 5.11), although it should be remembered that some other authors do not require CO_2 to achieve an equally good fit, depending on how volcanic effects are included. Vinnikov and Groisman (1981) also indicate, for example, that use of the stratospheric aerosol series of Lamb (1970) and Mitchell (1970) instead of that of Budyko (1969) leads to a qualitatively different conclusion that there is no significant influence of CO_2 on climate change over the last 100 years. Thus, even among those considering just volcanic and CO_2 effects, there is disagreement on the role of each factor.

Hansen et al. (1981) and Gilliland (1982) have achieved what appear to be excellent fits to the temperature record by considering CO_2, volcanic injections, and solar variations (Figure 5.12). While the similar conclusions of these two studies may appear to be re-enforcing (see CO_2/Climate Review Panel, 1982), it is disturbing that their data bases and relationships are quite dissimilar and in some instances contradictory.

(a) The maximum variation of Hansen et al.'s global temperature record is about 0.5°C, whereas Gilliland's northern hemisphere temperature record has a range of about 0.9°C.

(b) Hansen et al.'s volcanic record (from Lamb) has major peaks for Krakatoa (1883), Soufriere/Santa Maria (1902-1904), and Agung (1963). Gilliland's volcanic record appears to have major peaks for Askja (1875), unidentified (1885-1886), Katmai (1912), and Surtsey (1963-1965), most of which are high-latitude or local volcanoes.

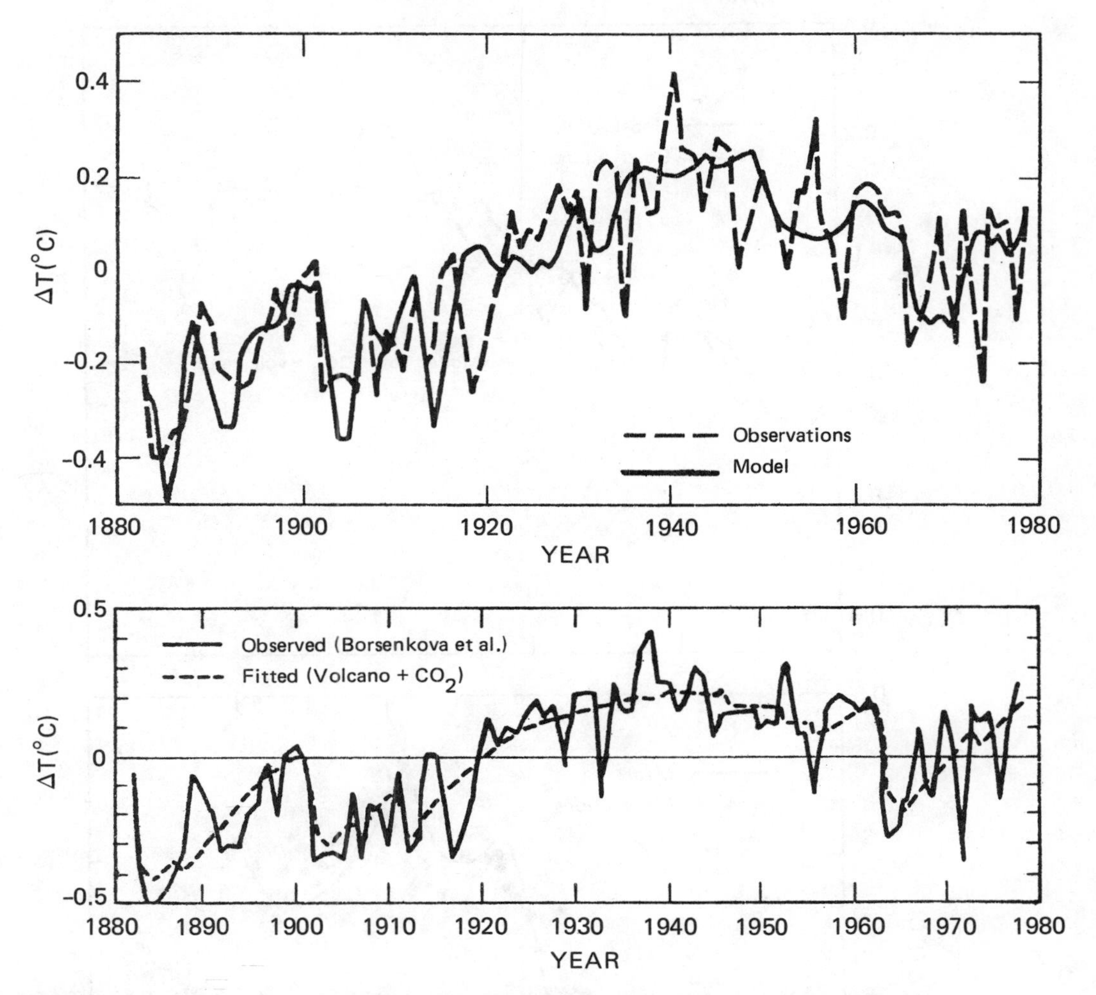

FIGURE 5.11 Comparison of the observed change in surface air temperature and model predictions of the change in temperature when considering the increase of CO_2 concentrations and changing stratospheric aerosol loading. (Top) Vinnikov and Groisman (1982) compare their results with data for the entire northern hemisphere from Vinnikov and Groisman (1981); (bottom) Mitchell (1983) compares with data north of 17.5° N.

(c) Hansen et al. show a CO_2 effect beginning in the 1880s, whereas Gilliland's CO_2 effect does not start until 1925. (The initial concentration of CO_2 assumed by each of these investigators is also 5-10% above recent estimates of CO_2 concentrations in the last century.)

(d) Hansen et al. use Hoyt's quite variable umbral/penumbral record as a measure of changes in solar radiance. Gilliland depends strongly on a smooth solar radius cycle of about 76 years and also includes solar variations having cycles of 12.4 and 22 years; phase and amplitude of the solar cycles were arbitrarily determined to provide the best fit with the temperature record.

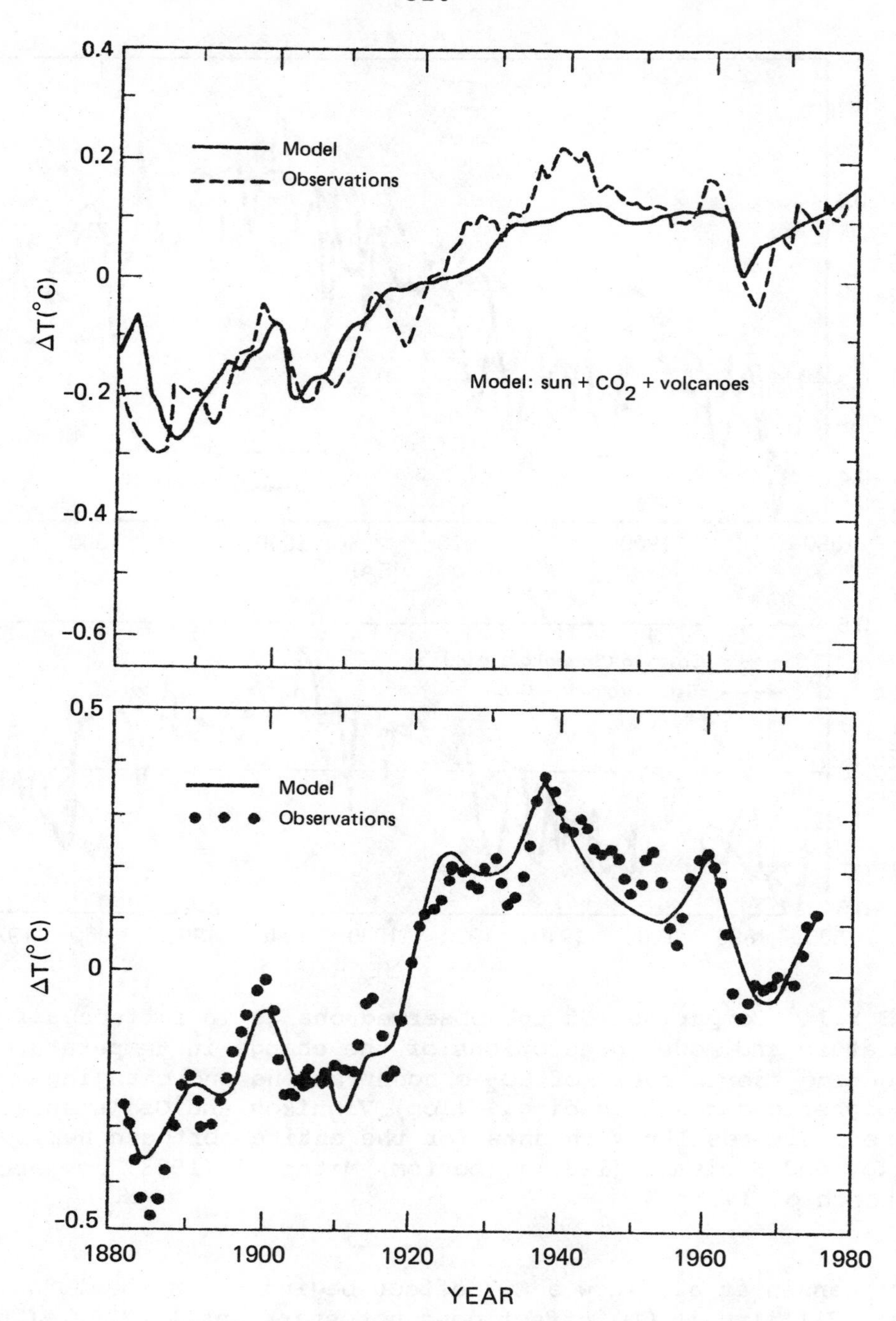

FIGURE 5.12 Comparison of the observed change in surface air temperature and model predictions of the change in temperature when considering the increase of CO_2 concentrations and changes in solar irradiance and stratospheric aerosol loading. (Top) Hansen (1980); (bottom) Gilliland (1982).

These and other similar studies (Hansen et al., 1981; Vinnikov and Groisman, 1982; Gilliland, 1982; for example) acknowledge uncertainties in their presentation of data and their formulation of conclusions. They have been very helpful in raising questions, suggesting relationships, and identifying gaps in our data bases and observational approach, and we cannot preclude the chance that at least one may be correctly relating causal factors and temperature changes. However, contrasting causal components of the climate change and differences in data bases make it difficult to accept the results as reinforcement of the general hypothesis of the CO_2-induced climate shift.

5.2.4 Steps for Building Confidence

Are results to date sufficient basis for declaring that a climate change due to increased CO_2 has already occurred? Given the potential importance of such a finding, we must require a high standard of agreement in attributing climatic effects to causal factors.

Methodologies for determining the statistical significance of suggested climatic changes have recently been discussed by Madden and Ramanathan (1980), Wigley and Jones (1981), Klein (1982), Epstein (1982), Hayashi (1982), Murphy and Katz (1982), Katz (1980, 1981), and an international group (World Meteorological Organization, 1982a). Epstein, for example, decomposes the time series of observations of a climatic variable into three components: a "natural" climatic mean, a possible climatic change induced by some extrinsic factor such as CO_2, and a random variability. The natural mean may be taken to be constant, or it may include the estimated influences of some forcing factors, e.g., solar or volcanic activity. Various hypotheses regarding the climatic change may now be stated and tested.

The conclusions to date have depended crucially on the assumptions made regarding the underlying climatic trend, the increase of CO_2 between the mid-nineteenth century and recent times, and the expected CO_2 influence. Once the hypothetical CO_2 signal is prescribed on the basis of estimated CO_2 concentrations and model simulations, the remaining variance of the climatic record is partitioned between the fluctuations explained by other external factors and the unexplained fluctuations deemed to be "noise." Reduction in the unexplained variance relative to the variance explained by CO_2 is taken as supporting the hypothesis of CO_2 influence. It is difficult to draw unambiguous conclusions from studies of this type for the following reasons:

1. The natural variability of global mean temperature is imperfectly known because of the relatively short period of instrumental record. Moreover, the spectrum of observed climate variability exhibits considerable power at low frequencies. In other words, natural climate variations have been observed on time scales commensurate with the time scale of CO_2 increase. At least in part, these may be attributable to external factors whose influence might, in principle, be removed. However, an unknown and possibly large amount of natural variability or

noise might well remain at low frequencies, making it difficult indeed to distinguish over short periods of record between slow trends induced by increasing CO_2 and equally slow trends that reflect the natural low-frequency variability of the climate system. For this reason, the hypothesis that the observed variability is entirely due to natural causes cannot be unequivocally rejected on the basis of the studies conducted so far.

2. The global mean temperature record has been reconstructed from a relatively short and geographically limited set of observations, primarily over land areas of the northern hemisphere. The period of record spans a period of great change and expansion in the human societies that make and record the observations. While investigators have conscientiously worked to remove or correct human influences such as urbanization on, for example, temperature records, it is difficult to assess their degree of success. The accuracy and representativeness of the data are thus open to question, although the broad trends are believed to be reliable.

3. Solar radiance variations have been estimated from surrogate observations, e.g., sunspots, umbra/penumbra ratios, and solar radius measurements, whose relationship to solar irradiance is by no means clear. Moreover, the relationship between solar variability and climate has been determined only empirically from a limited data base.

4. Atmospheric turbidity has been estimated from data on volcanic activity (e.g., individual eruptions or the "Dust Veil Index," or acidity measurements in ice cores) or from actinometric observations. None can be considered very reliable. As with solar variations, turbidity influences have been only empirically fitted to a limited record.

5. The history of atmospheric CO_2 concentrations before 1958 is poorly known, with estimates of mid-nineteenth-century concentrations ranging from 250 to 290 ppm. The expected temperature change is thus correspondingly uncertain.

6. Other neglected factors, e.g., surface albedo, aerosols, and--in recent years--anthropogenic trace gases, may have influenced the temperature record.

7. The choice of the appropriate signal of CO_2-induced climate change is by no means clear; many results from models of different degrees of physical completeness are available (Schlesinger, 1982, 1983). Moreover, model results are available only for equilibrium conditions, while the real world is presumably exhibiting a transient response to increasing CO_2. The transient response will be greatly complicated by the thermal inertia of the ocean and the distribution of ocean and land on the Earth's surface. Thus, real-world transient responses might be quite different in their regional details from those that might be inferred from equilibrium models (see, Thompson and Schneider, 1979; Schneider and Thompson, 1981; and Bryan et al., 1982).

8. Such studies are prone to a familiar pitfall of statistical inference, namely, the testing of multiple hypotheses. As noted by Epstein (1982), "If enough different hypotheses are examined, then, by chance, it is likely that statistics supporting one of them will be found."

9. The "noise" accounted for by both external factors omitted from the analysis and unexplained sources is not independent from one time of observation to another, nor can it be reasonably modeled as a first-order Markov or autoregressive process. (If it could be so modeled, then our best prediction of next year's climate would only involve the present single year's data corrected for changes in the included external factors; in fact, however, averages over many preceding years prove to be the best climatological estimates.) Thus, statistical techniques that are more careful and sophisticated than those for white noise or first-order autoregression will be required to deal with the increase of fluctuation energy at lower frequencies.

There is a methodological point that should be made here with regard to claims to have detected a CO_2-induced warming based on a "successful" model fit to the climatological record (e.g., Hansen et al., 1981) as opposed to the simpler approach of identifying a warming signal rising above the "noise" of intrinsic climate variability (e.g., Madden and Ramanathan, 1980). Tests of statistical significance are required in both approaches. However, these tests have usually been made only in the latter, partly because significance tests for model fits are not generally to be found in textbooks and, indeed, must be developed separately for each model type. Until such tests have been devised and carefully applied, the scientific community will remain skeptical of claims to have detected with statistical confidence a CO_2-induced signal in a single parameter such as temperature.

Notwithstanding methodological difficulties, earlier detection may be sought in two ways. First, we may attempt to improve the objectivity and physical basis of the estimates of external influences on the past climatic record, thus constraining the range of plausible hypotheses that could be tested. Needs are the following:

1. Better determination of the natural variability of temperature, particularly at low frequencies by extending the period of record back in time through use of proxy records and by distinguishing between ocean and land records.
2. Improvement in the accuracy and representativeness of the temperature record through incorporation of marine data and continued attention to influences such as urbanization.
3. Better data bases on possible changes in solar output and atmospheric aerosol loading.
4. Better reconstruction of the changes in CO_2 concentration over the last hundred years.
5. Objective, physically based, and observationally validated relationships between solar variability, volcanic aerosols, other possible factors and climate.

A second recently suggested approach is to attempt to isolate a pattern of changes specifically attributable to increasing CO_2 concentrations (MacCracken and Moses, 1982). This approach is discussed in the following sections of this chapter.

5.3 A STRATEGY FOR MONITORING CO_2-INDUCED CLIMATE CHANGE

5.3.1 The "Fingerprinting" Concept

Proposals for monitoring programs to detect the effects of increasing CO_2 date back to the SCEP (1970) and SMIC (1972) reports. Recently, participants at a Department of Energy sponsored workshop on First Detection of Carbon Dioxide Effects (Moses and MacCracken, 1982) proposed a three-part framework for detection of CO_2 effects involving:

1. Identification of changes,
2. Identification of possible causative factors,
3. Isolation of the parts of the changes attributable to increasing CO_2.

With respect to the last of these, a suggestion was made ". . . to develop a unique CO_2-specific 'fingerprint' for the CO_2 response involving a set of several parameters, distinctive from responses that would be caused by all other known influences, and to search for this correlated pattern of changes, not just for a change in one isolated parameter." (MacCracken and Moses, 1982.)

The concept of fingerprinting is based on the notion that a composite index based on multivariate statistical analysis of several parameters in space and time might enable us to attribute more positively climatic changes to increased CO_2. Indeed, changes in some climatic elements might help to distinguish the effects of CO_2 (or CO_2 in combination with other radiatively active gases) from those due to variations of some of the other factors or external conditions that could influence climate. For example, one might anticipate that the pattern of tropospheric temperature changes caused by turbidity variations due to volcanic aerosols would differ from those caused by the more globally uniform variations of CO_2.

While the notion of an index that would unequivocally reveal the influence of CO_2 on climate is indeed enticing, its practical application does not appear immediately feasible. The relationships between atmospheric CO_2 and climate variables must be deduced from model simulations. Simulations of the equilibrium response to highly elevated levels of CO_2 show considerable scatter in results (see Schlesinger, 1982, 1983), and simulations of the transient response to slowly changing CO_2 have not yet been accomplished. Even if a plausible index could be deduced from models, its statistical characteristics in the appropriate frequency range would need to be assessed through study of past data and through model simulations. Appropriate multivariate statistical tests would then have to be designed and applied. These difficulties must be overcome before a scientifically rigorous monitoring and detection strategy based on this approach can be devised and applied to provide clear guidance to policymakers.

Nevertheless, the concept of a "CO_2 fingerprint" provides useful guidance in the design of a geophysical monitoring program that will both provide data for research and help us to follow the course of climate. We thus suggest that monitoring programs designed to shed light on the effects of increasing CO_2 should focus on the climatic variables whose patterns of change in space and time are indicated by models to respond most strongly to CO_2 increases, together with those external factors that may also influence climate--particularly those external factors with greatest influence on the selected climatic variables. The remainder of this chapter presents some specific suggestions based on this approach.

5.3.2 Considerations in Climate Monitoring

5.3.2.1 Statistical Variability and Expectations of Change

All climatic parameters are highly variable in space and time. Their variability or "noise" arises both from systematic physical processes whose effect could in principle be calculated and from the random fluctuations of a turbulent atmosphere. The characteristics of the variability, including its preferential occurrence in one frequency range or another, can in some cases be determined from observed data, especially for climatic parameters at the Earth's surface such as air temperature and precipitation. Direct estimates of variability may in other cases have to be supplemented from the data simulated by comprehensive climate models (Manabe and Hahn, 1981).

Estimation of the climatic change expected to result from increased CO_2 depends on climate models that address the processes and scales of interest. Estimation of the changes to be expected from increased CO_2 should therefore include their seasonal and geographical characteristics, their longer-term and arealy averaged properties, and their evolution in time as CO_2 concentrations increase. Some preliminary information of this sort is already available; extended simulations are required to provide more complete information.

A schedule of expected climatic changes is especially necessary. Most experiments that have been made to date have estimated the response of climate from the difference between two simulated equilibrium climates of a model employing normal and twice the normal concentrations of atmospheric CO_2. From the results of such experiments the climatic changes expected at other (and generally lower) levels of CO_2 are then found by simple interpolation, usually logarithmic. This approach overlooks the facts that all the elements of the climate system do not interact either at the same rate or in the same places, especially insofar as they involve the oceans, and that the CO_2 concentration will not "wait" at any level for the climate to reach an equilibrium. In this connection, model experiments treating an exponential growth of CO_2, and hence a linear growth of its effects, may be particularly helpful. The information necessary for a realistic schedule of changes, including their seasonal and geographical distribution, in a variety of climatic parameters that are expected to occur as the CO_2 concentration reaches progressively higher levels, can be provided only by comprehensive general circulation model (GCM) simulations; such studies are important in the design of monitoring strategies.

5.3.2.2 Initial Selection of Parameters

A program for monitoring and detection of CO_2 effects can be developed on the basis of our expectations of the effects of increasing CO_2, estimates of the accompanying climatic noise, and a schedule of the expected responses to increased CO_2. Existing knowledge is adequate to formulate an initial monitoring strategy that can be revised as our understanding improves. The first step should be to identify the climatic parameters whose responses to increased CO_2, individually or in combinations, are likely to be significant. Atmospheric model studies suggest that likely candidates are the tropospheric and surface air temperature (which should rise), sea temperature (which should rise), stratospheric temperature (which should fall), and atmospheric water vapor or specific humidity (which should increase). In addition, the downward flux of infrared radiation at the surface should increase, while at the top of the atmosphere the spectral distribution of outgoing infrared radiation should shift with respect to the primary CO_2 emission bands. In the ocean and cryosphere, the amount of snow and ice in polar regions should decrease over the long term and the global sea level should rise.

In addition to large-scale responses, GCM climate simulations are expected to indicate the geographical and seasonal characteristics of such changes, and their evolution over time, all of which may provide additional indications. For example, the expected polar warming may occur principally in winter because of the increase of heat conduction through thinner sea ice, while soil moisture in midlatitudes may be reduced in summer (cf., Chapter 3). Other changes whose characteristics are yet to be definitively shown by climate model simulations may occur in the meridional circulation and hydrologic cycle.

The statistical significance of changes in individual variables can be judged against appropriate measures of their natural variability. Assessment of the significance of changes in some composite index could be judged similarly.

5.3.2.3 Revision and Application of a Monitoring Strategy

A multiple-effects monitoring strategy for CO_2-induced climatic change can be implemented even though all the information required for its rigorous design is not available. Using a combination of model simulations and observational data, at least partial information on the expected changes and levels of variability can be assembled. As further information becomes available from both new observations and new model simulations, additional variables can be included and a revised schedule or timetable of the effects expected over future years can be prepared.

The interpretation of the climatic record should also include an assessment of the possible effects of changes in other external factors in the climate system, apart from CO_2 and apart from internally generated noise. Such possibly competing factors include the large-scale injection of aerosols into the atmosphere by volcanic eruptions and possible variations of the solar constant on both interannual and decadal time scales. As with CO_2, our best hope for a knowledge of the climatic effects of these events rests with the use of a hierarchy of models, including comprehensive GCMs. Some experiments have suggested, for example, that the annual average (equilibrium) tropospheric warming resulting from an increased solar constant resembles in some respects that found with increased CO_2, although the transient response might differ in some respects (Schneider and Thompson, 1981). Since such factors have been used in conjunction with changes in CO_2 in simple climate models to simulate the historical variations of the northern hemispheric surface temperature (Hansen et al., 1981), it is important that simulations be made with comprehensive GCMs in which the significance of any climatic changes due to changes in solar radiation can be determined and, if possible, differentiated from those due to other factors. This information will help us to interpret monitored climate records in terms of signals due to variations in incoming radiation, aerosol loading, or greenhouse gas concentrations and residual changes that may contain a signal attributable to increased CO_2.

5.3.3 Candidate Parameters for Monitoring

Previous sections have discussed the problems of identifying CO_2-induced warming from present data and showed that there are weaknesses in the interpretations, caused by the inadequacy of the existing data sets, the methods of analysis used, and the underlying theory. Data employed, except in recent times, have rarely been collected with long-term monitoring of CO_2 effects in mind, so that development of a coherent and integrated monitoring strategy becomes essential at this time. Monitoring of CO_2 effects is a challenging task if the number of parameters to be monitored is to be kept low, in order to reduce costs, and if at the same time unambiguous cause-effect relationships are to be established.

The following criteria were chosen to assess the suitability of a number of parameters for long-term monitoring:

1. Sensitivity. How does the effect exerted on climate by the variable or the changes experienced by the variable on decadal time scales compare with that associated with corresponding changes in CO_2?
2. Response characteristics. Are changes likely to be rapid enough to be detectable in a few decades?
3. Signal-to-noise ratio. How large are the relevant changes in relation to the variability due to measurement errors and causes not accounted for?

TABLE 5.7 Primary Parameters for Monitoring the Causes and Effects of Climate Change

A. Primary parameters for monitoring the causes of climate change
   - Carbon dioxide
   - Stratospheric aerosols
   - Solar radiance
   - Other "greenhouse" gases and ozone
     - Nitrous oxide (N_2O)
     - Methane (CH_4)
     - Chlorocarbons ($CFCl_3$, CF_2Cl_2, for example)
     - Stratospheric ozone
     - Tropospheric ozone

B. Primary parameters for monitoring the effects of climate change
   - Atmosphere
     - Global temperature
       - Mean surface air temperature
       - Tropospheric temperature distribution
       - Stratospheric temperature distribution
     - Radiation
       - Upward terrestrial and reflected solar radiation at the top of the atmosphere
     - Cloud and water vapor
       - Precipitable water content of the atmosphere
       - Equivalent emission temperature (cloudiness)
   - Cryosphere
     - Sea ice cover
     - Snow cover
     - Ice cap mass balance changes
   - Oceans
     - Sea level
     - Sea temperature

4. Past data base. Are data on the past behavior of the variable adequate for determining both a base level and its natural variability?
5. Spatial coverage and resolution of required measurements.
6. Required frequency of measurements.
7. Feasibility of technical systems. Can we make the required measurements?

Cost is another important consideration, but meaningful assessments of technical feasibility and estimates of costs were beyond the capabilities of the present panel. As has been pointed out in other sections of this report, the detection of climate change is largely a signal-to-noise ratio problem, and it is therefore important to compile comprehensive data sets and to apply the best statistical techniques available to the data. The technical feasibility of making the measurements and the associated cost, however, will put restrictions on any monitoring strategy that can be put into effect. However, a number of primary parameters for monitoring have been tentatively selected (Table 5.7), based on the above criteria. They are organized under the headings of "causes" of climatic change and "effects" of CO_2. In the latter category, parameters to be monitored have been grouped under the headings of atmosphere, cryosphere, and oceans. This is not a comprehensive list of candidate parameters but includes only those that have high aggregate ratings in the seven criteria listed above.

5.3.3.1 Causal Factors

External factors that have been suggested as influencing the climate have time scales, i.e., periodicities or e-folding times, ranging from less than 1 to more than 10^9 years. Focus on the climatic effects of anthropogenic CO_2 emissions allows us to limit our attention to those that have time scales of one to several hundred years. Within this time range, the external factors that may influence the climate include the composition of the atmosphere, volcanic activity, land surface modification, and solar variations. Table 5.8 lists trace gases and aerosols that affect atmospheric composition and that are influenced by man's activities. Table 5.9 lists the trace gases with potentially important radiative, climatic, or chemical effects. We further limit our consideration to external factors that might induce hemispheric or global temperature changes of magnitude 0.1 K or greater over about a century and to a lesser extent according to feasibility of monitoring. According to these criteria, monitoring of the following factors is particularly important:

1. Carbon dioxide
2. Solar radiance
3. Stratospheric aerosols
4. Other greenhouse gases* and ozone
   (a) Nitrous oxide
   (b) Methane
   (c) Chlorocarbons
   (d) Stratospheric ozone
   (e) Tropospheric ozone

Although ozone is not strictly an external factor, its chemistry is affected by anthropogenic emissions of a number of gases (e.g., nitrogen oxides and chlorocarbons); by monitoring ozone directly, we avoid dependence on knowledge of the details of the chemistry. Nonvolcanic sulfur (OCS and SO_2, for example) and volcanic emissions both contribute to stratospheric aerosol.

Several other potential external influences have not been included at this stage. The possible effects of modifications in surface albedo,

---

*This list will no doubt be revised as research continues. See Machta (this volume, Chapter 4, Section 4.3).

TABLE 5.8 Principal Anthropogenic Sources of Trace Gases and Aerosols

	Anthropogenic Source	Comments
Gas		
CO_2	Fossil fuel combustion	Possibly large biospheric component
CO	Internal combustion engines	
Hydrocarbons C_2H_4, for example	Internal combustion engines	
Chlorocarbons	Refrigerants, solvents, propellants	Those of concern entirely man-made
CH_4	Internal combustion engines, industry, change in land use	Large component from biological activity
N_2O	Combustion, fertilizer manufacture	Large natural component from biological activity
NO, NO_2	Internal combustion engines, aircraft	High-flying aircraft are an upper-tropospheric and lower-stratospheric source
Sulfur Compounds		
OCS, CS_2	Fossil fuel conversion	Volcanoes are an intermittent source of sulfur
SO_2	Combustion	
Ozone	Anthropogenic contribution is from chemical reaction of other trace gases	
Aerosols		
Sulfate	Conversion from SO_2 and other sulfur-bearing compounds	Most important for stratospheric aerosols
Silicate or carbon-containing	Combustion, soil erosion	Diesel engines especially, closely tied to land use

roughness, and surface thermal characteristics by desertification, urbanization, and extension of agriculture (e.g., Charney, 1975; Sagan et al., 1979; Potter et al., 1975, 1980) are not included because effects of such changes appear to be primarily regional and because of the difficulty of developing a data base. Tropospheric aerosols are not included because there are large regional variations in aerosol amount, composition, and characteristics, so that an accurate monitoring network would require many stations as well as detailed analysis of the samples. These neglected effects contribute, of course, to the

TABLE 5.9 Trace Gases with Potentially Important Radiative and Chemical Effects on the Global Atmosphere

Chemical	Composition	Surface Concentration (ppmv)	Effects of Increased Concentrations: Change in Surface Temperature[a]	Effects of Increased Concentrations: Change in Ozone Column[b]
Carbon dioxide	CO_2	340	Increase	Increase
Chlorocarbons	$CFCl_3$	1.9×10^{-4}	Increase	Decrease
	CF_2Cl_2	3.0×10^{-4}	Increase	Decrease
	CH_3CCl_3	1.5×10^{-4}		Decrease
	etc.			
Stratospheric ozone	O_3 (strat.)	(0.1-10.)[c]	Increase[d]	Increase
Tropospheric ozone	O_3 (trop.)	(0.02-0.1)[c]	Increase	Increase
Methane	CH_4	1.7	Increase	Increase
Nitrogen oxides	NO_x (=NO, NO_2)	(3×10^{-5}-0.015)[c]		Decrease (strat.) Increase (trop.)
Nitrous oxide	N_2O	0.3	Increase	Decrease
Carbon monoxide	CO	0.12		Increase (trop.)
Hydrocarbons	C_2H_4, etc.	$\lesssim 1 \times 10^{-3}$	[e]	Increase (trop.)
Sulfur Compounds				
Carbonyl sulfide	OCS	5.0×10^{-4}	Decrease[f]	?
Carbon disulfide	CS_2	0.3×10^{-4}		
Sulfur dioxide	SO_2	1.0×10^{-4}		
Stratospheric water vapor	H_2O (strat.)	(3-5)[c]		Increase

[a] The change in surface temperature as a measure of the climatic effect.
[b] The change in the ozone column as a measure of the chemical effect.
[c] Range in concentration with altitude.
[d] Assumes a uniform percentage change with altitude. Changes in the vertical distribution may cause an increase or a decrease in surface temperature.
[e] Could increase surface temperature if concentrations increase drastically.
[f] Effects result from formation of stratospheric aerosols.

unexplained variance in the climatic record. Clearly, continuing research on the possible influences on climate and on the means for their measurement is essential.

The following subsections provide more detailed discussion for those external parameters listed above and present monitoring requirements. In addition to recommending improved monitoring of some of the factors, in several cases it will be essential to continue the records of surrogate indicators that have been developed in order to be able to calibrate these methods and extend the data bases back in time.

5.3.3.1.1 Carbon Dioxide

Sensitivity. Numerical model studies indicate that doubling atmospheric carbon dioxide concentrations would increase global average temperatures (ΔT_d) by 3 $\pm$ 1.5°C (Climate Research Board, 1979; CO_2/Climate Review Panel, 1982). Changes in radiative flux are approximately logarithmically related to CO_2 concentration. Although water vapor and albedo feedback mechanisms are a function of global mean temperature, for a moderate warming, temperature change (ΔT) is nearly linearly proportional to radiative flux, so that

$$\Delta T \sim \frac{T_d}{\ln 2} \cdot \ln\left(\frac{[CO_2]}{[CO_2]_0}\right),$$

where ΔT_d is the expected temperature change for doubling, and $[CO_2]_0$ and $[CO_2]$ are the concentrations for the base period and the current period, respectively.

Rate of Change. Since 1958, the annual increase in the atmospheric concentration of CO_2 has ranged between 0.6 and 2.2 ppmv (Machta, this volume, Chapter 3, Section 3.4). The concentration in 1982 was slightly above 340 ppmv. The concentration before the industrial revolution probably was in the range 250-290 ppmv. Releases of CO_2 to the atmosphere result from both fossil fuel emissions and changing land use. The mean growth rate of fossil fuel CO_2 production over the period 1861-1980 was 3.5% per year (Elliott, 1982). In 1980 total CO_2 emissions from fossil fuels are estimated to have been about 5.2 x 10^9 tons of carbon (Marland and Rotty, 1982). Uncertainty associated with calculations of fossil fuel CO_2 emissions is estimated to be 10-13.5%. CO_2 emissions from cement manufacture may add an additional 2% to the estimate of fossil fuel emissions. The history of biospheric emissions is poorly known and controversial. Global net release of carbon to the atmosphere from the biosphere since 1860 has been estimated to be as high as 180 x 10^9 tons of carbon. The release from biospheric sources in 1980 was given as 1.8 x 10^9 tons of carbon according to the FAO Production Yearbook, the most comprehensive and apparently detailed data source, and has been estimated to be as high as 4.7 x 10^9 tons of carbon according to other sources (see Woodwell, this volume, Chapter 3, Section 3.3). Regrowth and stimulation of photosynthesis from increased atmospheric CO_2 may have compensated for some of these emissions (see Machta, this volume, Chapter 3,

Section 3.4). Nordhaus and Yohe (this volume, Chapter 2, Section 2.1) project that the atmospheric CO_2 concentration will most likely increase by an average of 0.3% per year up to the year 2000, reaching a concentration at that time of 367 ppmv.

Signal-to-Noise Ratio. The CO_2 concentration in a single sample can be measured to within 0.1 ppm, but at most stations in the global network the accuracy is about 1 ppm. The seasonal cycle and annual increase of atmospheric CO_2 can be easily detected at remote locations where effects of local biospheric, anthropogenic, and geologic sources are avoidable. The atmosphere, averaged on a zonal and annual basis, appears to be within 4-6 ppm of being well mixed, and gradients of this magnitude can be explained by longer times for transport from oceanic or biologic sources and sinks (see Machta, this volume, Chapter 3, Section 3.4).

Adequacy and Availability of Data Base. Continuous, accurate observations of CO_2 began at Mauna Loa Observatory in 1958. Additional research and measurements are needed to reconstruct past concentrations. Before 1958, most measurements are of uncertain accuracy and may not be representative of the global average concentration. For purposes of detection of climate change, the present network provides adequate measures of global average concentrations, and data are readily available.

Efforts are under way to estimate past CO_2 concentrations by, for example, analysis of solar spectral measurements, carbon isotopic ratios in tree rings, reconstructed pCO_2 in former oceanic surface waters, and CO_2 concentration in air bubbles trapped in ice. Efforts to reconcile the results of these various approaches, including the effects of local station bias and different temporal characteristics, are needed.

Spatial Coverage and Resolution of Additional Measurements Required. The present global network provides adequate coverage and resolution of CO_2 concentration for purposes of detection of climatic changes.

Frequency of Measurements Required. For purposes of studying the climatic effect, the record of monthly average CO_2 concentration now being made is sufficient. Climate model studies of the climatic effects of increasing CO_2 concentrations are most unlikely to use more frequent data than annual average concentration.

Feasibility and/or Existence of Technical Systems; Continuity. Measurements made by the present global network of surface stations are satisfactory for determining the trend of CO_2 concentrations. Although satellite measurement of atmospheric CO_2 would provide improved global coverage, it is not evident that sufficient accuracy could be achieved to determine the interannual variations in the trend or in the range of the seasonal cycle.

5.3.3.1.2 Stratospheric Aerosols

Sensitivity. Stratospheric aerosols affect climate through their net effects on the components of the Earth's radiation budget. The delicate balance that exists between the solar and thermal components of the radiation budget can be altered by the range of possible stratospheric aerosol (or cirrus cloud) scattering and absorption properties such that either a net cooling or warming of the lower atmosphere may result. Radiative transfer calculations (e.g., Pollack et al., 1976b) indicate that the sign and magnitude of the temperature change depend critically on the composition (refractive indices), size distribution, scattering phase function, and optical depth of the stratospheric aerosols. Surface-temperature changes associated with potential radiative perturbations of climate were calculated by various research groups. For example, Hansen et al. (1981), using a simplified one-dimensional model, calculated that a persistent increase in stratospheric (sulfuric acid) aerosol optical depth of 0.2 (representative of a large volcanic event) would produce a cooling of 1.9°C in the Earth's surface temperature. In reality, however, effects over land and sea might be quite different.

Natural volcanic eruptions throughout history have long been thought to constitute the primary source of aerosols in the stratosphere through the direct injection of large quantities of gases and ash. Although the Mount St. Helens eruption was powerful and injected large quantities of material into the stratosphere (~0.5 km^3), its climatic impact was insignificant (Robock, 1981, 1983). Climatic effects of volcanic eruptions, in general, depend on the chemical composition of the aerosols injected into the stratosphere and, in particular, on the amount of sulfur contained. The climate effects are not a simple function of the volume of solid material ejected or the magnitude of eruption.

Rate of Change. Several empirical and statistical studies offer a history of volcanic impact on climate. A discussion of climatological evidence from past volcanoes is given in the recent report of the NASA Workshop on the Mount St. Helens Eruption of 1980 (Newell and Deepak, 1982). Lamb (1970) proposed an empirical dust veil index (DVI) in correlating atmospheric optical depth perturbations of the atmosphere with volcanic eruptions over the past century. Thus, the historical record of volcanic eruptions must account for composition factors in assessing climatic effects. This holds for future extrapolations. There is no strong basis for projecting aerosol perturbations or change of composition. No clear trend or pattern emerges from the historical record.

Signal-to-Noise Ratio. Satellite measurements of peak background (pre-eruption) stratospheric infrared extinction (1.0 μm) are typically on the order of 10^{-4} km^{-1} (McCormick et al., 1981). The corresponding optical depth between the tropopause and 30-km altitude is approximately 10^{-3}. Immediately following the Mount St. Helens eruption, local extinction values as high as 10^{-1} km^{-1} (optical depths near 1) were observed, although the average stratospheric optical depth was typically 10^{-3}. Volcanic eruptions, which are believed to have had an impact on climate, have produced stratospheric optical depths of about $\tau \sim 10^{-1}$ for several years following the eruptions.

Adequacy and Availability of Data Base. Volcanic eruptions have been related to climatic changes since a suggestion by Benjamin Franklin in 1784 (cf. Pollack et al., 1976a). Extended inventories of volcanic eruptions have been prepared by Lamb (1970) and Simkin et al. (1981). Pollack et al. (1976a) have assembled and reviewed data on transmissivity. Actinometric data, which may also include the effects of tropospheric aerosols, have been assembled by Pivovarova (1968, 1977) and Bryson and Goodman (1980). Hammer (1977) suggests that acidity in a Greenland ice core may be a measure of northern hemisphere volcanic activity. In recent years lidar data have become available, but time histories exist over only limited parts of the globe for about 10 years.

Observations of stratospheric aerosol extinction have been made by the Stratospheric Aerosol Measurement II (SAM II) instrument aboard the Nimbus-7 research satellite (since October 1978) and by the Stratospheric Aerosol and Gas Experiment (SAGE) aboard the Applications Explorer Mission 2 (AEM-2) (February 1979-March 1982). A follow-on SAGE mission, SAGE II, is planned for launch in 1984 in conjunction with the Earth Radiation Budget Experiment (ERBE). Aerosol information may also be inferred from operational Advanced Very High Resolution Radiometer (AVHRR) data from National Oceanic and Atmospheric Administration (NOAA) satellites.

Stratospheric aerosol data products are routinely archived at the National Space Science Data Center (NSSDC) at the National Aeronautics and Space Administration's (NASA) Goddard Space Flight Center, Greenbelt, Maryland.

Spatial Coverage and Resolution of Additional Measurements Required. Future high-latitude coverage beyond SAM II will be needed for investigating the importance of polar stratospheric clouds (PSCs) on climate. SAGE covered the geographical regions between 72° N and 72° S latitude, while SAM II observes both polar regions. Planned SAGE II measurements will provide continuing aerosol extinction measurements at low and midlatitudes through the mid-1980s. No corresponding measurements are planned for the polar regions. Remote measurements must be accompanied by in situ measurements of aerosol radiation properties and size distribution, as part of a coordinated ground-truth program. For example, lidars and aircraft flights are needed to calibrate the satellite measurements. In particular, additional lidars are needed in the southern hemisphere. Comprehensive data, including latitudinal variation, are needed in aerosol impact model studies.

Frequency of Measurement Required. Measurements are needed to establish the monthly variation of stratospheric aerosols.

Current and planned satellite observational programs are believed to be adequate for establishing the seasonal and interannual variability of the present background stratosphere. Provision must be made to respond to major volcanic eruptions as they occur and to continue observations beyond the lifetime of current programs.

Feasibility and/or Existence of Technical Systems; Continuity. Existing satellite systems supplemented by a few key lidar systems for calibration purposes are adequate.

5.3.3.1.3 Solar Radiance

Sensitivity. Solar radiation is the fundamental energy source that drives the motions of the atmosphere and oceans. The associated storage and transport of heat and mass then establish the Earth's temperature and climate on regional and global scales. Temporal variations in the sun's output have long been thought to be a major cause of recorded climate change. Although a great number of researchers have sought clear evidence in linking solar variations to specific weather and climate responses, the results have thus far been inconclusive. A recent report by the Geophysics Study Committee (1982) concludes that the role of the sun in producing global circulation, climatic zones, seasonal changes, and the recurrence of periods of glaciation is well recognized, but intrinsic solar variability is neither implied nor required to account for these phenomena. The simple changes in insolation responsible for these effects are predictable, far in advance, on the basis of known parameters of the orbit, figure, and motions of the Earth.

Although numerous mechanisms exist that can cause climate change, it is important, nevertheless, to understand to what extent the basic solar forcing of the climate system is changing. Sensitivity studies reported by Hansen et al. (1981) suggest a 0.5°C global average surface temperature rise associated with a hypothetical 0.3% increase in the solar constant. Systematic changes of only 0.5% per century could explain the entire range of past climate from tropical to ice-age conditions (Eddy, 1977).

Rate of Change. The integral of the broad spectrum of solar electromagnetic radiation incident on the Earth has historically been termed the solar constant, as a consequence of the lack of evidence to the contrary. Until recently, detailed information on short-term natural variability in the solar constant was restricted mainly by the lack of adequate instrumentation and by the variability in atmospheric transmission, which limits the accuracy of surface-based observations of the sun.

Since late 1978, independent accurate measurements of solar radiance by two NASA research satellites have confirmed that the solar constant is indeed variable. Both the Solar Maximum Mission (SMM) and Nimbus 7 spacecraft carry advanced active cavity radiometers that overcome the basic deficiencies in prior measurements. SMM results reported by Willson et al. (1981) yield a mean solar radiance at 1 AU of 1368 W m^{-2} with an absolute uncertainty of less than $\pm 0.5\%$. Several observed large decreases (dips) in radiance of up to 0.2%, lasting about 1 week, are highly correlated with the development of sunspot groups whose below-average temperatures reduce the total output of solar radiation. Solar faculae, with above-average temperatures and, therefore, radiative output higher than normal, cause a much smaller variability appearing as a radiative excess. The facular effects correlate with irradiance peaks before and after some observed dips. A measured 0.05% per year downward trend in the 4-year record may be related to the general pattern of solar activity variation over an 11-year cycle.

Signal-to-Noise Ratio. The solar radiance monitor carried aboard the SMM spacecraft consists of a combination of three independent electrically self-calibrated cavity pyrheliometers having nearly uniform wavelength sensitivity from the far-ultraviolet through the far-infrared regions. This sensor combination has provided daily observations of the sun with an estimated measurement precision of less than 0.005% since launch. Such precision should be sufficient to detect solar variability of a magnitude able to affect climate.

Adequacy and Availability of Data Base. Routine satellite measurements of solar radiance have been made since October 1978. This measurement capability, if sustained, is believed adequate for meeting current climate requirements for broadband solar radiance monitoring.

Spatial Coverage and Resolution of Additional Measurements Required. Current measurements should be continued to provide a data base encompassing at least one 11-year solar cycle.

Frequency of Measurements Required. The wide range of natural variability found in the current data set supports the need for daily measurements over the next few years. Beyond then, less frequent (monthly?) measurements may be adequate.

Feasibility and/or Existence of Technical Systems; Continuity. Although current technology can meet the scientific requirements for broadband solar radiance monitoring, no future satellite missions carrying these instruments have yet been approved. The proposed Upper Atmosphere Research Satellite (UARS) would have this capability.

5.3.3.1.4 Nitrous Oxide (N_2O)

Sensitivity. Donner and Ramanathan (1980) and Wang et al. (1976) calculated a surface warming of 0.3 K and 0.4 K, respectively, for a doubling of the N_2O concentration. A doubling of N_2O concentrations would also decrease the total ozone column by as much as 15% (World Meteorological Organization, 1981) through the catalytic reactions of N_2O-produced nitrogen oxides with ozone; the change in O_3 concentration would also have climatic effects.

Rate of Change. Current tropospheric N_2O concentrations are approximately 300 ppbv. N_2O levels appear to be approximately 0.8 ppbv higher in the northern hemisphere than in the southern hemisphere.

Atmospheric measurements (Weiss, 1981) indicate that tropospheric concentrations of nitrous oxide (N_2O) have been increasing at a rate of approximately 0.2% per year since the first measurements were made in 1963. Weiss suggests that a substantial fraction of this increase may be explained by combustion of fossil fuels. Nitrogen fertilizers may also be an important source of N_2O. It would be desirable to find means for determining the pre-industrial N_2O concentrations in the atmosphere.

The N_2O concentration decreases with altitude in the stratosphere. The primary sink for N_2O is photodissociation in the stratosphere. The primary source of nitrogen oxides in the present stratosphere is the reaction of N_2O with excited oxygen atoms.

Signal-to-Noise Ratio. The N_2O concentration at the surface can now be measured with a relative precision of less than 0.5% (Weiss, 1981) and to an absolute accuracy of 3%. Because of its long lifetime, N_2O in the troposphere is well mixed, to within 5 ppbv, i.e., 1.7% of its mean (World Meteorological Organization, 1981). This range is explainable in terms of dynamics and variations in strong local sources and sinks (Levy et al., 1979).

Adequacy and Availability of Data Base. Measurements of N_2O exist since 1961, with more extensive measurements beginning in 1976. Several groups have undertaken programs to monitor tropospheric N_2O (Weiss, 1981; Pierotti and Rasmussen, 1977, 1978). Weiss (1981) reports measurements at three NOAA Geophysical Monitoring for Climatic Change (GMCC) sites (Mauna Loa, Point Barrow, and South Pole) plus ship data.

Spatial Coverage and Resolution of Additional Measurements Required. The surface sources and sinks of N_2O are still not well understood. Additional research is needed, and continued surface monitoring should be supported.

Frequency of Measurements Required. A monthly record at several sites in each hemisphere is desirable.

Feasibility and/or Existence of Technical Systems; Continuity. Current ground-based observations should continue in order to monitor global concentrations, and satellite measurements of N_2O are likely to become more available within the next decade. The latter may be better able to determine tropospheric and stratospheric variations in N_2O.

5.3.3.1.5 Methane (CH_4)

Sensitivity. For a doubling of the present CH_4 concentrations, Wang et al. (1976) estimated a surface warming of 0.2-0.4 K, depending on model assumptions. Donner and Ramanathan (1980) and Lacis et al. (1981) calculated a warming in surface temperature of 0.3 K for the same change in CH_4 concentrations. Changes in methane concentration may also influence the global ozone distribution due to the reactivity of methane with hydroxyl radicals and other trace gases.

Rate of Change. Major sources of methane (CH_4) are believed to be mining and production of fossil fuels; anaerobic fermentation of organic material due to microbial action in rice paddies, swamps and marshes, tropical rain forests, and tundra; and enteric fermentation in mammals and the activity of termites. The primary atmospheric sink is by reaction with hydroxyl radical (OH) in the troposphere and stratosphere. Methane degradation is an important source of atmospheric carbon monoxide. Water vapor and odd hydrogen species (OH, HO_2) are important products of CH_4 oxidation in the stratosphere. The atmospheric lifetime of methane is approximately 7 years. Current tropospheric concentrations are about 1.7 ppmv. Methane concentrations have increased 1-2% per year since careful observations intended to determine presence of a trend were started in 1978 (Rasmussen and Khalil, 1981; see Machta, this volume, Chapter 4, Section 4.3).

Signal-to-Noise Ratio. At this time it is difficult to assess the extent to which the recorded increase represents a short-term fluctuation in the methane cycle or a long-term trend. Historical studies of methane concentrations are desirable.

Methane can be measured with a precision of 0.01 ppmv and an accuracy within a few percent. CH_4 is reasonably well mixed in the troposphere but decreases in concentration with altitude in the stratosphere owing to reaction with OH and other radical species. More CH_4 is found in the northern hemisphere than in the southern hemisphere.

Adequacy and Availability of Data Base. Regular measurements of CH_4 have been made for only a few years.

Spatial Coverage and Resolution of Additional Measurements Required. A few sites in each hemisphere should be adequate, if well chosen. The existing programs may be sufficient, although additional measurements may be needed for a better understanding of surface sources and sinks of CH_4 and variations with altitude.

Frequency of Measurements Required. A monthly record at several sites in each hemisphere is desirable.

Feasibility and/or Existence of Technical Systems; Continuity. Ground-based measurements may need to be expanded to improve monitoring of globally averaged tropospheric concentrations. Satellite measurements may become available within a decade.

5.3.3.1.6 Chlorocarbons (e.g., $CFCl_3$, CF_2Cl_2)

Sensitivity. A number of chlorocarbons and chlorofluorocarbons have strong IR bands (Ramanathan, 1975; Wang et al., 1976). Ramanathan (1975) estimated that increasing the concentrations of both $CFCl_3$ and CF_2Cl_2 to 2 ppbv could raise the surface temperature by 0.9 K. For the same chlorocarbon abundances, Lacis et al. (1981) calculated a surface temperature change of 0.65 K. Other chlorocarbons with known absorption features in the same region are CCl_4, $CHCl_3$, CH_2Cl_2, and CH_3Cl. Chlorocarbon emissions are also of concern because they dissociate in the stratosphere, and the resulting ClO_x species might significantly affect stratospheric ozone concentration through ozone-destroying catalytic reactions.

Rate of Change. Large quantities of these industrially produced chemicals are made for a variety of uses, such as solvents, refrigerants, and spray-can propellants. The chlorocarbons of primary current concern are $CFCl_3$ (also referred to as CFC-11) and CF_2Cl_2 (CFC-12) because of large amounts produced, long atmospheric lifetimes, and increasing stratospheric and tropospheric concentrations. Concentrations of these two chlorocarbons continue to increase by approximately 10% per year, although rates of emission have not increased significantly since 1976.

The lifetime of $CFCl_3$ and CF_2Cl_2 are approximately 60 years and 100 years, respectively. These species are well mixed in the troposphere. Once in the stratosphere, these species eventually photodissociate or sometimes react with excited oxygen atoms. Some of the other chlorocarbons react with hydroxyl in the troposphere. These species have much shorter tropospheric lifetimes.

Signal-to-Noise Ratio. $CFCl_3$ and CF_2Cl_2 are well mixed in the troposphere with mixing ratios of approximately 190 and 320 pptv, respectively. Estimated precision for $CFCl_3$ and CF_2Cl_2 measurements are 2-5% (World Meteorological Organization, 1981). Ten percent is the estimated accuracy for measurements of $CFCl_3$ and CF_2Cl_2, with the limit of detection being about 1 pptv. Approximately 10% larger concentrations of $CFCl_3$ and CF_2Cl_2 exist in the northern hemisphere than in the southern hemisphere. Errors are larger for other chlorocarbons.

Adequacy and Availability of Data Base. The present programs sponsored by the federal government and by the Chemical Manufacturers Association, if continued, are probably adequate to monitor the increase in the most important of these species. Additional data would be useful for such other chlorocarbons as CH_3CCl_3, $CFCl_2$, CF_2Cl, and CCl_4.

Spatial Coverage and Resolution of Additional Measurements Required. Because radiatively important chlorocarbons are long-lived species and relatively well mixed in the troposphere, measurements at only several sites in each hemisphere are required. This can be accomplished by assuring that the record from sites now sponsored for research purposes (or similarly located sites) is continued for monitoring purposes.

Frequency of Measurements Required. A monthly record of the concentration of a number of chlorocarbons at several sites in each hemisphere is desirable. Because of the accuracy required to detect changes in chlorocarbons with time, additional sites would have to be chosen carefully, so that long records can be developed.

Feasibility and/or Existence of Technical Systems Required; Continuity. Present ground-based measurements are probably adequate for determination of global concentrations. Satellite measurements may come available within the next decade.

5.3.3.1.7 Stratospheric Ozone

Sensitivity. Stratospheric ozone can affect climate through its influence on dynamic and radiative coupling mechanisms between the stratosphere and troposphere. The absorption of solar radiation by stratospheric ozone is primarily responsible for the increase in stratospheric temperatures with altitude and is thus linked to the dynamics of the stratosphere. Bates (1977) and Geller and Alpert (1980) examined the effect on tropospheric planetary waves from changes in the zonal mean wind and temperature structure in the stratosphere. Bates (1977) found that dramatic changes in the northward flux of sensible heat resulted from changes in stratospheric structure. Geller and Alpert (1980) found that the stratospheric structure had to be altered below about 35 km before any significant changes in the structure of tropospheric planetary waves resulted. Many uncertainties remain in determining the effect of changes in stratospheric ozone on tropospheric dynamics. However, the modeling studies described above suggest that planetary wave coupling may link tropospheric weather and climate to changes in the stratosphere.

A reduction in stratospheric ozone can modify the surface temperature through two competing radiative processes (solar and long wave). With less ozone, more solar radiation is transmitted through the stratosphere, thereby enhancing solar heating of the troposphere and Earth's surface. On the other hand, the reduced absorption of solar radiation in the stratosphere cools the stratosphere, thereby reducing long-wave emission from the stratosphere downward into the troposphere by all long-wave emitting species. The changes in solar and long-wave fluxes have opposing effects on surface temperature. Ramanathan et al. (1976) calculated a surface cooling of 0.1 K for a 10% uniform reduction in stratospheric ozone. However, Ramanathan (1980) has shown that changes in the vertical ozone distribution can have a significant effect on surface temperature, even if the total ozone column does not change. Since the change in ozone concentration with altitude differs, depending on the source of this perturbation (e.g., CFCs, NO_x), the change in surface temperature may differ depending on the perturbation even if the change in total ozone is the same (irrespective of the radiative effect of the perturbing species).

Rate of Change. Several analyses have been made of measurements of total ozone from 37 Dobson stations. The studies by Bloomfield et al. (1981), St. John et al. (1981), and Reinsel (1981) give the following 95% confidence intervals for global increase in total ozone in the 10-year period 1970-1979:

Bloomfield et al.	(1.7 $\pm$ 2.0)%
St. John et al.	(1.1 $\pm$ 1.2)%
Reinsel	(0.49 $\pm$ 1.3)%

Total ozone variations differed significantly with geographical location of the instrument. Total ozone increased at some stations and decreased at others.

Satellite measurements of ozone concentration in the upper stratosphere between 1970 and 1979 indicated a reduction of up to 0.46%/yr at altitudes between 33 and 43 km (Heath and Schlesinger, 1982). The satellite data are not consistent concerning whether ozone increased or decreased at higher altitudes. Recent analyses of surface-based Umkehr data are consistent with a decrease in upper stratospheric ozone of 0.3-0.4%/yr during the 1970s (Reinsel et al., 1983).

Signal-to-Noise Ratio. There is a great deal of variation in the ozone record. Ozone concentrations fluctuate on a variety of spatial and temporal scales owing to natural causes; these fluctuations tend to mask possible systematic changes due to manmade perturbations. Observed ozone changes averaged over the northern hemisphere and the world suggest that a time interval of approximately 10 years is the shortest period that is meaningful for calculating ozone trends. Ozone varies with the 11-year solar cycle, and there is also a strong quasi-biennial oscillation (World Meteorological Organization, 1981) that affects trend analyses for shorter averaging times.

Some variation is attributed to differences in instruments, in instrument operators, or in adjustment/calibration procedures. Those Dobson stations with the longest records (several decades) are generally considered to be the most reliable. Unfortunately, most of the stations began operating after 1957. The noise in the data contributes to the uncertainty in the trend analyses, which is considerable (as indicated above).

Satellite records are relatively short, with the first ozone measurements beginning in mid-1970.

Adequacy and Availability of Data Base. An extensive global network of stations measuring total ozone exists. However, most of the observing stations are on continents and in the northern hemisphere. Satellite measurements providing more uniform spatial sampling could eventually lead to better trend measurements. Also, since only one instrument is used, the present intercalibration errors between surface devices are eliminated.

Recent modeling studies (e.g., Wuebbles et al., 1983) suggest that total ozone measurements may not be a sensitive indicator of the impact of human activities on the global atmosphere. These studies suggest that monitoring of changes in the distribution of ozone with altitude should be more useful. However, major uncertainties exist in both the existing Umkehr network and satellite measurements. Only about 18 stations currently make regular Umkehr measurements; only 3 of these stations are in the southern hemisphere. In addition to calibration drift, the Umkehr method is also subject to errors from the effect of dust and aerosols and the inability to make measurement under cloudy conditions. As with the total ozone measurement, the satellite data are limited by the short duration of continuous and homogeneous global data coverage.

The ozone data are available from several central sources. The Dobson data are compiled and checked prior to public release by the Center for Ozone Data for the World located in Toronto, Ontario, Canada. The data are usually available about a year after they are taken by the Dobson stations.

NASA satellite data are available from the National Space Science Data Center at the NASA Goddard Space Flight Center, Greenbelt, Maryland. Data from the NOAA satellites are not available from this center and must be obtained directly from NOAA. It usually takes more than a year for the satellite data to be processed, reviewed, and released to the Data Center.

Spatial Coverage and Resolution of Additional Measurements Required. The ozone data based on Dobson and Umkehr stations have limited geographic coverage. Most stations are located in continental areas at middle latitudes in the northern hemisphere. Consequently, it is difficult to estimate global total ozone when there are vast areas of the world (mostly in the southern hemisphere) where measurements are not taken. The existing total ozone and ozone distribution networks need to be enlarged, particularly by adding stations in the southern hemisphere.

The spatial coverage on ozone from satellites is very good, but the satellite data are limited in temporal coverage. Global satellite measurements of ozone began in 1970, and the longest single record of data is 7 years. Although a continuous ozone record is available owing to overlap of various satellite systems, there is a significant bias between systems that makes ozone trend analysis using satellite data difficult.

Frequency of Measurement Required. Because of the large temporal and spatial variations of ozone, frequent measurements are needed. With a limited surface network, daily measurements of ozone should be taken (and usually are). The best stations take several measurements each day to monitor diurnal variations and to reduce the noise in the measurements.

Satellite systems monitor stratospheric ozone continuously with the objective of covering the entire global area at least once each day. Solar UV instruments usually are in polar orbits and take measurements at the same local time at each geographic location. IR instruments are capable of taking day and night measurements. Because of the operating cycle of satellite systems, a satellite usually has gaps in its temporal and spatial coverage. Daily measurements are adequate for obtaining monthly-mean distribution of ozone.

Feasibility and/or Existence of Technical Systems; Continuity. The Dobson instruments are considered to be excellent in quality and are often used as a reference for comparison with satellite measurements. NOAA, in cooperation with Environmental Protection Agency and the Chemical Manufacturers Association, is currently developing a new Umkehr network with improved instrumentation and extensive intercalibration program. Improvements are being made in satellite instrumentation, so the data have improved in spatial resolution and in consistency with other ozone data. Future NOAA satellites will carry operational ozone-measuring instruments.

5.3.3.1.8 Tropospheric Ozone

Sensitivity. Although only 5-10% of the total column amount of ozone is in the troposphere, a uniform percentage change in tropospheric ozone can have about the same effect on surface temperature as the same percentage change in stratospheric ozone. This results from the fact that owing to pressure broadening of the lines in the 9.6-μm band of ozone, the total long-wave opacity of tropospheric ozone is nearly the same as that of stratospheric ozone. The solar effect of a change in tropospheric ozone is different than that of stratospheric ozone in that both the solar and long-wave effects are in the same direction in this case. An increase in tropospheric ozone increases solar absorption plus enhances the long-wave "greenhouse" effect of ozone; both effects tend to increase surface temperature. Fishman et al. (1979) calculated that a doubling of tropospheric ozone concentrations would lead to an increase in surface temperature of 0.9 K.

Rate of Change. An analysis of ozonesonde data (World Meteorological Organization, 1981) shows that tropospheric ozone in the layer from 2 to 8 km increased in northern middle latitudes by about 7% during the last decade. An increase has also been observed in ozone in the layer between 8 and 16 km, but the magnitude is less than that for the 2- to 8-km layer. The cause of this increase is not known, but it is consistent with model calculations that predict an increase in upper tropospheric ozone due to increased NO_x emissions from aircraft engines. An increase in tropospheric ozone is not observed at all latitudes nor at all stations in middle latitudes. Some model calculations also suggest large photochemical production and destruction rates for ozone in the troposphere, implying night-day shifts in concentrations.

Signal-to-Noise Ratio. The annual-mean values of the ozonesonde measurements in the 2- to 8-km layer have an uncertainty of ±2-3%. Ozonesonde measurements in the troposphere have higher relative accuracy than in the stratosphere. As the ozonesonde reaches high altitudes, pump correction factors have to be applied to the measurements to account for reductions in pump efficiency.

Adequacy and Availability of Data Base. Both Umkehr and ozonesonde data bases are limited in spatial coverage. Although Umkehr measurements are made of tropospheric ozone, the Umkehr technique is more accurate for ozone in the upper stratosphere. While ozonesonde measurements have high vertical resolution, there are only a small number (<20) of stations. Only one ozonesonde station is in the southern hemisphere. Many ozone measurements are taken near the ground (in the boundary layer) in urban areas to monitor pollution. To study climatic effects, measurements are needed of the ozone distribution through the troposphere.

Spatial Coverage and Resolution of Additional Measurements Required. The ozonesonde data give detailed information about the vertical profiles of ozone through the troposphere. Because these balloon measurements are expensive, the number of launches from each station per year is limited. Budget cutbacks have led to a reduction in the number of operating stations. Since a long data record is needed for trend analysis, keeping or restoring the ozonesonde network to its peak level is important. Although adding additional stations would be desirable, maintaining what has been operational is a reasonable objective during this time of limited resources.

The Umkehr network provides an independent check on the ozonesonde measurements, and vice versa. Because Umkehr measurements are made with ground-based instruments, there is a lower cost per measurement, so the frequency of measurements is greater than that of the ozonesonde network.

Frequency of Measurement Required. Monthly mean profiles of ozone are needed for trend analysis. Consequently, daily measurements are desirable. This is feasible for the Umkehr network, but costs prohibit ozonesonde measurement at this frequency. At least one ozonesonde measurement per week would be desirable at each location.

Feasibility and/or Existence of Technical Systems; Continuity. The spatial and temporal coverage of the existing ozonesonde network is limited, and the network should be maintained and expanded, if possible. New, highly automated, ground-based ozone monitoring instruments have been designed, allowing greater frequency of measurements and less dependence on operator skill. Because the Dobson instruments are well established, there will be some reluctance to replace them with the new instruments until the quality and reliability of the instrument is fully demonstrated and adequate overlap calibrations are available and have been carefully studied.

5.3.3.2 Atmospheric Parameters

The atmosphere is the part of the climate system with the least thermal inertia and, therefore, the part that can respond most quickly to the radiative effects of increasing CO_2 concentrations. Climate models have focused most attention on the changes that increasing CO_2 concentrations will induce in the atmosphere, and it is therefore quite appropriate to look first to the atmosphere for any evidence of CO_2-induced climate changes.

Since the greenhouse effect on the radiation balance of the atmosphere affects temperature directly, a first change to look for is in the temperature field. Global and hemispheric mean surface temperatures have been studied extensively in this regard--regional temperatures cannot be considered as sufficiently representative. Arctic and

Antarctic temperatures may respond more than the global mean, but unfortunately the variability of polar temperatures from year to year is also larger, so it is unclear whether the signal-to-noise ratio will be less there (Kelley and Jones, 1981). Similarly, at middle and high latitudes wintertime surface temperatures may show a slightly larger increase than summertime temperatures, but the variability of wintertime temperatures, or noise, is more than twice as large as that in summer. This suggests that summertime temperature response may have a larger signal-to-noise ratio. It will also be important to distinguish between temperatures over land and over the ocean.

It has been argued that temperature averaged over the lower troposphere, rather than surface temperature, may be more representative of changes in the radiation balance because of the greenhouse gases. Unfortunately, good upper-air coverage does not go back beyond about 1950, and even now it is deficient over the oceans. So far, satellite indirect soundings have not been used to fill in the gaps in the climatological record. The situation is worse in the stratosphere where the global rocketsonde sounding system has been in operation for only a relatively short time and is on the point of being dissolved altogether. In view of the difference between tropospheric and stratospheric temperature responses to increased CO_2, both stratospheric and tropospheric temperatures should be monitored.

Atmospheric fluxes of radiation should also be monitored. The monitoring of upward shortwave (reflected and scattered solar) radiation and emitted infrared radiation at the top of the atmosphere deserves special attention. Satellites are potentially powerful tools for monitoring these parameters. However, interpretation of most satellite measurements to the accuracy required to explain temperature changes of a few tenths of a degree appears difficult, as this requires that absolute changes in the radiation budget of the order of 0.2% over a period of years be detected. Detailed analysis of the spectrum of outgoing infrared radiation may, however, provide useful information.

Clouds may also influence long-term climate variations through their radiative effects. To illustrate their potential importance, changes in global cloud cover of a few percent could mask the warming effect of doubled carbon dioxide. Unfortunately, it is difficult to determine with confidence the three-dimensional distribution of clouds because of their ill-defined boundaries and complicated configurations. Instead, one can monitor the equivalent blackbody temperature of the cloud ensembles by measuring their outgoing radiation in one or more atmospheric "windows." From this it should be possible to calculate not only the upward flux of thermal radiation but also the radiatively effective cloud height when the vertical distribution of temperature in the atmosphere is given. This, and knowledge of the planetary albedo, allows one to evaluate important aspects of the influence of the cloud cover on the radiation balance of the Earth-atmosphere system.

Finally, the precipitable water content of the atmosphere will change with CO_2 increases. Model calculations indicate that the atmospheric precipitable water will increase 5-15% as the climate warms in response to doubled CO_2 concentrations (Manabe and Stouffer, 1980; Wetherald and Manabe, 1981). Observations of this parameter may, therefore, help confirm whether the numerical models are properly simulating the role of water-vapor processes in contributing to climate change, although changes of precipitable water in the near future will, of course, be much less than for doubled CO_2.

5.3.3.2.1 Global Mean Surface Air Temperature

Sensitivity. When it is assumed that the CO_2 content of the atmosphere is doubled and statistical thermal equilibrium is achieved, the more realistic of climate modeling efforts predict a global surface warming of between 2 and 3.5°C, with greater increases at high latitudes; if one allows for more feedback mechanisms, a range of 1.5 to 4.5°C is suggested (Climate Research Board, 1979; CO_2/Climate Review Panel, 1982). The results from climate models also suggest that the CO_2-induced warming is approximately proportional to the increase of the logarithm of the CO_2 concentration in the atmosphere.

Rate of Change. The large thermal inertia of the oceans will probably delay the response of the atmospheric temperature to an increase of the CO_2 concentration in the atmosphere. Since the CO_2-induced warming penetrates deeper into the ocean as time passes, the effective thermal inertia of the ocean depends on the time scale under consideration. Therefore, the length of the delay in the climatic response depends on the rate of the CO_2 increase and can change with time. By use of a simple model of the atmosphere-ocean system, Hoffert et al. (1980) recently estimated the temporal variation of the atmospheric temperature in response to a predicted increase of the CO_2 concentration. According to their results, the delay of response is approximately 10-20 years during the period of A.D. 1980-2000. Transient responses may also vary zonally and regionally because of the ocean's thermal inertia (Schneider and Thompson, 1981; Thompson and Schneider, 1982).

Signal-to-Noise Ratio. It is expected that CO_2-induced change of surface air temperature will be positive over most of the Earth's surface, whereas the trend of the natural temperature variation changes sign from one geographical location to another. Therefore, the signal-to-noise ratio of area-averaged surface air temperature should increase as the area for the averaging increases. For this reason, the global mean surface air temperature is one of the most promising quantities for the early detection of the CO_2 climate signal.

According to the recent study of Jones et al. (1982), the standard deviation of the annually averaged area mean temperature over the entire northern hemisphere is about 0.86°C, while surface temperature anomalies for the northern hemisphere are typically $\pm$0.2-0.4°C (Clark et al., 1982). Depending on assumptions of temperature sensitivity to increases in CO_2 concentration and ocean thermal lag, an increase of atmospheric CO_2 concentration to between 400 and 450 ppm should raise the global average surface air temperature distinctly beyond the expected natural fluctuations of surface temperature.

Adequacy and Availability of Past Data Base. Long, reliable, and representative temperature records are needed. Continuing analysis and estimates of error of the temperature record are called for; dissemination of relevant records is generally good. Analyses of the temporal variation of the annually averaged global (or hemispheric) mean surface air temperature during the past 100 years have been made by many authors by use of instrumental data (see Section 5.2.2.1), with significantly differing results.

Spatial Coverage and Resolution of Additional Measurements Required. Estimates of hemispheric (or global) mean surface air temperature are subject to large errors because of the sparseness of the observational network, particularly over the oceans. Improved observational coverage would not only improve the representativeness of hemispheric or global averages but would also permit monitoring of regional averages and differentiation between changes over ocean and over land. For example, model simulation (e.g., Manabe and Stouffer, 1980) indicate larger CO_2-induced temperature changes in polar regions than in lower latitudes, and transient responses over the ocean may be expected to differ from those over land (Schneider and Thompson, 1981; Ramanathan, 1981). For these reasons, the observational network of surface air temperature measurements employed in monitoring and detection studies should be supplemented by determinations of sea-surface temperatures from satellite observations, despite the fact that sea-surface temperature is not identical with surface air temperature.

Frequency of Measurements Required. One or preferably two observations per day will suffice.

Feasibility. Satellite measurements are currently available. However, further improvement of the accuracy of sea-surface temperature determination is required in order to improve estimates of surface air temperature.

5.3.3.2.2 Tropospheric Temperature Distribution

Sensitivity. The CO_2-induced change of tropospheric temperature has been estimated by use of general circulation models of climate (e.g., Manabe and Wetherald, 1975; Manabe and Stouffer, 1980). According to these studies, the warming of the surface layer of the atmosphere is expected to be particularly large in high latitudes (i.e., 2-3 times larger than the increase of global average temperature). The high-latitude warming decreases sharply with increasing altitude. In low latitudes, the CO_2-induced warming is predicted to be somewhat smaller than the global average. The warming of the tropical troposphere will increase significantly with increasing altitude.

Rate of Change. Since the troposphere is closely coupled to the Earth's surface through dry and moist convective transfer of heat, the response time of the tropospheric temperature is not very different from that of global surface air temperature discussed in the preceding subsection. It is, however, probable that the response time of surface air temperature over continents is somewhat shorter than that over oceans because of the differences in effective thermal inertia between ocean and continent. The response time may also have latitudinal variation because the speed of the penetration of thermal anomaly into the ocean depends on latitude.

Signal-to-Noise Ratio. The signal-to-noise ratio for the CO_2-induced warming of zonal mean surface air temperature was recently estimated by Madden and Ramanathan (1980) and Wigley and Jones (1981). The results from these studies suggest that the signal-to-noise ratio of the CO_2-induced warming of zonal mean surface air temperature is at a maximum in middle latitudes in summer when the magnitude of natural temperature fluctuation is relatively small. It is desirable to conduct similar analyses for the zonal mean temperature of the troposphere.

Adequacy and Availability of Past Data Base. Tropospheric temperatures have been observed by radiosondes on a large scale since the Second World War. More recently, satellites have provided a dense coverage of remotely sensed soundings, although their accuracy is inadequate to determine trends. Angell and Korshover (1978a) estimated the temporal variation of the globally averaged temperature in the surface to 100-mbar layer for the period from 1958 to 1976. They used radiosonde data from 63 stations that are fairly evenly distributed. The recent compilation of upper-air data by Oort (1982) should also prove valuable for the determination of the past trend of tropospheric temperature.

Spatial Coverage and Resolution of Additional Measurements. The coverage over the oceans of radiosonde observations of atmospheric temperature is sparse. To overcome this deficiency, the network of radiosonde observation should be augmented with satellite measurements of infrared or microwave radiances (to be converted to temperature through appropriate inversion techniques).

A spatial resolution of 250 km is desirable.

Frequency of Measurements Required. Two observations per day are desirable.

Feasibility and/or Existence of Technical Systems; Continuity. Currently, radiance data from satellites are routinely inverted to vertical temperature distributions of the atmosphere, which are used as input data for the daily numerical weather forecasts. These data might be used for the long-term monitoring of tropospheric temperature; careful assessment of the accuracy of current and proposed temperature determination from satellites should be made in the light of expected CO_2-induced change of climate.

5.3.3.2.3 Stratospheric Temperature Distribution

Sensitivity. Results from both radiative-convective models and general circulation models indicate that a cooling of the stratosphere could result from an increase of the CO_2 concentration in the atmosphere. This cooling contrasts with the warming expected to occur in the troposphere. Therefore, the simultaneous monitoring of temperature not only of the troposphere but also of the stratosphere should yield valuable information for distinguishing CO_2-induced climate change from climate change caused by other factors.

According to Fels et al. (1980), cooling due to the doubling of the atmospheric CO_2 concentration would be about 7°C and 11°C at altitudes of 30 km and 45 km, respectively.

Rate of Change. Since the thermal coupling between the stratosphere and troposphere is relatively weak, the response time of stratospheric temperature is not prolonged by the large thermal inertia of oceans as is the case with respect to the tropospheric temperature. According to results from time integration of a one-dimensional, radiative-convective model of the atmosphere, the response time (e-folding time) of the stratosphere is on the order of a few months (see, for example, Manabe and Strickler, 1964).

Signal-to-Noise Ratio. The standard deviation of natural temperature fluctuation in the stratosphere has a large seasonal variation, being small in summer and large in winter. In contrast, the CO_2-induced change of stratospheric temperature estimated by model experiments has relatively small dependence on season. Therefore, it is expected that the signal-to-noise ratio of the CO_2-induced change of stratospheric temperature will be at a maximum in summer. In assessment of signal-to-noise ratio, it is necessary to know the standard deviations of the temperature fluctuation in the middle and upper stratosphere.

Adequacy and Availability of Past Data Base. Recently, Angell and Korshover (1978b) estimated global temperature variation in the 100-30-mbar layer between 1958 and 1977. It is desirable to extend similar analyses to the 10-mbar level where the CO_2-induced cooling is expected to be substantial. In addition, the data from chopper radiometer measurements from satellite and rocketsonde observations of temperature in the stratosphere and mesosphere should be analyzed in order to determine the recent trend of temperature in the middle atmosphere.

Spatial Coverage and Resolution of Additional Measurements Required. It is recommended to monitor the future variation of temperature in the middle and upper stratosphere where the magnitude of the CO_2-induced cooling is expected to be large. A spatial resolution of 500 km is desirable. This could be accomplished by observation of stratospheric temperature from satellite (e.g., by limb measurements), augmented by observations by radiosonde and rocketsonde. It is expected that these observations will yield a global distribution of stratospheric temperature at various altitudes.

Frequency of Measurements Required. One observation per day is adequate.

Feasibility and/or Existence of Technical Systems; Continuity. Current observation by radiosonde and satellite should be continued. In addition, rocketsonde observation of stratospheric temperature should be continued at least at a few key locations in order to calibrate the temperature determined by inversion technique from radiance data obtained from satellites. Careful assessment of the accuracy of the determination of stratospheric temperature from satellites should be made.

5.3.3.2.4 Upward Terrestrial Radiation and Reflected Solar Radiation at the Top of the Atmosphere

Sensitivity. Infrared emission to space and reflected solar energy from Earth depend primarily on clouds and surface characteristics. Models indicate that total precipitable water will increase, sea-surface temperature will increase, sea ice and snow cover will change, and stratospheric temperature will decrease as CO_2 rises. Climate models cannot reliably tell us, at present, the impact of CO_2 change on cloud amount, height, and type, although we know that key parameters for cloud formation, such as the water content of the atmosphere and the atmospheric lapse rate, will change.

Rate of Change. Since most solar energy is absorbed in the oceans, the response time is linked to the thermal inertia of the oceans. Regional feedbacks, especially over continents, could be possible in terms of time periods as short as seasons or months. The secular rates of change of these two "top-of-the-atmosphere" parameters is likely to be very low. An annual cycle of the two radiation parameters has been observed on a global scale (Ellis et al., 1978) and is under study. The CO_2 impact on the amplitude and phase of these annual cycles should be studied as well.

Signal-to-Noise Ratio. Natural variability of the emitted and reflected radiation on seasonal time scales has been estimated by Stephens et al. (1981). In addition to the annual amplitudes, the tropical latitudes have a semiannual wave forced by Earth-sun geometry, and certain regions have quasi-random variations in emitted and reflected radiation likely due to cloud-cover changes. For the globe as a whole, the annual cycle of emitted IR radiation is ± 4 W/m^2 and for albedo it is $\pm 1.5\%$. For regions of the Earth the natural "noise" is large, with albedo variations of $\pm 15\%$ (absolute units) about a mean of 50% common in northern polar regions, for example. The CO_2-induced signal is currently unknown but is likely to be small in the next decade or two, suggesting a low signal-to-noise ratio.

Adequacy and Availability of Data Base. The existing record of Earth radiation budget measurements begins in the 1960s with intermittent data from various satellite experiments until 1975. Then, the Nimbus 6 and Nimbus 7 experiments provide a continuous data set to the present. If Nimbus 7 lasts two more years, it will overlap with the improved ERBE experiment planned for 1984-1986. These data are adequate for the

CO_2-related task. However, it is urgent to ensure continued ERBE-type measurements beyond 1986. Plans to do so have not yet been finalized.

Spatial Coverage and Resolution of Additional Measurements Required. Satellite coverage is global. A few high-quality surface radiation stations, as in the U.S. Geophysical Monitoring for Climatic Change (GMCC) network, provide ground checks to complement the satellite observations.

Frequency of Measurements Required. Daily satellite observations are desirable at as many local times as possible in order to account for diurnal effects (i.e., non-sun-synchronous satellite orbits).

Feasibility and/or Existence of Technical Systems; Continuity. The Nimbus 6 and 7 experiments demonstrated high-precision proof of concept. Data are also being provided by the AVHRR instruments on NOAA operational satellites. The ERBE will carry improved instrumentation and offer needed improvements in local time sampling and in-orbit calibration. As noted above, it is important that ERBE-type measurements be planned now to continue beyond 1986.

5.3.3.2.5 Precipitable Water Content of the Atmosphere

Sensitivity. Model calculations predict that atmospheric precipitable water will increase 5-15% as the climate warms in response to doubled CO_2 concentrations (Manabe and Stouffer, 1980; Wetherald and Manabe, 1981). The radiative effects resulting from this increase are a central aspect of the projected climatic response leading to the predicted warming due to increasing CO_2 concentrations. Observation of this parameter will, therefore, help to confirm whether the numerical models are properly simulating the role of water vapor processes in contributing to climatic change.

Rate of Change. By the Clausius-Clapeyron equation, the fractional change in saturation water vapor mixing ratio (and therefore in precipitable water vapor if relative humidity remains constant, which is usually assumed) is approximately proportional to the fractional change in atmospheric temperature. Since most water vapor is present in the troposphere, the response time for the increase in water vapor should be about the same as for tropospheric temperature (except for the possible influence of year-to-year water storage in the land surface layers). If constant relative humidity is assumed, the response times would be identical.

Signal-to-Noise Ratio. The intra-annual range in 5-year averages of mean monthly northern hemisphere specific humidity is from 2.17 g/kg in February to 3.81 g/kg in August (Oort and Rasmusson, 1971). The inter-annual variation in annual average northern hemisphere specific humidity is about 0.05 g/kg (Oort, 1982). Climate models project that a doubling of CO_2 concentrations would increase the specific humidity by about 0.3 g/kg (Manabe and Wetherald, 1980), so that the effect should clearly become evident. For a more complete analysis of signal to noise, information on the latitudinal distribution of the expected signal from models and noise from observations is needed.

Adequacy and Availability of Past Data Base. The new 15-year global data base of radiosonde-derived atmospheric specific humidity by Oort offers the potential to develop important baseline data. Development of a global or hemispheric data base extending back before about 1950 is doubtful, however, owing to the limited number of vertical profiles. Because the vertical soundings are taken on a fixed 12-hourly (or 24-hourly) basis only at radiosonde stations, adequate definition of the global integral may not be possible; however, the relatively long lifetime of atmospheric water (~10 days) and recent evidence that the radiosonde network temperature record is in quite good agreement with the more extensive surface network may ameliorate these difficulties and permit at least the detection of changes.

Spatial Coverage and Resolution of Additional Measurements Required. A data set based on satellite measurements of atmospheric water vapor (whether direct or indirect) would be useful in order to provide adequate global coverage and to reduce variations introduced because of sampling errors in the present surface network. Prabhakara et al. (1982), for example, have developed a global map of precipitable water based on Nimbus 7 microwave measurements. Monthly and latitudinal averages should be developed since model results will probably indicate that the fractional increases in precipitable water will be a function of latitude.

An alternative approach to direct or indirect measurement of water-vapor amount would be to measure the changes in the radiative flux expected to result from the projected changes in CO_2 and, primarily, water-vapor amount. At the surface, downward infrared radiation is projected to increase 15-20 W/m^2 when climate has reached equilibrium after a doubling of CO_2 concentrations; this is about a 5% increase in total downward radiation. At the top of the atmosphere, total upward infrared radiation will likely change only about 1% under similar conditions in response to small planetary albedo changes. To improve the signal-to-noise ratio, consideration of expected changes in the spectral distribution of the infrared radiation must be undertaken. Here the operational AVHRR on current NOAA satellites and other satellite programs may provide some useful information. Until detailed investigations are undertaken, however, the preferred option is to monitor atmospheric water vapor amount directly (from radiosondes and/or high-resolution spectral measurements on satellites).

Frequency of Measurements Required. Measurements once or twice daily should be sufficient to develop monthly averages. Special care will have to be taken, however, so that account is taken of the presence and extent of clouds in the soundings.

Feasibility and/or Existence of Technical Systems; Continuity. The continuation of the radiosonde network is assured for the purposes of weather prediction. Surface-based microwave radiometry may provide additional data in the future (Skoog et al., 1982). Improved coverage over the oceans, particularly in the southern hemisphere, would be helpful. Nimbus 7 and Seasat have demonstrated the feasibility of satellite water-vapor measurements over the ocean (Prabhakara et al., 1982; Alishouse, 1983). Research into satellite detection systems is currently under way and should be continued. Improvements will likely require either detailed spectral measurements by new satellites or clever computation of the ratios of broadband fluxes now being measured (e.g., Rosenkranz et al., 1982). Compilation of the integral water vapor amounts used to initialize daily global weather forecasts should be started and compared with integrals based on the radiosonde network. Monitoring of precipitable water content will be a challenging task, but we believe that the problem should be addressed; an approach commencing with a feasibility study, evaluation of instrument errors, and similar questions is appropriate.

5.3.3.2.6 Equivalent Emission Temperature (Cloudiness)

Sensitivity. One of the important factors that can control the long-term variation of climate is cloud cover. For example, an increase in cloud cover exerts two opposing influences on climate. On one hand, it tends to produce cooling of the Earth by reflecting a large fraction of insolation. On the other hand, it contributes to the warming of the Earth by reducing the outgoing terrestrial radiation at the top of the atmosphere. Small percentage changes in global cloudiness could thus accentuate or counteract a CO_2-induced warming, especially at the regional scale. Unfortunately, it is difficult to determine with confidence the needed information on the three-dimensional distribution of cloud cover, because cloud cover often has ill-defined boundaries and complicated configuration. As an alternative to direct monitoring of cloud cover, one can monitor the equivalent temperature for the upward window radiation. From equivalent emission temperature it is possible to compute both the upward flux of blackbody radiation from a cloud top and the effective cloud height when the vertical distribution of temperature in the atmosphere is given. By monitoring the long-term variation of both effective cloud height and planetary albedo, one can evaluate the net effect of the changes in cloud cover and optical properties of the Earth's surface.

Rate of Change. Changes should occur in close association with CO_2 and climate changes.

Signal-to-Noise Ratio. Since the temperature change for doubled CO_2 is expected to be about 3°C, one might estimate that global mean equivalent emission temperature should be measured to an accuracy of a few tenths of a degree. It has not been possible, however, to obtain a reliable estimate of the signal-to-noise ratio for the CO_2-induced change of the equivalent emission temperature (or effective height of the source of upward window radiation) for the following reasons: (a)

In view of poor performance of current general circulation models of the atmosphere in simulating the global distribution of cloud cover and its seasonal variation, it is premature to trust the CO_2-induced changes of cloud cover and effective emission temperature as determined by such models. (b) The analysis of the natural variability of effective emission temperature for upward window radiation is not available.

Availability of Past Data Base. Over a decade of full-disk geostationary satellite visible images has been archived at the University of Wisconsin. Window radiance data from NOAA operational satellites are routinely archived by the National Environmental Satellite Service, which regularly publishes analyses including both images and tabulated data.

Spatial and Temporal Resolution of Additional Measurements Required. The spatial and temporal resolution of future satellite observation should be determined in the light of careful assessment of sampling error involved in the time (or space) averaging of window radiance as obtained from a satellite. The past observation of this variable by a geosynchronous satellite provides ideal data for such an assessment.

Frequency of Observations Required. The frequency of observations will be determined by the sampling error studies described above.

Feasibility and/or Existence of Technical Systems; Continuity. This is an area for further study. Plans and feasibility studies should be made in conjunction with the recently initiated International Satellite Cloud Climatology Project (ISCCP) (World Meteorological Organization, 1982b).

5.3.3.3 Cryospheric Parameters

The features of the cryosphere include snow, sea ice, glaciers, ice sheets, permafrost, and river and lake ice. Perennial ice at present covers about 7% of the world oceans and 11% of the land surface, almost entirely in the polar regions. Seasonal snow and ice, however, occupy 15% of the Earth's surface in January, at a time when Antarctic sea ice is near its minimum extent, and 9% in July when there is almost no snow cover in the northern hemisphere. Table 5.10 shows the global distribution of ice and snow. The volume of water locked up in the Antarctic and Greenland ice sheets could potentially raise the world's mean sea level by 77 m.

The general problems of possible snow and ice responses to a carbon dioxide-induced warming were reviewed by Barry (1978, 1982) and Kukla (1982) in terms of the sensitivity of individual components of the cryosphere. The response of snow and ice covers to climatic change varies greatly in terms of time scale. Typical residence times of solid precipitation in the various reservoirs are approximately 10^{-1} to 1 year for seasonal snow cover, 1-10 years for sea ice, and 10^3-10^5 years for ground ice and ice sheets. In each case certain

TABLE 5.10 Distribution of Ice and Snow[a]

	Area (106 km^2)	Volume (106 km^3 water)	Sea-Level Equivalent (m)
Land Ice			
Antarctica[b,c]	12.2	25	70
Greenland	1.8	2.7	7
Small ice caps and mountain glaciers	0.5	0.12	0.3
Ground Ice (excluding Antarctica)			
Continuous	7.6		
Discontinuous	17.3		1
Sea Ice			
Arctic: max.	15		
min.	5		
Antarctic: max.	20		
min.	2.5		
Total Land Ice, Sea Ice, and Snow			
Jan: N. hemisphere	58		
S. hemisphere	18		
July: N. hemisphere	14		
S. hemisphere	25		
Global mean annual	59		

[a]SOURCE: Hollin and Barry (1979).
[b]Excludes peripheral, floating ice shelves (which do not affect sea level).
[c]Roughly 10% of the Antarctic ice is in West Antarctica and 90% is in East Antarctica.

phases of the seasonal regime are particularly critical for the occurrence of snow and ice and their response to climatic variations. Of primary importance are the times of seasonal temperature transition across the 0°C threshold (-1.8°C in the case of seawater). Other threshold effects that influence radiative and turbulent energy exchanges arise as a result of the large albedo differences between snow cover (about 0.80) and snow-free ground (0.10-0.25) or between ice (0.65) and water (0.05-0.10).

Climate research indicates that doubling the CO_2 concentrations will lead to a significant reduction in the extent of snow cover and sea ice with perhaps, if the warming persists, melting and deterioration of the major polar ice sheets. Sea-ice and snow-cover extent can be monitored routinely from satellites, and it has been shown that the mass balance of the large continental ice sheets of Antarctica and Greenland could also be monitored from satellites. These three cryospheric parameters are the most promising ones to monitor. Others, like smaller glaciers and river and lake ice, for example the dates of freezeup of the latter, are in most cases probably too noisy to be used as good indicators of climate change. Several circulation models suggest that increased CO_2 concentrations will lead to winter warming in the polar regions that is several times as large as in middle and low latitudes. Cryospheric conditions, particularly those responding rapidly to climatic change, may thus be excellent early indicators of CO_2-induced effects. It is, nevertheless, important also to mention other, slower-responding phenomena, because of the environmental significance of potential changes that may be detected in them.

5.3.3.3.1 Sea-Ice Cover

Sensitivity. Model calculations by Manabe and Stouffer (1980) show sea-ice cover, both in areal extent and thickness, to be a sensitive indicator of climate change. They assume an increase of four times the present CO_2 concentration and therefore project a very large warming in the polar region. This results in the sea-ice cover of the Arctic Ocean being reduced to a seasonal ice cover, which reforms in winter. Budd (1975) calculates from empirical data in Antarctica that an annual change of 1°C in mean temperature corresponds to a 70-day variation in the duration of sea ice at the margin and a 2.5° latitude variation in maximum extent. Observations (Vinnikov et al., 1981) and paleoclimatic reconstructions using sediment data (Hays, 1978) confirm the relationship between temperature and sea-ice extent. Despite the thicker ice in the Arctic (~3 m) compared with that in the Antarctic (~2 m), empirical and modeling results seem to indicate that the ice extent responds rapidly (on a seasonal time scale) to climate changes.

Rate of Change. The natural variability of ice extent is large, and Zwally et al. (1983b) have not detected a systematic decrease of ice extent in Antarctica since 1973. Data for the Arctic also so far do not indicate any clearcut effects due to a CO_2 warming there in the past 25 years.

Signal-to-Noise Ratio. Sea-ice extent is a noisy parameter when considered over short time scales or small space scales, since it is determined by numerous environmental parameters and by both dynamic and thermodynamic processes (Pritchard, 1980). In the Antarctic, where sea ice displays a wide seasonal variation in extent, recent studies (Zwally et al., 1983b) have shown no systematic trend. Decreases between 1973 and 1980 were within 1 standard deviation of the long-term mean (Budd, 1980) and have been followed by increases since 1980. In the largely enclosed Arctic Ocean, variations in the ice extent are more limited on a seasonal basis, and ice thickness changes may be the first indicator of a climate change.

Adequacy and Availability of Past Data Base. Accurate data on the extent of sea ice in both hemispheres are limited to the satellite records of the last two decades or so, although historical data in isolated instances, e.g., Iceland (Vilmundarson, 1972) and northern Europe (Vinnikov et al., 1981) date back several centuries. Sediment core data (Hays, 1978) have extended this data base by many thousands of years. The satellite records are too short at present to determine definite trends, but continued monitoring over the next 10-15 years should establish whether incipient or proposed trends are significant. Data on sea-ice extent are also not yet archived routinely in digital form.

Spatial Coverage and Resolution of Additional Measurements Required. Satellite measurements allow routine integration of the areal extent of sea ice in both hemispheres. It would be useful also to have sea-ice thickness distributions in both polar regions, but these measurements are at present not feasible for satellites.

Frequency of Measurements Required. Weekly averages for each hemisphere, as determined at present, are adequate, from which monthly and annual means as well as maxima and minima can be derived.

Feasibility and/or Existence of Technical Systems; Continuity. Technical systems exist to carry out routine sea-ice monitoring from spacecraft. All-weather and night capability is essential, since both polar regions are dark for prolonged periods, and seasonal cloud systems are extensive. Microwave or radar systems are needed, but there is likely to be a hiatus in the launch of U.S. spacecraft with such systems that can perhaps be filled only by using European or Japanese satellites.

5.3.3.3.2 Snow Cover

Sensitivity. For snow cover, the CO_2 signal is more difficult to interpret than for sea ice, since the effects of CO_2-induced warming on snowfall and snow cover will vary with latitude. In low and middle latitudes, where the occurrence of snow rather than rain is frequently marginal, warming will decrease the frequency of the snowfall and the duration of snow cover on the ground. In high latitudes, snowfall is limited by the frequency of cyclonic incursions and the moisture content of the air, and there is a tendency for warm winters to be snowy, as for example at Barrow, Alaska (Barry, 1982). The year-to-year variability of snow cover in the northern hemisphere is large, but global warming could eliminate the occurrence of snow completely in broad areas of low snowfall frequency (Dickson and Posey, 1967), increase it at higher latitudes, but also possibly result in an overall increase in the length of the snow-free season in the higher latitudes due to warmer summer temperatures.

Rate of Change. The generally thin snow cover of the Arctic requires little energy for melting and can therefore respond rapidly to changes in the energy balance triggered by CO_2-induced warming. The duration of snow cover at high latitudes is determined primarily by summer temperatures, since the depth of snow is not highly variable from year to year (Barry, 1982). Typically, a 30-40-cm snow cover in the Arctic disappears in about 10 days from the start of melting and requires about 2-3 kJ cm^{-2} (Weller and Holmgren, 1974).

Signal-to-Noise Ratio. The signal-to-noise ratio of snow cover extent in the Arctic is likely to be high, because snowfall at high latitudes is highly variable in space and time. For example, one day's precipitation amount may be a large percentage of the total precipitation in some areas (Maxwell, 1980). As discussed above, snow cover may either increase or decrease, depending on latitude, geographical location, and change in circulation patterns caused by CO_2 effects.

Adequacy and Availability of Past Data Base. Information on the duration of snow cover and the last date of snow on the ground is available for Canada (Potter, 1965) and in maps of probability for the northern hemisphere (Dickson and Posey, 1967). Kukla (1981) and Matson and Wiesnet (1981) have recently compiled satellite information on monthly snow limits. Snow depth data are not available in convenient archives, although they are recorded in written synoptic weather reports, and selected mapping has been performed by the British Meteorological Office since 1962 for Eurasia and since 1971 for North America (Taylor, 1980).

Spatial Coverage and Resolution of Additional Measurements Required. A regular program of mapping global snow cover extent and depth with a higher time and space resolution is needed for present purposes. A 50-km grid is required. Frequent satellite observations with a horizontal resolution of 1-4 km would make this possible for snow-cover extent. Snow-depth data from synoptic weather reports currently have a coarse resolution, and refinement is desirable here as well.

Frequency of Measurements Required. Snow maps, which at present are compiled weekly, are adequate for long-term monitoring studies, although studies of synoptic-scale interactions would require daily maps.

Feasibility and/or Existence of Technical Systems; Continuity. Satellite systems like those currently in use are adequate for purposes of measuring snow cover extent, but at present snow depth cannot be measured from space.

5.3.3.3.3 Ice-Cap Mass Balance Changes

Sensitivity. The effects of a warming on the Greenland and Antarctic ice sheets are likely to be complex (Bentley, 1983; Revelle, this volume, Chapter 8). In the short-run CO_2-induced climate changes could result in either positive or negative transient mass balance changes of the ice sheets, depending on regional shifts in temperature and precipitation. The potential of CO_2-induced changes in the next few decades to initiate disintegration of the West Antarctic ice sheet is very small. On the other hand, Ambach (1980) states that a temperature increase of only 1.5°C will cause a decisive negative change in the mass balance of Greenland. Such mass balance changes will in turn slowly affect sea level. For a 3°C CO_2-induced warming, Revelle (this volume, Chapter 8) calculates a 60-70-cm rise in global sea level, about half due to ablation of the Greenland and Antarctic ice caps and about half due to increase in the specific volume of seawater resulting from an increase in temperature. We thus need to measure the mass balance of these ice sheets in order to understand long-term sea-level changes.

Rate of Change. Because of the large mass and long residence time (about 10^3-10^5 years) of the ice in the Greenland and Antarctic ice sheets, their responses to CO_2-induced warming will be slow. Based on our present knowledge, it appears that a CO_2-induced warming on the century time scale will have only minor consequences for ice sheets, but changes in their thickness may be detectable at intervals of 5-10 years. Detectable shorter-term changes could include higher melting on ice shelves, changes in iceberg calving rates, or changes in surface gradient on ice shelves near ice rises. Ice shelves have shorter response times and provide a first indication of the state of health of the ice sheet.

Signal-to-Noise Ratio. The signal-to-noise ratio is probably quite high, since the interiors of the large ice sheets are stable and relatively inactive. Noise may be increased by the limits of present measurement techniques.

Adequacy and Availability of Past Data Base. The mass balance of the Greenland ice sheet is well known (Ambach, 1980), and satellite measurements, using airborne radio echo sounding (Robin et al., 1977) and radar altimetry have begun to give us similar data for Antarctica. Previous estimates of the mass balance changes of Antarctica are unreliable, varying from positive to negative values. Few data are available on the extent of melt features on ice shelves, on calving rates of iceberg, or on the response of ice shelves to climatic changes.

Spatial Coverage and Resolution of Additional Measurements Required. Satelliteborne radar altimeter flights in a polar orbit are required for at least every 10° of longitude. The vertical resolution of height of the ice sheet should be ± 5 cm. Areas in which to look for changes are not only the elevation of the interior of ice sheets but also gradient changes near ice shelves and ice rises.

Frequency of Measurements Required. Changes in the ice-sheet mass balance of Greenland and Antarctica should be monitored at 5-year intervals. Seasonal features, such as the extent of melt features on ice shelves, should be surveyed annually.

Feasibility and/or Existence of Technical Systems; Continuity. The feasibility of making such measurements has been demonstrated by using the radar altimeter on the GEOS-3 satellite launched in 1975 (Brooks et al., 1978) and by studies based on Seasat data (Zwally et al., 1983a) and simulations of possible future satellite systems (Zwally et al., 1981). No U.S. satellite currently carries either radar or laser altimeters, however, but plans for future satellites should include them.

5.3.3.4 Oceanic Parameters

The oceans are key elements in the Earth's climate system. However, there are still major uncertainties in our knowledge of how the coupled ocean-atmosphere system works and, therefore, how it may change when CO_2 is added to the atmosphere. The available, though sketchy, evidence points toward the fact that the ocean is most probably delaying the temperature signal of increasing CO_2 by mixing heat downward. It is clear from model studies, which up to now have treated the oceans quite simply (e.g., Gates et al., 1981; Manabe and Stouffer, 1980), that the response of the atmosphere is paced by that of the ocean.

Thompson and Schneider (1979) and Schneider and Thompson (1981) discussed the question of the transient response of the atmosphere to CO_2 using models based on the thermal inertia of the upper layers of the oceans in combination with their interaction with deeper waters. Similar conclusions about the delay of a warming, i.e., as long as a few decades, were reached in reports of the National Academy of Sciences (Climate Research Board, 1979; CO_2/Climate Review Panel, 1982). Ramanathan (1981) emphasizes the inadequacy of present coupled models for examining the transient response. Work of Fine et al. (1981) on tritium penetration into the ocean suggested that the delay time may have been underestimated. Regional climate changes will also be strongly associated with climatic variation in the ocean (Schneider and Thompson, 1981; Bryan et al., 1982; Thompson and Schneider, 1982).

In order to identify changes in the ocean due to CO_2 warming, a long-term measurement program is required. Brewer (this volume, Chapter 3, Section 3.2) examines the changes in ocean chemistry to be expected from increasing CO_2. Baker and Barnett (1982) describe the physical oceanographic variables that would be expected to respond.

Of the parameters identified by these authors, which include sea level, sea temperatures, salinity, and ocean circulation patterns, the first two seem most appropriate for monitoring the possible effects of CO_2-induced warming. Changes in sea level, though not driven by thermal expansion alone, may be the best indicator of the global change in ocean temperature, because an observational network exists, at least in the northern hemisphere, and sea-level data are representative of integrated, rather than point, measurements.

5.3.3.4.1 Sea Level

Sensitivity. Sea level should be a sensitive indicator of CO_2 effects. A change of 0.1% of the global land ice cover will result in a sea-level change of about 5 cm (Flint, 1971), and increases in ocean temperature will presumably accompany increases in atmospheric temperature; a change of 0.5°C in the upper 200 m would increase sea level by roughly 2 cm (Baker and Barnett, 1982). Given a 3°C atmospheric warming, Revelle (this volume, Chapter 8) estimates a rise in sea level about 100 years from now of at least 30 cm, resulting from ocean warming, and a probable rise of between 60 and 70 cm, if ice ablation is included. Both of these effects will be global in nature, and it may be far easier to detect a signal that is coherent in all the oceans than to identify one that is rather regional in character. Measurement problems exist, however, which are discussed below.

Rate of Change. Rises in sea level in response to projected CO_2-induced warming will be slow, but much more rapid than recent historic rates. Estimates of the rate of sea-level rise so far during this century range from 1 to 3 mm/year (see Revelle, this volume, Chapter 8). Problems of tectonic movement and poor station distribution in terms of location, offshore current, and wind systems, for example, leave one uncertain as to the reality and meaning of these numbers.

Signal-to-Noise Ratio. The relatively long time series of sea level provide opportunities to estimate signal-to-noise ratios and hence make detection of global changes more feasible (cf., Madden and Ramanathan, 1980). However, there are relatively few sea-level stations in the southern hemisphere that possess a long record, and, further, sea-level data in huge ocean regions must be reconstructed from limited hydrographic data These deficiencies will make detection of a truly global signal somewhat more difficult.

Adequacy and Availability of Past Data Base. Long time series of sea-level data are available (e.g., Emery, 1980; Revelle, this volume, Chapter 8), but their interpretation is complicated by the problems listed above. Nevertheless, these studies suggest a coherent rising of sea level on scales of oceanic dimensions. The statistical significance of the changes and their relationship to CO_2-induced warming are hard to estimate.

Spatial Coverage and Resolution of Additional Measurements Required. It appears feasible to achieve spatial coverage sufficient to ameliorate some of the difficulties mentioned above, although problems in the interpretation of individual records will remain. Measurements are needed at all open ocean island locations; primarily lacking at present are islands in the Atlantic and Indian Oceans. Also needed are stations around the Antarctic continent. Global sea-level coverage will be available from the various altimetric satellites now being proposed for the late 1980s. The estimated accuracy of these, ± 10 cm, will be too low for useful estimation of CO_2 effects in the next one or two decades; however, over the long term (i.e., a century) such satellite measurements will be helpful.

Frequency of Measurements Required. Since most sea-level measurements are made for tidal prediction, the frequency of measurement is more than adequate for long-term sea-level change. It is critical to keep the measurement going for a long time (decades).

Feasibility and/or Existence of Technical Systems; Continuity. Tide-gauge measurements are simple and have been carried out for a long time. The existing technology that permits unattended operation for months to years with data recording on tape cassettes is entirely adequate for the purpose.

5.3.3.4.2 Sea Temperature

Sensitivity. Most simulations of increased atmospheric CO_2 show a substantial warming of surface air temperature over the globe. The effect is most pronounced in high latitudes. As with sea level, change in sea-surface temperature resulting from this air temperature increase should be global in nature. However, its magnitude will have a strong regional character (Baker and Barnett, 1982), owing to regional variation in vertical mixing and diffusion and the relative importance of different physical processes in the ocean heat budget; detection of a CO_2-induced temperature signal in the oceans will thus be difficult. Certainly, the nature and magnitude of such a signal merits further study.

Rate of Change. Response time of sea temperatures to CO_2 effects could be relatively slow owing to the thermal inertia of the oceans. Data on sea-surface temperature (SST) for the Indian, North Atlantic, and North and Tropical Pacific Oceans show perhaps a 1°C rise over the last 80 years in all oceans (Baker and Barnett, 1982). The change may be smaller because of possible errors due to the gradual conversion from measurements by bucket thermometer to ship's injection thermometers; the latter consistently showing SSTs warmer by 0.4°C.

Signal-to-Noise Ratio. Current data archives on the surface temperature field of the world oceans are quite extensive and should make proper determinations of signal-to-noise ratios relatively simple. The historical record of surface temperature and subsurface temperature needs to be analyzed to determine the type of signal that could be detected over the background. Estimates of regional signal-to-noise ratios also need to be made; these have not been performed so far. The effects of mesoscale eddies add a large noise to subsurface ocean temperature that is generally not so prominent in the SST field.

Adequacy and Availability of Past Data Base. Measurements of sea-surface temperature have been made in many regions of the world for the past 100 years from ships, islands, and coastal stations, although many of these data are crude and unrepresentative. More complete analyses of this record should be undertaken. Analyzed data covering the North Atlantic and Pacific are available back to about 1948 (Walsh and Sater, 1981; Bunker, 1980). Subsurface temperature fields are much less well known. The historical record of subsurface ocean temperature is unfortunately relatively short, and the data are subject to errors introduced by changes in the instrumentation (bathythermographs), which may introduce bias into the data sets.

Spatial Coverage and Resolution of Additional Measurements Required. It is now possible to collect data on sea-surface temperatures for all oceanic regions of the world. Infrared satellite sensors have been used since the early 1970s, and improved satellite analyses are now produced daily for all oceanic regions. Programs of direct measurement must, however, be maintained for their better accuracy, and new methods (e.g., acoustic tomography) should be explored for measuring integrated ocean temperature.

Frequency of Measurements Required. Daily analyses are available from satellite infrared sensors; their frequency is adequate.

Feasibility and/or Existence of Technical Systems; Continuity. It is necessary to evaluate rigorously how well the sea-surface temperature signal can be extracted from the satellite microwave radiometer. The demonstrated accuracy of the spaceborne sea-surface temperature systems still remains inadequate to resolve the signals as observed by direct measurement. For studies of warming, accuracies of from 0.1 to 0.5°C are needed, but this accuracy has not yet been achieved. Much of this uncertainty comes from surface foam, water vapor, and liquid water in the path of the sensor; the new multichannel sensors may show marked improvement. Acoustic tomography (Munk and Wunsch, 1982) also offers promise of providing integrated ocean temperatures.

5.3.4 Conclusions and Recommendations

5.3.4.1 Priority of Parameters to be Monitored

To help determine the current and projected effects of increasing atmospheric CO_2 on climate, we recommend further elaboration and, if sound, the development of a monitoring strategy in which many measures of the state of the climate systems are monitored and analyzed as an ensemble.

If recent climatic trends are sustained, it seems likely that there will be an increasing number of claims to have distinguished a significant warming. The problem will then become increasingly one of attribution of cause and effect. The most promising means to achieve convincing attribution will be development of reliable records of several parameters in addition to temperature. Clearly, for technical or cost reasons adequate monitoring of some parameters will be much more readily achievable than others. To accomplish early attribution, initial emphasis should be given to these.

While initial emphasis may be placed on parameters that may be monitored immediately, cheaply, or easily, it is important over the long run to build up a rather complete data base, not only for reasons of detection but also for research and for calibration of models of the climate system. It may take until 2010 or 2020 to begin to have useful data bases on some of the parameters mentioned here, but they should not be neglected. Instead, we should anticipate that a monitoring program will gradually evolve into a program to verify and calibrate crucial aspects of model calculations, especially the numerous projected effects of increasing atmospheric CO_2, for example, sea-level rise and changes in rainfall in midlatitudes.

Based on this initial survey, we summarize our recommendations for monitoring in Table 5.11.

TABLE 5.11 Priority in Monitoring Variables for Early Detection of CO_2 Effects

Priority	Monitoring Causal Factors by Measuring Changes in	Monitoring Climatic Effects by Measuring Changes in
First	CO_2 concentrations Volcanic aerosols Solar radiance	Troposphere/surface temperatures (including sea temperatures) Stratospheric temperatures Radiation fluxes at the top of the atmosphere Precipitable water content (and clouds)
Second	"Greenhouse" gases other than CO_2 Stratospheric and tropospheric ozone	Snow and sea-ice covers Polar ice-sheet mass balance Sea level

5.3.4.2 Measurement Networks

The key to a successful monitoring strategy is a global observation system. Satellites are a major component of such a system, and it is essential to be able to continue monitoring without interruption on a long-term basis the radiative fluxes, the planetary albedo, snow and ice extent, and sea-surface temperatures and to improve the spaceborne measurements of tropospheric and stratospheric temperatures, precipitable water content of the atmosphere, mass balance of the polar ice sheets and sea level, as well as aerosols, ozone, and other atmospheric constituents.

Many of the satellite measurements that are being made at present are difficult to calibrate. Of particular concern are the vertical soundings made from spacecraft and the lack of supporting surface-based or surface-launched profiling systems. A concern, for example, is the dismantling of the global rocketsonde system of stratospheric temperature soundings. Some key stations should be retained to calibrate the instruments flown on satellites. Similarly, the only reliable method of characterizing stratospheric aerosols at present is by the deployment of lidar systems. By adding one or two such systems in the southern hemisphere to the existing network, and through occasional aircraft flights to calibrate the satellite soundings, an adequate amount of data could be collected. Other parameters, for example ozone, are also inadequately measured at present. Total ozone values derived from Dobson spectrometer and satellite profile measurements are not enough in themselves, and the existing ozonesonde network must be maintained and augmented in data-sparse regions.

The southern hemisphere presents a particular problem in monitoring climatic changes that can only be solved by improving and perfecting the present satellite-based sounding techniques.

Table 5.1 summarizes requirements and technical systems for monitoring high-priority variables.

5.3.4.3 Modeling and Statistical Techniques

The internal physical consistency and relative ease of diagnosis of simulated climatic data make the construction of realistic and comprehensive models a prerequisite for the development of a successful fingerprinting strategy for the detection of CO_2-induced climatic change. In addition, climate models are needed in order to determine the accuracy that is required in monitored climatic variables. Unfortunately, the CO_2-induced climatic changes calculated from the various current climate models continue to show substantial differences. In order to develop an effective monitoring strategy, it is essential that further intensive efforts be made to improve climate models by validating them against the observed structure and behavior of the ocean-atmosphere system and to make effective use of model improvements.

Another important element is the development of methods for the statistical identification of a CO_2-induced climate signal against the background of natural climatic fluctuations. Statistical techniques applied so far focus on the significance of the signal-to-noise ratio, assuming that the data at individual points may be modeled as a first-order autoregressive process. Further use should be made of significance tests that consider longer-term dependence in the climatic time series and that provide estimates of confidence limits. An essential ingredient of a successful detection strategy will be the development of techniques that take into account not only the temporal correlation but also the spatial correlation that is characteristic of nearly all climatic variables and that lead to more careful and sophisticated statistical tests of a possible CO_2 climatic signal. Here again, model-simulated data can be used effectively to begin to develop and test the statistical procedures that will ultimately have to be applied to monitored observational data. However, purely statistical inferences will have to be buttressed to the greatest extent possible by physical reasoning.

5.3.4.4 Objective Evaluation of Evidence

The differing interpretations of the effects of likely CO_2-induced climate changes, as arrived at by different authors (see Section 5.2), often using identical data sets, underline the need for objective evaluation of evidence. It is wise to anticipate the need at national and international levels for periodic efforts to evaluate evidence and arbitrate between divergent opinions, where necessary. Some centers that can perform such functions are already in existence as part of national and international programs. At these centers climatic indices should be collected and compared, and evaluation and improvement of techniques for identification of climatic changes should be encouraged.

REFERENCES

Alishouse, J. C. (1983). Total precipitable water and rainfall determinations from SEASAT scanning multichannel microwave radiometer. J. Geophys. Res. 88:1929-1935.

Ambach, W. (1980). Anstieg der CO_2-Konzentration in der Atmosphäre und Klimaänderuug: Mögliche Auswirkungen auf dem Grönländischen Eisschild. Wetter und Leben 32:135-42.

Angell, J. K., and J. Korshover (1978a). Global temperature variation; surface--100 mb: an update into 1977. Mon. Wea. Rev. 106:755-770.

Angell, J. K., and J. Korshover (1978b). Estimate of global temperature variations in the 100-30 mb layer between 1958 and 1977. Mon. Wea. Rev. 106:1422-1432.

Augustsson, T., and V. Ramanathan (1977). A radiative-convective model study of the CO_2-climate problem. J. Atmos. Sci. 34:448-451.

Baker, Jr., D. J., and T. P. Barnett (1982). Possibilities of detecting CO_2-induced effects: ocean physics. In U.S. Department of Energy (1982), pp. 301-342.

Baldwin, B., J. B. Pollack, A. Summers, O. B. Toon, C. Sagan, and W. Van Camp, (1976). Stratospheric aerosols and climatic change. Nature 263:551-555.

Barry, R. G. (1978). Cryospheric responses to a global temperature increase. In Carbon Dioxide, Climate and Society, J. Williams, ed. Pergamon, Oxford, pp. 169-180.

Barry, R. G. (1982). Snow and ice indicators of possible climatic effects of increasing atmospheric carbon dioxide. In U.S. Department of Energy (1982), pp. 207-236.

Bates, J. R. (1977). Dynamics of stationary ultra-long waves in middle latitudes. Q. J. Roy. Meteorol. Soc. 103:397-430.

Bentley, C. R., ed.(in press). CO_2-induced Climate Change and the Dynamics of Antarctic Ice. Proceedings of AAAS Symposium, Toronto, January 1981, American Association for the Advancement of Science, Washington, D.C..

Bloomfield, P., M. L. Thompson, G. S. Watson, and S. Zeger (1981). Frequency Domain Estimation of Trends in Stratospheric Ozone. Technical Report 182. Dept. of Statistics, Princeton U., Princeton, N.J.

Borzenkova, I. I., K. Ya. Vinnikov, L. P. Spirina, and D. I. Stekhnovskii (1976). Variation of northern hemisphere air temperature from 1881 to 1975. Meteorol. Gidrol. 7:27-35.

Broecker, W. S. (1975). Climatic change: are we on the brink of a pronounced global warming? Science 188:460-463.

Brooks, R. L., W. J. Cambell, R. O. Ramseier, H. R. Stanley, and H. J. Zwally (1978). Ice sheet topography by satellite altimetry. Nature 274:539-43.

Bryan, K., F. G. Komro, S. Manabe, and M. J. Spelman (1982). Transient climate response to increasing atmospheric carbon dioxide. Science 215:56-58.

Bryson, R. A. (1980). CO_2, Aerosols, and Modeling Climate of the Recent Past. In U.S. Department of Energy (1980), pp. 139-143.

Bryson, R. A., and G. J. Dittberner (1976). A non-equilibrium model of hemispheric mean surface temperature. J Atmos. Sci. 33:2094-2106.

Bryson, R. A., and B. M. Goodman (1980). Volcanic activity and climatic changes. Science 207:1041-1044.

Bryson, R. A., and W. M. Wendland (1970). Climatic Effects of Atmospheric Pollution. Global Effects of Environmental Pollution, S. F. Singer, ed. Springer-Verlag, New York, pp. 130-138.

Budd, W. F. (1975). Antarctic sea ice variations from satellite sensing in relation to climate. J. Glaciol. 73:417-427.

Budd, W. F. (1980). The importance of the antarctic region for studies of the atmospheric carbon dioxide concentration. In Carbon Dioxide and Climate: Australian Research, G. I. Pearman, ed. Australian Academy of Science, Canberra, pp. 115-28.

Budyko, M. I. (1969). The effect of solar radiation variations on the climate of the earth. Tellus 21:611-619.

Budyko, M. I. (1974). Climatic Changes. Hydrometeoizdat, Leningrad (47 pp.); transl., American Geophysical Union, 1977.

Budyko, M. I., and K. Ya. Vinnikov (1973). Modern climate variations. Meteorol. Gidrol. 9:3-13.

Bunker, A. F. (1980). Trends of variables and energy fluxes over the Atlantic Ocean from 1948 to 1972. Mon. Wea. Rev. 108:720-732.

Callendar, G. S. (1938). The artificial production of carbon dioxide and its influence on temperatures. Q. J. Roy. Meteorol. Soc. 64:223-240.

Callendar, G. S. (1961). Temperature fluctuations and trends over the earth. Q. J. Roy. Meteorol. Soc. 87:1-12.

Charney, J. G. (1975). Dynamics of deserts and drought in the Sahel. Q. J. Meteorol. Soc. 101:193-202.

Clark, W. C., ed. (1982). Carbon Dioxide Review: 1982. Oxford U. Press, New York, 469 pp.

Clark, W. C., K. H. Cook, G. Marland, A. M. Weinberg, R. M. Rotty, P. R. Bell, L. J. Allison, and C. L. Cooper (1982). The carbon dioxide question: a perspective for 1982. In Clark (1982).

Climate Research Board (1979). Carbon Dioxide and Climate: A Scientific Assessment. National Research Council, National Academy of Sciences, Washington, D.C., 22 pp.

CO_2/Climate Review Panel (1982). Carbon Dioxide and Climate: A Second Assessment. National Research Council, National Academy Press, Washington, D.C., 72 pp.

Dickson, R. R., and J. Posey (1967). Maps of snow-cover probability for the northern hemisphere. Mon. Wea. Rev. 95:347-53.

Donner, L., and V. Ramanathan (1980). Methane and nitrous oxide: their effects on terrestrial climate. J. Atmos. Sci. 37:119-124.

Eddy, J. A. (1976). The Maunder minimum. Science 192:1189-1202.

Eddy, J. A. (1977). Climate and the changing sun. Climatic Change 1:173-190.

Eddy, J. A., R. L. Gilliland, and D. V. Hoyt (1982). Changes in the solar constant and climatic events. Nature 300:689-693.

Elliott, W. P. (1982). A note on the historical industrial production of carbon dioxide. NOAA Air Resources Laboratories, Rockville, Md.

Ellis, J. S., T. H. Vonder Haar, S. Levitus, and A. H. Oort (1978). The annual variation in the global heat balance of the earth. J. Geophys. Res. 83:1958-1962.

Emery, K. O. (1980). Relative sea level from tide gauge records. Proc. Nat. Acad. Sci. 77:6968-6972.

Epstein, E. S. (1982). Detecting climate change. J. Appl. Meteorol. 21:1172.

Fels, S. B., J. D. Mahlman, M. D. Schwarzkopf, and R. W. Sinclair (1980). Stratospheric sensitivity to perturbations in ozone and carbon dioxide: radiative and dynamical response. J. Atmos. Sci. 37:2265.

Fine, R. A., J. L. Reid, and H. G. Ostlund (1981). Circulation of tritium in the Pacific Ocean. J. Phys. Oceanog. 11:3.

Fishman, J., V. Ramanathan, P. J. Crutzen, and S. C. Liu (1979). Tropospheric ozone and climate. Nature 282:818-820.

Flint, R. F. (1971). Glacial and Quarternary Geology. Wiley, New York.

Gates, W. L., K. H. Cook, and M. E. Schlesinger (1981). Preliminary analysis of experiments on the climatic effects of increased CO_2 with an atmospheric general circulation model and a climatological ocean model. J. Geophys. Res. 86:6385.

Geller, M. A., and V. C. Alpert (1980). Planetary wave coupling between the troposphere and the middle atmosphere as a possible sun-weather mechanism. J. Atmos. Sci. 37:1197-1215.

Geophysics Study Committee (1982). Solar Variability, Weather, and Climate. National Research Council, National Academy Press, Washington, D.C., 106 pp.

Gilliland, R. L. (1981). Solar radius variations over the past 265 years. Astrophys. J. 248:1144-1155.

Gilliland, R. L. (1982). Solar, volcanic, and CO_2 forcing of recent climatic changes. Climatic Change 4:111-131.

Hammer, C. U. (1977). Past volcanism revealed by Greenland ice sheet impurities. Nature 270:482-486.

Hansen, J. (1980). CO_2, solar variations, recent climate, and model predictions. In U.S. Department of Energy (1980), pp. 144-153.

Hansen, J., D. Johnson, A. Lacis, S. Lebedeff, P. Lee, D. Rind, and G. Russell (1981). Climate impact of increasing atmospheric carbon dioxide. Science 213:957-966.

Hayashi, Y. (1982). Confidence intervals of a climatic signal. J. Atmos. Sci. 39:1895.

Hays, J. D. (1978). A review of the late quarternary climatic history of Antarctic seas. In Antarctic Glacial History and World Paleoenvironments: August 17, 1977, Birmingham, United Kingdom. F. M. van Zinderen Bakker, ed. A. A. Balkema, Rotterdam, The Netherlands, pp. 57-71.

Heath, D. F., and B. M. Schlesinger (1982). Secular and periodic variations in stratospheric ozone from satellite observations--1970-1979. NASA/Goddard Space Flight Center Preprint.

Hirschboeck, K. (1980). A new worldwide chronology of volcanic eruptions. Palaeogeogr. Palaeoclim. Palaeoecol. 29:223-241.

Hoffert, M. I., A. J. Callegari, and C.-T. Hsieh (1980). The role of deep sea heat storage in the secular response to climatic forcing. J. Geophys. Res. 85:6667.

Hollin, J. T., and R. G. Barry (1979). Empirical and theoretical evidence concerning the response of the earth's ice and snow cover to a global temperature increase. Environ. Int. 2:437-444.

Hoyt, D. V. (1979a). An empirical determination of the heating of the earth by the carbon dioxide greenhouse effect. Nature 282:388-390.

Hoyt, D. V. (1979b). Variations in sunspot structure and climate. Climatic Change 2:79-92.

Jones, P. D., T. M. L. Wigley, and P. M. Kelly (1982). Variations in surface air temperatures: Part 1. Northern hemisphere, 1881-1980. Mon. Wea. Rev. 110:59-70.

Katz, R. W. (1980). Statistical evaluation of climate experiments with general circulation models: inferences about means. Climatic Research Institute Report 15. Oregon State U., Corvallis.

Katz, R. W. (1982). Statistical evaluation of climate experiments with general circulation models: a parametric time series modeling approach. J. Atmos. Sci. 39:1446-1455.

Kelley, P. M., and P. D. Jones (1981). Annual temperatures in the Arctic, 1881-1981. Climate Monitor 10:122-124.

Klein, W. H. (1982). Detecting climate effects of increasing atmospheric carbon dioxide. In U.S. Department of Energy (1982), pp. 175-194.

Kraus, E. B. (1955). Secular changes of tropical rainfall regimes. Q. J. Roy. Meteorol. Soc. 81:108.

Kukla, G. J. (1981). Climatic role of snow covers. In Sea Level, Ice and Climatic Change, I. Allison, ed. International Assoc. of Hydrological Sciences Publ. 131. IAHS, Wallingford, Oxfordshire, U.K., pp. 79-107.

Kukla, G. J. (1982). Carbon dioxide and polar climates. In U.S. Department of Energy (1982), pp. 237-288.

Lacis, A., J. Hansen, P. Lee, T. Mitchell, and S. Lebedeff (1981). Greenhouse effect of trace gases, 1970-1980. Geophys. Res. Lett. 8:1035-1038.

Lamb, H. H. (1970). Volcanic dust in the atmosphere with a chronology and assessment of its meteorological significance. Phil. Trans. Roy. Phil. Soc. London:425-533.

Levy, H., J. O. Mahlman, and W. J. Moxim (1979). A preliminary report on the numerical simulation of the three-dimensional structure and variability of atmospheric N_2O. Geophys. Res. Lett. 6:155-158.

MacCracken, M. C. (1983). Have we detected CO_2-induced climate change?--Problems and prospects. In U.S. Department of Energy (1983), Vol. 5, pp. 3-45.

MacCracken, M. C., and H. Moses (1982). The first detection of carbon dioxide effects: workshop summary. Bull. Am. Meteorol. Soc. 63:1164-1178.

Madden, R. A., and V. Ramanathan (1980). Detecting climate change due to increasing carbon dioxide. Science 209:763-68.

Manabe, S., and D. G. Hahn (1981). Simulation of atmospheric variability. Mon. Wea. Rev. 109:2260-2286.

Manabe, S., and R. J. Stouffer (1980). Sensitivity of a global climate model to an increase of CO_2 concentration in the atmosphere. J. Geophys. Res. 85:5529-5554.

Manabe, S., and R. F. Strickler (1964). Thermal equilibrium of the atmosphere with a convective adjustment. J. Atmos. Sci. 21:361-385.

Manabe, S., and R. T. Wetherald (1975). The effects of doubling the CO_2 concentration on the climate of a general circulation model. J. Atmos. Sci. 32:3.

Manabe, S., and R. T. Wetherald (1980). On the distribution of climate change resulting from an increase in CO_2 content of the atmosphere. J. Atmos. Sci. 37:99.

Marland, G., and R. Rotty (1982). Carbon dioxide emissions from fossil fuels, 1950-1981. Institute for Energy Analysis, Oak Ridge, Tenn.

Mass, C., and S. H. Schneider (1977). Statistical evidence on the influence of sunspots and volcanic dust on long-term temperature records. J. Atmos. Sci. 34:1995-2004.

Matson, M., and D. R. Wiesnet (1981). New data base for climate studies. Nature 289:451-456.

Maxwell, J. B. (1980). The Climate of the Canadian Arctic Islands and Adjacent Waters. Environment Canada, Canadian Government Publishing Centre, Hull, Quebec, 531 pp.

McCormick, M. P., W. P. Chu, L. R. McMaster, G. W. Grams, B. M. Herman, T. J. Pepin, P. B. Russell, and T. J. Swissler (1981). SAM-II aerosol profile measurements, Poker Flat, Alaska, July 16-19, 1979. Geophys. Res. Lett. 8:3-4

Miles, M. K., and P. B. Gildersleeves (1978). A statistical study of the likely influence of some causative factors on the temperature changes since 1665. Meteorol. Mag. 107:193-204..

Mitchell, J. M., Jr. (1961). Recent secular changes of global temperature. Ann. N.Y. Acad. Sci. 95:235-250.

Mitchell, J. M., Jr. (1970). A preliminary evaluation of atmospheric pollution as a cause of the global temperature fluctuation of the past century. In Global Effects of Environmental Pollution, S. F. Singer, ed. Springer-Verlag, New York, pp. 139-155.

Mitchell, J. M., Jr. (1983). An empirical modeling assessment of volcanic and carbon dioxide effects on global scale temperature. American Meteorological Society, Second Conference on Climate Variations, New Orleans, Louisiana, January 10-14, 1983.

Munk, W, and C. Wunsch (1982). Observing the ocean in the 1990s. Phil. Trans. Roy. Soc. London Ser. A 307:439-464.

Murphy, A. H., and R. W. Katz (1982). Statistical methodology for first detection of carbon dioxide effects in the atmosphere. In U.S. Department of Energy (1982), pp. 165-174.

Newell, R. E., and A. Deepak, eds. (1982). Mount St. Helens Eruptions of 1980--Atmospheric Effects and Potential Climatic Impact. NASA report SP-458. National Aeronautics and Space Administration, Washington, D.C.

Newhall, C. G., and S. Self (1982). The volcanic explosivity index (VEI): an estimate of explosive magnitude for historical volcanism. J. Geophys. Res. 87:1231-1238.

Oliver, R. C. (1976). On the response of hemispheric mean temperature to stratospheric dust: an empirical approach. J. Appl. Meteorol. 15:933-950.

Oort, A. H. (1982). Global atmospheric circulation studies, 1958-1973. NOAA Prof. Paper. U.S. Govt. Printing Office, Washington, D.C. (in preparation).

Oort, A. H., and E. M. Rasmusson (1971). Atmospheric circulation statistics. NOAA Prof. Paper 5. U.S. Govt. Printing Office, Washington, D.C.

Paltridge, G., and S. Woodruff (1981). Changes in global surface temperature from 1880 to 1977 derived from historical records of sea surface temperature. Mon. Wea. Rev. 109:2427-2434.

Pierotti, D., and R. A. Rasmussen (1977). The atmospheric distribution of nitrous oxide. J. Geophys. Res. 37:5823-5832.

Pierotti, D., and R. A. Rasmussen (1978). Inter-laboratory calibration of atmospheric nitrous oxide measurements. Geophys. Res. Lett. 5:353-355.

Pivavarova, Z. I. (1968). The long-term variation of the intensity of solar radiation according to the observations of actinemetric stations [in Russian]. Glavnaya Geofiz. Obs. Trudy 233:17-37.

Pivavarova, Z. I. (1977). Radiation Characteristics of the Climate of the USSR (in Russian). Gidrometeoizdat, Leningrad.

Plass, G. N. (1953). Some problems in atmospheric radiation. Proc. Toronto Meteorol. Conf.:53. Royal Meteorological Society, London.

Plass, G. N. (1956). The carbon dioxide theory of climatic change. Tellus 8:140-154.

Pollack, J. B., O. Toon, C. Sagan, A. Summers, B. Baldwin, and W. Van Camp, (1976a). Volcanic explosion and climatic change: a research assessment. J. Geophys. Res. 81:1071-1083.

Pollack, J. B., O. B. Toon, A. Summers, B. Baldwin, C. Sagan, and W. Van Camp, (1976b). Stratospheric aerosols and climate change. Nature 263:5578

Potter, J. G. (1965). Snow cover. Climatological Studies 3. Meteorological Branch, Dept. of Transport, Toronto, Canada.

Potter, G. L., H. W. Ellsaesser, M. C. MacCracken, and F. M. Luther (1975). Possible climatic impact of tropical deforestation. Nature 258:697-698.

Potter, G. L., H. W. Ellsaesser, M. C. MacCracken, J. S. Ellis, and F. M. Luther (1980). Climate change due to anthropogenic surface albedo modification. In Interactions of Energy and Climate, W. Bach, J. Pankrath, and J. Williams, eds. Reidel, Dordrecht, pp. 317-326.

Prabhakara, C., H. D. Chang, and A. T. C. Chang (1982). Remote sensing of precipitable water over the oceans from Nimbus 7 microwave measurements. J. Atmos. Sci. 59.

Pritchard, R. S., ed. (1980). Sea Ice Processes and Models. Proceedings of the Arctic Ice Dynamics Joint Experiment, U. of Washington Press, Seattle, 474 pp.

Ramanathan, V. (1975). Greenhouse effect due to chlorofluorocarbons: climatic implications. Science 190:50.

Ramanathan, V. (1980). Climatic effects of anthropogenic trace gases. In Interactions of Energy and Climate, W. Bach, J. Pankrath, and J. Williams, eds. Reidel, Boston, Mass., pp. 269-280.

Ramanathan, V. (1981). The role of ocean-atmosphere interactions in the CO_2-climate problem. J. Atmos. Sci. 38:918.

Ramanathan, V., L. B. Callis, and R. E. Boughner (1976). Sensitivity of surface temperature and atmospheric temperature to perturbations in the stratospheric concentration of ozone and nitrogen dioxide. J. Atmos. Sci. 33:1092-1112.

Rasmussen, R. A., and M. A. K. Khalil (1981). Increase in the concentration of atmospheric methane. Atmos. Environ. 15:883.

Reinsel, G. C. (1981). Analysis of total ozone data for the detection of recent trends and the effects of nuclear testing during the 1960s. Geophys. Res. Lett. 8:1227-1230.

Reinsel, J. C., J. C. Tiso, A. J. Miller, C. L. Mateer, J. J. DeLuisi, and J. E. Frederick (1983). Analysis of upper stratospheric Umkehr ozone profile data for trends and the effects of stratospheric aerosols. American Geophysical Union Spring Meeting, Baltimore, Maryland, May 30-June 3, 1983. Abstract in EOS 64:199.

Revelle, R., and H. E. Suess (1957). CO_2 exchange between atmosphere and ocean, and the question of an increase of atmospheric CO_2 during the past decades. Tellus 9:18.

Robin, G. de Q., D. J. Drewry, and D. T. Meldrum (1977). International studies of ice sheet and bedrock. Phil. Trans. Roy. Soc. London Ser. B 279:185-96.

Robock, A. (1978). Internally and externally caused climate change. J. Atmos. Sci. 35:1111-1122.

Robock, A. (1979a). The performance of a seasonal global climatic model. Report of the JOC Study Conference on Climate Models: Performance, Intercomparison, and Sensitivity Studies, Vol. 2. GARP Publ. 22. World Meteorological Organization, Geneva, Switzerland, pp. 766-802.

Robock, A. (1979b). The Little Ice Age: the northern hemisphere average observations and model calculations. Science 206:1402-1404.

Robock, A. (1981). The Mount St. Helens volcanic eruption of 18 May 1980: minimal climatic effect. Science 212:1383-1384.

Robock, A. (1983). Ice and snow feedbacks and the latitudinal and seasonal distribution of climate sensitivity. J. Atmos. Sci. 40:986-997.

Rosenkranz, P. W., M. J. Komichak, and D. H. Staelin (1982). A method for estimation of atmospheric water vapor profiles by microwave radiometry. J. Appl. Meteorol. 21:1364.

Sagan, C., O. B. Toon, J. B. Pollack (1979). Anthropogenic albedo changes and the earth's climate. Science 206:1363-1367.

St. John, D. S., S. P. Bailey, W. H. Fellner, J. M. Minor, and R. D. Snee (1981). Time series search for a trend in total ozone measurements. J. Geophys. Res. 86:7299-7311.

SCEP (1970). Man's Impact on the Global Environment, Study of Critical Environmental Problems. MIT Press, Cambridge, Mass.

Schlesinger, M. E. (1982). A Review of Climate Model Simulations of CO_2-Induced Climatic Change. Report No. 41, Climatic Research Institute, Oregon State U., Corvallis, 135 pp.

Schlesinger, M. E. (1983). Simulating CO_2-induced climate change with mathematical climate models: capabilities, limitations, and prospects. In U.S. Department of Energy (1983), III.3-III.140.

Schneider, S. H., and S. L. Thompson (1981). Atmospheric CO_2 and climate: importance of the transient response. J. Geophys. Res. 86:3135.

Schönwiese, C. D. (1981). Solar activity, volcanic dust, and temperature: statistical relationships since 1160 A.D. Arch. Meteorol. Geophys. Bioclimatol. Ser. A 30:1-22.

Simkin, T., L. Seibert, L. McClelland, W. G. Nelson, D. Bridge, C. G. Newhall, and J. Latter (1981). Volcanoes of the World. Hutchinson, Ross, New York.

Siquig, R. A., and D. V. Hoyt (1980). Sunspot structure and the climate of the last one hundred years. In Ancient Sun: Proceedings of the Conference on the Ancient Sun: Fossil Record in the Earth, Moon, and Meteorites. Boulder, Colorado, Oct. 16-19, 1979. R. O. Pepin, J. A. Eddy, and R. B. Merrill, eds. Pergamon, Elmsford, N.Y., pp. 63-67.

Skoog, B. G., J. I. H. Askne, and G. Elgered (1982). Experimental determination of water vapor profiles from ground-based measurements at 21.0 and 31.4 GHz. J. Appl. Meteorol. 21:394.

SMIC (1971). Inadvertent Climate Modification, Study of Man's Impact on Climate. MIT Press, Cambridge, Mass.

Stephens, G., G. Campbell, and T. Vonder Haar (1981). Earth radiation budgets. J. Geophys Res. 86:9739-9760.

Taylor, R. A. H. (1980). Snow Survey of the Northern Hemisphere. Glaciological Data, Report GD-9. World Data Center-A for Glaciology, Boulder, Colo., pp. 75-76.

Thompson, S. L., and S. H. Schneider (1979). A seasonal zonal energy balance climate model with an interactive lower layer. J. Geophys. Res. 84:2401-2404.

Thompson, S. L., and S. H. Schneider (1982). Carbon dioxide and climate: the importance of realistic geography in estimating the transient temperature response. Science 217:1031-1033.

U.S. Department of Energy (1980). Proceedings of the Carbon Dioxide and Climate Research Conference. Washington, D.C., April 24-25, 1980. Prepared by The Institute for Energy Analysis, Oak Ridge Associated Universities, L. E. Schmitt, ed., Report No. CONF-8004110, UC-11, December 1980, 287 pp.

U.S. Department of Energy (1982). Proceedings of the Workshop on First Detection of Carbon Dioxide Effects. Harper's Ferry, West Virginia, June 8-10, 1981. Harry Moses and Michael C. MacCracken, Coordinators. Prepared by The Institute for Energy Analysis, Oak Ridge Associated Universities, N. B. Beatty, ed., Report No. DOE/CONF-8106214, UC-11, May 1982, 546 pp.

U.S. Department of Energy (1983). Proceedings: Carbon Dioxide Research Conference: Carbon Dioxide, Science and Consensus. September 19-23, 1982. Berkeley Springs, West Virginia. Compiled by The Institute for Energy Analysis, Oak Ridge Associated Universities. Report No. CONF-820970, 506 pp.

Vilmundarson, T. (1972). Evaluation of historical sources of sea ice near Iceland. In Sea Ice, Proceedings of an International Conference at Reykjavik, Iceland, May 1971, pp. 159-169.

Vinnikov, K. Ya., and P. Ya. Groisman (1981). The empirical analysis of CO_2 influence on the modern changes of the mean annual Northern Hemisphere surface air temperature. Meteorol. Gidrol. 11:35-45.

Vinnikov, K. Ya., and P. Ya. Groisman (1982). The empirical study of climate sensitivity. Atmos. Oceanic Phys. 18 (in press).

Vinnikov, K. Ya., G. V. Gruza, V. F. Zakharov, A. A. Kirillov, N. P. Kovyneva, and E. Ya. Ran'kova (1980). Current climatic changes in the Northern Hemisphere. Meteorol. Gidrol. 6:5-17. (English translation by Allerton Press: Sov. Meteorol. Hydrol. 6:1-10.)

Vinnikov, K. Ya., G. V. Gruza, V. F. Zakharov, A. A. Kirillov, N. P. Kovyneva, and E. Ya. Ran'kova (1981). Modern changes in climate of the northern hemisphere. Meteorol. Gidrol. 6:5-17.

Walsh, J. E., and J. E. Sater (1981). Monthly and seasonal variability in the ocean-ice-atmosphere systems of the North Pacific and North Atlantic. J. Geophys. Res. 86:7425-7445.

Wang, W. C., Y. L. Yung, A. A. Lacis, T. Mo, and J. E. Hansen (1976). Greenhouse effects due to man-made perturbations of trace gases. Science 194:685.

Weiss, R. F. (1981). The temporal and spatial distribution of tropospheric nitrous oxide. J. Geophys. Res. 86:7185-7195.

Weller, G., and B. Holmgren (1974). The microclimates of the Arctic tundra. J. Appl. Meteorol. 13:854-62.

Wetherald, R. T., and S. Manabe (1981). Influence of seasonal variation upon the sensitivity of a model climate. J. Geophys. Res. 86:1194.

Wigley, T. M. L., and P. D. Jones (1981). Detecting CO_2-induced climatic change. Nature 292:205-208.

Willett, H. C. (1950). Temperature trends of the past century. Centenary Proceedings 195, Royal Meteorological Society.

Willett, H. C. (1974a). Recent statistical evidence in support of the predictive influence of solar-climate cycles. Mon. Wea. Rev. 102:679-686.

Willett, H. C. (1974b). Do recent climate fluctuations portend an imminent ice age? Geofis. Intern. 14:265-302.

Willson, R. C., S. Gulkis, M. Janssen, H. S. Hudson, and G. A. Chapman (1981). Observations of solar irradiance variability. Science 211:700-702.

World Meteorological Organization (1981). Joint WMO/ICSU/UNEP Meeting of Experts on the Assessment of the Role of CO_2 on Climatic Variations and Their Impact (Villach, Austria, November 1980). World Climate Programme Report No. 3. World Meteorological Organization, Geneva, 35 pp.

World Meteorological Organization (1982a). Report of the WMO(CAS)/JCS Meeting of Experts on Detection of Possible Climate Change (Moscow, October 3-6, 1982). World Climate Programme Report No. 29. World Meteorological Organization, Geneva, 43 pp.

World Meteorological Organization (1982b). The International Satellite Cloud Climatology Project (ISCCP) Preliminary Implementation Plan. World Climate Programme Report No. 35. World Meteorological Organization, Geneva, 85 pp.

Wuebbles, D. J., F. M. Luther, and J. E. Penner (1983). Effect of coupled anthropogenic perturbations on stratospheric ozone. J. Geophy. Res. 88:1444-1456.

Yamamoto, R. (1980). Change of global climate during recent 100 years. Proceedings of the Technical Conference on Climate--Asia and Western Pacific, Guangzhou, China, 15-20 Dec. 1980. World Meteorological Organization, Geneva, pp. 360-375.

Zwally, H. J., R. H. Thomas, and R. A. Bindschadler (1981). Ice-sheet Dynamics by Satellite Laser Altimetry. Technical Memorandum 82128. National Aeronautics and Space Administration, Goddard Space Flight Center, Greenbelt, Md., May 1981, 11 pp.

Zwally, H. J., R. A. Bindschadler, A. C. Brenner, T. V. Martin, and R. H. Thomas (1983a). Surface elevation contours of Greenland and Antarctic ice sheets. J. Geophys. Res. 88:1589-1596.

Zwally, H. J., C. L. Parkinson, and J. C. Comisco (1983b). Variability of antarctic sea ice and CO_2 change. Science 220:1005-1012.

6 Agriculture and a Climate Changed by More Carbon Dioxide

Paul E. Waggoner

6.1 INTRODUCTION

The crops that feed us stand outdoors in the wind, rain, and frost. Except for 17 pounds of fish in the 1400 pounds each American eats yearly, all the food for us and the feed for our animals, too, grows on a third of a billion acres of cropland and vast rangelands and pastures, exposed to the annual lottery of the weather. Although about 50 million acres of American crops are protected from drought by irrigation, even these depend on precipitation in the long run, and sheltering crops in greenhouses from temperature extremes is too expensive for staples. Thus, decade after decade grocery prices for all and hunger of the poor puncture the arrogance of our technology and remind us that a few chance degrees of warmth or drops of rain, properly timed, protect us. If it were not so, "Then all the concern about future climate that has been so widespread in recent years would be much ado about nothing" (McQuigg, 1979).

6.1.1 Concentrating on a Critical, Susceptible, and Exemplary Subject

Agriculture is broad. It is the science and art of the production of plants and animals useful to man and in varying degrees their preparation for man's use and their disposal by marketing. This chapter on agriculture could, therefore, encompass forests and fisheries as well as farmers, hunters, and grocers as well as farmers, logs, and hogs

---

Several individuals have assisted the author. Clarence Sakamoto, Norton Strommen, and Tom Hodges assembled the data and used the regression and simulator models to predict changes in yield. Herbert Enoch, Gary Heichel, Robert Loomis, Israel Zelitch, and James Tavares contributed to the section on the effects of CO_2 on photosynthesis and plant growth. Marvin Jensen and Glenn Burton advised on water and breeding. The author gratefully acknowledges these contributions, but it is the author alone who takes responsibility for any errors or omissions.

as well as crops. It could encompass the Ukraine and the Humboldt current, Iowa and the Hoboken docks, California and the Amazonian forest. Although weather surely affects all of these, their survey in a single chapter would be incomplete or superficial, and we have concentrated on a critical, susceptible, and exemplary part: American crop production.

Crops are critical because without their photosynthetic conversion of solar to food energy there would be neither bread nor meat. American crops are critical to Americans because they feed us and bring in $40 billion of our foreign exchange (USDA, 1982). And, they are critical to others: for example, in 1979 the United States provided 42% of the wheat and 19% of the rice exported by the nations of the world, and fully 43% of the world's corn crop is American (USDA, 1982).

The susceptibility of crops, rooted in place and exposed outdoors, makes them biologic indicators of weather. American crops growing from 35 to 49 degrees latitude are within the zone that meteorologists predict will experience a change in the weather as CO_2 increases, and thus these critical and susceptible crops are also exemplary. Often we shall concentrate further, examining wheat, corn, and soybeans, which outdistance in value any other American crop.

6.1.2 Agriculture and Past Changes in the Weather

Later paragraphs will show technical calculations and projections of changes in American crops matching projections of the weather. Some were derived from historical statistics. Changes in the weather cause changes in agriculture that are too complex to be distilled into a few statistics, however, and at this point a background of real life stories is painted. History is, after all, the laboratory notebook where the results of experiments performed by Nature herself are recorded.

Nature has not, of course, actually experimented less than 100 ppm with CO_2 within the human era. Nevertheless, her experiments with temperature, rain, and snow show in general how farmers are affected by atmospheric change, and then CO_2 may bring changes in temperature and water themselves.

When rain fell abundantly on Italy from 450 to 250 B.C., intensive agriculture was fruitful and the Roman Republic was vigorous. From 100 B.C. to A.D. 50 rain again fell abundantly and Roman civilization was high. After A.D. 80 rain was light, vines and olives replaced cereal, and the Roman Empire declined and fell (Brooks, 1970). This demonstrated the adaptation of agriculture by changing crops--but the new crop did not sustain the Empire.

The years 1301-1350 experienced a change in climate, and Brooks (1970) calculated a maximum of raininess for the half century. Tuchman (1978) wrote that the medieval people were unaware that

> ...owing to the climatic change, communication with Greenland was gradually lost, that the Norse settlements there were being extinguished, that cultivation of grain was disappearing from Iceland and being severely reduced in Scandinavia. But they

could feel the cold, and mark with fear its result: a shorter growing season.

This meant disaster, for population increase in the last century had already reached a delicate balance with agricultural techniques. Given the tools and methods of the time, the clearing of productive land had already been pushed to its limits. Without adequate irrigation and fertilizers, crop yield could not be raised nor poor soils be made productive. Commerce was not equipped to transport grain in bulk from surplus-producing areas except by water. Inland towns and cities lived on local resources, and when these dwindled, the inhabitants starved.

In 1315, after rains so incessant that they were compared to the Biblical flood, crops failed all over Europe, and famine, the dark horseman of the Apocalypse, became familiar to all. The previous rise in population had already exceeded agricultural production, leaving people undernourished and more vulnerable to hunger and disease. Reports spread of people eating their own children, of the poor in Poland feeding on hanged bodies taken down from the gibbet.

In this ancient experiment were demonstrations that colder as well as warmer weather can damage; also the length of the season as well as the mean temperature is critical, transportation can alleviate hunger, and the impact of changed climate is sharp at the poleward margin of farming.

In 1886-1893 an American experiment was performed in the Middle Border, and the Wayne Township (Kansas) Farmer's Club recorded the results (Malin, 1936). Problems were met with full force because movement into the region was swift during the most favorable weather of 30 years. After the farmers were settled, however, "Out of 471 acres of fall wheat there is not wheat enough to cover 15 acres. All winter killed." Although flooding was sometimes a problem, drought was the major force. In 1886 Turkey wheat was not favored because millers did not like it, but 4 years of hot summers and cold winters made the hardy Turkey wheat the favorite. Although in 1885 the acreage of corn per farm was four times that of wheat, by 1905 there were more than 3 acres of wheat for every acre of corn. The predominance of grazing land over cropland ran in cycles with the weather. A final adaptation was flight: in 1895 a traveler across the County reported that it was practically deserted. Enough people were left, however, for the subject of "Rainfall and the Populist Party in Nebraska" (Barnhart, 1925).

The Dust Bowl of the 1930s was a natural experiment with results dramatized in John Steinbeck's <u>Grapes of Wrath</u> (1939), and drifting soil is vividly remembered. The catastrophe was not only wilting of crops: pests encouraged by the drought amplified the damage (National Research Council, 1976).

The red spores of wheat rust were blown over the Wheat Belt and infected wheat; rust fungus erupted from stems and leaves. The greatest losses of wheat during 1921-1950 were in the 1930s: 3.4 million metric tons in 1935 and 1937 were much greater than the runner-up, 1 million metric tons in 1923.

Weeds invaded. The prickly pear cactus headed east, invading 1.6 million hectares in western Kansas. More young jackrabbits survived in drier weather, and productive rangeland was turned into pastures of prickly pear with a plague of jackrabbits.

Grasshopper plagues are generally associated with drought. Like a chapter in Exodus, the report of 1936 was of a loss of $106 million, more than the total gross income from all farm products in Arizona, Nevada, New Mexico, Utah, and Wyoming combined. The hoppers made such a clean sweep in South Dakota that jackrabbits, faced with starvation, escaped into Nebraska (Schlebecker, 1953).

Finally, an experiment in direct effect on cattle is added. A cattle boom began in Dakota in 1883, The Bad Lands Cow Boy claimed, "We have never heard of a solitary head ever having died in the Bad Lands because of exposure," and Theodore Roosevelt invested. In 1886-1887 storm piled on storm, children were lost and froze within a hundred yards of home, and cattle, desperate for shelter, smashed their heads through ranch house windows. The average loss of cattle was 75%, and Roosevelt rode for 3 days the following summer without seeing a live steer. Extreme weather rather than the average affected farming. A change of weather and the consequent boom to bust took only 3 or 4 years (McCullough, 1981).

Nature's experiments demonstrated the following:

- Farmers fit husbandry and crops to the weather.
- Swift change disrupts.
- Colder as well as warmer and wetter as well as drier can damage.
- Pests amplify effects of bad weather.
- The very soil can be changed by weather.
- Even warm-blooded animals are affected by weather.
- Occasional extremes destroy agriculture.
- Impact of changed weather is sharp in marginal climates.
- Migrations and political upheaval are blown by bad weather.
- Empires rise as well as fall as yields wax and wane in changing climates.

Against this background we shall now examine the range of change meteorologists set before us and then calculate or speculate from science on the changes in crops that would follow.

6.1.3 The Range of Change in the Atmosphere

Carbon Dioxide. The CO_2 concentration of the atmosphere has risen from below 300 parts per million by volume (ppmv) in the 1800s to about 340 ppmv at present. It is estimated that by the year 2025 it will likely reach 425 ppmv, and by 2065 it will probably pass 600 ppmv (Nordhaus and Yohe, this volume, Chapter 2, Section 2.1). Nordhaus and Yohe's upper estimate for CO_2 concentration in the year 2000 is 400 ppmv (Chapter 2, Figure 2.23). We shall see that CO_2 has a direct effect on crops.

Temperature. Available climate models indicate that a doubling of the CO_2 content could raise the global annual average surface temperature by 3°C (this volume, Chapter 4). If we assume the rapid rise in atmospheric concentrations to 400 ppmv in A.D. 2000, the temperature rise by A.D. 2000 would be about 1°C. In polar latitudes a doubling of the atmospheric CO_2 concentration would cause a 5 to 10°C warming (Mitchell, 1977). The polar and higher latitudes are sensitive because of summer changes in the albedo of regions normally covered by snow and ice throughout the year. For middle latitudes, Manabe and Stouffer (1980) calculated a lesser warming: 3°C at the U.S.-Canadian border after a doubling of CO_2.

Length of Growing Season. Kellogg (1977) pointed out a simple relation between summertime mean temperature and the length of the growing season at middle and high latitudes: A 1°C change in mean temperature for the summer corresponds approximately to a 10-day change in the length. We must beware that the predicted annual average temperatures include winter, and conceivable changes in daily amplitudes or variability (Neild, 1979) could modify the simple relation. Using this 10-day rule of thumb, however, one can reasonably consider the effect on crops of a lengthening of the growing season by 10 days in the northern United States where a 1°C increase in average annual temperature has been predicted by A.D. 2000.

Precipitation and Moisture. Manabe and his colleagues (1981) have projected the changes in soil moisture that will accompany the predicted rise in temperature with three models of increasing geographic detail. They estimated the change from present CO_2 concentrations to four times as much. The three models all predicted drier summers at middle and high latitudes. Snow would melt earlier at high latitudes, causing an earlier transition from spring to summer and less rain. The earlier onset of summer would evaporate and transport more moisture from the soil.

Conclusion. Looking toward a horizon near A.D. 2000, the agriculturalist may be uncertain how the climate of a precise place will change, but he can reasonably consider how crop production would be changed by an increase to about 400 ppmv, a mean warming of about 1°C in the northern United States with a growing season about 10 days longer, and more frequent drought in the United States caused by somewhat less rain and slightly more evaporation.

6.2 EFFECTS OF CO_2 ON PHOTOSYNTHESIS AND PLANT GROWTH

Carbon dioxide is a major substrate for photosynthesis and, therefore, can directly affect plant growth if CO_2 is limiting. The current 340 ppmv appears to be limiting (Figure 6.1) and, therefore, a rise in atmospheric CO_2 levels should increase photosynthesis. However, most effects of CO_2 on photosynthesis and plant growth have been studied and measured during short periods where other factors such as light, water, temperature, and nutrients are optimal. In addition, growth habits and adaptations to different environments might mitigate or alter the effects of changing CO_2 concentration. Yield is the important integration of physiology, and it is difficult to predict the changes in yield that might follow a rise to 400 ppm of CO_2. This section summarizes the current understanding of the direct effects of CO_2 on plant growth and the factors that must be considered in projecting these effects to yields. More discussion of direct effects can be found in the proceedings of the International Conference on Rising Atmospheric Carbon Dioxide and Plant Productivity, Athens, Georgia, May 1982 (Lemon, 1983).

In addition to showing the direct effect of CO_2 on plants, the curves of Figure 6.1 are critical in calculating the rise in CO_2 itself. The percent change in photosynthesis per percent change in CO_2 in the air, which is typically covered by the term "beta" in carbon cycle models (c.f. Woodwell, this volume, Section 3.3.3), must be about 0.25 to account for the rise caused by burning fossil fuel since the Industrial Revolution (Bjorkstrom, 1979). The beta of the illuminated wheat leaf in 400 ppmv CO_2 (Figure 6.1) is 0.8 but of the corn leaf is 0.1. These botanical estimates from the laboratory certainly do not conflict with the estimate of 0.25 from the global fuel consumption and atmospheric CO_2.

6.2.1 Photosynthesis

About 90% of the dry weight of plants derives from the reduction of CO_2 to carbohydrates by photosynthesis. In single leaves in bright light, net photosynthesis increases with CO_2 above the current atmospheric level of 340 ppmv, and this is confirmed in whole plants by greater crop yields (see Section 6.2.6).

On the basis of photosynthetic properties plants are classified into C_3 plants, C_4 plants, and crassulacean acid metabolism (CAM). Although photosynthesis occurs in all plants by the C_3 pathway, the C_4 and CAM plants have specialized steps to sequester CO_2 into the leaf. In C_4 plants CO_2 is incorporated into C_4-dicarboxylic acids that can be transported within the leaf to sites where CO_2 is released for photosynthesis. In specialized desert plants of the CAM type the pores in the leaves (stomates) open at night to collect CO_2 into organic acids that later regenerate the CO_2 inside the leaf for photosynthesis during daylight. This allows CAM plants to keep their stomates closed during the heat of daylight and thereby save water.

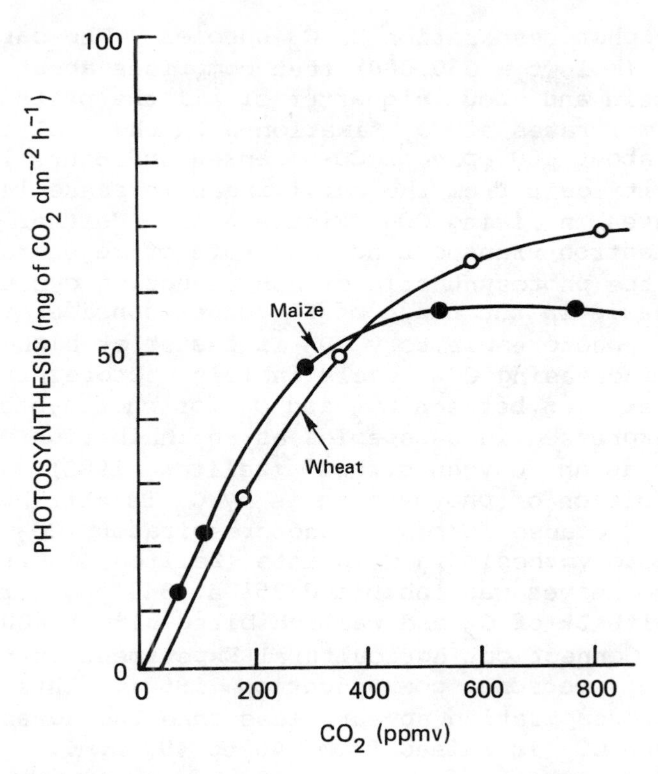

FIGURE 6.1 Typical photosynthesis response of plants to CO_2. Net photosynthesis of wheat is about 70 mg of CO_2 dm^{-2} h^{-1} compared with maize (about 55 mg of CO_2 dm^{-2} h^{-1}) for equivalent light intensity (0.4 cal cm^{-2} min^{-1}). Maize is saturated at a lower CO_2 concentration (~450 ppm) than wheat (~850 ppm). CO_2 in ppmv is percent $CO_2 \times 10^4$. (Adapted from Akita and Moss, 1973.)

The same enzyme that catalyzes the first step in the reduction of CO_2 to carbohydrates can also oxidize the first product. This oxidation occurs in light also but uses O_2 to oxidize carbon back to CO_2. It is called photorespiration, and its rate is determined by the ratio of CO_2 to O_2 within the leaf. Photorespiration rates are high in C_3 plants. By increasing CO_2 levels from 340 to 400 ppmv, CO_2 uptake is enhanced about 20% in C_3 (high-photorespiration) species and about 7% in C_4 (low-photorespiration) species (Hesketh, 1963; Akita and Moss, 1973). Thus, increasing CO_2 directly benefits C_3 species more than C_4 species. Rising CO_2 can be expected to alter leaf carbon metabolism and thereby affect the rate and duration of photosynthesis and the fate and partitioning of the photosynthate.

6.2.1.1 Rate of Photosynthesis

The faster CO_2 fixation per leaf area at higher CO_2 may be explained by the kinetic properties of ribulose bisphosphate carboxylase, the primary CO_2-fixing enzyme in the chloroplast, and to some extent by the consequently slower photorespiration in C_3 species. The carboxylase is a large molecule (mol wt = 560,000) that comprises about half the total chloroplast protein and about a quarter of all the protein in the leaf. Half-maximal rates of CO_2 fixation with the isolated enzyme are obtained at about 600 ppmv of CO_2 (Jensen and Bahr, 1977), which is not greatly difficult from the curvilinear increase in net photosynthesis of leaves in rising CO_2 (Figure 6.1). Neither the photochemistry and electron transport nor the rate of regeneration of the CO_2 acceptor by the photosynthetic carbon reduction cycle appear to limit photosynthesis in the range of CO_2 under consideration.

The release of photorespiratory CO_2 is faster at higher O_2 concentrations, while increasing CO_2 levels inhibit photorespiration (Zelitch, 1971). The competition between CO_2 and O_2 for photosynthesis and photorespiration is expressed in C_3 species as an inhibition of photosynthesis by O_2, or as an "oxygen stress" (Zelitch, 1982). A large proportion of the inhibition of photosynthesis by O_2 is attributable to photorespiration, because losses of photorespiratory CO_2 can be as high as 50% of net photosynthesis in C_3 plants (Zelitch, 1982). Photosynthesis in tobacco leaves was inhibited 35% at 340 ppmv of CO_2 at 21% of O_2 compared with 3% of O_2 and was inhibited 31% at 400 ppmv of CO_2 (R. B. Peterson, Connecticut Agricultural Experiment Station, New Haven, Connecticut, personal communication, 1982). Thus, the benefit of regulated photorespiration appears less than the advantage of faster carboxylation when CO_2 is raised from 340 to 400 ppmv.

6.2.1.2 Duration of Photosynthesis

Prolonged and faster photosynthesis caused by increased CO_2 in bright light produces more sucrose, sometimes increasing starch accumulation in the chloroplasts, and excessive starch can deform chloroplasts and decrease photosynthesis (Guinn and Mauney, 1980). Besides increasing sucrose, faster photosynthesis may change the levels of phosphorylated compounds and decrease orthophosphate levels in chloroplasts, which feed back to inhibit photosynthesis and decrease CO_2 assimilation (Walker, 1976). Such negative feedback might be bred against to obtain the full benefits of increased CO_2 on yield.

Yield depends on an adequate storage or sink to accept the products of photosynthesis. If the sink is inadequate, feedback will decrease photosynthesis. Sink capacity and yield tend to increase in parallel until they reach the limit set by the photosynthetic capacity (Evans, 1975).

6.2.1.3 Fate and Partitioning of Photosynthate

The products of photosynthesis and the efficiency of their translocation to the sites of conversion to starch, protein, and lipids may affect photosynthesis itself as well as the accumulation of carbon in the storage organ. Higher CO_2 decreases photorespiration and thus indirectly will affect nitrogen metabolism, since ammonia turns over rapidly in leaves during photorespiration, and the balance of amino acids available for storage will change (Lawyer et al., 1981). Although less lipids might be made in leaves, more may be produced in storage organs that synthesize lipids from translocated sucrose. These speculations are based on sound biochemistry, but data are lacking.

6.2.2 Drought

CO_2 directly affects the water in plants because CO_2 affects the pores or stomata through which water is transpired. The epidermis of leaves is generally perforated by countless microscopic stomata, where the size of the opening is regulated by the shrinking and swelling of two guard cells that border it. The CO_2 for photosynthesis is acquired through these pores, and since the pores open into the moist interior of the leaves, water is transpired. Most stomata close in the dark, conserving water when photosynthesis stops. In fact, mere narrowing of the pores conserves water, even in a crop with several acres of leaf surface area per acre of land (Waggoner et al., 1964). By regulating the size of stomata, plants can seek a balance between the necessary uptake of CO_2 and the stresses caused by excessive loss of water. We noted earlier the unique method used by CAM plants to acquire CO_2 under arid conditions.

In bright light, maize stomata narrow when CO_2 concentration increases from 300 to 600 ppmv, and transpiration decreases about 15% from the potted plants. Transpiration from pots of wheat decreased only about 5%, however, and the difference between wheat and corn is assumed typical of the difference between C_3 and C_4 plants (Akita and Moss, 1973). In crops in the field, Baker (1962, 1965) observed that transpiration of corn and cotton in bright light decreased by 20 and 35% between 300 and 600 ppmv.

The loss of water from the soil includes evaporation from the soil surface itself, especially when it is moist and unshaded by foliage, and from wet foliage as well as transpiration through stomata. Thus the loss of water from a field, month in and month out, will be affected less by CO_2 than by the percentages of 5 to 35% given above.

Although we know of no experiments showing that crops can tolerate drought in CO_2-rich air, there is some foundation to expect an increase to 400 ppmv of CO_2 would decrease transpiration from a crop with abundant foliage by the order of 5%, somewhat more in C_4 and somewhat less in C_3 crops. Since reduced transpiration would deplete soil water somewhat more slowly, drought would logically be somewhat less frequent.

6.2.3 Nutrients

The demand for nitrogen and mineral nutrients in plant growth is tied to photosynthesis. As photosynthesis increases with increasing CO_2, the carbohydrate available for plant growth will increase and, in turn, impose demands for increased fertilizer or available soil nutrients. Since plant biomass typically has a minimum nitrogen content of between

2 and 3% of dry matter and a phosphorus content near 0.2 to 0.3%, nutrients may be needed to realize any increase in production.

CO_2 may improve the availability of nutrients: Photosynthates are utilized as energy by symbiotic nitrogen-fixing organisms. The nitrogen in plants with such symbionts associated with their roots may increase. For the same reason more photosynthates might increase uptake of N, P, and K by roots in their association with bacteria and fungi. And finally, increased photosynthesis and growth could enlarge the pool of decomposing soil organic matter that can serve as a reservoir of soil nutrients.

6.2.3.1 Nitrogen Metabolism

Nitrogen fixation consumes much photosynthate for its energy. Nitrogen-fixing symbionts such as *Rhizobium* and *Frankia*, therefore, are greatly affected by the photosynthate available in the plant. While quantitative data are limited, experiments with CO-enriched plants show increased N_2 fixation. In field experiments, Hardy and Havelka (1977) showed more N_2 fixed in soybeans in air enriched with CO_2 to 800-1200 ppmv, and the total kilograms of N_2 fixed per hectare over the growing season was much higher with CO_2-enrichment. In terms of harvestable product, Rogers et al. (1980) found a 28% increase in weight of seeds harvested per soybean plant at 520 ppmv of CO_2 in comparison with 340 ppmv.

Of all the essential elements for plant growth it is usually nitrogen that is limiting. Thus the increased photosynthesis in a CO_2-enriched atmosphere will increase demand for nitrogen. There also might be a shift to plants capable of N_2 fixation and to plants that have a beneficial association with free-living, N_2-fixing microbes. The increased nitrogen in these plants would permit increased growth and, in turn, increased photosynthesis, more photosynthate, and even more N_2 could be fixed until some other factor became limiting.

6.2.3.2 Organic Matter and Rhizosphere Association

Increased atmospheric CO_2 will not likely affect the rhizosphere. The respiration of roots and soil microbes maintain CO_2 concentrations in the air spaces in soil that are 10 to 50 times higher than in the atmosphere. Doubling atmospheric CO_2 would likely have little direct effect on roots and soil microbes.

If CO_2 increased plant growth it could increase the plant remains incorporated as soil organic matter and influence the cycling of minerals in the soil and other soil properties. Soil organic matter is composed of two major factions that occur in roughly equal amounts. One factor is rapidly turning over and is composed of readily metabolized organic molecules such as cellulose; the other is composed of slowly accumulating, stabilized aggregates of phenols, other aromatic molecules, and inorganic particles with turnover times of at least 100 years (Van Veen and Paul, 1981). Although increases in degradable biomass from green manuring generally raise the activity of soil microbes and change soil organic matter only slightly, CO_2 will be recycled to the atmosphere faster in response to the increase of degradable organic matter. With increased substrates in the form of root exudates or degradable soil organic matter there may be increased nitrogen fixation by bacteria, both free-living and symbiotic, and increased mycorrhiza.

Many minerals required for growth, such as P, Zn, and Cu, are unavailable to roots because they are predominantly in immobile and insoluble forms in the soil. Micorrhizal fungi may increase their availability to host plants by penetrating more soil.

Thus, increased photosynthesis and degradable biomass are likely to increase soil nitrogen levels, perhaps by 5 to 10%, and may slightly decrease the phosphorus and soil nutrients not tied up in living biomass (Lemon, 1983).

6.2.4 Phenology

The development of new organs is distinct from their mere enlargement or growth, and one can reasonably ask if this development is affected by CO_2 in the air.

Although one plant, cucumber, flowered earlier and the fruit was ready for market two weeks earlier when the air was enriched with CO_2, another plant, pepper, proceeded at the same rate (Enoch et al., 1970). CO_2 did not change the period from pruning to harvest of roses (Zieslin et al., 1972), and a tenfold increase in CO_2 scarcely shortened the time from flowering to ripe strawberries (Enoch et al., 1976).

Thus changed timing of flowering and other stages in the life of crops is not likely to be an important consequence of the rise in CO_2 that we are considering.

6.2.5 Weeds

Environment affects crops, their pests, and the relation between them. Among the insect, disease, and weed pests of plants, it is the weeds fueled by their own photosynthesis that can be directly affected by changing atmospheric CO_2. We shall examine the effects of insects and disease on crops later in this chapter in the context of changes in temperature and moisture.

Annually weeds exact a toll of some $18 billion in the United States by competing with crops for light, water, and fertilizer (Chandler, 1981). Because C_3 and C_4 plants respond differently to changes in atmospheric CO_2 concentrations, the generalities of weeds versus crops have been examined in those terms. Will the competition tip toward the weed if it is C_3 and benefits from an increasing CO_2 concentration while the crop is C_4 and responds less to increasing CO_2?

Twelve of the 15 crops that feed the world are C_3 plants (Harlan, 1975), whereas 14 of the 18 most damaging weeds are C_4 (Patterson,

1982). Although that seems good fortune, 19 of the 38 major weeds of American maize, a C_4 plant, are C_3 plants (USDA, 1972).

The generality of C_3 gaining relative to C_4 plants in rising CO_2 has been demonstrated by crops and weeds. The C_3 weed velvet leaf (Abutilon theophrasti) increased its growth more than maize when CO_2 was increased. On the other hand, the C_4 weed, itch grass (Rottboellia exaltata), gained less than the C_3 crop, soybean (Glysine max) (Patterson and Flint, 1980).

In addition to a gradual increase of a few percent in growth, CO_2 can conceivably remove a limitation to the spread of a weed. Thus, okra, which is a crop becoming a weed, can grow at lower temperature and presumably at higher latitudes if CO_2 is enriched (Sionit et al., 1981b).

Limitations of fertilizer, water, or light might of course limit the realization of an advantage of CO_2 to a weed or crop. In fact, however, limited nutrients and water have failed to nullify the benefits of CO_2, including those to the height and leaf area that will affect competition for light (Patterson and Flint, 1982). Thus increases in atmospheric CO_2 may affect the competition of weeds and crops, sometimes to the advantage of the crop and sometimes to the weed.

6.2.6 Direct Effects of CO_2 on Yield

The integration and the practical outcome of all the effects enumerated above is yield. Calculating the direct effect of increased CO_2 on yield is chancy because of lack of experimental data. Few food, feed, or fiber crops have been grown at elevated CO_2 from sowing to harvest. Most experiments have been brief, with emphasis on a specific stage of growth, and have been conducted with flowers and ornamentals rather than crops. Kimball (1982) recently summarized the results of 70 CO_2 enrichment experiments conducted during the past 64 years. In Table 6.1 the results of several experiments are presented for the entire life cycle of crops.

All experiments were in growth chambers and greenhouses; no results from fields were available. In all experiments plants were grown in optimum environments without pests and with abundant water and nutrients. We cannot claim that light was optimum because plants seldom enjoy this status when grown in chambers and greenhouses. Although all experiments were performed in equable environments, all experiments were, nevertheless, dissimilar. Some experiments included stressful conditions as treatments. As these stresses might reflect a situation encountered in the field, their effect on response of yield to CO_2 was important to our evaluation.

First beta, here taken as the percent change in yield per percent change in CO_2, is examined. The range is a 0.1 to 0.9% increase in dry weight per percent increase in CO_2. As in the estimation of beta from the simple response to CO_2 shown in Figure 6.1, the global estimate of 0.25 for beta (see Woodwell, this volume, Chapter 3, Section 3.3) is not denied by these additional botanical estimates. In addition, there is evidence in Table 6.1 that drought or lack of

TABLE 6.1 Changes in Yields of Crops in Optimum and Stressful Environments Anticipated from Atmospheric Enrichment to 400 ppmv of CO_2

Crop	Change in Yield (%)	Component Harvested	Yield Increment/ CO_2 Increment (%/ppmv of CO_2)	Yield Change by Enrichment (%/60 ppmv of CO_2)	Reference
Optimum Environments					
Barley	0.9[a]	Grain	0.18	11	Gifford et al. (1973)
Corn	0.28	Young shoots	0.03	1.9	Wong (1979)
Cotton	0.6	Lint	0.34	20	Mauney et al. (1978)
Soybean	0.4[a]	Grain	0.04	2	Hardman and Brun (1971)
Wheat	0.4	Grain	0.13	8	Gifford (1979)
Wheat	0.3	Grain	0.07	4	Sionit et al. (1980)
Wheat	0.6	Grain	0.13	8	Sionit et al. (1981a)
Stressful Environments					
Corn (1/3 normal N)	0.28	Young shoots	0.03	1.9	Wong (1979)
Wheat (water limited)	0.6	Grain	0.44	26	Gifford (1979)
Wheat (1 H_2O stress cycle)	0.5	Grain	0.10	6	Sionit et al. (1980)
Wheat (2 H_2O stress cycles)	0.2	Grain	0.05	3	Sionit et al. (1980)
Wheat (1/8 normal nutrient)	0.1	Grain	0.02	1	Sionit et al. (1981a)

[a]Calculated from shoots only.

fertilizer will not prevent any increase in photosynthesis and a sequestering of more CO_2 as its concentration increased.

Proceeding to the yield of grain or cotton lint, one sees increases of 0.07 to 0.34% per ppmv increase in CO_2 in optimum growing environments. There is no clear evidence that this relative change is less in wheat that lacks water, but the wheat deprived of nutrients did respond little to CO_2.

We conclude from the examples in Table 6.1 and from the extensive survey of Kimball (1982) that CO_2-enrichment to 400 ppmv by A.D. 2000 may increase the annual yields of well tended crops by, say, 5%. In comparison, the yield of Illinois corn roughly quadrupled during the past half century, and about three fourths of the change is attributable to improved husbandry (Haigh, 1977). Although the evidence of Table 6.1 for poor growing conditions is equivocal about quantity, some increase in yield--even in poor circumstances--is indicated.

6.3 PREDICTING THE CHANGES IN YIELD THAT WILL FOLLOW A CHANGE TO A WARMER, DRIER CLIMATE

Two orderly means are at hand to calculate how the yields of corn, soybeans, and wheat will change if rising CO_2 makes the climate warmer and drier. In one method, the statistical regression model, history is distilled to obtain the change in yield for a specified change in weather, for example, the decrease in quintals of wheat/hectare in Kansas after a 10% decrease in March precipitation. In the other method, the physiology of wheat and the physics of evaporation are assembled and used to compose a computer program or "simulator" of wheat to calculate the change in wheat yields per change in environment. We shall employ both these methods, comparing and verifying them as well as making predictions.

More must be said of the climatic change to be caused by an increasing concentration of atmospheric CO_2 than simply "1°C warmer and a bit drier." The direct effects of CO_2 on plant growth have already been dealt with in the preceding section and are not incorporated here. To approximate the CO_2-induced weather change for the year 2000, we shall employ the actual weather data for a location, which includes the natural correlations between temperature and rain, and then increase the temperature by 1°C and subtract 10% of the precipitation. The reader must remember the vagueness of predictions of changes and realize the 1°C and 10% drier is merely a rational example and not a prediction. The corresponding change in the length of the growing season will not affect the historical models unless spring and fall temperatures are employed and are proxies for frost; the growing season will be explicitly changed in the simulations. The change in evaporations with 1°C warming will implicitly enter the historical models and be explicitly calculated in the simulations from the increased warmth.

How shall the results be presented? The historical models provide coefficients of quintals/hectare/millimeter and so forth that we can examine. The simulators allow us to calculate the frequency distributions, year by year, of yields in the present climate and the distributions, year by year, if the climate, on the average, warms by 1°C and lacks 10% of the precipitation. This is a more informative result than a single normal or average, which so rarely occurs.

6.3.1 History

For decades the yield of a "crop reporting district" has been estimated after each growing season. A crop reporting district often includes several counties. For example, Iowa is composed of nine crop reporting districts. The Cooperative Climatological Service of the National Weather Service has been recording temperature and precipitation daily at about a dozen stations in each district. Using these data, Thompson (1969) estimated the effect of weather by correlating the yields with the weather.

Crop yield, temperature, and precipitation for a district are used in statistical regression models. The model for a particular crop in a particular area is generally expressed as

$$Y_i = a + b_1 t_i = b_2 X_{2i} + \dots + b_n X_{ni},$$

where

Y_i	= estimated yield in the ith year;
a	= intercept;
t_i	= surrogate for technology in the ith year;
b_1	= coefficient representing the effect of technology in quintals/hectare/year;
b_2 to b_n	= coefficients representing the effect in quintals/hectare/unit change in the weather;
X_{2i} to X_{ni}	= weather variables such as precipitation, temperature, potential evapotranspiration (PET), and evapotranspiration (ET) in the ith year.

Thus, by multiple linear regression the effect of weather, factor by factor, on the yields of crops is distilled from a history of weather and yields, accumulated for decades by faithful observers.

Beginning in the mid-1970s a comprehensive set of estimates of the b coefficients was made by the National Oceanic and Atmospheric Administration, the U.S. Department of Agriculture, the National Aeronautics and Space Administration, and the University of Missouri for the three major crops in the states of the nation's grain belt. In some cases the temperature and precipitation were combined into variables representing evaporation or periods of excessive heat (Sakamoto, 1978).

Our results are presented state by state. The results for a state, however, were calculated from regressions for the several districts in a state and then aggregated to the whole state because this provides a better estimate for a state than directly relating state yield to state weather (Sakamoto, 1982). For example, yields estimated for the nine districts in Iowa and then aggregated are within about 8% of the actual state yield (LeDuc, 1980). The parameters that we shall show for each state are aggregations of the districts of a state.

The effect of weather on wheat is seen in Table 6.2. Look at the column for the Red River Valley, which is northern and likely to experience a change in climate from rising CO_2. The valley includes eastern North Dakota and western Minnesota. The increase in yield for a 1°C warming in April, the planting season, is represented by a b coefficient of 0.21 quintal/ha. On the other hand, a 1°C warming in June or July will increase the frequency of hot days over 32°C and yield will be decreased by 0.25 and 0.42 quintal/ha for each hot day added to June and July. In the Red River Valley, spring wheat is heading in June and July, and hot spells during heading decrease yield. In Kansas, heading occurs in winter wheat around May; therefore hot spells in Kansas in May decrease yields there. South Dakota shows the effect of weather in the transition region between winter and spring wheat where spring wheat is relegated to unfavorable sites where

TABLE 6.2 The Effect of Weather on Yields (in quintals/ha) of Winter Wheat (WW) and Spring Wheat (SW)[a]

	Crop Region, State					
	Red River Valley (SW)	North Dakota (SW)	South Dakota (SW)	Nebraska (WW)	Kansas (WW)	Oklahoma (WW)
A. Variable						
Yield, Average 1978-1980	18.2	14.9	12.0	21.3	21.3	19.7
Temperature, °C						
Jan to Feb	--	--	--	-0.24	--	--
Apr	0.21	--	--	--	--	--
May	--	--	--	-0.74	--	--
July	--	--	-0.69	--	--	--
Temperature, No. of days above 32°C						
May	--	--	--	--	-0.30	-0.16
June	-0.25	-0.42	-0.29	-0.22	--	--
July	-0.42	-0.21	--	--	--	--
Precipitation, mm						
Sept to Nov	--	--	0.01	0.02	--	--
Aug to Nov	0.01	0.02	--	--	0.02	--
Sept to Dec	--	--	--	--	--	0.01
Jan to Feb	--	--	--	--	--	0.01
Mar	--	--	--	--	0.06	0.03
May	--	0.02	--	-0.02	--	-0.02
June	--	0.02	--	-0.04	-0.01	-0.02
Combined Variables						
Mar Prec minus PET[b]	--	-0.01	--	--	--	--
May Prec minus PET[b]	--	--	--	--	-0.01	--
Apr Prec/PET[c]	--	--	0.71	--	--	--
$(\text{Apr Prec/PET})^2$	--	--	-0.322	--	--	--
B. Calculated Estimated Changes						
Yield, quintals/ha	-1.32	-1.77	-1.36	-1.04	-1.04	-0.37
Change from 1978-1980 average	-7%	-12%	-11%	-5%	-5%	-2%

[a]A. The effects are given as b coefficients in quintals/ha/unit of variable, i.e., mm of precipitation, °C of temperature, or fraction of a ratio (after Sakamoto, 1978). B. Estimates of the change in yield with a 1°C increase in temperature and a 10% decrease in precipitation from the historic average temperature and precipitation recorded for the regions.

[b]PET, potential evapotranspiration in millimeters, a measure of the demand for water estimated from temperature.

[c]The variable Apr Prec/PET is calculated as the deviation from normal. This combined variable and its square produce a curvilinear response with both too little and too much rain relative to the demand, PET. Either extreme in rainfall is harmful.

the higher yielding winter wheat will not prosper. Hence, in South Dakota higher temperatures in July decrease wheat yields 0.69 quintal/ha for each degree centigrade of warming.

The effect of precipitation on yield also can be seen in Table 6.2. In Kansas, a decrease of 1 mm in March precipitation causes a decrease in yield with a b coefficient of 0.06 quintal/ha for each millimeter decrease in rain. In addition, precipitation (Prec) can be combined with a measure of the demand for water, potential evapotranspiration (PET), in the combined variable April Prec/PET as is shown in Table 6.2 for South Dakota. A decrease in precipitation would also mean an increase in potential evapotranspiration. Thus a decrease in April precipitation in South Dakota of about 10% would decrease the ratio April Prec/PET about 16%, causing a very small decrease in yield.

To appraise the consequences of climatic change caused by rising CO_2 we must, however, examine the consequences of both warmer and drier weather. For example, to obtain an estimate of the change in wheat yield in South Dakota if weather becomes 1°C warmer and 10% drier one needs only to multiply the b coefficients in Table 6.2 by the change in the recorded weather variables normal for the region. A 1°C increase in temperature might mean two more days in June with temperatures over 32°C. A 10% decrease in the September to November precipitation is a decrease of 8 mm of rain for this 3-month time period. Combining a 1°C increase in July temperature, two more days in June over 32°C, an 8-mm decrease in September to November precipitation and a 16% decrease in April Prec/PET would reduce the average yield by 11% from 14 to 12.53 quintals/ha. The calculation is (+1°C)(-0.69 yield/°C) + (2 days)(-0.29 yield/day) + (-8 mm)(0.01 yield/mm) + (-16% ratio)(0.71 yield/ratio) + (-16% ratio)(2)(-0.32 yield/ratio). Because the relation between the ratio April Prec/PET and yield is curvilinear, the effect of decreasing it 16% is negligible. In short, a reasonable consequence of a CO_2 increase and climate change is an 11% decrease in the wheat yield for South Dakota of 1.36 quintals/ha.

At a southern location, Kansas, the combined changes in temperature and rainfall produce a somewhat smaller relative decrease. Two more very hot days would occur in May; a 10% decrease in the August to November precipitation is a decrease of 22 mm, and the 10% decrease in March and June precipitation would be 4.5 mm and 10 mm, respectively. The outcome of all this is a 1.04 quintal/ha loss or about a 5% decrease in yield in comparison with the average yield for the 3 years 1978-1980.

The effect of weather on corn is seen in Table 6.3. In Illinois a warming of 1°C in July decreases yield 1.56 quintals/ha or about 2% of the 3-year average yield of 68.8 quintals/ha. July is the time of pollination, and at this stage corn is most sensitive to heat. In August, after pollination, the effect of heat is only half that in July. Warmth in October raises yields 0.57 quintal/ha/°C because October is frost time.

July is a critical time for precipitation as well as heat. A 10% decrease in July rain and correspondingly an increase in potential evapotranspiration would decrease July Prec minus PET by about 11 mm and yield by 0.28 quintal/ha. A decrease in precipitation during planting in May, on the other hand, increases yield.

TABLE 6.3 The Effect of Weather on Yields (in quintals/ha) of Corn in Three States[a]

	Crop Region, State		
	Iowa	Illinois	Indiana
A. Variable			
Yield, Average 1978-1980	72.7	68.8	65.3
Temperature, °C			
July	--	-1.56	--
Aug	--	-0.64	--
Oct	--	0.57	--
July to Aug Average	--	--	-2.34
Precipitation, mm			
May	--	--	-0.017
Sept to June	0.013	--	--
Sept to June SDFN[b]	-0.0001	-0.00006	--
July	--	--	0.045
Combined Variables			
Apr and May PET[c]	-0.12	--	--
May Prec/PET	--	-1.49	--
July Prec minus PET	0.076	0.025	--
(June ET/ET[d] + July ET/ET)/2	--	--	2.27
B. Calculated Estimated Change			
Yield, quintals/ha	-2.36	-1.72	-2.80
Change from 1978-1980 average	-3%	-3%	-4%

[a]A. The effects are given as b coefficients in quintals/ha/unit of variable, i.e., mm of precipitation, °C of temperature or fraction of a ratio (after Leduc, 1980). B. Estimates of the change in yield with a 1°C increase in temperature and a 10% decrease in precipitation from the historic average temperature and precipitation recorded for the regions.
[b]SDFN, departure from normal precipitation, squared.
[c]PET, potential evapotranspiration in millimeters, a measure of the demand for water.
[d]ET/ET, evapotranspiration divided by average evapotranspiration.

To appraise the consequences of a combined warming and drying of Illinois by rising CO_2, we decreased the ratios of May Prec/PET by 12%, decreased precipitation by 10%, and raised temperatures 1°C. The net effect is a decrease of 1.7 quintals/ha, or 3%. The decrease is moderated because less rain is favorable for corn planting in May but unfavorable for grain development and growth during July to September. Likewise, a favorable, warmer October moderates the effect of an unfavorable increase in temperatures during July and August.

Thus a steady warming and drying throughout the year can produce a modest effect on yield because warmer and drier is beneficial at planting and harvest, whereas it is harmful at the height of the summer.

In Iowa, unlike in Illinois, the warmer spring is unfavorable as shown by the negative coefficient for April and May PET, the decrease in September to June precipitation is unfavorable, and the warmer summer decreases July Prec minus PET. The net is a greater decrease in Iowa, 2.4 quintals/ha, than in Illinois, showing geographical variations in the consequences of the same change in climate.

The effect of weather on soybeans is seen in Table 6.4. Yields are normally about 22 quintals/ha in the three states of Iowa, Illinois, and Indiana. In Iowa a wet winter makes a late spring and a high September-to-May precipitation; a decrease of 10% or 47 mm in September-to-May precipitation would increase yield by 0.2 quintal/ha. A warming of 1°C in June would increase yield of the warmth-loving soybeans by 0.43 quintal/ha and in September by 0.35 quintal/ha.

The effect of summer rain in Iowa is in the ratios of actual evapotranspiration as a fraction of the long-term average. Assuming that evapotranspiration by the plants is limited by the water available, a 10% decrease in rain would decrease evapotranspiration (ET) by 0.1 and yields by 1.9 or 0.6 quintals/ha, respectively, if the decrease were in July or September.

The effect of changes in both temperature and rainfall in Iowa were calculated for a 10% decrease in precipitation and a decrease of 10% in the ratio of ET to ET with a 1°C warming. The net effect is a loss of 1.5 quintals/ha or a 7% decrease below 1978-1980 average yields.

Now we see how history can foretell the outcome of a changed climate. A couple of remarks are in order: "multiple collinearity" can mislead us, and simpler parameters might be obtained.

If two climatic factors are correlated, their correlation changes the regression coefficient, say, quintals per hectare per degrees Celsius in the tables. Sometimes the correlation will cause the coefficient to overstate and sometimes understate the effect of the factor, say, temperature on yield. It may even produce a nonsensical value. This is the problem of multiple collinearity. Fortunately in Tables 6.2, 6.3, and 6.4 the coefficients generally make good agricultural sense.

In fact, the problem of multiple collinearity has been minimized in the tables by combining variables. For example, July precipitation and temperature are correlated, and both affect Iowa corn. The two were combined in a single variable, July Prec minus PET, by calculating PET from temperature and subtracting it from rainfall, making a single factor expressing the harm to corn from drought in July or the benefits from timely rain during tasseling.

The invention of combined variables to avoid collinearity problems and the importance of different seasons in different states with different crops produced the plethora of variables in the tables. Likely, repeating the calculations with an eye for simplicity and standardization could reduce the plethora. Nevertheless, history has been distilled into parameters in the tables for an objective forecast of the agricultural consequences of a climate changed by more CO_2.

TABLE 6.4 The Effect of Weather on Yields (in quintals/ha) of Soybeans in Three States[a]

	Crop Region, State		
	Iowa	Illinois	Indiana
A. Variable			
Yield, Average 1978-1980	23.6	21.9	22.0
Temperature, °			
May	--	--	0.13
June	0.43	--	0.35
Sept	0.35	--	--
Precipitation, mm			
Sept to May	-0.0042	-0.0031	--
July	--	--	0.019
July Prec SDFN[b]	--	-0.0003	-0.0002
August	--	0.016	--
Combined Variables			
July Prec minus PET[c]	--	0.039	--
July ET/ET[d]	19.24	--	--
Aug ET/ET	--	--	11.83
Sept ET/ET	6.06	1.99	--
B. Calculated Estimated Change			
Yield, quintals/ha	-1.55	-0.82	-1.25
Change from 1978-1980 average	-7%	-4%	-6%

[a]A. The effects are given as b coefficients in quintals/ha/unit of variable, i.e., mm of precipitation, °C of temperature or fraction of a ratio (after Motha, 1980). B. Estimates of the change in yield with a 1°C increase in temperature and a 10% decrease in precipitation from the historic average temperature and precipitation recorded for the regions.
[b]SDFN, departure from normal precipitation, squared.
[c]PET, potential evapotranspiration in millimeters, a measure of water demand.
[d]ET/ET, evapotranspiration divided by average evapotranspiration.

In the large, the tables and our calculations have shown some geographical variation in the outcome of climatic change. For example, there are substantial differences in the average yield between spring and winter wheat, therefore, a decrease of 1.36 quintals/ha in South Dakota is an 11% decrease in yield, whereas a decrease of 1.04 quintals/ha is a 5% drop in yield for Kansas. Also, the season of sensitivity to weather is earlier in Kansas than northward in South Dakota. Although we cannot appraise the consequences, it is well known that the relative variability of precipitation and hence variability of yield increases as the amount decreases.

The effect of a change in weather is a net of several biological impacts and hence b coefficients. Thus the effect of 1°C warming may be advantageous in the spring and fall and disadvantageous in the summer. This tempers the net outcome of the changed weather.

Finally, the b coefficients distilled from history indicate that the assumed climate change caused by CO_2 would reduce yields of the three crops in several states by 2 to 13%. However, a caution must be made concerning the calculated estimates of yield changes given in Tables 6.2, 6.3, and 6.4. These values are estimated on the basis of a uniform weather change of 1°C warmer and 10% drier throughout the whole year. It is unlikely that the warmer and drier climate brought about by an increase in atmospheric CO_2 concentrations will be so evenly modulated through all the seasons nor from place to place. Also, we must assume that the killing extremes of weather are related to means in the same way in the changed climate as they are now. Nevertheless, the small percentage decreases estimated are remarkably consistent for the three major crops throughout the American grain belt.

6.3.2 Simulation

The physiology of the crop and the physics of evaporation provide an alternative to history as a foundation for predicting the consequence of changed temperature and moisture.

A well-known simulator of crop growth was composed by Duncan et al. (1967). Essentially, they assembled such things as the photosynthesis of maize leaves as a function of radiation and temperature, the architecture of the canopy of leaves, and the calendar of crop development. Other simulators have incorporated more details of metabolism, including the division of plant stuff between foliage and grain (Maas and Arkin, 1980). Still others have incorporated the transpiration of water and hence the depletion of soil water and occasional drought (Richardson and Ritchie, 1973). Simulators have been composed for wheat (Maas and Arkin, 1980) and soybeans (Curry, 1971). Although simulators are not perfect, they are logical assemblies of current physiology and physics of crops, and we shall examine the consequences of climatic change with the wheat simulator.

The wheat simulator begins a season with parameters specifying a variety, date of planting, the water in the soil, and the soil's ability to hold water. Day by day the simulator uses the temperature, rainfall, and solar radiation to calculate use of water, plant growth, and its allocation to leaves and grain. If it becomes cold, evaporation, growth, and progress through the life cycle slows. If it rains, soil moisture is replenished. And if it becomes hot, growth and development decrease, evaporation increases, and soil moisture may be depleted, inhibiting growth. At the end of the season, the sum of the growth of grain is the yield, which reflects the variety and, especially, the weather.

To perform our simulations we must have some weather observations. We began with the actual observations of nine crop reporting districts in North Dakota, each district with 15 to 20 stations. We chose the

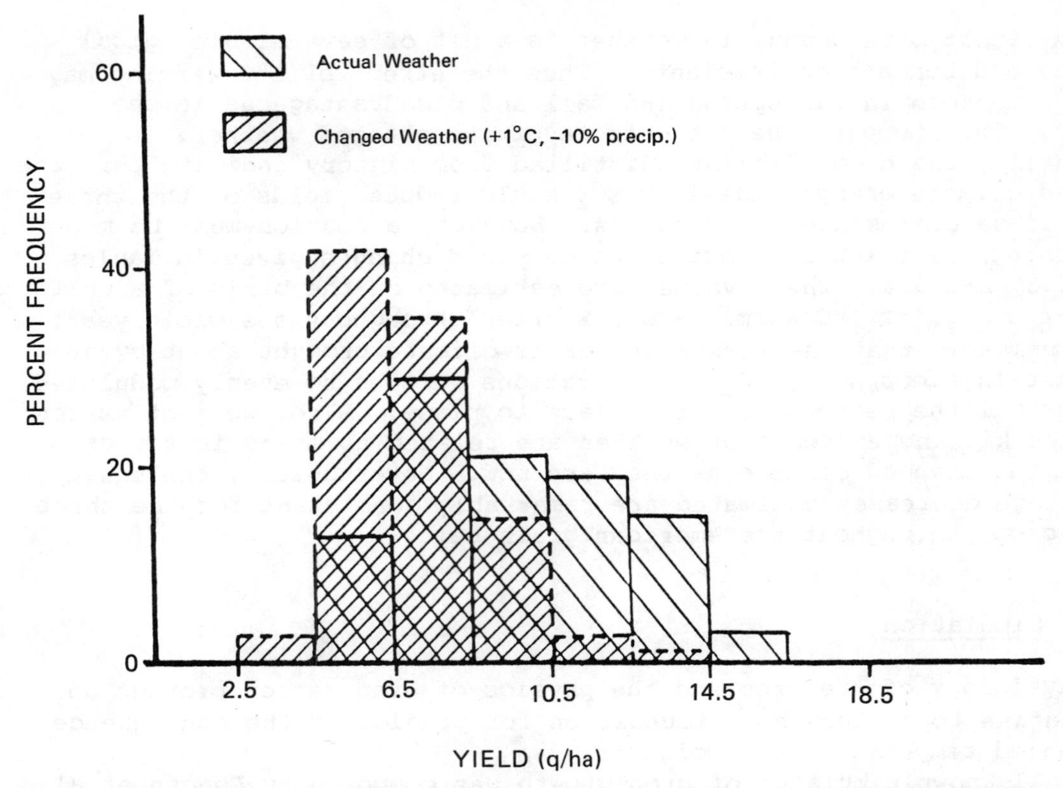

FIGURE 6.2 North Dakota simulated spring wheat yield (1949-1980).

years 1949 to 1980. Because only temperature and rainfall were observed at these stations, the solar radiation required by the simulator was estimated from the temperature and precipitation (Richardson, 1981). We also needed a planting date for the simulated wheat yield, and we estimated it by the method of Hodges and Artley (1981) in soil with a capacity for 175 mm of water. For the first spring of our simulations we assumed that the soil was filled to capacity, and in later springs the content was calculated from the rain and evaporation of the preceding months. We used the parameters for a common variety of spring wheat in North Dakota. In the simulator an improvement in technology would be reflected in a change in some of the parameters. Thus, a better variety would change the productivity for the North Dakota variety.

For each year we simulated the yield for each crop reporting district and added these to make a yield for North Dakota. The simulated yields for the 32 years, 1949-1980, form a frequency distribution of 4.5 to 16.5 quintals/ha for the present climate (Figure 6.2). The distribution is skewed with a median yield of 8.5 quintals/ha. This is less than the actual average yield of 12.0 quintals/ha for the years 1978-1980 (see Table 6.2). To eliminate the underestimation of the simulated yield would require tuning the model to provide values close to existing yields. We can, however, use the untuned model to calculate the relative change if rising CO_2 makes the climate warmer and drier.

The weather for each crop reporting district was changed for increased CO_2 in the atmosphere by adding 1°C to the daily temperatures and decreasing each rain or snow fall by 10%. The 1°C warming increased evaporation and also the solar radiation, which we calculated from the temperature. The distribution of yields in the changed climate has a median yield of 6.5 quintals/ha (Figure 6.2). Both the median and range are 2 quintals/ha less than for the unchanged weather.

The simulated change of 2 quintals/ha is more than the 1.36 quintals/ha estimated from the b coefficients of Table 6.2. In relative terms the 2 is, of course, a substantially greater fraction of the median of 8.5 than 1.36 is of its mean of 12 quintals/ha. Having to choose between the two, one chooses the actuality incorporated in the b coefficients and the 11% decrease and is pleased that the simulation is in the same direction and range. In addition, we have gotten a frequency distribution of yields from the simulator. In some years, the changed climate provided larger yields than the median for the normal climate, but in general the warmer and drier climate from a changed CO_2 would decrease yield.

6.3.3 Summary

The changes in yield that would follow a change to the warmer, drier climate assumed to follow an increase in atmospheric CO_2 have now been forecast by two orderly, explicit, and objective means. The reader knows we have not even employed every coefficient in our tables nor simulated wheat yields in every important region. Neither have we simulated corn or soybean yields. We have not combined the beneficial direct effect of CO_2 on plant growth with the disadvantages of warmer and drier, considered possible geographic shifts of the grain belt nor incorporated the changes breeders will make in varieties to exploit a new climate. We have considered only the single change of warmer and drier.

Despite the fog created by the foregoing list of qualifications and warnings, a clear conclusion shines through. The warmer and drier climate assumed to accompany the increased CO_2 will decrease yields of the three great American food crops over the entire grain belt by 5 to 10%, tempering any direct advantage of CO_2 enhancement of photosynthesis.

6.4 PATHOGENS AND INSECT PESTS

Sometimes a change in the weather that has only a modest direct effect on a crop is amplified into a disaster by a third party--a pest.

A concrete example of a plant disease (Tatum, 1971) makes the point. The spring of 1970 was wet in the Southeast. An early hurricane striking Biloxi, Mississippi, exemplified the weather. The summer was also wet and warm in the Corn Belt. Summer moisture in the Corn Belt generally benefits corn, and a bumper crop should have been produced. Thanks to a third party, a fungal pest, the parties of hybrid corn and moist weather did not produce the expected crop. It was, in fact, an eighth less than the year before, and yield per acre was off by one sixth. That was 535 million bushels of corn for the United States.

Before a change in the weather can evoke an epidemic, a susceptible crop and a virulent pest must be on stage. In the United States, 70 million to 80 million acres are planted to a single host, corn; about 1 acre of cropland in 5 is planted to the single crop of corn.

The host was uniform. When hybrid corn was introduced in the 1930s, seed was produced by detasseling or emasculating the rows that were to be the female parent and allowing pollen to fly only from the rows designated as male parent. In 1951, geneticists invented a labor-saving scheme to obviate detasseling in the hot sun. They found in Texas a cytoplasm, T-cytoplasm, that could sterilize the tassels of the female parent, and they also found a restorer gene that the male parent could transmit to the offspring, letting it produce fertile pollen and a crop in the next year. The method was so profitable that by 1970 most of the 67 million acres were planted to seed that inherited its cytoplasm from a single male-sterile plant from Texas. Despite the expanse of host, however, it was healthy. In a sampling of years from 1935 to 1969 only half as many reports were published about corn disease as about diseases of a crop grown on a much smaller acreage (Ullstrup, 1972).

The year 1970 changed things. A new pest came on stage: Race T of *Helminthosporium maydis*.

The microscopic fungi of the genus *Helminthosporium* have long caused only minor diseases of corn. Because all the *Helminthosporium* diseases were accused of decreasing corn yield only about 2%, Americans naturally paid little attention when Philippine scientists reported in 1961 an especial susceptibility of T-cytoplasm corn to *Helminthosporium maydis*.

During 1969, however, greater-than-normal susceptibility to *Helminthosporium* was observed in seed and test fields, and over winter the association between T-cytoplasm corn and Southern corn leaf blight, the disease caused by *Helminthosporium maydis*, was confirmed (Tatum, 1971).

The pathogenic fungus had done what many shifty pathogens had done before: it had produced a new race that was selected for the vast acreage of uniform host, and it burst forward in the moist weather of 1970. An urban society studied agriculture, the Chicago *Tribune* alone publishing 37 articles in 1970 (Ullstrup, 1972). A mutated microscopic fungus had amplified a change to a warm, moist summer, and the little fungus began a series of calamities that brought wheat deals, expensive groceries, and beef boycotts to the early 1970s.

Other diseases, like the wheat rust mentioned in Section 6.1, also amplify the effect of changed weather.

The pests of insects and weeds are other amplifiers. Whereas the encouragement of grasshoppers by drought cited in Section 6.1 is straightforward, some amplifications are more convoluted. Aphids carry viruses that infect crops, and because aphids usually multiply faster and move more frequently in warm weather than in cool, viruses spread faster in southern potato fields, and seed certified to be virus-free is grown in northern Maine.

Biological control provides another convoluted case. The walnut aphid attacked orchards in both the warm, dry San Joaquin Valley and cooler coastal valleys. A parasitic insect from France controlled the walnut aphid in the cool coastal valleys, but an Iranian biotype of the parasite was required to control the aphid pest on walnuts in the warmer Valley (Fraser and van den Bosch, 1973).

Being photosynthesizing plants, weeds can be directly affected by CO_2 as described above. They can also be affected by changes in the weather and amplify the effect on crops. A "pestograph" (U.S. Dept. of Agriculture, 1970; National Research Council, 1976) shows how. Field bindweed and Canadian thistle compete well with crops in cool, dry climates. If the climate is moister, the thistle persists and is joined by quackgrass. If instead the climate is both moister and warmer, both thistle and bindweed suffer and foxtail, coffeeweed, and nutsedge prosper.

Adding a convolution, we mention that the pesticides for controlling diseases, insects, and weeds are affected by the weather, especially moisture. Moisture may affect the absorption of a pesticide, particularly an herbicide, or rain may simply wash it away.

Because pest outbreaks require the threesome of virulent pest, susceptible crop, and favorable weather, outbreaks cannot be predicted from weather alone. Many seasons of favorable weather passed before the Southern corn leaf blight epidemic of 1970 or the tobacco blue mold epidemic of 1979 (Lucas, 1980) because another factor was missing.

Nevertheless, week-to-week changes in pests can be anticipated from the weather, and mathematical models of the effect of weather on pests have been composed. These range from statistical regressions between weather and disease on wheat (i.e., Burleigh et al., 1972) to simulations on a computer of how weather affects the stages in the life of a pathogen (Waggoner et al., 1972). Some simulations even unite a simulator of a weevil with one of alfalfa (Miles et al., 1974).

The practical goal of such models is generally more precise timing of pesticides, no needless application, and fewer pests. Given a pest to consider, however, the models can show how a pest will change with the weather--if new actors do not enter. History shows, however, that pests are so varied and variable that new and surprising pests will appear so frequently that a quantitative calculation for a given pest would give a misleading certainty. Instead, agriculture must expect novel pests, keep its research system in trim, and control new pests as they arise or the direct and indirect effects discussed above will be rendered trivial by a shifty and aggressive pest.

6.5 IRRIGATION IN A WARMER AND DRIER CLIMATE

Irrigation is both important and susceptible. Its importance derives from its expanse and from the value of irrigated crops. Fully 50 million acres or about 1 in 7 American acres of cropland are irrigated. The quarter trillion cubic meters of irrigation water withdrawn from American streams and groundwater represents about half of all withdrawals of this natural resource. Averaging wheat yields over all

American fields, humid as well as arid, one sees an average of 1.9 ton/ha on unirrigated versus 3.7 ton/ha on irrigated land. Valuable crops are grown on irrigated fields because irrigation reduces variability of water and produces consistently high yields. Thus, most of the irrigated cropland (44 million acres) occurs on only 12% of the farms, and these farms produce fully 40% of the market value of the crops from all American cropland! (Jensen, 1982b.)

Since irrigation uses runoff and runoff is a fraction of precipitation, the effect of a prolonged change in the amount of precipitation will usually be a greater change in the runoff to supply irrigation. For example, a simulation of a 10% decrease in precipitation over a 2000-km^2 watershed with an annual mean runoff of 400 mm showed a decrease in runoff of 25% to 300 mm per year. Furthermore, a 10% decrease in precipitation on a 9000-km^2 watershed with only 11-mm mean runoff would decrease runoff 40% to 7 mm per year (J. Nemec and J. Schaake, 1982, "Sensitivity of Water-Resource Systems to Climate Variation," unpublished manuscript, WMO Secretariat, Geneva).

This reasoning cannot, however, be applied blindly to a large watershed. For example, the runoff in the bulk of the Colorado River Basin is little, but in the high Rocky Mountains, which contribute most of the water, runoff is a larger portion of the precipitation, and a 10% reduction in precipitation could decrease runoff more nearly 10% than the preceding calculations suggest. This is treated in Chapter 7.

The agricultural effects from changes in precipitation patterns reducing runoff for irrigation might be mitigated by transferring water from one river basin to another. Southern California's agriculture is largely sustained by such large diversions of water. But such large water storage and transfer projects take many decades to accomplish and depend on an accurate prediction of the changes in the total runoff from the effected river basins (Cooper, 1982).

The decline of irrigation systems, often caused by salinity and rising water table, is not a new experience for mankind and intrigues archaeologists studying the extinction of cultures. Even without drastic changes in the availability of water, irrigated land faces a limited period of production because of the buildup of residual salts carried into the land by the irrigation water and left behind as water evaporates or is transpired into the atmosphere. Of the 1450×10^6 ha of world cropland under agricultural cultivation about 16% (230×10^6 ha) is irrigated and about 2×10^5 ha of this irrigated land is removed from production annually because of buildup of salt concentrations that inhibit crop production (Hodges et al., 1981). Without increased availability of water to wash away salts, irrigated land is ultimately lost to crop production. Salt-tolerant species can be developed as new crops or standard crops can be bred to be salt tolerant; but finally all crops fail in excessively salty soil.

Right now in America, we foresee another sort of decline without a change in the weather: Extraction of groundwater exceeds recharge by 25 billion to 30 billion cubic meters annually. About 60% of this overdraft is in the Ogallala aquifer, and by A.D. 2000 the conversion from irrigated to dryland farming is expected with the accompanying years of fallow and changes in crops and yields (Jensen, 1982b).

The loss in irrigated yield accompanying the change in weather specified earlier can be envisaged in two ways. For grain one might simply and roughly say that some areas can no longer be irrigated, and on those acres the yield will be halved at least because the subsequent dryland crops will be grown in alternate years.

The value of truck crops would sustain their irrigation. In 1977, California and Texas produced about half the value of the principal commercial vegetable crops. These two states have a third of the irrigated cropland, and a decrease in water and irrigated area in those states could reduce yields to zero on many acres and thus decrease the fresh vegetables in the produce section of the supermarket, especially in the winter.

6.6 ADAPTING TO THE CHANGE TO A WARMER, DRIER CLIMATE

The predictions in the preceding sections tell from a foundation of physiology and history how a change in climate would change the yields of crops if farmers persisted in planting the same varieties of the same species in the same way in the same place, ignoring the weather. The histories related in Section 6.1, however, are filled with stories of adaptations; and with more mobility, technology and knowledge, the future farmers will surely adapt even faster than the ones of the histories. The safest prediction of any we shall make is: Farmers will adapt to a change in climate, exploiting it and making our preceding predictions too pessimistic.

If the climate changes, farmers will move themselves, change crops, modify varieties, and alter husbandry. The loss of cropland to the margins of the desert, for example, may be replaced in the national production by higher yields and more cropland at the frigid margins. Seeking higher yields and more profit, they will correct their course annually, and they may even adapt to a slowly changing climate unconsciously and successfully.

The seasonal adaptation by migration between the marginal climates near deserts or mountains is familiar, although the movable animals are not the stationary crops that have been the chief concern in this chapter. In Section 6.1 permanent migrations were exemplified by the flight from the Middle Border in the 1890s and from the Dust Bowl in the 1930s. With the facility of modern communication and the scope of a nation that would still span many climates, farmers will surely move as promptly as the Okies, saving themselves while abandoning the cropland to some other use.

6.6.1 Breeding New Varieties

The change of crop species was exemplified in Section 6.1 by the change from corn to wheat as Kansas grew drier from 1885 to 1905. Although the market rather than the weather was the impetus, the rapidity of the adoption of new species of crops was recently demonstrated in Minnesota and North Dakota (Frey, 1982): from 1976 to 1980 the acreage of sunflowers in the two states increased nearly sixfold to 2 million ha, and then in 2 more years it decreased by a third as dry beans and corn became more profitable than sunflowers in the Red River Valley. In the process, the area of small grains decreased and has not recovered. Adaptation by a change of crop can be swift, and a change in CO_2 and climate will make an opportunity for introducing new species in farming and make the preservation of species and the introduction of plants more important.

Changing the variety of a crop by planting a different seed can be even swifter than changing crops because so little needs to be altered from the dealer who supplies chemicals to the farmer who must finance equipment on to the consumer who may scarcely realize the change. The question is whether breeders can make new varieties adapted to a climate as fast as it changes.

The adaptability of crops was long ago demonstrated. In 1857, Wendeling Grimm brought his family and 7 kg of alfalfa seed from The Grand Duchy of Baden to Minnesota. Although the Minnesota winters, more severe than the German winters, killed most of Grimm's alfalfa, some survived. Planting seed from the survivors again and again, he gradually developed an alfalfa that could become relatively dormant in the fall and resist Minnesota cold (Burton, 1980). Given time for population improvement, Grimm produced a new variety for a sudden change to much colder winters. Winter wheat is another example (Rosenberg, 1982). Since the 1920s the growing zone for hard red winter wheat has been expanded through breeding and improved agronomic technologies to include climates with a range of temperatures and rainfall comparable with or greater than those predicted to occur with a doubling of the atmospheric CO_2 concentration.

Because new varieties of crops are bred and tested outdoors, they are indirectly adapted. For example, the yield of modern hybrids of corn in seven states during 1972-1976 decreased only a third in bad years, whereas the yield of open-pollinated corn in the seven states during 1928-1936 fell fully two thirds in bad years (Harvey, 1977). When one open-pollinated cultivar and 24 corn hybrids developed during 1930-1970 were compared, the new hybrids were especially superior in environments that caused low yields because they were less sensitive to drought (Russell, 1974).

The breeding of adapted varieties does have limits and costs. The forage called Bermuda grass tolerates heat and drought, and its tolerance of cold has been improved. Still, it is restricted to central and southern states (Burton, 1980). Tolerance for one stress may bring susceptibility to another as when plants selected for heat and drought tolerance do not have early vigor in cold soil (Jensen, 1981). Still, some costs can become benefits: Since cold tolerance is correlated with low yields (Jensen, 1981), it must also be said that adapting varieties to a warmer climate will bring higher yields.

Plant breeding must keep pace with the change in climate if yields and the food supply are not to decline. Although knowing the factor that has changed in the environment, learning the mechanism of resistance to the changed factor, and then rationally engineering just that change in the genes of the crop is appealing, the process is still beyond our capabilities. We are not, however, vulnerable.

Since plant breeding and testing is performed in the open and in the region of the intended farming, the logical but lengthy process of measuring the weather, discovering the mechanism of resistance, and engineering the genes is truncated and accelerated. For example, a single producer of seed corn has 22 testing sites in the United States and 17 abroad where a quarter million rows of breeding material and 400 inbred lines are yield tested. If the climate shifts, some of this multitude will perform well in the new environment and be shared among sites and advanced without time lost to follow the logical but lengthy chain. Although a complete breeding cycle is approximately a decade from the beginning of inbreeding to the marketing of a product, the objectives can be shifted during the first 5 years, and new hybrids emerging are those adapted to the final 3 years (Frey, 1982).

For maize and other major crops, other seed producers and experiment stations provide the insurance of other programs and diversity. Although fads and excursions present some hazards, steady work by the plant-breeding establishment will surely produce varieties of the life-sustaining, major crops that sustain man that are well adapted to the environment as it shifts gradually to 1°C warmer and 10% drier than the present.

6.6.2 Adapting to Less Water

The narrowing of stomata by rising CO_2 may reduce the transpiration of water. Reducing the harm of drought is, however, a well-known game played by storing more precipitation and getting more yield from the stored water. Much that follows is taken from Greb (1979).

Increasing storage is exemplified by a history of cropping in the Great Plains (see Table 6.5). Crops are planted every other year to permit soil moisture to build up during the fallow year to support such alternate year cropping. Tillage during the fallow year controls weeds and decreases the loss of soil moisture while stubble captures more

TABLE 6.5 Progress in Fallow Systems and Wheat Yields, Akron, Colorado (Greb, 1979)

Years	Fallow System	Fallow Efficiency Storage/ Precipitation (%)	Yield (tons/ha)	Water-Use Efficiency (kg/m^3)
1916-1930	Plow harrow	19	1.1	0.12
1931-1945	Shallow disk, rod weeder	24	1.2	0.14
1946-1960	Improved conventional tillage, begin stubble mulch 1957	27	1.7	0.21
1961-1975	Stubble mulch, begin minimum tillage with herbicides 1969	33	2.2	0.28
1976-1990	Estimate. Begin no-till 1983	40	2.7	0.32

precipitation. Together these doubled the proportion of precipitation stored, which is called the fallow efficiency. Other means of increasing storage include barriers that catch snow, leveling and terracing to decrease runoff, and harvesting water from nearby acreage. Controlling nonbeneficial plants along western rivers could increase water supply by 31 billion m^3 or about the same as the overdraft of groundwater in America (Jensen, 1982a). Matching irrigation to need can decrease pumping, but if the former excess water was flowing back to the water supply, the saving of water will be moderated.

Getting more yield per acre increases the yield of marketable product per unit of water consumed, which is called water-use efficiency. Generally, evaporation from a productive crop is little more than from a crop that merely shades the ground. The increase in water-use efficiency can be seen in Table 6.5 where yields as well as storage increase with time, increasing water-use efficiency even more than fallow efficiency.

Changing the crop as well as increasing fertility can increase water-use efficiency, Table 6.6. The use of water in this example is changed insignificantly by either plant or fertilizer. Since the yield of grass is tripled from the least fertilizer on the less productive grass to the most fertilizer on the more productive grass, the water-use efficiency is also tripled. In the same way, breeding more productive varieties and controlling pests increase water-use efficiency.

A final and large adaptation is moving to a different place. From 1944 to 1978 the irrigated cropland in California increased 1480 thousand ha. At about the same time the California cropland of commercial vegetables for fresh market increased 84 thousand ha, while that in the Atlantic states from North Carolina to Georgia decreased 20 thousand ha. During the same period, the area in New York and New Jersey declined by 57 thousand ha. This was an adaptation by moving to the warmth, sun, and irrigation of California. If water becomes short in California, the nation might adapt by increasing its vegetable fields near the Atlantic where the rain falls.

TABLE 6.6 The Effect of Plant and Fertilizer on Water-Use Efficiency, Akron, Colorado (Greb, 1979)

Plant	Nitrogen (kg/ha)	Water Use (mm)	Yield (tons/ha)	Water-Use Efficiency (kg/m^3)
Russian wild rye	0	159	0.95	0.56
	28	194	1.54	0.75
	56	161	1.84	1.08
Crested wheatgrass	0	158	1.76	1.06
	28	159	2.22	1.33
	56	156	2.72	1.64

6.7 CONCLUSION

We have called the roll of direct effect of CO_2 on plants and of side effects of warmer and drier, pests, and adaptation. What will be the net or integration of all of these on yield?

Although answering seems foolhardy rather than courageous, some large matters seem clear. The direct effects of more CO_2 in the air are beneficial: increase CO_2 around a prosperous leaf and it will assimilate more carbon and lose less water. The indirect effects of warmer and drier, on the other hand, are slightly harmful in the American grain belt as calculations from both past statistics and physiological simulators show. Although pests will change, the direction that they will change is imponderable. While CO_2 is directly narrowing stomata and the need for water, it is also decreasing rainfall.

This conservative forecast of countervailing effects influences the prediction of CO_2 in the air. A portion of each ton of CO_2 injected into the atmosphere from a stack or exhaust can shortly dissolve in the ocean or become sugar and eventually wood or soil organic matter, tempering the increase in atmospheric CO_2. If real plants in corn fields and forests increase their net photosynthesis along the curves observed in the laboratory, the increase in atmospheric CO_2 would be tempered by plants to about 500 ppmv. Unfortunately, to the extent that indirect effects of warmer and drier prevent the rise in yield expected from the laboratory experiment with CO_2, vegetation will temper the rise in CO_2 to a lesser extent.

Thus in the end one sees that the effects on plants of the gradual changes in CO_2 and gradual changes in weather foreseen for A.D. 2000 are modest, some positive and some negative. The wise forecast of yield, therefore, seems a continuation of the incremental increases in production accomplished in the past generation as scientists and farmers adapt crops and husbandry to an environment that is slowly changing with the usual annual fluctuations around the trend.

REFERENCES

Akita, S., and D. N. Moss (1973). Photosynthetic responses to CO_2 and light by maize and wheat leaves adjusted for constant stomatal apertures. Crop Sci. 13:234-237.

Baker, D. N. (1962). The effect of air temperature on the rate of photosynthesis in corn. Ph.D. Thesis, Cornell University, Ithaca, N.Y.

Baker, D. N. (1965). Effects of certain environmental factors on net assimilation in cotton. Crop Sci. 5:53-56.

Barnhart, J. D. (1925). Rainfall and the Populist party in Nebraska. Am. Political Sci. Rev. 19:527-540.

Bjorkstrom, A. (1979). Model of CO_2 interactions between atmosphere, oceans and land biota. In The Global Carbon Cycle, B. Bolin et al., eds. Wiley, New York, pp. 403-457.

Brooks, C. E. P. (1970). Climate Through the Ages. 2d ed. Dover, New York.

Burleigh, J. R., M. G. Eversmeyer, and A. P. Roelfs (1972). Development of linear equations for predicting wheat leaf rust. Phytopathology 62:947-953.

Burton, G. W. (1980). Breeding programs for stress tolerance in forage and pasture crops. Paper presented at Workshop on Crop Reactions to Water and Temperature Stress Tolerance in Forage and Pasture Crops, Duke U., Durham, N.C.

Chandler, J. M. (1981). Estimating losses of crops to weeds. Handbook of Pest Management in Agriculture, Vol. 1, D. Pimentel, ed. CRC Press, Boca Raton, Fla., pp. 95-109.

Cooper, C. F. (1982). Food and fiber. In Carbon Dioxide Review: 1982, W. C. Clark, ed. Oxford U. Press, New York, pp. 297-320.

Curry, R. B. (1971). Dynamic simulation of plant growth, Part I. Development of a model. Trans. Am. Soc. Agric. Eng. 14:946-949.

Duncan, W. G., R. S. Loomis, W. A. Williams, and R. Hanau (1967). A model for simulating photosynthesis in plant communities. Hilgardia 38:181-205.

Enoch, H. Z., I. Rylski, and Y. Samish (1970). CO_2 enrichment to cucumber, lettuce and sweet pepper plants grown in low plastic tunnels in a subtropical climate. Israel J. Agric. Res. 20:63-69.

Enoch, H. Z., I. Rylski, and M. Spigelman (1976). CO_2 enrichment of strawberry and cucumber plants grown in unheated greenhouses in Israel. Sci. Hortic. 5:33-41.

Evans, L. T. (1975). The physiological basis of crop yield. Crop Physiology, L. T. Evans, ed. Cambridge U. Press, Cambridge, pp. 327-335.

Fraser, B. D., and R. van den Bosch (1973). Biological control of the walnut aphid in California: the interrelationship of the aphid and its parasite. Environ. Entomol. 2:561-548.

Frey, N. M. (1982). Plant breeding under increasing CO_2 concentration and in a changing climate. Paper presented at AAAS Conference on Plants under Increasing CO_2. Athens, Ga.

Gifford, R. M. (1979). Growth and yield of CO_2-enriched wheat under water-limited conditions. Aust. J. Plant Physiol. 6:367-378.

Gifford, R. M., P. M. Bremner, and D. B. Jones (1973). Assessing photosynthetic limitation to grain yield in a field crop. Aust. J. Agric. Res. 24:297-307.

Greb, B. W. (1979). Reducing drought effects on croplands in the West Central Great Plains. USDA Agr. Info. Bull. 420.

Guinn, G., and J. Mauney (1980). Analysis of CO_2 exchange assumptions: feedback control. Predicting Photosynthesis for Ecosystem Models, Vol. 2, J. D. Hesketh and J. W. Jones, eds. CRC Press, Boca Raton, Fla., pp. 1-16.

Haigh, P. A. (1977). Separating the Effects of Weather and Management on Crop Production. C. F. Kettering Foundation (ST 77-4), Dayton, Ohio.

Hardman, L. E., and W. A. Brun (1971). Effect of atmospheric CO_2 enrichment at different developmental stages on growth and yield components of soybeans. Crop. Sci. 11:886-888.

Hardy, R. W. F., and U. D. Havelka (1977). Possible routes to increase the conversion of solar energy to food and feed by grain legumes and cereal grains (crop production): CO_2 and N_2 fixation, foliar fertilization, and assimilate partitioning. In Biological Solar Energy Conversion, A. Mitsui et al., eds. Academic, New York, pp. 299-322.

Harlan, J. R. (1975). Crops and Man. American Society of Agronomy, Madison, Wisc.

Harvey, P. H. (1977). ARS National Research Program, NEP No. 20040 breeding and production: corn, sorghum and grain millets. Am. Seed Trade Assoc. 1977 Yearbook and Proceedings 94th Annual Conference, pp. 178-182.

Hesketh, J. D. (1963). Limitations to photosynthesis responsible for differences among species. Crop Sci. 3:493-496.

Hodges, C. N., R. M. Fontes, E. P. Glenn, S. Katzen, and L. B. Colvin (1981). Sea water-based agriculture as a food production defense against climate-variability. In Food-Climate Interactions, W. Bach et al., eds. Reidel, Dordrecht, Holland, pp. 81-100.

Hodges, R., and J. A. Artley (1981). Spring Small Grains Planting Date Distribution Model. AgRISTARS Rep. SR-L1-040332. March 1981, JSC 16858. Lyndon B. Johnson Space Center, Houston, Tex.

Jensen, M. E. (1982a). Water resource technology and management. Proceeding RCA Symposium: Future Agricultural Technology and Resource Conservation. December 5-9, 1982, Washington, D.C.

Jensen, M. E. (1982b). Overview-irrigation in U.S. arid and semiarid lands. Prepared for the Office of Technology Assessment, Water-Related Technologies for Sustaining Agriculture in Arid and Semiarid Lands.

Jensen, R. G., and J. T. Bahr (1977). Rubulose 1,5-bisphosphate carboxylase-oxygenase. Ann. Rev. Plant Physiol. 28:379-400.

Jensen, S. D. (1981). Discussion. In Plant Breeding II, K. J. Frey, ed. Iowa State U. Press, Ames, p. 171.

Kellogg, W. W. (1977). Effects of Human Activities on Global Climate. Technical Note No. 156. WHO-No. 486, World Meteorological Organization, Geneva.

Kimball, B. A. (1982). Carbon dioxide and agricultural yield: an assemblage and analysis of 430 prior observations. WCL Report 11. U.S. Water Conservation Laboratory, Phoenix, Ariz., pp. 1-48.

Lawyer, A. L., K. L. Cornwell, P. O. Larsen, and J. A. Bassham (1981). Effects of carbon dioxide and oxygen on the regulation of photosynthetic carbon metabolism by ammonia in spinach mesophyll cells. Plant Physiol. 68:1231-1236.

LeDuc, S. K. (1980). Corn models for Iowa, Illinois and Indiana. NOAA Center for Environmental Assessment Services, Columbia, Mo.

Lemon, E. R., ed. (1983). CO_2 and Plants: The Response of Plants to Rising Levels of Atmospheric Carbon Dioxide. AAAS Selected Symposium No. 84. Westview Press, Boulder, Colo.

Lucas, G. D. (1980). The war against blue mold. Science 210:147-153.

Maas, S. J., and G. F. Arkin (1980). TAMW: A wheat growth and development simulation model. Program in Model Documentation Report No. 80-3, October 1980. Blackland Research Center, Texas Agricultural Experiment Station, Temple, Tex.

Malin, J. C. (1936). The adaptation of the agricultural system to sub-humid environment. Agric. History 10:118-141.

Manabe, S., and R. J. Stouffer (1980). Sensitivity of a global climate model to an increase of CO_2 concentration in the atmosphere. J. Geophys. Res. 85:5529.

Manabe, S., R. T. Wetherald, and R. J. Stouffer (1981). Summer dryness due to an increase of atmospheric CO_2 concentration. Clim. Change 3:347-386.

Mauney, J. R., K. E. Fry, and G. Guinn (1978). Relationship of photosynthetic rate to growth and fruiting of cotton, soybean, sorghum, and sunflower. Crop Sci. 18:259-263.

McCullough, D. (1981) Mornings on Horseback. Simon & Schuster, New York, 445 pp.

McQuigg, J. D. (1979). Climatic variability and agriculture in the temperate regions. In WMO Proc. World Climate Conference, WMO-No. 537, World Meteorological Organization, Geneva, pp. 406-425.

Miles, G. E., R. M. Peart, R. J. Bula, M. C. Wilson, and T. R. Hintz (1974). Simulation of plant-pest relationships with GASP IV. ASAE Paper 74-4022. American Society of Agricultural Engineers, St. Joseph, Mich.

Mitchell, J. M., Jr. (1977). Carbon dioxide and future climates. EDS March. Environmental Data Service, National Oceanographic and Atmospheric Administration, Washington, D.C., pp. 3-9.

Motha, R. P. (1980). Soybean models for Iowa, Illinois and Indiana. NOAA Center for Environmental Assessment Services, Columbia, Mo.

National Research Council (1976). Climate and Food: Climatic Fluctuation and U.S. Agricultural Production. Board on Agriculture and Renewable Resources, Commission on Natural Resources, National Academy of Sciences, Washington, D.C.

Neild, R. E., H. N. Richman, and M. W. Seeley (1979). Impacts of different types of temperature change on the growing season for maize. Agric. Meteorol. 20:367-374.

Patterson, D. T. (1982). Effects of light and temperature on weed/crop growth and competition. In Biometeorology in Integrated Pair Management, J. L. Hatfield and I. J. Thomason, eds. Academic, New York, pp. 407-420.

Patterson, D. T., and E. P. Flint (1980). Potential effects of global atmospheric CO_2 enrichment on the growth and competitiveness of C_3 and C_4 weed and crop plants. Weed Sci. 28:71-75.

Patterson, D. T., and E. P. Flint (1982). Interacting effects of CO_2 and nutrient concentration. Weed Sci. 30:389-394.

Richardson, C. W. (1981). Stochastic simulations of daily precipitation, temperature, and solar radiation. Water Resources Res. 17:182-190.

Richardson, C. W., and J. T. Ritchie (1973). Soil-water balance for small watersheds. Trans. Am. Soc. Agric. Eng. 16:72-77.

Rogers, H. H., et al. (1980). Response of Vegetation to Carbon Dioxide. No. 001. Field Studies of Plant Responses to Elevated Carbon Dioxide Levels. Research Progress Report from the U.S. Department of Energy and the U.S. Department of Agriculture, DOE Carbon Dioxide Research Division, Washington, D.C. 20545.

Rosenberg, N. J. (1982). The increasing CO_2 concentrations in the atmosphere and its implications on agricultural productivity, II. Effect through CO_2-induced climate change. Clim. Change 4:239-254.

Russel, W. A. (1974). Comparative performance for maize hybrids representing different eras of maize breeding. Proceedings 29th Annual Corn and Sorghum Res. Conf., Am. Seed Trade Association, pp. 81-101.

Sakamoto, C. (1978). Reanalysis of CCEAI, U.S. Great Plains Wheat Yield Models. Center for Climatic and Environmental Assessment Technical Note No. 78-3. National Aeronautics and Space Administration, Houston, Tex.

Sakamoto, C. (1982). Application of Process Models for Large Area Crop Yield Estimation. Level I AgRISTARS Project Review, April 19, 1982. Lyndon B. Johnson Space Center, Houston, Tex.

Schlebecker, J. T. (1953). Grasshoppers in American agricultural history. Agric. History 27:85-93.

Sionit, N., H. Helmers, and B. R. Strain (1980). Growth and yield of wheat under CO_2 enrichment and water stress. Crop Sci. 20:687-690.

Sionit, N., D. A. Mortensen, B. R. Strain, and H. Helmers (1981a). Growth response of wheat to CO_2 enrichment and different levels of mineral nutrition. Agron. J. 73:1023-1027.

Sionit, N., B. R. Strain, and H. A. Beckford (1981b). Environmental controls on the growth and yield of okra, I. Effects of temperature and of CO_2 enrichment at cool temperature. Crop Sci. 21:885-888.

Steinbeck, J. (1939). Grapes of Wrath. Viking, New York.

Tatum, L. A. (1971). The Southern corn leaf blight epidemic. Science 171:1113-1116.

Thompson, L. M. (1969) Weather and technology in the production of corn in the U.S. Corn Belt. Agron. J. 61:453-456.

Tuchman, B. (1978). A Distant Mirror; The Calamitous 14th Century. Knop, New York.

Ullstrup, A. J. (1972). Impacts of Southern corn leaf blight epidemics of 1970-1971. Ann. Rev. Phytopathol. 10:37-50.

U.S. Dept. of Agriculture (USDA) (1970). Selected weeds in the United States. Agricultural Handbook 366. U.S. Government Printing Office, Washington, D.C.

U.S. Dept. of Agriculture (USDA) (1972). Extent and Cost of Weed Control with Herbicides and an Evaluation of Important Weeds, 1968, USDA ARS-H-1. U.S. Department of Agriculture, Washington, D.C.

U.S. Dept. of Agriculture (USDA) (1982). Agricultural Statistics, 1981. U.S. Government Printing Office, Washington, D.C.

van Veen, J. A., and E. A. Paul (1981). Organic carbon dynamics in grassland soils. I. Background information and computer simulation. Can. J. Soil Sci. 61:85.

Waggoner, P. E., J. L. Monteith, and G. Szeicz (1964). Decreasing transpiration of field plants by chemical closure of stomata. Nature 201:97-98.

Waggoner, P. E., J. G. Horsfall, and R. J. Lukens (1972). EPIMAY: A simulator of Southern corn leaf blight. Conn. Agric. Exp. Stn. Bull. 729. New Haven, Conn.

Walker, D. A. (1976). Plastids and intracellular transport. In Transport in Plants III, C. R. Stocking and U. Heber, eds. Encyclopedia of Plant Physiology. Springer-Verlag, Berlin, pp. 85-136.

Wong, S. C. (1979). Elevated atmospheric partial pressure of CO_2 and plant growth. Oecologia 44:68-74.

Zelitch, I. (1971). Photosynthesis, Photorespiration, and Plant Productivity. Academic, New York.

Zelitch, I. (1982). The close relationship between net photosynthesis and crop yield and the critical role of oxygen stress. BioScience 32:796-802.

Zieslin, N., A. H. Halavy, and H. Z. Enoch (1972). The role of CO_2 increasing the yield of "Baccara" roses. Hortic. Res. 12:92-100.

7 Effects of a Carbon Dioxide-Induced Climatic Change on Water Supplies in the Western United States

Roger R. Revelle and Paul E. Waggoner

In this chapter we show that warmer air temperatures and a slight decrease in precipitation would probably severely reduce both the quantity and the quality of water resources in the western United States. Similar effects can be expected in many water-short regions elsewhere in the world. We have not attempted to estimate these, primarily because we do not know enough to be able to do so. But we hope that hydrologists of other countries will be stimulated by our calculations to investigate the probable consequences of a CO_2-induced climate change on water resources in their own countries. In all countries, planning and construction of large-scale water-resource systems takes many decades. The time involved is of the same order of magnitude as the time over which a significant change in climate from increase of carbon dioxide and other greenhouse gases can be expected. Thus, we believe that planners and managers of water systems throughout the world should be able to make good use of forecasts of the hydrologic consequences of a warmer climate and of possible changes in precipitation.

7.1 EMPIRICAL RELATIONSHIPS AMONG PRECIPITATION, TEMPERATURE, AND STREAM RUNOFF

To assess the effects on the United States' water resources of probable climatic change we used the empirical relationship found by Langbein et al. (1949) among mean annual precipitation, temperature, and runoff. This was based on representative data from 22 drainage basins in the conterminous United States. Their relation in Table 7.1 gives the estimated annual runoff for different values of mean annual precipitations and weighted mean annual temperatures. The latter were computed for each catchment basin by dividing the sum of the products of average monthly temperature and precipitation by the mean annual precipitation. In this way, the average temperature during each month is weighted by the precipitation during that month.

The catchments studied by Langbein and his colleagues were distributed over climates from warm to cold and from humid to arid, but in Table 7.1 we have shown the relations among runoff, temperature, and precipitation only for relatively arid areas. In these arid areas, the value of actual evapotranspiration is less than the potential evapo-

TABLE 7.1 Runoff (mm yr^{-1}) as a Function of Precipitation and Temperature[a]

Weighted Average Temperature (°C)	Precipitation (mm yr-1) 200	300	400	500	600	700
- 2	54	92	154	230	330	440
0	40	74	124	190	275	380
2	28	57	95	154	225	320
4	17	40	78	125	190	265
6	9	25	60	100	155	220
8	0	17	42	82	128	185
10		8	29	64	103	155
12		0	19	47	80	130
14			10	32	65	105
16			0	20	50	85

[a]Source: W. B. Langbein et al. (1949).

transpiration that would occur if sufficient water were present and evapotranspiration was controlled mainly by temperature. For example, at a temperature of 4°C, potential evapotranspiration is about 450 mm yr^{-1}. Yet Langbein's data show that even when annual precipitation is only 300 mm, there is still significant runoff, about 13% of the precipitation. Correspondingly, average actual evapotranspiration must be only 260 mm yr^{-1}.

From Table 7.1 we observe that for any given annual precipitation, runoff diminishes rapidly with increasing temperature. Similarly, for any given temperature, the proportion of runoff to precipitation increases rapidly with increasing precipitation. For example, at a weighted mean annual temperature of 4°C and annual precipitation of 200 mm, runoff is only 8.5% of precipitation, whereas for the same temperature and an annual precipitation of 700 mm, runoff is 38% of precipitation. At an annual temperature of 8°C, runoff is zero when precipitation is 200 mm or less and is 185 mm--or 26.4% of precipitation--when the average annual precipitation is 700 mm.

For any particular region, the relations shown in Table 7.1 are rather crude approximations because many physical factors, including geology, topography, size of drainage basin, and vegetation, may alter the effect of climate on runoff. We believe, nevertheless, that these relationships can be used without serious error to describe the effects of relatively small changes in average temperature and precipitation on mean annual runoff.

In Table 7.2, we have used the data in Table 7.1 to compute the approximate percentage decrease in runoff for a 2°C increase in temperature. Climate models (e.g., Manabe and Wetherald, 1980) indicate that a temperature change of this magnitude or greater is likely as a result of the doubling of carbon dioxide and increased concentration of "greenhouse gases" expected during the next century. We see that for a

TABLE 7.2 Approximate Percent Decrease in Runoff for a 2°C Increase in Temperature[a]

Initial Temperature (°C)	Precipitation (mm yr-1) 200	300	400	500	600	700
- 2	26	20	19	17	17	14
0	30	23	23	19	17	16
2	39	30	24	19	17	16
4	47	35	25	20	17	16
6	100	35	30	21	17	16
8		53	31	22	20	16
10		100	34	22	22	16
12			47	32	22	19
14			100	38	23	19

[a]Computed from Table 7.1.

present weighted mean annual temperature of 4°C and annual precipitation of 300 mm, a 35% diminution in runoff would follow a 2°C warming. The percentage decrease in runoff from a warming diminishes with increasing precipitation and becomes greater for successively higher values of the initial temperature.

Table 7.3 shows the approximate percentage decreases in runoff for a 10% decrease in precipitation. According to the results of climate models (see Chapter 4), such a diminution in precipitation is likely, at least in certain regions of the United States, with a doubling of atmospheric CO_2. Again we see that the effect becomes larger with higher average annual temperatures. There is a relatively small difference in the percentage decrease in runoff at a given temperature over the range of initial precipitation values shown in the table. Comparison of Tables 7.2 and 7.3 shows that below an initial mean annual precipitation of 500 mm, the effects of a 2°C warming are larger than those caused by a 10% decrease in precipitation. The reverse is true when mean annual precipitation is 500 mm or more.

7.2 EFFECTS OF CLIMATE CHANGE IN SEVEN WESTERN U.S. WATER REGIONS

Stockton and Boggess (1979) have used Langbein's empirical relation to estimate the effects of a climatic change on water in the 18 water regions of the conterminous United States defined by the U.S. Water Resources Council (1978). They find that a 2°C warming and a 10% reduction in precipitation would not have serious effects in the humid regions east of the 100th meridian. In the West, however, the impact would be severe on seven water regions: the drainage basins of the Missouri, Arkansas-White-Red, Rio Grande, and Colorado rivers; the river basins draining into the Gulf of Mexico from the northern two thirds of Texas; and the rivers of California. The only western water

TABLE 7.3 Approximate Percentage Decrease in Runoff for a 10% Decrease in Precipitation[a]

Temperature (°C)	Initial Precipitation (mm yr-1) 300	400	500	600	700
- 2	12	16	17	18	18
0	14	16	17	19	19
2	15	16	19	19	20
4	17	19	19	21	21
6	23	23	21	21	21
8	30	24	24	22	22
10		24	27	23	23
12		40	30	25	25
14			34	30	27
16			50	36	29

[a]Computed from Table 7.1.

regions that would not be severely affected are the water-rich Pacific Northwest and the Great Basin (parts of Nevada, Utah, and Idaho), where demand is relatively small and groundwater reserves are large.

In estimating the impact of climate change, Stockton and Boggess assumed:

1. The region-by-region variation in annual runoff is predominantly influenced by climate, although other factors such as geology, topography, vegetation, and many other variables may be important, especially in smaller drainages.
2. The empirical curves associating total annual precipitation and total annual runoff with weighted mean annual temperature are appropriate for all 18 regions although derived from a relatively small (22 drainage basin) sample.
3. Changes in land use have relatively small influences on regionwide annual runoff.
4. Annual runoff is not greatly affected by large-scale groundwater overdraft.
5. Evapotranspiration is controlled solely by temperature.
6. The postulated climatic change does not modify the present monthly distribution of temperature and precipitation; only the amplitude of the present distribution is increased or decreased.
7. Selection of a few meteorological stations for each region adequately establishes the relation of the weighted mean temperature and annual precipitation to annual runoff.

In Table 7.4 we have summarized the estimates by Stockton and Boggess of the effects of a 2°C increase in temperature and a 10% reduction in precipitation in the seven water regions of the western United States in which climate change would have the most serious

TABLE 7.4 Comparison of Water Requirements and Supplies for Present Climatic State and for a 2°C Increase in Temperature and 10% Reduction in Precipitation in Seven Western U.S. Water Regions[a]

	Present Climate						Warmer and Drier Climate		
Water Region[b]	Area ($10^{10}/m^2$)	Mean Annual Runoff (10^{10} m^3 yr^{-1})	Mean Annual Runoff (mm)	Mean Annual Supply (10^{10} m^3 yr^{-1})	Mean Annual Requirements[c] (10^{10} m^3 yr^{-1})	Ratio of Requirement to Supply	Mean Annual Supply (10^{10} m^3 yr^{-1})	Percent Change in Supply	Ratio of Requirement[d] to Supply
Missouri	132.4	8.50	64	8.50	3.63	0.43	3.07	-63.9	1.18
Arkansas-White-Red	63.2	9.35	148	9.35	1.67	0.18	4.32	-53.8	0.39
Texas Gulf	44.9	4.92	110	4.92	1.74	0.35	2.47	-49.8	0.70
Rio Grande	35.2	0.74	21	0.74	0.67	0.91	0.18	-75.7	3.72
Upper Colorado	29.6	1.64[e]	55	1.64	1.63[f]	0.99	0.99	-39.6	1.65
Lower Colorado	40.1	0.38	10	1.15[g]	1.37	1.19	0.50[g]	-56.5	2.68
California	42.9	9.56	222	10.18[g]	4.22	0.41	5.71[g]	-43.9	0.74
For the 7 regions together	388.3	35.09	90.4	35.09[h]	14.93	0.43	16.53[h]	-53	0.90

[a]Source: Stockton and Boggess (1979) and calculations in this paper for Upper Colorado Basin.
[b]As defined by the U.S. Water Resources Council (1978).
[c]Projected through year 2000 A.D.
[d]Assuming no increase in requirement because of increased evapotranspiration from irrigated farms or reservoirs.
[e]Average "virgin flow" of the Colorado River at Lee Ferry from 1931 to 1976.
[f]Includes allocation to Lower Basin States, California included, of 0.93×10^{10} m^3 yr^{-1}.
[g]Includes water received from Upper Colorado Basin, but not mined groundwater.
[h]Total is less than sum of the column because of flow of Lower Colorado derived from Upper Colorado (g).

impact. These regions cover about half the area of the conterminous United States, but they produce only about 15% of the mean annual stream runoff. The table shows the present mean annual water supply for each region, not including mined groundwater, in millions of hectare meters (10^{10} m^3 yr^{-1}) and the estimated mean annual requirement in the year 2000. The mean annual requirements listed in Table 7.4 represent "consumptive" use of water plus evaporation from reservoirs, that is, the total quantity of water that is evapotranspired in the course of beneficial human use. Although actual withdrawals from streams and underground aquifers are considerably larger, portions of these withdrawals are returned to the streams or back into the ground where the water may be reused. In the present climate the ratio of estimated requirements to supplies is less than one for all regions except the Lower Colorado River. In this region today, the deficit of supply is presently made up by extensive mining of groundwater.

For the postulated climatic change, supplies would greatly diminish in all regions, ranging from almost a 76% reduction in the Rio Grande region to nearly 40% in the Upper Colorado, with the result that estimated requirements would exceed supplies in the Missouri, Rio Grande, and Upper and Lower Colorado regions. Mean annual requirements would still be less than future mean annual supplies in the Arkansas-White-Red, Texas Gulf, and California regions. But requirements would almost certainly exceed supplies in the Texas Gulf and California regions during future prolonged droughts. Conditions are highly variable in different parts of the Arkansas-White-Red region, with the western part tending to be deficient in water supplies and the eastern part having a surplus. To maintain the present pattern of water use, large-scale transfers between basins might be necessary here even under average conditions, let alone to meet water requirements during prolonged droughts. A serious deterioration in water quality would follow from climatic change in all seven regions.

The ratio of future requirements to supply would probably be even less favorable than indicated in Table 7.4 because evapotranspiration from irrigated farms and reservoirs would undoubtedly increase with a rise in temperature. On a global basis this would be compensated for by an increase in precipitation, but this might or might not occur in the regions we are considering.

At present, California depends for about 15% of its water on imports from the Colorado River. These imports might be eliminated entirely with the postulated climatic change, in which case the ratio of mean annual requirements to mean annual runoff would increase to 0.83, more than double the present ratio.

In all seven regions, irrigation is by far the largest user. Its share of water withdrawals ranges from 68% in the Texas Gulf region to 95% in the Rio Grande region. Total water withdrawals for agriculture are now 13.2 million hectare meters, and in the seven regions the irrigated area is (very approximately) 13 million hectares (Rogers, 1983) so that, on the average, the annual depth of irrigation is about 1 m. Consumptive water use in irrigation is much smaller; a large share of the water that is not consumed reappears as return flows that can be used downstream. Reduction in the irrigated area and an increase in the efficiency of water use in irrigation would significantly lower the overall water requirements in the seven western regions. A 15% increase in water-use efficiency is probably feasible (Jensen, 1982). Reduction in the irrigated area might come about automatically if an interstate economic market for water were to develop, because economic returns to irrigation are relatively low in large parts of the seven western regions. On the other hand, potentially very large Indian claims for irrigation water for their reserved lands must eventually be settled, and this could result in a major reallocation of water rights (Back and Taylor, 1980).

The effects of future droughts in the Arkansas-White-Red, Texas Gulf, and California regions from the assumed climatic change could be significantly mitigated by construction of additional reservoirs for water storage, but increases in storage would help little in the Missouri, Rio Grande, or Upper and Lower Colorado regions because their storage reservoirs are already so large compared to the annual runoff. In these regions strict water conservation would be essential.

The mean annual requirements listed in Table 7.4 represent "consumptive" use of water plus evaporation from reservoirs, that is, the total quantity of water that is evapotranspired in the course of beneficial human use. Although actual withdrawals from streams and underground aquifers are considerably larger, portions of these withdrawals are returned to the streams or back into the ground where the water may be reused.

As Stockton and Boggess show, the one western region where a large surplus would still exist after their postulated climatic change would be the Pacific Northwest. Their estimated ratio of requirements to supplies following a 2°C warming and a 10% reduction in precipitation would be 0.10. The annual supply would then be 23.7 million hectare meters. Transfer of 20% of this total supply to water-short regions through large, long-distance conveyance could increase future supplies in the seven western regions shown in Table 7.4 by nearly 30%, thereby compensating for much of the estimated shortages from climatic change. The ratio of requirements to supplies in the Pacific Northwest region would still be a comfortable 0.18.

From an economic standpoint, however, such a transfer would probably not be desirable. The value of the hydroelectric energy that could be generated from a hectare meter of water in the Pacific Northwest, assuming a total head of 500 m and a price of 5¢ per kilowatt-hour, would be over $600, considerably in excess of the value of a hectare meter of water for irrigating the fodder and cereal crops grown in most of the western regions (Rogers, 1983).

7.3 THE COLORADO RIVER

Except for the Rio Grande, the waters of the Colorado River are more intensively used than those of any other major stream in the United States. Half the estimated "normal" flow of 18,500 million cubic meters per year at Lee Ferry in Arizona has been allocated by interstate compact, confirmed by federal law, to the lower basin states of California and Arizona, with minor amounts going to Nevada. Nearly all of the runoff originates from snow in the high-mountain area of western Colorado, southwestern Wyoming, and eastern Utah.* (As defined, the Upper Basin also includes northwestern New Mexico.)

To check the probable effect of a climatic change on the flow of the Colorado River, we calculated a multiple regression of the relation between annual averages of precipitation and temperature and the "virgin flow" of the Colorado at Lee Ferry, Arizona, which is the southernmost point on the river in the Upper Colorado Basin.

The "virgin flow" is the measured flow at Lee Ferry plus estimated depletions within the upper basin, evaporation from reservoirs, and changes in reservoir storage. Estimates for the annual virgin flows were furnished to us by Myron B. Holbert, chief engineer of the Colorado River Board of California. The flow for each water year--October 1 to September 30--from 1931 to 1976 is tabulated in Table 7.5. It ranges from 9,450 to 25,490 million cubic meters. The average for the 46-year period is 16,430 million cubic meters (13.5 million acre feet). This average value is 2065 million cubic meters, or 11% less than the assumed "normal flow" on which the Colorado River Compact is based, but it agrees well with Langbein's relationship, shown in Table 7.1, among temperature, precipitation, and runoff. Interpolating between the values given in the table for temperatures of 4 and 6°C and precipitation of 300 and 400 mm yr^{-1}, we arrive at a runoff of 53 mm, corresponding to an annual river flow for the 296,000 km^2 in the Upper Colorado drainage of 15,700 million m^3.

The flow of the Colorado is mainly collected in five catchments corresponding to the five climatic divisions in the Upper Colorado Basin outlined in Figure 7.1. Mean annual precipitation and temperature for each drainage area and water-year from 1931 to 1976 were provided to us by Daniel Cayan of the Scripps Institution of Oceanography, who obtained the original data from the National Climate Center, Asheville, North Carolina. The annual data for each division are the areally weighted averages of the records of individual weather stations. As Cayan has pointed out, the number of weather stations was not constant but gradually increased from 1931 to 1976. Between 1951 and 1980, continuous records were obtained from 45 stations in the western Colorado division, 10 stations in the Green and Bear drainages in Wyoming, and 10 in the three drainage basins of eastern Utah (North Mountains, Uinta Basin, and Southeast Utah). By 1980, there were 65 stations in western Colorado, 26 in the three Utah divisions, and 16 in southwestern Wyoming.

We weighted the data from these five drainage basins in proportion to their areas: Colorado drainage in western Colorado, 0.430; Green and Bear drainages in Wyoming, 0.184; North Mountains in Utah, 0.110;

---

*See Dracup (1977) and Howe and Murphy (1981) for useful discussions of the economic, social, political, and international problems of the Colorado River compact.

FIGURE 7.1 Basins or drainages of the western United States.

Uinta Basin, 0.053; and southeast Utah, 0.223. The weighted sum of the precipitation in millimeters for the five drainages is called PRECIP in the following equation. It varies from 262 to 439 mm yr^{-1}, with an average for the 46 years of 332 mm yr^{-1}. Weighted annual average temperatures obtained in the same manner are called CELSIUS. They range from 2.52 to 6.75°C and average 4.18°C. The weighted annual averages for 1931 to 1976 are tabulated in Table 7.5.

The relation of the virgin flow to PRECIP and CELSIUS is assumed to fit the equation:

$$\text{FLOW} = b_0 + (b_1 \times \text{PRECIP}) + (b_2 \times \text{CELSIUS}).$$

The regression coefficients, b_i, and their standard errors relating the virgin flow at Lee Ferry to the mean annual precipitation and the mean annual temperature in the watershed from 1931 to 1976 are as follows:

b_0 in millions of cubic meters = 9274 $\pm$ 3838,
b_1 in millions of cubic meters/mm = 52 $\pm$ 7,
b_2 in millions of cubic meters/°C = −2400 $\pm$ 507.

The fit of the equations to the data is represented by the square of the correlation coefficient, R^2, which is 0.73. About 75% of the variation of the flow about the mean is explained by the equations.

We see that a 2°C rise in temperature would decrease the virgin flow by 4800 $\pm$ 1015 million cubic meters yr^{-1}, or about 29% $\pm$ 6%. A 10% decrease in precipitation would reduce the flow by 1730 $\pm$ 230 cubic meters, or an additional 11% $\pm$ 1.4%. The combined effect would be a reduction in flow by 40% $\pm$ 7.4%, very close to the estimate of 44% given by Stockton and Boggess.

Estimates of the virgin flow between 1900 and 1930 were also provided by Holbert, and we attempted a similar analysis for these years, using data published in the Weather Bureau publication, *The Climates of the States*. Unfortunately, the data for these earlier years are sparse. Only 10 stations recorded precipitation more or less continuously during this period in western Colorado and even fewer in Utah and Wyoming. Temperature data were available from only Garnett, Colorado; Lander, Wyoming; and Modena, Utah, and none of these is in the high-mountain regions of heavy snowpack that contribute most of the runoff to the Colorado River.

The data for 1900 to 1930, Table 7.6, indicate that the average runoff was about 20% greater than between 1931 and 1976, while the average precipitation was about 7% less and the average temperatures for the three available stations were about 2°C higher. The low estimate of average precipitation may reflect a deficiency in estimating the quantities of water precipitated as snow. These estimates were very uncertain before the advent of the "snow courses" initiated by the U.S. Soil Conservation Service in the 1930s. The extreme variations in average annual temperature for the three stations were only 2.45°C, in contrast to the range of 4.23°C for the average of the much larger number of stations in 1931-1976.

TABLE 7.5 Annual Averages of Precipitation, Temperature, and Virgin Flow of the Colorado River at Lee Ferry, Upper Colorado Region, 1931-1976

Year	Precipitation (mm yr^{-1})	Temperature (°C)	Flow (10^6 m^3 yr^{-1})
1931	268	4.95	9583
1932	355	3.23	21270
1933	283	3.51	14009
1934	228	6.75	6958
1935	316	5.14	14247
1936	349	4.62	17023
1937	378	3.80	16948
1938	392	5.04	21643
1939	301	4.22	13666
1940	323	5.71	10609
1941	451	4.11	22386
1942	357	3.51	23592
1943	362	4.68	16164
1944	334	3.99	18693
1945	355	3.97	16542
1946	305	4.08	12860
1947	418	4.39	19083
1948	346	4.13	19258
1949	385	3.32	20199
1950	320	3.68	15904
1951	298	4.60	14366
1952	424	3.20	25490
1953	273	4.63	13119
1954	308	5.64	9450
1955	291	3.57	11333
1956	255	4.31	13259
1957	429	4.25	24787
1958	323	4.28	20339
1959	294	4.75	10619
1960	262	4.49	13893
1961	340	4.82	10432
1962	300	3.38	21338
1963	305	5.44	10423
1964	286	3.61	12527
1965	439	3.30	23329
1966	282	4.06	13825
1967	350	4.20	14687
1968	326	3.45	16854
1969	361	3.84	17745
1970	352	4.05	19002
1971	302	3.78	18645
1972	303	3.46	15021
1973	435	2.52	23893
1974	254	4.22	16355
1975	347	3.51	20447
1976	300	4.27	14064
Mean	333	4.18	16432

TABLE 7.6 Annual Averages of Precipitation, Temperature, and Virgin Flow of the Colorado River at Lee Ferry, Upper Colorado Region, 1901-1930

Year	Precipitation (mm yr^{-1})	Temperature (°C)	Annual Sum (m/sec)
1901	229	6.56	16753
1902	195	6.71	11586
1903	285	5.35	18264
1904	240	6.55	19297
1905	314	6.10	19769
1906	370	5.86	23585
1907	350	6.78	28865
1908	274	6.49	15857
1909	364	6.05	28709
1910	233	6.57	17574
1911	327	7.49	19770
1912	313	5.05	25311
1913	269	5.71	17852
1914	355	6.68	26176
1915	308	6.47	17303
1916	304	7.08	23684
1917	346	5.04	29650
1918	263	7.32	18951
1919	253	6.43	15372
1920	331	5.58	27076
1921	355	6.57	28389
1922	300	6.60	22580
1923	336	6.04	22535
1924	252	5.63	17517
1925	338	6.41	16077
1926	303	6.28	19554
1927	404	6.88	22963
1928	253	6.58	21314
1929	392	5.46	26432
1930	309	6.44	18361
Mean	306	6.29	21098

The multiple regression for the earlier period showed a smaller effect of temperature and a somewhat larger effect of precipitation than during 1931-1976. But during the earlier period the standard deviations for the effects of both temperature and precipitation were almost twice as great, and the square of the correlation coefficient was only 0.57.

We conclude tentatively that the weakness of the correlation and the relatively larger standard errors from 1901 to 1930 resulted from the paucity of precipitation and temperature data, the probable inaccuracy of estimates of the quantity of water precipitated as snow, and the probably unrepresentative character of the stations used for average temperatures. In both the earlier and the later periods, variations in precipitation were reflected almost linearly in variations in runoff without the amplification that might be expected from the results obtained on smaller watersheds (Schaake and Kaczmarek, 1979) or from Table 7.3.

After the postulated climatic change, the mean annual flows at Lee Ferry computed from our multiple regression equation for 1931 to 1976 would be only 9900 million cubic meters. This last amount is 2060 million cubic meters, or 17%, less than the historically lowest 10-year average annual flow of 11,960 million cubic meters, calculated by Stockton from tree-ring records for the decade from A.D. 1584 to A.D. 1593 (Dracup, 1977). A similar prolonged drought in the middle of the twenty-first century with the same percentage decrease in runoff as in 1584-1593 could bring the 10-year annual average flow down to 7200 million cubic meters, or about 44% of the annual average virgin flow from 1931 to 1976.

Although a 2°C warming is probably a conservative estimate of the effect over the next hundred years of increase of greenhouse gases for the northern United States, the magnitude and even the sign of possible changes in precipitation are uncertain. According to our regression equation, a 10% increase in average annual precipitation combined with a 2°C rise in average temperature would result in an 18% decrease in runoff. To counteract the effects of a 2°C warming completely, a 28% increase in precipitation would be required. Clearly, higher spatial resolution in climate models is needed for more credible forecasts of the effects of increasing atmospheric carbon dioxide and other greenhouse gases on the quantity of water supplies. Possibly, also, seasonal variations from year to year in precipitation and temperature may be more critical in determining Colorado River runoff than annual averaged variations. This possibility could be investigated by statistical analysis of monthly averages of temperature and precipitation for each year against the annual "virgin flow" at Lee Ferry.

7.4 CLIMATE CHANGE AND WATER-RESOURCE SYSTEMS

Planning and construction of major water-resource systems have a time constant of 30 to 50 years. In the past, these activities have been based on the explicit assumption of unchanging climate. The probably serious economic and social consequences of a carbon dioxide-induced climatic change within the next 50 to 100 years warrant careful consideration by planners of ways to create more robust and resilient water-resource systems that will, insofar as possible, mitigate these effects.

REFERENCES

Back, W. D., and J. S. Taylor (1980). Navajo water rights: pulling the plug on the Colorado River. Natural Resources Journal, January 1980, pp. 70-90.

Dracup, J. A. (1977). Impact on the Colorado River Basin and Southwest water supply. In Climate, Climatic Change, and Water Supply. National Academy of Sciences, National Academy Press, Washington, D.C.

Howe, C. W., and A. H. Murphy (1981). The utilization and impacts of climate information on the development and operations of the Colorado River system. In Managing Climatic Resources and Risks. Panel on the Effective Use of Climate Information in Decision Making, Climate Board, National Academy of Sciences, Washington, D.C., pp. 36-44.

Jensen, M. E. (1982). Water resources technology and management. Paper presented at Soil and Water Resources Conservation Act Symposium on Future Agricultural Technology and Resource Conservation, December 1982.

Laney, N. (1982). Does Arizona's 1980 Groundwater Management Act violate the Commerce Clause? Arizona Law Review 24:108-131.

Langbein, W. B., et al. (1949). Annual Runoff in the United States. U.S. Geological Survey Circular 5. U.S. Dept. of the Interior, Washington, D.C. (reprinted, 1959).

Manabe, S., and R. T. Wetherald (1980). On the horizontal distribution of climate change resulting from an increase of CO_2 content of the atmosphere. J. Atmos. Sci. 37:99-118.

Rogers, P. (1983). Water resource technology and management in the future of U.S. agriculture. Paper presented at the Soil and Water Resources Conservation Act Symposium on Future Agricultural Technology and Resource Conservation, December 1982.

Schaake, Jr., J. S., and Z. Kaczmarek (1979). Climate variability and the design and operation of water resource systems. In Proceedings of the World Climate Conference, pp. 290-312. World Meteorological Organization, Geneva, Switzerland.

Stockton, C. W., and W. R. Boggess (1979). Geohydrological implications of climate change on water resource development. U.S. Army Coastal Engineering Research Center, Fort Belvoir, Virginia.

U.S. Water Resources Council (1978). The Nation's Water Resources, The Second National Water Assessment. U.S. Govt. Printing Office, Washington, D.C.

8 Probable Future Changes in Sea Level Resulting from Increased Atmospheric Carbon Dioxide

Roger R. Revelle

Many processes can cause an apparent change in sea level at any particular location. They include local or regional uplift or subsidence of the land; changes of atmospheric pressure, winds, or ocean currents; changes in the volume of the ocean basins owing to volcanic activity, marine sediment deposition, isostatic adjustment of the Earth's crust under the sea or changes in the rate of seafloor spreading; changes in the mass of ocean water brought about by melting or accumulation of ice in ice sheets and alpine glaciers; and thermal expansion or contraction of ocean waters when these become warmer or colder. Only the last two processes are of primary interest in considering worldwide changes in sea level resulting from climatic change, such as the warming that may be induced by increasing atmospheric carbon dioxide. (Melting or formation of sea ice and floating ice shelves have no effect on sea level--a glass of ice water filled to the brim does not overflow while the ice melts.) But the other processes contribute to the "noise" that afflicts all sea-level records and may make their interpretation over periods of a few decades difficult or impossible.

For orientation, it is useful to keep in mind that sea level has risen 150 m in the 150 centuries since the peak of the last glacial period. Hence, the present rate of 10-20 cm per century is small compared with the average rate of 1 m per century over the past 15 millennia and very much smaller than the inferred maximum rise of perhaps 5 m per century immediately following the glacial period. Indeed, judging by Figure 8.1, the present is a time of quiet sea level compared with the violent oscillations that occurred during most of the last 100,000 years.

8.1 THE OBSERVED RISE IN SEA LEVEL DURING PAST DECADES

Barnett (1983d) examined all available records of annual averages of relative sea level. These were obtained from tide gauges around most of the world's continental margins and on a few islands. Barnett found 155 stations that had 30 or more years of record. Averaging over small areas of unusually high station density reduced this number to 82 usable data series. The longest covers the period from 1881 to 1980, and nearly all series extend from 1930 to 1980. The data are very unevenly

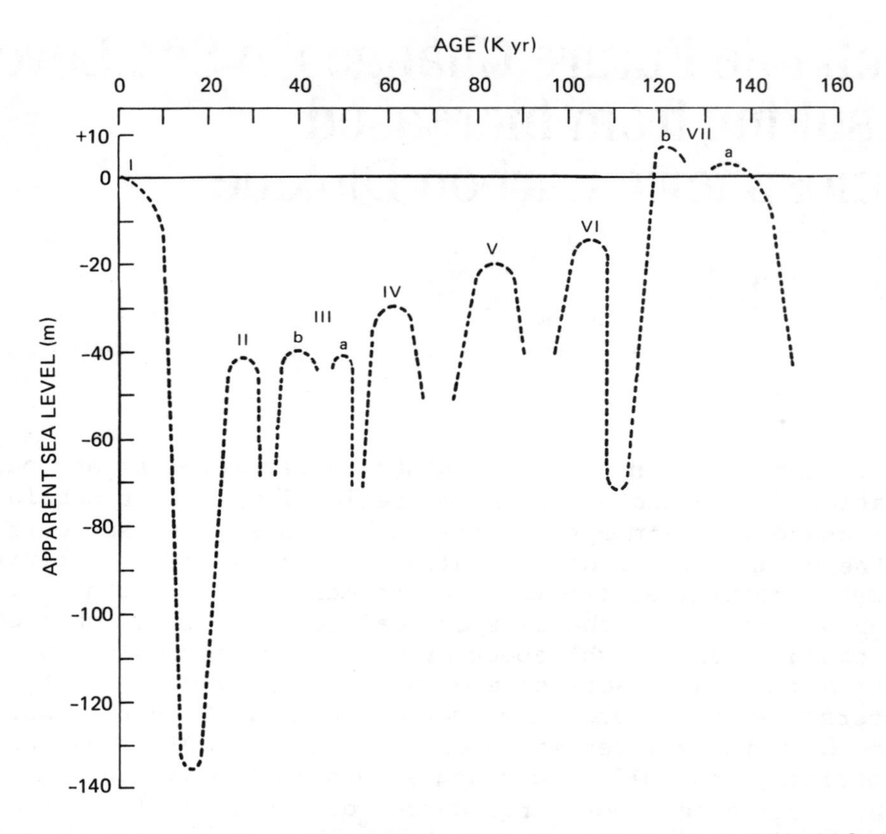

FIGURE 8.1 High stands of sea level during the past 150,000 years. Note 6-m terrace above present sea level 120,000-125,000 years ago, and 2-m terrace 135,000 years ago. (Source: Moore, 1982.)

distributed; there are only six stations in the southern hemisphere and only two (Hawaii and Bermuda) in the central ocean gyres.

To obtain some semblance of a global average, Barnett divided the stations, using empirical orthogonal function analysis (Barnett, 1978), into six oceanic regions in which there was strong coherence among different stations. For the 100 years from 1881 to 1980, the average sea level for all regions combined, weighted by the area of each, can be fitted by a linear trend of rising sea level of 14.3 $\pm$ 1.4 cm per century. During the 51 years from 1930 to 1980, a linear trend of 22.7 $\pm$ 2.3 cm per century was observed. The trend for the last 100 years represents a compromise between a time of little change (1881-1920) and a time of steady increase (1920-1980). Stations from Alaska and Scandinavia, where isostatic rebound is known to cause strong uplift of land, were omitted from the averages.

Other recent estimates of the rate of global sea-level rise, using geographically distributed selected stations or different methods of combining groups of stations, have been made by Fairbridge and Krebs (1962), Lisitzin (1974), Gornitz et al. (1982), Barnett (1983a), and Emery (1980). Their estimated rates were, respectively, 12, 11.2, 12, 15.1, and 30 cm per century. These different results reinforce

Barnett's conclusion (1983d) that estimates of global sea-level change with data available at present can vary substantially, depending solely on the method of analysis. He believes that the data do not support Emery's contention that the rate of rise has accelerated during recent decades.

Barnett (1983a,b,c) investigated the probable causes of the sea-level rise. He concluded from analysis of changes of dynamic heights in the upper water layers measured by repeated hydrographic casts, together with measurements, corrected for instrumental bias, showing an average rise in ocean surface temperature of about 0.4°C in the northern hemisphere since the beginning of the twentieth century, that warming and expansion of the upper water layers can account for less than 5 cm of the estimated rise in sea level. Gornitz et al. (1982) agree that only a small part of this observed rise is due to thermal expansion of the upper ocean waters. Barnett tentatively inferred that perhaps three fourths of the rise represents water added to the ocean from ablation or melting of the Greenland and Antarctic ice caps and alpine glaciers.

Observed changes in the length of the day since 1900 and a displacement by 730 cm toward 70° W in the position of Earth's mean pole of rotation are consistent with this explanation, which would mean that between 30,000 and 40,000 km^3 of ice (less than 0.2% of the land-ice volume) has melted, mainly from Greenland and Antarctica. Similar conclusions with respect to the length of the day were reached by Etkins and Epstein (1982). Estimates of mass balance of the Antarctic Ice Sheet (Bentley, 1983) suggest that the mass is stable and perhaps even increasing, but the noise level of the estimates is so high that a small net loss corresponding to a rise in sea level of 0.5 mm per year is not forbidden. Perhaps two thirds of the ice loss, or about 390 km^3 per year (1 mm/year of sea-level rise) may have come from Greenland. This idea is supported by the observations of Brewer et al. (1983) and Swift (1983) that the salinity of the deep water north of 50° in the North Atlantic has diminished over the past 20 years or so, as would be expected if freshwater from a melting ice cap were being added to the northern ocean.

8.2 THE FUTURE RISE IN SEA LEVEL

The projected climatic warming from increasing atmospheric carbon dioxide will lead to an increased transfer of water mass to the sea from continental and alpine glaciers. As shown below, the resulting rise in sea level could be about 40 cm over the next century. Increased downward infrared radiation will also lead to a warming and, therefore, expansion of the upper ocean waters, which can contribute another 30 cm for a total of 70 cm. Assuming the correctness of the figure of 4 W m^{-2} for the increased downward infrared flux with a doubling of carbon dioxide (Ramanathan et al., 1979),* both the estimates for ice

*Higher concentrations of other infrared gases will further increase the downward infrared flux (see Machta, this volume, Section 4.3).

melting and ocean thermal expansion still have large error bars--at least ±25%. These are due to our uncertainty over the causes of the present rise in sea level, our inability to predict whether changes in atmospheric circulation will cause more or less snow to fall on the ice caps, our ignorance of the conditions for advance or retreat of alpine glaciers, and our lack of understanding of the physical processes associated with the flux of heat to the ocean. Of even greater uncertainty is the potential disintegration of the West Antarctic Ice Sheet, most of which now rests on bedrock below sea level. This could cause a further sea-level rise of 5 to 6 m in the next several hundred years. We shall discuss each of these sources in turn.

8.2.1 Melting of Greenland Ice Cap and Alpine Glaciers

Ambach (1980) has investigated the effects of changes in atmospheric temperature on the Greenland ice cap. He shows that the altitude at which ice ablation occurs increases by about 100 m for each 0.6°C rise in air temperature, and he calculates that the area of ice ablation will increase by 340,000 km^2 for a 3°C rise in temperature. There would be an equal diminution in the area of ice accumulation, with the result that the ice volume would decrease by the equivalent of 480 km^3 each year, equal to an annual rise in sea level of 1.2 mm (12 cm per century). For a temperature rise of 6°C or more, which the general circulation models of Manabe and Wetherald (1975) indicate at the latitude of Greenland for a CO_2 doubling, the rate of rise would be at least 24 cm per century. The average rate over the next 80-100 years would be half of this, or 12 cm per century.

We are unable to calculate the rise in sea level due to a possible retreat of alpine glaciers. According to Barnett (1983a), glaciers contain about 240,000 km^3 of ice, and a 20% reduction in their volume during the next 100 years would bring about a rise of 12 cm in sea level.

Adding these figures for accelerated ice loss from Greenland and alpine glaciers to the estimated rate of loss from ice melting over the past 50 years of 17 cm per century, we arrive at a probable rise in sea level resulting from an increase in the mass of ocean water during the next century of 41 cm.

The volume of ice lost from ice caps and alpine glaciers over the century would be approximately 160,000 km^3, between 5 and 10% of the Greenland ice and around 0.5% of the total ice above sea level. The average latent heat of melting would be 1.2×10^{20} cal yr^{-1} or 1.5×10^{13} W. This is about 1% of the calculated increased downward flux to the ocean surface of infrared radiation of 4 W m^{-2} when atmospheric carbon dioxide has doubled and other greenhouse gases have increased by the expected amounts. Clearly, a major fraction of the downward infrared flux should be available to heat the upper ocean waters.

8.2.2 Heating of the Upper Oceans

This effect is strongly focused in the upper oceans for two reasons: it is where the heating is greatest and where the coefficient of thermal expansion is largest. For simplicity, the ocean can be divided into two layers: a mixed surface layer with an average thickness of 70 m in which the temperature change is uniform throughout and equal to the change in atmospheric temperature at the surface and the remainder of the ocean in which heat is transported by advection and turbulent diffusion.

Unfortunately, there is no good understanding of what happens in the deeper layer; major ongoing research efforts such as the projected World Ocean Circulation Experiment of the World Climate Research Program are directed at remedying this unsatisfactory situation. The evidence points toward the importance of heat transport along isopycnal (quasi-horizontal constant-density) surfaces from surface regions in higher latitudes toward the midlatitude interior. In comparison, crosspycnal (quasi-vertical) mixing processes, such as double diffusion (i.e., diffusion of heat and salinity at different molecular diffusion rates) and intermittent internal wave breaking, appear to be of lesser importance. This means that the heat transport needs to be considered as at least a two-dimensional (latitude-depth) problem and probably as a three-dimensional one. Nevertheless, one-dimensional mixing models persist, if for no other reason than that they can be solved. Here we sketch a one-dimensional treatment in the hope that, with the traditional empirically based choice of mixing parameters, we may arrive at some feeling as to the magnitude of the effect.

Let $\theta(z,t)$ represent the temperature change at different times and depths below the mixed layer relative to an assumed steady state in the year 1860. Then, in a simple two-layer system,

$$\frac{d\theta}{dt} = k \frac{d^2\theta}{dZ^2} + w \frac{d}{dZ} , \qquad (1)$$

where k is the coefficient of vertical eddy diffusion and w is the vertical component of advective velocity taken positive downward. Both k and w are assumed to be constant with depth down to 1000 m and zero below 1000 m. Cess and Goldenberg (1981) found that the time rate of change in temperature resulting from radiative heating of the atmosphere caused by observed and projected increases in atmospheric CO_2 is proportional to the temperature change, divided by an appropriate time constant, τ_c

$$\frac{d\theta}{dt} = \frac{\theta}{\tau_c} , \qquad (2)$$

and hence the solution of Equation (1) is

$$\theta(z,t) = \theta(0,t) \exp \frac{-(z - m)}{H} , \qquad (3)$$

where m = 70 m and

$$\frac{1}{H} = \frac{w}{2k} + \frac{1}{2}\left(\frac{w^2}{k^2} + \left(\frac{4}{k\tau_c}\right)\right)^{1/2} . \quad (4)$$

Using the past and projected future changes in atmospheric CO_2 since 1860, estimated by Machta and Telegadas (1974) and Keeling (1977), and the radiative heating at the surface computed by Ramanathan et al. (1979) for atmospheric CO_2 concentrations of 320, 426, 534, and 640 ppm, Cess and Goldenberg computed a value of 33 years for τ_c. If, as seems likely, the future proportional rate of increase of atmospheric CO_2 is lower than in the past, τ_c will be longer than 33 years, and the temperature changes at different depths below the mixed layer will be a larger fraction of the temperature changes at the surface. But the uncertainties in the calculation are so large that it is not very useful to attempt a more precise estimate of τ_c. By using a value of 33 years, we are probably computing a minimal rise in sea level from ocean warming resulting from a doubling of atmospheric carbon dioxide.

For the vertical component of advective velocity and the vertical-eddy diffusion coefficient in latitudes between 30° N and 30° S, we use Munk's (1966) suggested values,

$$w = 1.4 \times 10^{-5} \text{ cm sec}^{-1} ,$$

$$k < 30° = 1.3 \text{ cm}^2 \text{ sec}^{-1} .$$

The density stratification of the subsurface waters at higher latitudes is weaker than in the tropics, and consequently we assume that k north and south of the parallels of 30° N and 30° S, respectively, has twice its value in low latitudes, that is,

$$k > 30° = 2.6 \text{ cm}^2 \text{ sec}^{-1} .$$

Calculations of the increases in water temperature at different depths down to 1000 m were carried out for 10° latitude bands from 60° N to 80° S. Initial average surface ocean temperatures within each latitude band were taken from Sverdrup et al. (1942, p. 127). These were added to the change in average surface temperature for each latitude band projected by Flohn (1982) for the middle of the twenty-first century, assuming a doubling of atmospheric carbon dioxide and the probable increase in other greenhouse gases during the next 70-80 years.* Cess and Goldenberg show that a time delay of at least 20 years can be expected before the upper ocean waters come near to equilibrium with these projected values. Thus, our calculation gives the rise in sea level about 100 years from now.

---

*Flohn's projected global temperature rise is 4.2°C, somewhat higher than the value of 3°C used in the earlier discussion.

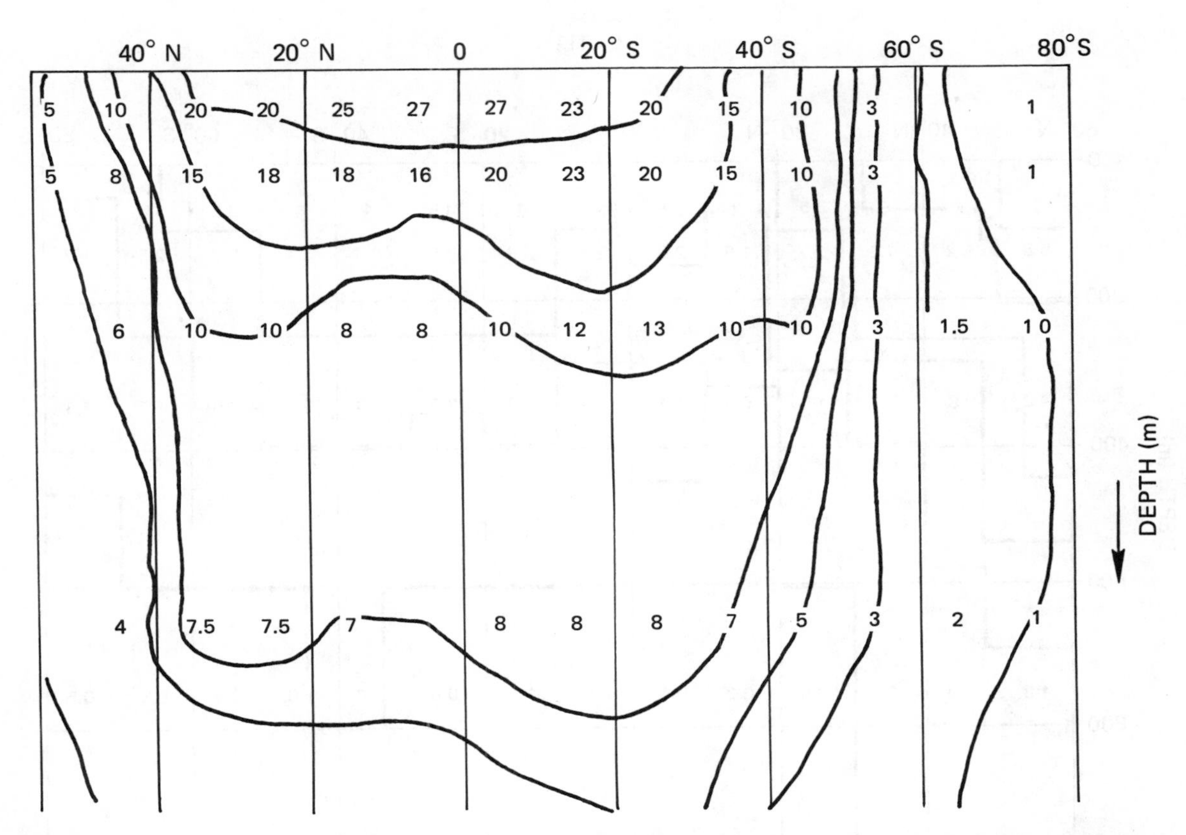

FIGURE 8.2 Present temperatures in the top 1000 m of the World Ocean, estimated from Atlantic and Pacific N-S sections (Figures 210 and 212, Sverdrup et al., 1942).

The average increases in temperature at depths between 0-100 m, 100-200 m, 200-500 m, and 500-1000 m were calculated from Equation (3) and added to the present values at these depth intervals. The latter were estimated from Figures 210 and 212 of Sverdrup et al. (1942, pp. 748 and 753). These figures give the distribution of temperature with depth for north-south sections in the Western Atlantic and Pacific Oceans, respectively. (See Figures 8.2 and 8.3.)

The increase in specific volume of seawater resulting from a given increase in temperature will vary markedly with temperature and depth. This quantity was calculated using the coefficients of thermal expansion at various temperatures for a salinity of $35^{o}/oo$ and depths of zero and 2000 m shown in Table 9 of Sverdrup et al. (1942, p. 60). The total change in specific volume (and thus in sea level) between zero and 1000 m for each 10° latitude band was multiplied by the percentage of ocean area in that band. We thus arrive at a rise in sea level about 100 years from now of at least 30 cm, resulting from ocean warming. A

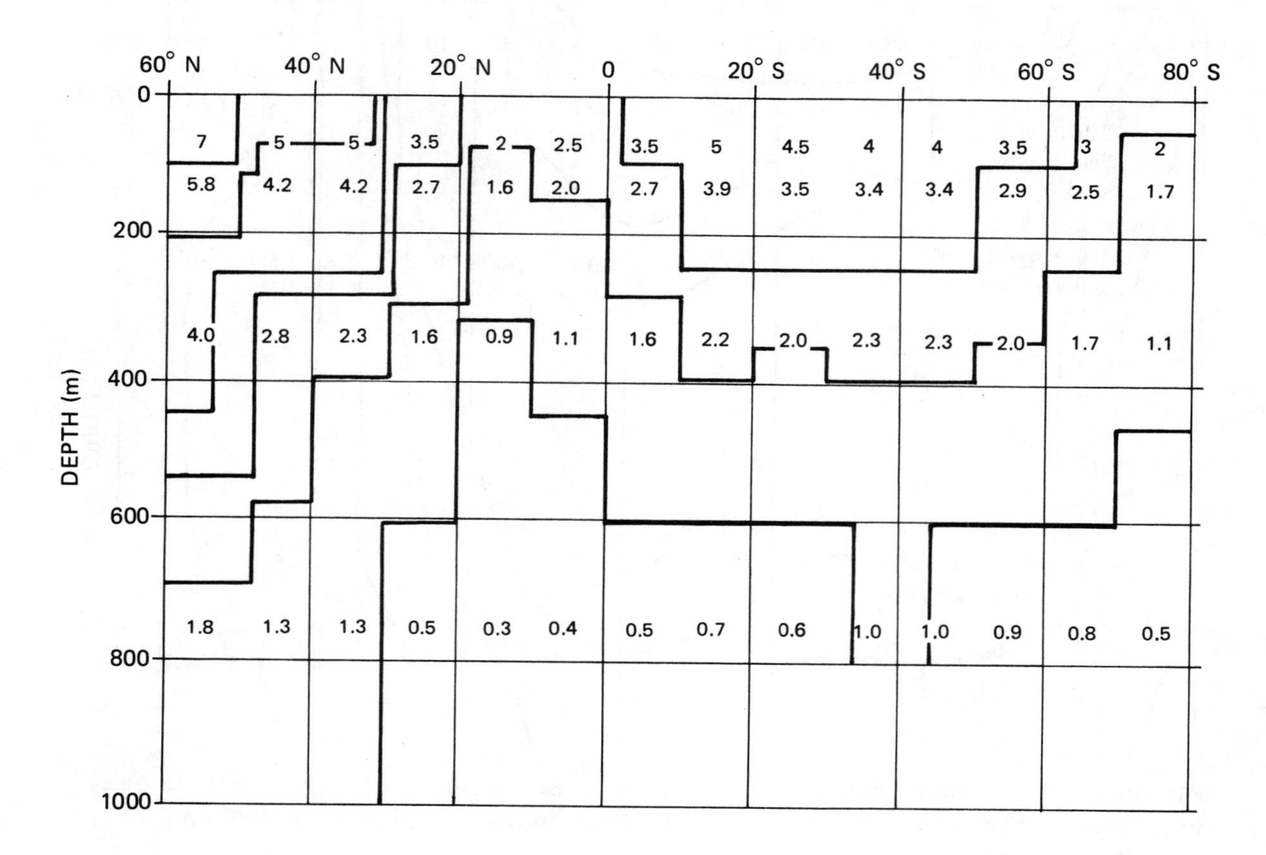

FIGURE 8.3 Computed near-equilibrium changes in ocean temperature for a doubling of atmospheric carbon dioxide and probable increase in other greenhouse gases, about 2080. The surface temperature increase is based on Flohn's (1982) prognosis.

similar calculation has been made by Gornitz et al. (1982); they estimate a rise of 20-30 cm during the next 70 years.

Adding this estimate for ocean warming to our estimates for melting in Greenland and Antarctica and in alpine glaciers, we arrive at a probable rise in sea level during the next 100 years of about 70 cm. But a much larger rate of rise is not unlikely during the following several centuries because of events in Antarctica.

8.2.3 Possible Disintegration of the West Antarctic Ice Sheet

West of the Transantarctic Mountains (approximately from the Meridian of Greenwich across the Antarctic Peninsula to 180° W), most of the Antarctic Ice Sheet rests on bedrock below sea level, some of it more than 1000 m beneath the sea surface. In its present configuration, this "marine ice sheet" is believed to be inherently unstable; it may be subject to rapid shrinkage and disintegration under the impact of a CO_2-induced climatic change (Mercer, 1978). (See Figure 8.4.)

A collapse of the West Antarctic Ice Sheet would release about 2 million km^3 of ice before the remaining half of the ice sheet began

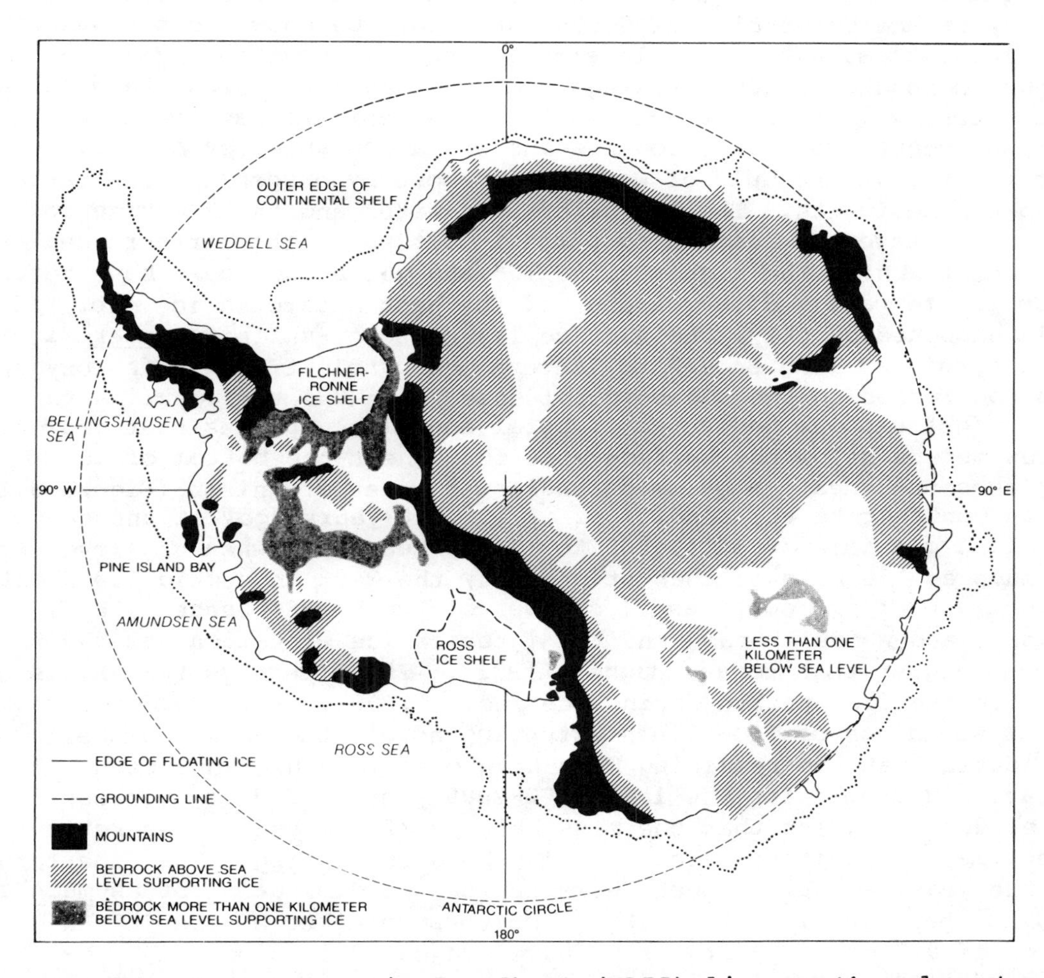

FIGURE 8.4 The West Antarctic Ice Sheet (WAIS) lies north and west of the Transantarctic Mountains (shown in black). It is believed to be unstable because most of it lies on rock below sea level. Disappearance of the ice above sea level would raise the world ocean by 5 to 6 m. At present, the ice sheet is held back by the Ross and Filchner-Ronne ice shelves, which, though mostly floating, are "pinned" by high places on the seafloor. From "Carbon Dioxide and World Climate" by Roger R. Revelle. Copyright © 1982 by Scientific American, Inc.

to float (Bentley, 1983). The resulting worldwide rise in sea level would be between 5 and 6 m. The oceans would flood all existing port facilities and other low-lying coastal structures, extensive sections of the heavily farmed and densely populated river deltas of the world, major portions of the state of Florida and Louisiana, and large areas of many of the world's major cities. Table 8.1 shows the percentage of the area of each coastal state of the United States that would be flooded by a 4.5-m rise and by a 7-m rise (Schneider and Chen, 1980).

Well-preserved fossil corals in late Pleistocene wave-cut or depositional terraces at 21 localities throughout the world have been dated by several groups of workers using measurements of the ratios of uranium and thorium isotopes. The results have been evaluated by Moore (1982) of the University of South Carolina. He shows that at almost all localities, a high sea level--perhaps 7 m above the present--occurred about 125,000 years ago. A somewhat lower high stand of the sea--about 2 m above present sea level--appears to have occurred around 135,000 years ago. This lower stand was accompanied by a major melt-water event in the Gulf of Mexico, as shown by oxygen/isotope ratios in deep-sea sediments. The influx of freshwater and the 2-m rise in sea level may have been caused by partial melting of the northern hemisphere ice sheet, while the higher stand 125,000 years ago could have resulted from the temporary disappearance of the West Antarctic ice cap. If this happened quickly, the surface layer of low-salinity water might have persisted long enough to leave a measurable trace in the oxygen isotope ratios in the oldest coral fossils on the terraces. A rapid rise might also be indicated if coral and algal species that grow best a few meters below the surface were found near the bottom of the terraces underneath remains of shallower water organisms (Figure 8.1).

In contrast to the conditions of 125,000 years ago, at the end of the last glacial period 17,000 to 12,000 years ago when sea level was as much as 150 m lower than at present, the West Antarctic Ice Sheet extended over the continental shelves of the Ross, Amundsen, Bellingshausen, and Weddell Seas, and the interior ice elevation was 500 to 2000 m higher than today (Hughes, 1983). Thus, perhaps two thirds of the ice has disappeared during the last 17,000 years. Evidence from radar soundings of flow lines extending across the Ross ice shelf indicates that the remaining West Antarctic Ice Sheet has been relatively stable for the last 1000-2000 years. Indeed, various lines of evidence suggest that the mass balance of the entire Antarctic Ice Sheet may be positive--i.e., ice may be accumulating. The uncertainty of the measurements is such, however, that a diminution of as much as 180 km^3 per year is not unlikely, corresponding to a rise in sea level of 0.5 mm per year, as we have suggested in discussing the present rate of sea-level rise.

The rate at which the remaining West Antarctic Ice Sheet could disappear under the impact of a CO_2-induced warming has recently been examined by Charles Bentley of the University of Wisconsin-Madison. He concludes that the mechanism for a relatively rapid disappearance would involve two processes occurring in sequence: accelerated melting on the underside of the Ross and Filchner-Ronne ice shelves by comparatively warm seawater (temperature between 2 and 3.5°C) circulating

TABLE 8.1 Summary of Estimated Geographic, Demographic, and Economic Impacts of 15- and 25-Foot (4.6- and 7.6-Meter) Rises in Sea Level for the Continental United States[a], [b]

Region/State	15-Foot Case					25-Foot Case				
	Flooded (%)	Population (millions)	Approx. % of State	EMV[c] ($ billions)	Approx. % of State	Flooded (%)	Population (millions)	Approx. % of State	EMV[c] ($ billions)	Approx. % of State
Florida	24.1	2.9	43	33.4	52	35.5	3.8	55	41.7	65
Gulf Coast	4.7	2.7	--	21.3	--	5.8	3.3	--	26.9	--
Texas	2.2	0.9	8	14.3	14	3.2	1.3	11	19.2	18
Louisiana	27.5	1.7	46	6.5	51	31.4	1.8	50	7.0	55
Mississippi	1.0	d	2	0.2	3	1.5	0.1	4	0.3	5
Alabama	0.8	d	2	0.3	2	1.1	0.1	2	0.4	2
Mid-Atlantic	5.3	1.8	--	11.6	--	7.6	2.5	--	16.6	--
Georgia	2.4	0.2	4	0.8	4	4.0	0.3	6	1.2	5
South Carolina	6.7	0.3	10	1.6	12	10.2	0.3	13	2.2	16
North Carolina	7.9	0.2	3	1.4	4	9.8	0.2	4	1.9	6
Virginia	3.1	0.7	16	4.1	12	4.8	1.0	21	5.8	16
Maryland	12.3	0.2	6	1.5	5	17.3	0.4	10	2.6	9
Dist. of Columbia	15.0	0.1	15	1.2	15	20.0	0.2	20	1.6	20
Delaware	16.0	0.1	19	1.0	18	25.0	0.1	25	1.3	25
North Atlantic	0.9	3.6	--	33.3	--	1.3	5.0	--	47.6	--
New Jersey	9.5	0.7	9	6.2	9	13.1	0.9	12	8.4	12
Pennsylvania	0.1	0.4	3	2.0	3	0.2	0.5	4	2.8	4
New York	0.6	2.1	12	22.0	12	0.8	3.0	16	30.9	16
Connecticut	1.2	d	2	0.6	2	2.2	0.1	3	1.1	3
Rhode Island	3.5	d	3	0.2	3	7.0	0.1	6	0.4	5
Massachusetts	2.1	0.3	4	2.2	5	3.4	0.4	8	3.7	8
New Hampshire	0.1	d	0	c	1	0.4	c	1	0.1	1
Maine	0.2	d	1	0.1	2	0.4	c	1	0.2	3
West Coast	0.6	0.8	--	7.8	--	1.2	1.4	--	14.2	--
California	1.0	0.7	4	7.3	3	1.7	1.2	6	12.7	6
Oregon	0.1	d	1	0.3	1	0.4	0.1	3	0.6	3
Washington	0.4	d	1	0.2	1	7.1	0.1	2	0.9	3
All Regions		11.6		107.5			15.7		146.9	
Continental U.S. (%)	1.5	5.7		6.2		2.1	7.8		8.4	

[a]SOURCE: Schneider and Chen (1980).
[b]Totals may not add exactly due to rounding errors.
[c]EMV equals Estimated Market Value derived by dividing the "locally assessed taxable real property" in each county by the "aggregate assessment sales price ratio," which is based on a sample of market values.
[d]Less than 0.1.

under the shelves, which could bring about a thinning of the shelves of one to several meters per year, followed by relatively rapid movement of ice streams and calving of icebergs along the ice front.

Thinning of ice shelves would result in their becoming "unpinned" from most high points of the seafloor and in a rapid retreat of the "grounding lines" (the boundaries between the floating ice shelves and the grounded inland ice), followed by still more rapid acceleration of the ice shelf thinning and grounding line retreat. This process of ice-shelf removal would not be necessary in the Pine Island Bay-Thwaites Glacier area of the Amundsen Sea because no ice shelves exist there.

After the removal of the ice shelves, "ice streams" in the inland ice could discharge directly into the ocean (Figure 8.5). According to Bentley, the width of the ice streams, including Thwaites and Pine Island Glaciers, would be about half the width of the new coastline, or 500-600 km. Removal of 2×10^6 km^3 of ice in 200 years would require a speed of discharge of 20 km/year, about 20 times faster than the present speed of movement. This is nearly 3 times faster than the fastest known ice stream/outlet glacier speed of 7 km/year in the Jakobshaven Glacier of Greenland. Bentley concludes that the ice streams cannot move at 20 km/year in their unbroken form but that this could occur by another process: calving and removal of icebergs, with the rate of the ice loss being equal to the width of the ice streams times their average thickness of 1 km, times the sum of the glacier speeds and the speed of retreat of the grounding lines. The limiting factor here would be the rate of removal of the icebergs by winds and ocean currents. Bentley calculates that the component of ocean current velocity at right angles to the glacial shoreline would have to be 2 cm sec^{-1} and that the current velocities in a series of gyres in the Ross, Amundsen, and Weddell Seas would have to be at least 10 cm sec^{-1}. He concludes that such rates of discharge and removal of icebergs "might barely be possible, although unlikely, and that the ice sheet could disappear in 200 years, but only after removal of the ice shelves. . . ."

Bentley considers the suggestion by some workers that all the ice could be discharged into the Amundsen Sea through Thwaites and Pine Island Glaciers in 200 years would require "unreasonably high" glacier speeds of at least 50 km yr^{-1} and ocean current speeds to carry away calved icebergs of 50 cm sec^{-1}. However, he suggests that one quarter of the West Antarctic ice could be discharged through Pine Island Bay in 200 years and half in 400 years. During this period the Ross and Filchner-Ronne ice shelves could have disappeared, and all the ice could be discharged within 500 years.

If the time required for the ice shelves to disappear is 100 years, Bentley's analysis would not be incompatible with a minimum time of 300 years for disintegration of the West Antarctic Ice Sheet. The corresponding average rate of rise of sea level would be slightly less than 2 m/100 years, beginning about the middle of the next century. Bentley's "preferred" minimum time of about 500 years would give a rate of sea-level rise of 1.1 m/100 years, which, as we have pointed out, is about the mean rate for the last 15,000 years. To either of these figures we must add a rise of 70 ± 18 cm between 1980 and 2080, which we have shown is likely to result from ocean warming and ice ablation in

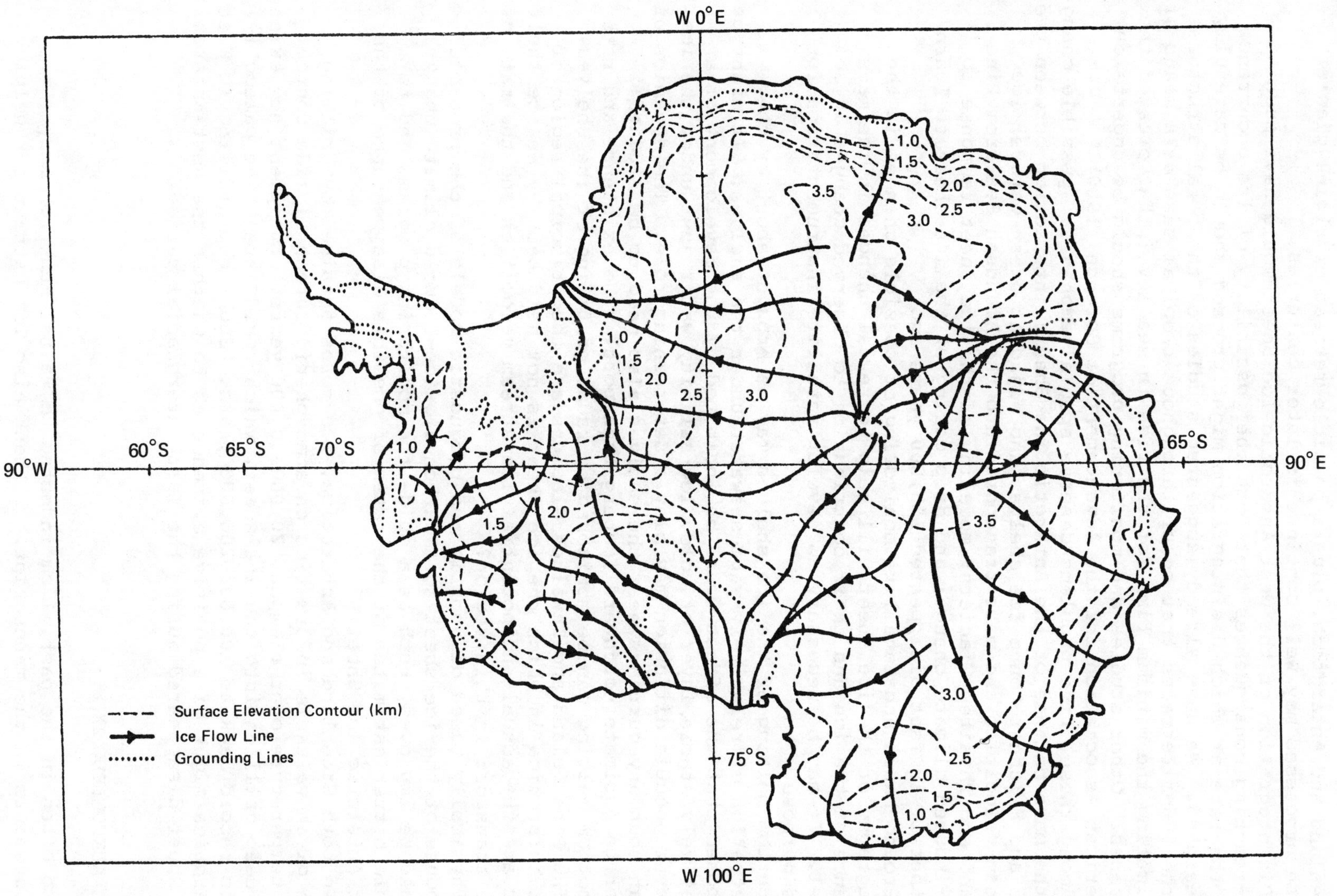

FIGURE 8.5 Generalized map of Antarctic surface elevations with schematic flow lines. Flow vectors diverge from ice divides into separate drainage systems. (Source: Bentley, 1983)

Greenland and Antarctica, plus a possible retreat of alpine glaciers. These processes may well continue in later centuries.

Disintegration of the West Antarctic Ice Sheet would have such far-reaching consequences that both the possibility of its occurrence and the rate at which disintegration might proceed should be carefully researched. We have already suggested studies of the coral structures in the 5-6-m terraces that occur throughout tropical seas as a means of estimating the minimum time for the rise in sea level that created the terraces. Other studies and monitoring programs should be undertaken (American Association for the Advancement of Science, 1980).

Among these, five problems deserve special emphasis: possible change in the mass balance of the Antarctic Ice Sheet; interaction between the Ross and Filchner-Ronne ice shelves and adjacent ocean waters; ice stream velocities and mass transport into the Amundsen Sea from Pine Island and Thwaites Glaciers; modeling of the ice sheet response to CO_2-induced climate change; and deep coring of the ice sheet to learn whether it in fact disappeared 125,000 years ago.

Ground-based and satellite monitoring of possible changes in the topography of the ice sheet will help to reveal changes in the mass balance of the ice and their contribution to observed changes in sea level. It may be necessary to develop satellite instrumentation for this purpose.

Bottom melting of the ice shelves may be brought about by the penetration of relatively warm seawater under the shelves. It might be supposed that the effect would be inhibited by the development of a relatively stable layer of cold, low-salinity water just under the ice. However, double diffusion may play a determining role. The dynamics of ocean boundary currents near the shelves also need to be studied in order to estimate northward iceberg transport. The existence and rate of bottom melting can be monitored by radar profiling of the shelves.

The Pine Island and Thwaites Glaciers are the principal region of West Antarctica in which the ice sheet is not held back by an ice shelf. What are the actual and potential ice stream velocities and the rate of mass transport in this region?

Physically based diagnostic and prognostic models of the probable response of the ice sheet to a carbon dioxide-induced climate change should be improved, both as a guide for monitoring programs and to enable better estimates of the rates of possible disappearance of the West Antarctic Ice Sheet.

In both Greenland and Antarctica, ice cores obtained by drilling have given valuable information on atmospheric carbon dioxide content and temperature over the past 20,000-30,000 years. Equipment now exists for deep drilling into the ice sheet, which could extend the paleoclimatic record back to 100,000-200,000 years. Such deep drilling in West Antarctica might give positive evidence as to whether the West Antarctic Ice Sheet disappeared during the last interglacial period.

8.4 ACKNOWLEDGMENTS

As so often in the past, I am indebted to Walter Munk for many improvements in the manuscript. I thank also Tim Barnett for helping me understand the uncertainties in existing data related to sea level and Jesse Ausubel for careful editing. The discussion of the West Antarctic Ice Sheet is partly based on an informal conference at the Scripps Institution of Oceanography attended by C. Bentley, A. L. Bloom, W. S. Broecker, E. S. Epstein, R. Etkins, V. Gornitz, T. Gross, D. Katcher, W. Nierenberg, J. S. Perry, R. Wetherald, and the author.

REFERENCES

Ambach, W. (translated by G. P. Weidhaas) (1980). Increased CO_2 concentration in the atmosphere and climate change: potential effects on the Greenland ice sheet. Wetter und Leben 32:135-142, Vienna. (Available as Lawrence Livermore National Laboratory Report UCRL-TRANS-11767, April 1982.)

American Association for the Advancement of Science (1980). Response of the WAIS to CO_2 Warming. Report of a meeting, Orono, Maine, April 1980 (mimeo.). (Made available by C. Bentley, University of Wisconsin-Madison.)

Barnett, T. P. (1978). Estimating variability of surface air temperature in the northern hemisphere. Mon. Wea. Rev. 106: 1353-1367.

Barnett, T. P. (1983a). Recent changes in sea level and their possible causes. Climatic Change 5:15-38.

Barnett, T. P. (1983b). Long-term trends in surface temperature over the oceans. Mon. Wea. Rev., in press.

Barnett, T. P. (1983c). Long-term changes in dynamic height. J. Geophys. Res., in press.

Barnett, T. P. (1983d). Some problems associated with the estimation of "global" sea level change. Climate Research Group, Scripps Institution of Oceanography, U. of Calif., San Diego, 40 pp. (mimeo).

Bentley, C. R. (1983). The West Antarctic Ice Sheet: diagnosis and prognosis. Proceedings, Carbon Dioxide Research Conference, Carbon Dioxide, Science and Consensus, Berkeley Springs, W. Va., September 1982. CONF-820970, NTIS, Springfield, Va.

Brewer, P. G., W. S. Broecker, W. J. Jenkins, P. B. Rhines, C. G. Rooth, J. H. Swift, T. Takahashi, and R. T. Williams (1983). A climatic freshening of the deep North Atlantic (north of 50° N) over the past 20 years. Science, to be submitted.

Cess, R. D., and S. D. Goldenberg (1981). The effect of ocean heat capacity upon global warming due to increasing atmospheric carbon dioxide. J. Geophys. Res. 86:498-502.

Emery, K. O. (1980). Relative sea levels from tide gauge records. Proc. Nat. Acad. Sci. 77:6968-6972.

Etkins, R., and E. S. Epstein (1982). The rise of global mean sea level as an indication of climate change. Science 215:287-289.

Fairbridge, R., and O. Krebs, Jr. (1962). Sea level and the southern oscillation. Geophys. J. Roy. Astron. Soc. 6:532-545.

Flohn, H. (1982). Climate change and an ice-free Arctic Ocean. In Carbon Dioxide Review: 1982, W. C. Clark, ed. Oxford U. Press, New York, pp. 145-199.

Gornitz, V., S. Lebedeff, and J. Hansen (1982). Global sea level trend in the past century. Science 215:1611-1614.

Hughes, T. (1982). The stability of the West Antarctic Ice Sheet: what has happened and what will happen. Proceedings, Carbon Dioxide Research Conference, Carbon Dioxide, Science and Consensus, Berkeley Springs, W. Va., September 1982. CONF-820970, NTIS, Springfield, Va.

Keeling, C. D. (1972). Impact of industrial gases on climate. In Energy and Climate. Geophysics Study Committee, National Research Council, National Academy of Sciences, Washington, D.C., pp. 72-95.

Lisitzin, E. (1974). Sea-Level Changes, Elsevier Oceanography Series, 8, Elsevier, New York.

Machta, L., and K. Telegadas (1974). Inadvertent large-scale weather modification. In Weather and Climate Modification, W. Hess, ed. Wiley, New York.

Manabe, S., and R. T. Wetherald (1975). The effects of doubling the CO_2 concentration on the climate of a general circulation model. J. Atmos. Sci. 32:3-15.

Mercer, J. H. (1978). West Antarctic Ice Sheet and CO_2 greenhouse effect: a threat of disaster. Nature 271:321-325.

Moore, W. S. (1982). Late pleistocene sea-level history. In Uranium Series Disequilibrium: Applications to Environmental Problems, M. Ivanovich and R. S. Harmon, eds. Clarendon, Oxford, p. 41.

Munk, W. H. (1966). Abyssal recipes. Deep Sea Res. 13:707-730.

Ramanathan, V. M., M. S. Lian, and R. D. Cess (1979). Increased atmospheric CO_2: zonal and seasonal estimates of the effect on the radiation energy balance and surface temperature. J. Geophys. Res. 84:4949-4958.

Revelle, R. (1982). Carbon dioxide and world climate. Sci. Am. 247:36-43.

Schneider, S. H., and R. S. Chen (1980). Carbon dioxide warming and coastline flooding: physical factors and climatic impact. Ann. Rev. Energy 5:107-140.

Sverdrup, H. U., M. W. Johnson, and R. H. Fleming (1942). The Oceans. Prentice-Hall, New York.

Swift, J. H. (1983). A recent O-S shift in the deep water of the northern Atlantic. MLRG, Scripps Institution of Oceanography, La Jolla, Calif. Submitted for Proc. Vol. 4th Biennial Ewing Symposium.

9 Climatic Change: Implications for Welfare and Policy

Thomas C. Schelling

9.1 INTRODUCTION

The protagonist of this study has been carbon dioxide. The research has been motivated by concern that atmospheric carbon dioxide is increasing and may increase faster as the use of fossil fuels continues to grow and by the known potential for a "greenhouse effect" that could generate worldwide changes in climate. The group responsible for the report is the Carbon Dioxide Assessment Committee; the study was authorized by an act of Congress concerned with carbon-intensive fuels; and the agency principally charged with managing the research is the Department of Energy. The topic is usually referred to as "the carbon dioxide problem," a global challenge to the management of energy resources.

In a chapter on the welfare and policy implications, however, there are reasons for taking climate change itself as our theme, not carbon dioxide or energy. First, over the span of time this report has to cover there could be changes in climate not due to human activity. Changes due to carbon dioxide have to be assessed against some projected background of natural change. The consequences of an increment in global temperature, for example, would depend on whether it were superimposed on a warming, a cooling, or no trend at all.

Second, carbon dioxide is not the only climate-affecting substance that people inject into the atmosphere. Other gases have greenhouse effects; and dust from farming and industrial activity and the myriad changes made in the Earth's surface can alter climate. Not only must the impact of carbon dioxide be assessed together with other climate-changing activities, but any policy response needs a focus wider than carbon dioxide.

Third, there is a natural tendency to define a problem by reference to the agent of change and to seek solutions in the domain suggested by the naming of the problem, "fossil fuels" or "CO_2." There is a legitimate presumption that in matters of the Earth's biosphere any drastic change may produce mischief. There is also a widespread methodological preference for preventive over meliorative programs and for dealing with causes rather than symptoms. But it would be wrong to commit ourselves to the principle that if fossil fuels and carbon dioxide are where the problem arises, that must also be where the solution lies.

Finally, although a precautionary attitude toward any drastic changes in world climates is prudent, we do not know that there actually is a problem until we have completed the investigation of what changes in climates may occur <u>and</u> what damages or blessings they may bring. In assessing something as complex as global meteorological dynamics, it is wise to avoid any formulation that speaks of "a problem" in the singular or that attempts to evaluate changes absolutely, without regard to the pace of change over time.

Until we reach the welfare and policy implications, research on atmospheric carbon dioxide generates its own agenda: sources of emissions, projected uses of fuels and patterns of land use, sources and sinks in the carbon cycle, radiative balance of the atmosphere and interchange of heat between oceans and air, changes in atmospheric temperature and induced changes in wind and precipitation, feedback through water vapor and reflecting ice, and the direct effect of carbon dioxide on photosynthesis. Which among the uncertainties are likely to yield to further research, on what time schedule, and how to deploy research resources are questions that have to be answered through a systems approach; but the main features of the system and its inputs are generated by a process that is mostly straightforward, even if not all of the relevant inputs can be anticipated before the research gets under way.

But the agenda for policy implications demands an iterative process. To illustrate, at certain CO_2 concentrations there are food-producing locations where reduced rainfall or snowfall is likely, and reduced food output, unless there are compensating improvements in water supply or water conservation or changes in crops or farming technology. To assess this change in climate requires projecting water resources and technologies and farming techniques; and to choose policies requires attending to whether it will be more economical to reduce production of CO_2 or to increase production of water.

We have a choice between conserving fuel and conserving water. While conserving fuel is an obvious policy option at the outset, the parallel importance of conserving water emerges only at the end of the scenario, enlarging the domain of required research. Defining the problem as "the CO_2 problem" can focus attention too exclusively on energy and fossil fuels, compared with calling it the water or the rainfall problem or, more evenhandedly, the issue of climate change.

How the issue is named can affect its apparent character. If the solution to foreseeable problems has to be reduced CO_2 emissions, both the problem and the solution are global in a severe sense. A ton of CO_2 produced anywhere in the world has the same effects as a ton produced anywhere else. Any nation that attempts to mitigate changes in climate through a unilateral program of energy conservation or fuel switching, or scrubbing of CO_2 from smokestacks, in the absence of some international rationing or compensation arrangement, pays alone the cost of its program while sharing the consequences with the rest of the world. Worldwide agreements involving the main consumers or producers of fossil fuels might be essential to programs for reducing CO_2 emissions; in contrast, water development and conservation is usually a national responsibility or involves a few neighboring countries.

So the organizing framework for welfare and policy implications of atmospheric CO_2 should be built around climate change not around the carbon dioxide.

9.1.1 Uncertainties

In approaching these welfare and policy implications there are two kinds of uncertainty. One is the subject of the preceding chapters: uncertainties about sources and uses of energy, which in turn embody uncertainties about population, per capita income, energy-using and energy-producing technologies, density and geographical distribution of populations, and the distribution of income; a multitude of uncertainties about the carbon cycle; and, finally, all the uncertainties in translating a growth curve for CO_2 in the atmosphere into appropriately time-phased changes in climate in all the regions of the globe.

The second is uncertainty about the kind of world the human race will be inhabiting as the decades go by, through the coming century and beyond. This overlaps the uncertainties just mentioned: per capita income, for example, both influences the use of fossil fuels and affects how readily the world's population can afford, or can adapt to, changes in climate. Similarly, the structures people inhabit, the ways people and goods are transported, the foods people eat, the ways countries defend themselves, and the geographical distributions of populations within and among countries all affect land use and the kinds and amounts of energy used and hence the production of CO_2; but they also affect the ways that climate impinges in living and earning, even on what climates are preferred. The mobility of people, capital, and goods--the readiness with which people can migrate, goods can be traded, and capital for infrastructure and productive capacity can flow among regions and countries--will also determine how much difference the changes in climate can make. The location and significance of national boundaries, and various international and supranational institutions, would have much to do with whether adverse climatic effects in some places can be offset, in a welfare assessment, by improvements in other places.

9.1.2 The Time Dimension

In addition to uncertainty, there is the problem of managing an indefinite succession of future times. There is a temptation to pick some arbitrary concentration of CO_2, like double the current level, and to experiment with variables to get an idea of when in the next century that particular concentration might occur, using the median of some probability distribution that yields a single date. Alternatively, we pick a date, like the middle of the next century, and estimate the likely change in "climate" from now to then; to do this we have to pick some arbitrary conventional figure, like the mean expected increment in global atmospheric temperature, as index of the seriousness of climate change.

These simplifications can be helpful as long as it is kept continuously in mind that they are merely shorthand expressions for a dynamic process of many dimensions. The trick is keeping that continuously in mind.

It is possible to do something more sophisticated. As in the approach to CO_2 emissions (this volume, Chapter 2, Section 2.1), we can make random selections of probabilistic values of key variables and generate time paths into the future for climate change. That generates more information than a person can keep in mind, especially if the several time paths are not roughly parallel; we then have the same problem of compressing all those results into a manageable number of dynamically descriptive parameters. Even then, what remains is a somewhat arbitrary intermediate variable with no direct climatic interpretation. Even if temperature change were all we really wanted to know in the different regions and localities, the global mean temperature is little help unless each of us knows how to extract from it the atmospheric temperatures at geographical locations of interest. But usually we are even more interested in winter and summer precipitation, humidity, cloud cover, fog, wind velocity, seasonal temperature variability, and the annual variations of temperature and rainfall around their local means. So even locally we have no single measure of what we might mean by climate change.

Whatever the way we translate a CO_2 concentration into some standard index like mean temperature (or temperature differential between pole and equator), and however we translate that locally into rain and snow and wind and sunshine and degree days, there is still no assurance that the changes that interest us in the things we call "climate" will vary proportionally with that numerical index (or even monotonically) in any given location. And finally, sports fans and farmers do not even agree on what is an improvement in climate and what is a worsening. For all these reasons it is difficult to keep in mind what it is that we want to be talking about as a "measure" of climate change in the future.

9.1.3 Discounting, Positive or Negative

Then there is how we think about the future. The climate changes anticipated are at an unaccustomed planning distance in the future. The troubling changes are probably beyond the lifetimes of contemporary decision makers but not beyond intimate association; our grandchildren will live into the span of time we have in mind.

There arises the issue of discounting costs and benefits that accrue in the future for comparison with costs incurred now. Should future populations count equally with present populations, or count more because there will be more of them, or count less because we feel less concern for people remote in any significant dimension like space, time, or nationality? Will future populations on average have vastly superior living standards, or somewhat better, or worse than today's generation? A century from now will large parts of the world's population be in extreme poverty, or will nearly everybody be better off, climate change notwithstanding, than most of the world's population is now? If we were to undergo economic sacrifices to improve the well-being of future generations, is what we might do about climate change likely to be worth more than alternative legacies that we might leave them? These are not questions to be answered here--nor are there universal answers; they will be answered differently by different people and different governments--but they are unavoidable in an actual judgment on the welfare and policy implications of climate change in the lives of our grandchildren and their children.

9.1.4 Perspective on Change

The most perplexing uncertainties are not in the train of events from the burning of coal to the changes in precipitation and temperature. Those at least have a certain structure. More open-ended and unrestricted, and demanding imagination as much as estimation, are questions about what the world will look like. How will people be living and working and moving and raising families and entertaining themselves in Peking or New York or on the plains of Kansas or Patagonia, in the Nile Valley, southern England, or northern India? A mistake hard to avoid is superimposing a climate change that would occur gradually in the distant future on life as we know it today--today's habitations and transport, today's agriculture and construction and fishing, today's urban complexes, today's working hours and living standards, diet and warmth, indoor and outdoor activity.

A useful exercise is to project ourselves 75 or 100 years into the past and imagine how different life is now, with its blessings and its problems, from what we might have expected had we been concerned, toward the end of the last century or the beginning of this one, with climate changes during the decades we are now experiencing.

Anyone can amuse himself with a list of technologies, political events, and demographic and environmental phenomena that would be most startling to students of social policy around the turn of the last century. Electronics was not dreamed of. Electric light would have been new in our lifetime and unknown to most of our countrymen. There was telephone but no radio. Nuclear energy for electricity or propulsion, let alone for weapons and medicine, was way over the horizon. (Transatlantic travel by zeppelin was a generation in the future.) Satellites in geosynchronous orbit would have been useless without today's electronics. Anesthesia was by ether, there were no antibiotics, bedbugs were a scourge, and yellow fever had caused abandonment of the Panama Canal. Air-conditioning was by ice, and the New York World Trade Towers would have seemed like first steps toward enclosing cities. Electric street railways were transforming our cities. California had half the population of Massachusetts. Soybeans were not grown in Iowa nor rice in California. The greatest military advance to come in the next war was an unbelievably lethal defensive combination of machine guns, barbed wire, and mud.

Russia was czarist. Africa belonged to Europe. Average weekly working hours in U.S. industry were 60, and there were no child labor laws. U.S. life expectancy at birth was 47 years (it is now 74). Only a third of the U.S. population lived in places with more than 5,000 inhabitants, barely a fifth in places with 50,000. More than a quarter of the horsepower of all prime movers in the United States came from animals.

Will there be changes in the way people live and work over the next hundred years that make as much difference as in the past hundred? Has our forecasting and assessing improved since the turn of the last century? If we had perfect climate forecasts for all the inhabited regions of the world for the century that begins, say, in the year 2025, there would undoubtedly be important parts of the world, and segments of populations in all parts of the world, where it would be difficult to put an algebraic sign on the apparent welfare impact, let alone assess the magnitude. For the world as a whole we might not be confident of the direction of change in some aggregate measure of welfare.

Undoubtedly there will be places where some predicted change in climate could have no foreseeable benefit and where some potential damages could be foreseen with clarity. But unless we impute to ourselves foresight much superior to what we might willingly claim for ourselves were we doing our work in 1900, it is likely that most of the identifiable changes in welfare due to climate change would be, for most parts of the world, swamped by other uncertainties. Even whether, on behalf of the world, we would prefer the mean global atmospheric temperature to rise by a couple of degrees over the next hundred years, or to fall by that amount, or to stay as it is, would depend on considerations other than assessments of what the specific changes in local and regional climate would do to life and welfare in the affected populations.*

9.1.5 Prudential Considerations

What are those other considerations? One is that drastic change in the environment is costly to adapt to. Our technologies, our homes, our habits, our crops, and our travel patterns and recreational activities--but most of all the knowledge and custom that people bring to bear in earning their livings, the technologies embedded in structures and equipment, the genetics of plants and animals--arose in evolutionary processes. Any change in the environment that makes the work habits and the technologies and the crops unsuitable requires migration, adaptation, and replacement. All of these are costly and uncertain.

To this it can be replied, inconclusively, that if the change is slow the adaptations and replacements, even the migrations, need not be traumatic or even especially noticeable against the ordinary trends of

---

*A personal observation: Every expression of concern that I have read or heard about the effect of rising temperatures on human health and comfort has been about summer heat; I conjecture that in 1900 the comments would have been about winter cold, autumn frost, and spring thaw, and the tone would have been positive.

obsolesence, movement, and change. The issue is suddenness and unexpectedness. Of course, where migration is politically impossible, or capital immobile, the movements and adaptations may not be forthcoming.

A powerful argument is that we are good at recognizing precisely those impacts that we are good at accommodating. The reason we recognize them may be that we already know enough about them to anticipate and to adapt. It is consequences we have not thought of that may find us no better at adapting than we were at anticipating. They could be of greater magnitude than the consequences we can foresee. An example might have been the potential collapse of the West Antarctic ice sheet; that one has been thought of, but there could be something comparably different from ordinary climate that we have not thought of.

It is wise to be concerned about any prospective change in some major index of climate, like the mean annual global atmospheric temperature, that goes beyond the boundary of values believed to have been experienced throughout the history of civilization. Certainly the temperate parts of Europe and North America would find bleak the prospect of a global cooling by two or three degrees, if the experience of the reported "Little Ice Age" of a few centuries ago is any guide. Still, it is not a cooling we are faced with but a warming, and there may be an asymmetry in our favor. Or we just look at things from the point of view of temperate climates that have more to lose from a cooling than from a warming. Or perhaps we just have not done the thought experiment of superimposing the colder winters on today's world and, by a series of approximations, thought of how we might adapt to such winters.

A fair conclusion from the preceding chapters is that there could be, within the next 100 or 200 years, a systematic global change in climate exceeding anything that has occurred in the last 10,000 years. To be specific, the mean annual global atmospheric temperature would be higher, and the temperature differential between poles and equator would be smaller. And in specific locations the major components of climate, like temperature and rainfall, may move outside the limits experienced during the past 10,000 years.

9.1.6 Variation in Human Environments

This prospect is sometimes described as a greater change in climate than people have undergone during recorded history. That is a dramatic way to describe the relative perturbation of some crucial global climate variables. But if we have in mind people and their climates, rather than climates in fixed localities, and especially if we think of mankind rather than individuals, that understates the climate changes to which people have been subjected individually and through which mankind has passed statistically.

The members of this committee undergo greater changes in climate whenever they meet than most populations will undergo if they remain stationary while climate changes during the next hundred years. People who moved north in the United States during the 1930s in search of jobs or west in the 1940s, who migrated southwest in the postwar period in search of pleasant environments, who left economically declining parts of New England, or who were recruited into the army, all experienced long-term or permanent changes of climate greater than any changes typically recorded over the past 10,000 years.

People experience climate change mainly by moving. People have been moving on continental and intercontinental scales at least since the time of the Roman Empire. If changes in climate during the past several thousand years have been small in comparison with the changes we must consider possible during the coming hundred years, the latter are similarly small in comparison with the changes that large parts of the world's population have undergone in populating the western hemisphere. Even if people did not move, differential population growth in regions with different climate would have meant that "mankind" underwent large changes in average or statistical climate--the fractions of the population living in different climates.

There is no indication that new kinds of climate will result from the rise in global average temperature. Climates will shift; some climates may become more and some less widespread. But the variation in climate is so great from desert to pine forest to rain forest, from the wide variation of summer and winter temperatures in the Dakotas to the narrow band of San Francisco, from permafrost on the North Slope to fog on the California coast, that the changes over _time_ will still be unremarkable compared with variations across _space_.

In 1860, 98% of the U.S. population lived in humid continental or subtropical climates; barely 2% were scattered among the tropical, semiarid, or steppe climates, the marine and Mediterranean climates, or the mountain climates. In 1980 the percentages in these latter zones had increased from 2% to 22%. There is continuous cross-migration among areas: in each of the two 5-year periods between 1970 and 1979, approximately 10 million people switched residence from one to another among the four large divisions--Northeast, North Central, South, and West. For the United States as a whole, slightly over half of the population has spent its life to date in the same state; the fraction who will spend, or have spent, an entire lifetime in the same state is much smaller. Climate variation even within the humid subtropical region, which had 32% of the population in 1860 and again in 1980 (but 29% in 1920) is perhaps as great as the likely changes due to CO_2, so movement _within_ the climate "region" adds to the amount of relevant mobility.

And this is still apart from the significant changes in microclimate--local rather than regional--entailed in movement from countryside to Pittsburgh, Jersey City, or Milwaukee. The actual climates numbers of Americans have experienced since 1800 are suggested by Table 9.1 and the subsequent maps (Figures 9.1-9.5) that illustrate it.

9.2 A SCHEMA FOR ASSESSMENT AND CHOICE

We now develop a framework for policy choices. The framework ought to be comprehensive. It should include theoretical possibilities that may

TABLE 9.1 U.S. Population by Climatic Zone[a,b] (Figures in Parentheses are Percentage of Total Population in that Climate Zone)

Climatic Zone[c]	Description	Population 1800	1860	1920	1980
Aw	Tropical wet and dry (Savannah)	0	2,996 (<1)	129,741 (<1)	2,793,140 (1)
BS_{Total}	Total semiarid and steppe	0	64,018 (<1)	4,291,664 (4)	21,000,465 (9)
BS_1	Southern California	0	15,657	1,138,802	11,801,232
BS_2	Central Valley of California	0	17,426	276,820	1,262,423
BS_k	Middle-latitude steppe	0	30,935	1,415,622	7,936,810
BW_h	Tropical and subtropical desert	0	28,029 (<1)	743,263 (<1)	4,955,742 (2)
Caf	Humid subtropical (warm summer)	2,034,536 (42)	9,426,517 (32)	32,360,561 (29)	71,932,014 (32)
Cb	Marine (cool summer)	0	39,246 (<1)	1,795,406 (2)	4,447,811 (2)
Cs	Dry-summer subtropical (Mediterranean)	0	202,420 (<1)	1,636,597 (2)	8,675,763 (4)
Daf	Humid continental (warm summer)	2,348,030 (49)	16,074,866 (54)	59,811,474 (54)	90,882,262 (40)
Dbf	Humid continental (cool summer)	435,665 (9)	3,586,555 (12)	9,394,792 (8)	13,710,636 (6)
H_{Total}	Total undifferentiated highlands	0	184,896 (<1)	1,559,963 (1)	9,147,733 (4)
H_1	Northwestern Mountain Region	0	158,012	1,343,146	8,150,762
H_2	Colorado Mountain Region	0	28,884	167,123	891,756
H_3	Arizona Mountain Region	0	0	49,694	105,215

[a]Source: U.S. Census Bureau, 1800, 1860, 1920, 1980; data compiled by Clark University Cartographic Service.
[b]Climatic zones shown in Index Map.

be of no contemporary significance, because we have to think about choices as they evolve through the century. The framework should make room for imagination, not just for options that currently look cost-effective.

The framework should lend itself to different levels of universality. While atmospheric CO_2 is a global condition, the consequences and many of the policy implications will be regional and local. Governments will assess consequences and choose policies according to the climatic impacts on their own populations and territory. At the same time, some national governments, including ours, need a framework for assessing consequences worldwide and policy options that are international in scope.

Just as governments will assess differently the implications of climate change for their own countries, some perceiving gains and others losses, so will interests be divided within countries. Not only are some countries, like our own, large enough to have diverse climates subject to different kinds of change, but people in the same climate are affected differently according to how they live and earn their living, their age and their health, what they eat, and how they take

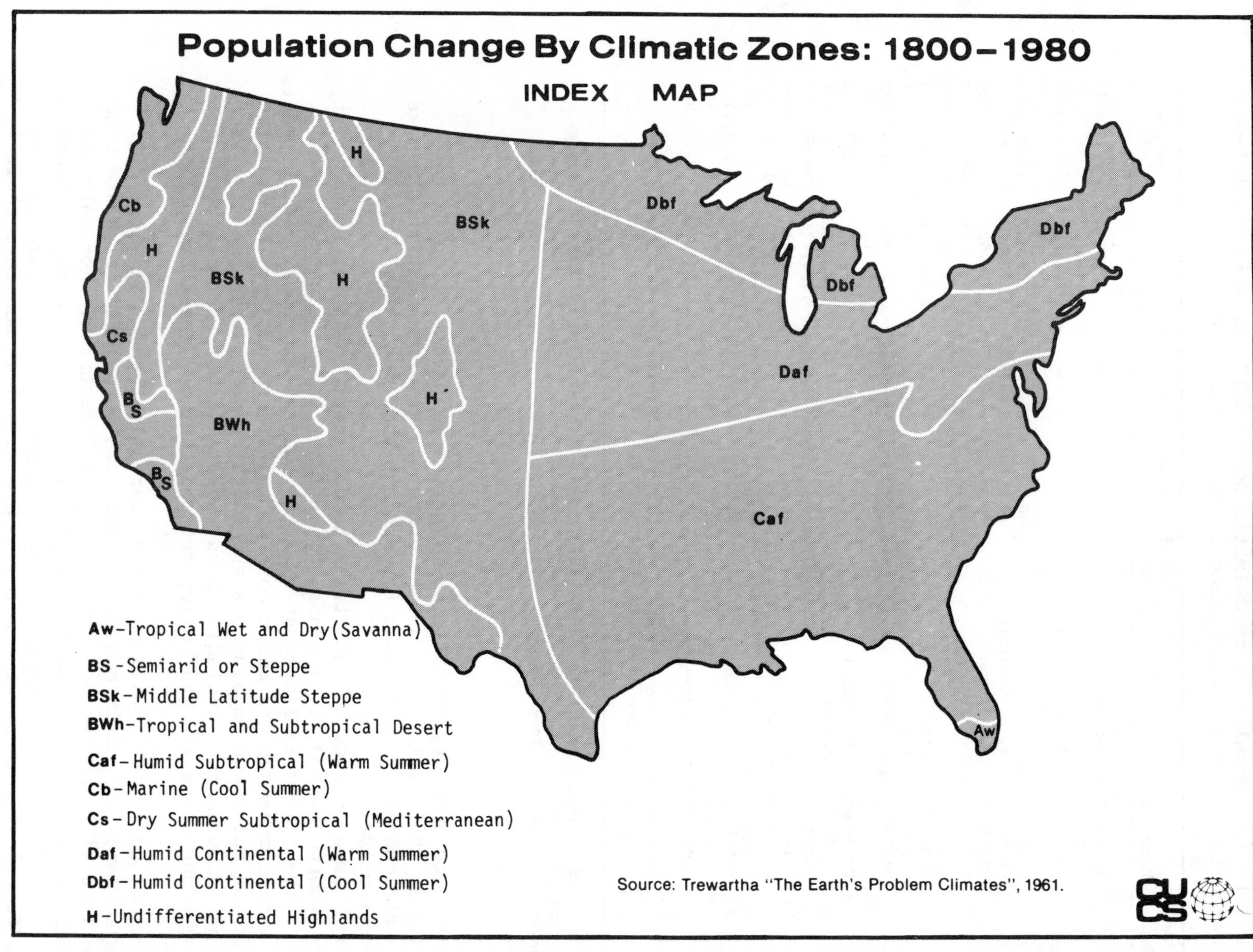

FIGURE 9.1 Population change by climatic zones: 1800-1980. (Prepared by Clark University Cartographic Service.)

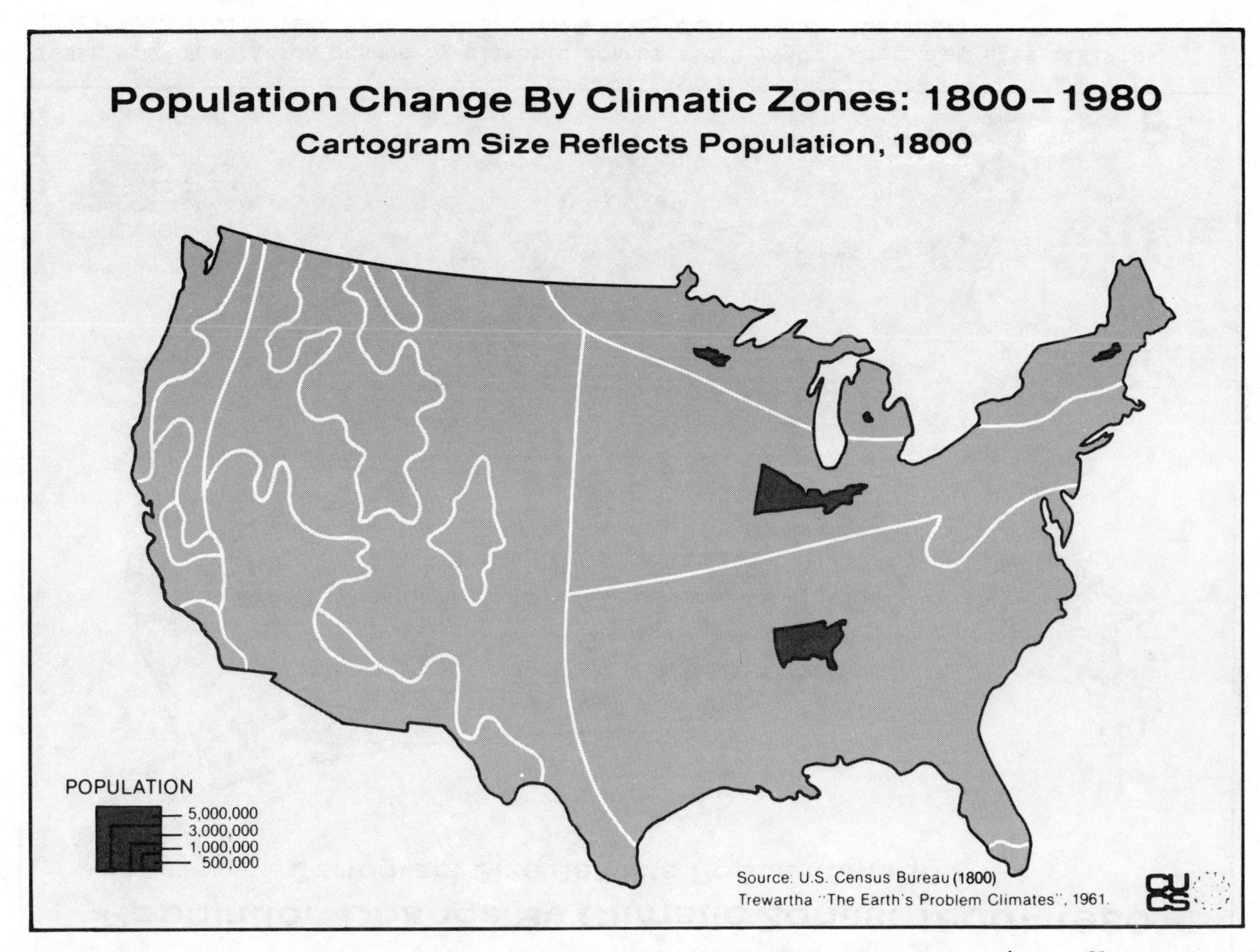

FIGURE 9.2 Population change by climatic zones: 1800-1980. Cartogram size reflects population, 1800. (Prepared by Clark University Cartographic Service.)

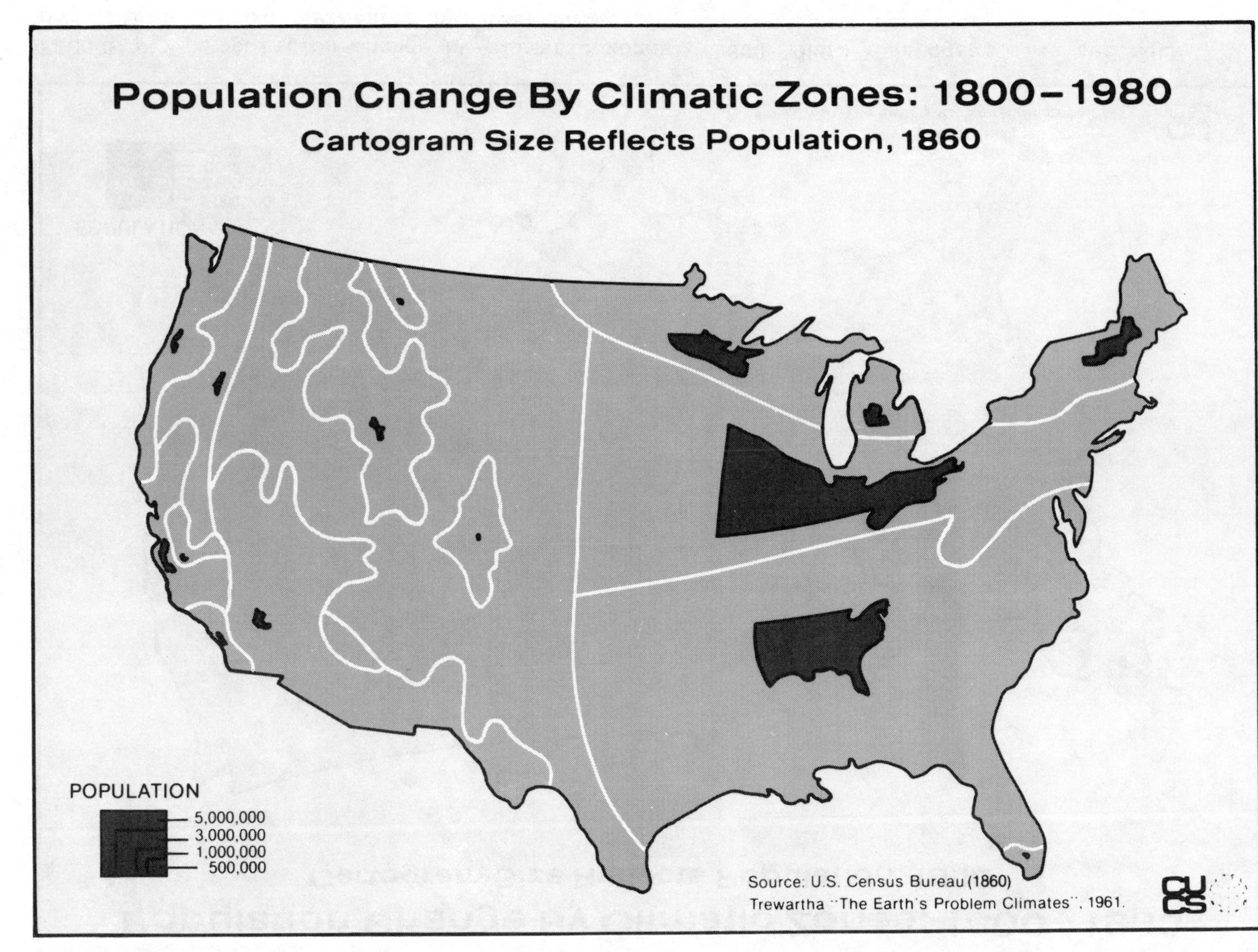

FIGURE 9.3 Population change by climatic zones: 1800-1980. Cartogram size reflects population, 1860. (Prepared by Clark University Cartographic Service.)

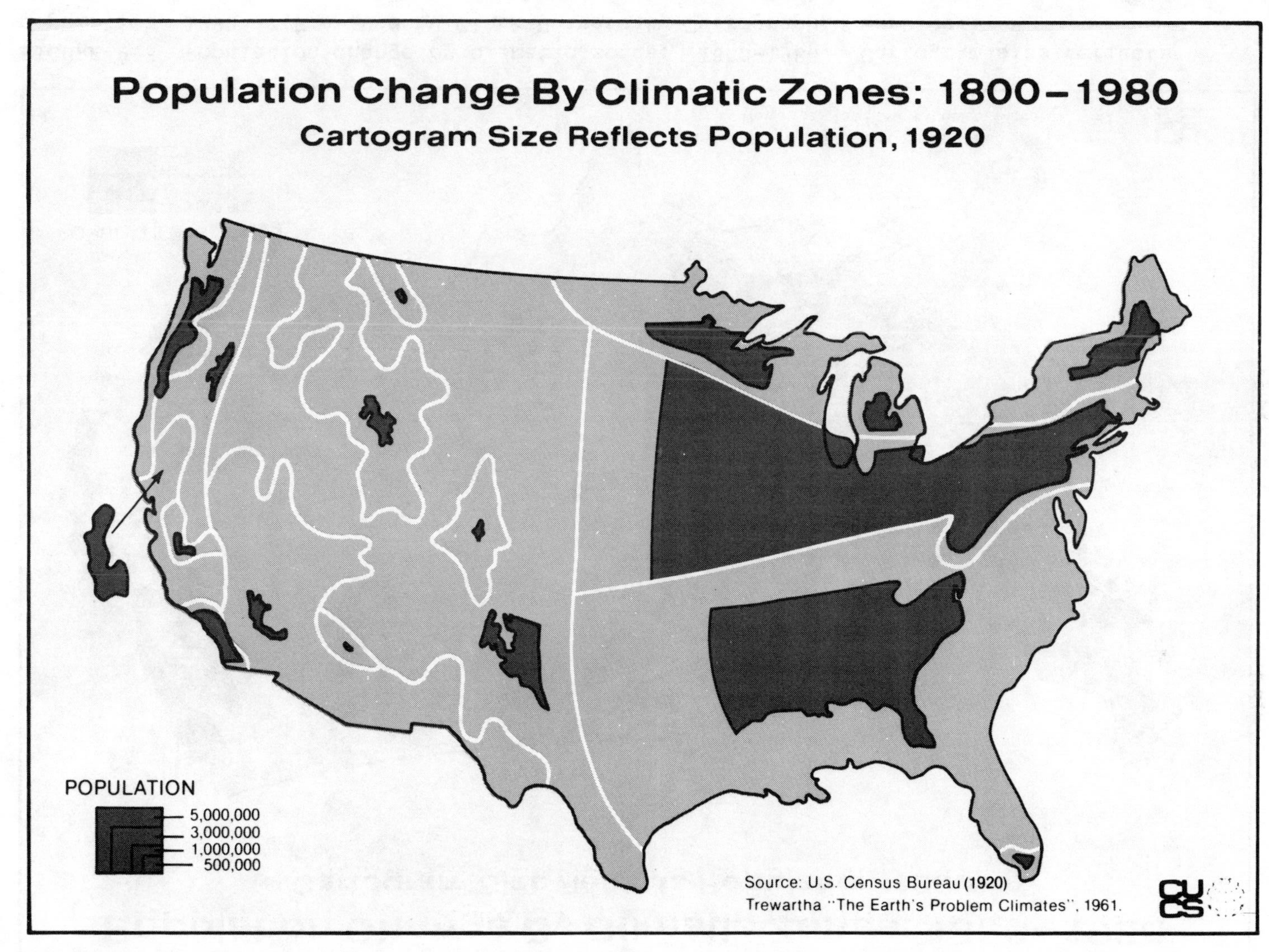

FIGURE 9.4 Population change by climatic zones: 1800-1980. Cartogram size reflects population, 1920. (Prepared by Clark University Cartographic Service.)

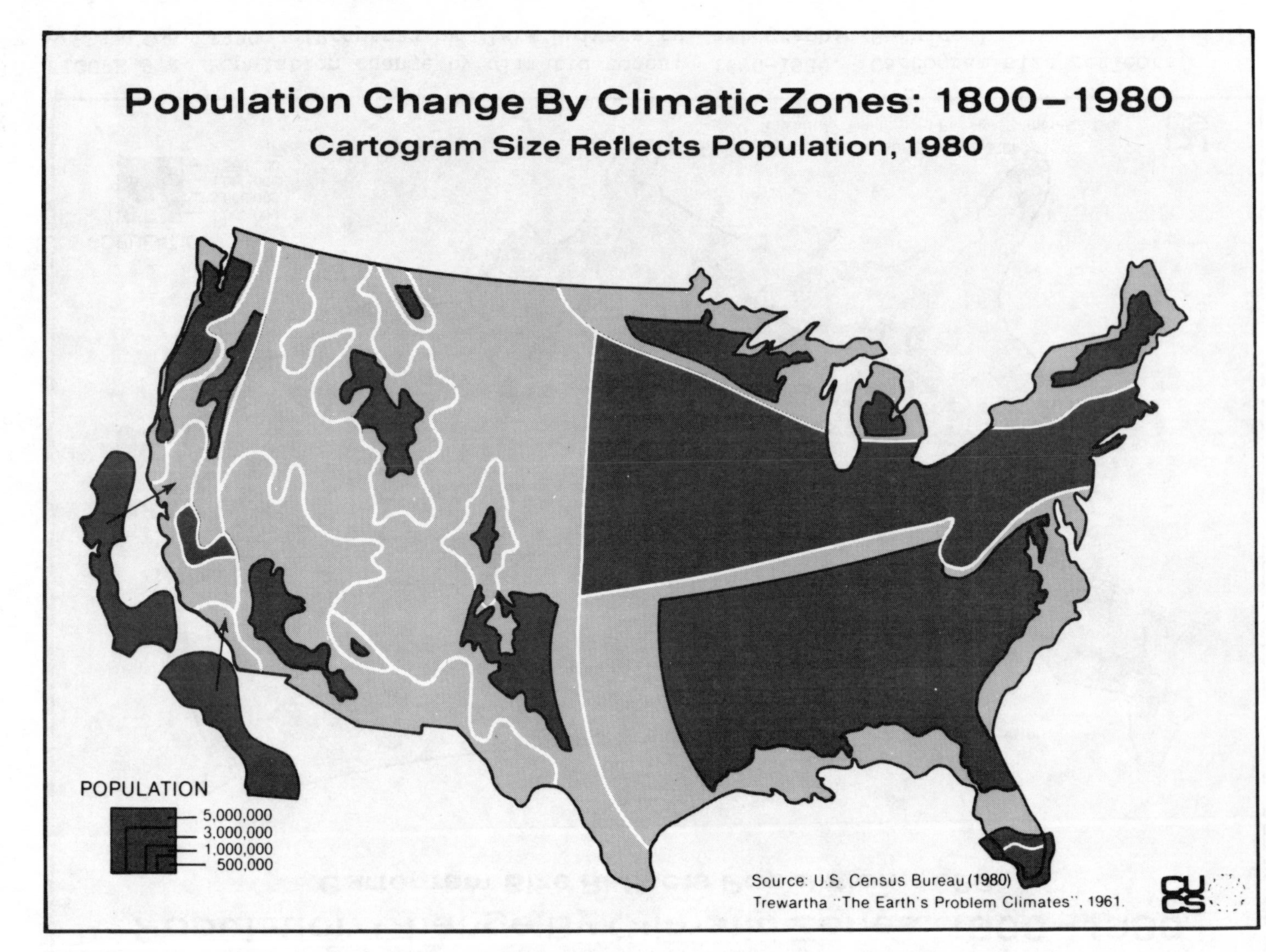

FIGURE 9.5 Population change by climatic zones: 1800-1980. Cartogram size reflects population, 1980. (Prepared by Clark University Cartographic Service.)

their recreation. Our framework has to be susceptible of disaggregation.

The framework should be construed as moving through time. The changes take time; the uncertainties unfold over time; policies and their effects have lead times, lag times, and growth rates. Governments and people will attach different discounts to events and conditions at different distances in the future. And a country that appears to be victim or beneficiary of a climate change forecast for the next 75 years would not be helped or hurt the same amount, or even in the same direction, by an additional 75 years of the same scenario. Uncertainties in the forecast CO_2 effects on climate are not merely uncertainties in some average magnitude but also, especially in local or regional change, uncertainties about the algebraic sign of some measure of net welfare and in the distribution of gains and losses among sectors of a population.

What we do not want, therefore, is a framework for assessing welfare and policy that is oriented toward some "bottom line." There will be as many bottom lines as there are users of the framework, according to their interests and responsibilities over space, time, and people.

9.2.1 Five Categories

The framework (Table 9.2) consists of a sequence of discrete categories. They can be thought of as arrayed from left to right. The point of departure, Category 1, contains the options for affecting the production of CO_2. This is taken as starting point not to prejudice the question whether fuel and energy policies are the preferred policies for anticipating climate change but because this is the most familiar category, the one that occurs first and most naturally to people concerned with CO_2 and climate change and the category least likely to be challenged as inappropriate, irrelevant, or improper. The framework itself entails no necessary presumption that the less CO_2 the better.

The last remark suggests a Category Zero preceding no. 1, the background, including any natural changes in climate, against which potential changes are to be judged. The sequence of these categories is generated by the premise that, taking the background into account, future emissions of CO_2 would probably prove nonoptimal: reducing CO_2 emissions is an obvious set of policies even if not the preferred set, and successive categories might be generated by following a sequence of "next steps." Thus Category 2, following production of CO_2, is removal. If we cannot help producing too much, can we remove some?

If too much is produced and not enough removed, so that the concentration is going to increase and climate is going to change in systematic fashion, can we do something about climate? Category 3 consists of policies to modify climate and weather.

Finally, Category 4 is all the policies or actions taken in consequence of anticipated or experienced climate change. It is important to include actions as well as policies because much adaptation will be by individuals and private bodies, and the term "policy" might exces-

TABLE 9.2 CO_2-Induced Climatic Change: Framework for Policy Choices

	Policy Choices for Response[a]			
Possibly Changing Background Factors	(1) Reduce CO_2 Production	(2) Remove CO_2 from Effluents or Atmosphere	(3) Make Countervailing Modifications in Climate, Weather, Hydrology	(4) Adapt to Increasing CO_2 and Changing Climate
.	.	.	.	.
.	.	.	.	.
.	.	.	.	.

sively imply that the reactions to foreseen climate change, or to climate uncertainty, were solely the responsibility of governments.

9.2.2 Background Climate and Trends

Absent a CO_2 effect, would global temperature rise or fall, or neither, or both, over the next hundred or two years, and by how much; and what uncertainty is there in our estimate? If the Earth were in for a cooling trend and prospects were universally gloomy, an offsetting greenhouse effect--the burning of fossil fuels--would offer some desired climate modification as an inadvertent but beneficial by-product. Indeed, as a by-product not easily controlled, it would be beyond diplomatic dispute and not dependent on agreement or cooperation. If the natural trend were toward a warmer atmosphere, added warming would aggravate concern.

Part of this background is the emission, and residence in the atmosphere, of other gases or particulates that affect radiation in either direction. Several have been identified that could have a significant greenhouse effect. Over the coming century other emissions or environmental effects might have similarly significant inadvertent effects on global climate.

Other background elements are the expected rise or fall in sea level, subsidence of land near sea level, and whether more land and more people, as in The Netherlands, will be below sea level already. An important background variable here is the fraction of the world's population and capital structures that would be expected over the coming century to populate low-lying areas that might have to be abandoned or protected.

More generally, for climate rather than sea level, what are the regional demographic trends and the likely (continuously changing) distribution of populations among the changing climate zones? Migration and relative growth expected in the normal course of events could aggravate or mitigate the damages due to climate changes and enhance or negate the benefits.

If the main effects of climate change will be on agriculture, background variables must include population and income, land use and agricultural productivity, dietary practices, and other determinants of whether decrements in food output due to climate change would be expensive to make up or to do without, at the income levels at that time. (The problems of assessing gains and losses to the people of a country, or to parts of the population of a country, are illustrated by the observation that for 50 years the government of the United States has had a policy of removing farmland from cultivation in the interest of higher farm incomes.)

The prospects for water supply--sources, uses, transport, storage, and conservation--are evidently of importance. Rainfall, snowfall, and evaporation are among the key elements in climate change. Diminishing groundwater or growing needs for irrigation would be main determinants of whether increased rain and snow would be welcome and how costly reduced precipitation would be.

This list is incomplete in two respects. First, it consists only of highlights and does not contain everything that a priori we could recognize as relevant background, especially in judging the needs and problems of particular countries and regions. But it is necessarily incomplete also because when we come to the fourth category, policies to adapt, to mitigate, or to take advantage of climate change, we are bound to uncover significant parts of the background whose significance was not apparent the first time around.

9.2.3 Production of CO_2

There are two subdivisions, energy and land use. Energy breaks into three: reduced energy, reduced fossil fuels, and less carbon-intensive fossil fuels.

The different sources of energy have different advantages and disadvantages. Petroleum is the most versatile, gas the cleanest (except for pure hydrogen); nuclear power is limited to large-scale production of steam (or, indirectly, hydrogen); hydroelectric is fixed in location. Reducing energy thus has different effects on CO_2 according to whether the savings are in transport, electric power, home heating, or industrial processes. And CO_2 per unit of heat varies among countries because of the mix of energy sources.

Among the fossil fuels gas produces the least and coal the most CO_2 per unit of heat (oil about 0.8 and gas 0.6 times the carbon of coal). Because their production is itself energy-intensive, coal-based synthetics entail the greatest injection of CO_2 per unit of heat, counting the emissions of a synthetics plant. Currently interesting coal-based synthetics appear to require about 40 units of coal to convert 100 into synthetics and thus would put 40% more carbon into the atmosphere than plain coal, per unit of heat. (For petroleum refining the corresponding figure is about 5%.)

Most of the known fossil resources are not gas and oil but coal and other currently uneconomical sources like shale, heavy oil, and tar sands. Shale is high in carbon per unit of heat because of the fuel used in retorting the shale.

Land use is the other subdivision in this category, but it cuts across Categories 1 and 2 symmetrically. There is no way to "unuse" fuel that has been burned, but forests can be grown or cut and the net effect can go either way. Preserving a forest rather than permanently removing it can be thought of as producing less CO_2 or removing it from the atmosphere.

The relevant land use is more than living biomass. Dead matter on and in the soil amounts to two or three times the carbon of the living vegetation. What happens to forests affects the release of carbon from the exposed soil and so does what happens to unforested land through cultivation and erosion.

9.2.4 Removal of CO_2

Removal shares with production the characteristic that it affects the global inventory. The incentives, therefore, are nonexistent for the individual, the firm, or the small nation, as long as removal is at any significant cost. The costs fall entirely on whoever undertakes removal; the benefits are shared indiscriminately.

Removal can be subdivided into processes that take CO_2 out of the atmosphere at large and those that "scrub" it from stack gases and other exhausts. The former are unrestricted by locale and operate on concentrations in the hundreds of parts per million (ppm); the latter operate at fixed sites and on much denser concentrations, in the tens or hundreds of thousands of ppm.

A natural way to remove carbon at large is through photosynthesis. In growing, trees remove CO_2. They release it on being consumed by fire, termites, bacteria, fungus, and in other forms of oxidation. Permanent removal requires that the global stand of forests increase in carbon content or that fallen and felled trees be preserved from oxidation.

Three apparent possibilities for permanently increasing the carbon content of standing forests would be to increase the forested area, to increase the carbon density of forest by enhancement of growth or longevity, and to enhance the density of carbon by cultivating alternative species. The first is "land management" and typically constitutes abstaining from deforestation or guarding against erosion. The second should occur to some extent as a consequence of CO_2 enhancement; we have no reliable extrapolation from greenhouse experiments with enhanced CO_2 to the long-term growth of natural forests. The third could provide a criterion for reforestation.

There are some orders of magnitude to keep in mind. The carbon in existing vegetation is roughly equal to the carbon in the atmosphere. Worldwide, forests cover a quarter or a third of the continents, and this amount is about three times the area of land in cultivation or pasture. Reforested land takes a few decades to a century to reach, say, 75% of the carbon density of mature forest, depending on whether the climate is warm and humid, temperate, or subpolar. The preponderant fraction of carbon in living vegetation is trees, not shrubs, grasses, and crops. A one seventh decrement in forest worldwide of average carbon density thus adds--and promptly, because of disposal for clearing--on the order of 100 gigatons of carbon (Gt of C); a one seventh increment in forest removes, more slowly, the same 100 Gt of C. A doubling of atmospheric CO_2 (from current levels) is an addition of about 700 Gt of C; increasing mature forest by a seventh would thus offset a seventh of a doubling, or, what is approximately the same thing, might retard the growth of atmospheric CO_2 by something like a decade during the second half of the next century.

Biomass is an exceedingly land-intensive form of carbon storage; it requires a permanent inventory of living trees. An alternative would be to grow trees for harvest and preserve them from decay. If reforestation is the reverse of deforestation, "refossilization" of harvested forest would be the reverse of fossil fuel combustion. ("Fossil" is Greek for "dug up.")

A few pertinent considerations. Preserving trees on site (coating them in plastic?) utilizes land, as biomass does. Floating logs downriver and sinking them in the deep ocean has been thought of; getting them to the rivers and towing them to sea would be expensive and fuel-intensive. Randomly stacked, 100 Gt of C in the original lumber would occupy something like 1000 km^3 on the ocean bottom.

One more technology might be increasing the transfer of carbon from ocean surface to deeper ocean. Pumping, perhaps by thermal gradient energy, has been proposed; the benefits with any currently conceived technology would be small compared with the cost.

"Scrubbing" from stack gases requires disposal. Possibilities are pumping into deep ocean, where ocean is accessible, or piling up sludge somewhere, unless usable construction materials can be extracted or formed. One hundred gigatons of carbon might be associated with 100-200 km^3 of sludge.

One conclusion is inescapable, irrespective of a hundred years' technological change: "sweeping" the atmosphere with trees can be no great part of any solution to a CO_2 problem. That does not mean that a strategy for the use of lands and forests should ignore CO_2, only that the role of trees, standing or fossilized, will be modest.

"Scrubbing" from stacks and "washing" by the oceans offer the possibility of yielding to technological advance.

9.2.5 Modification of Climate and Weather

The common distinction is between weather and climate. Both include the same descriptive components--temperature, humidity, rain and snow, clouds and fog, hours and intensity of sunlight, and other atmospheric details of special interest like visibility and ceiling for aircraft. Weather is what is experienced at a given time, where time is longer than instantaneous, as in "occasional showers" or "gusting winds," but no longer than the few days covered by short-range forecasts. The exact definition is not necessary to make the distinction with "climate." The latter is a probabilistic description of weather throughout the seasons of the year. Climate differs from weather in describing not a state but a pattern, including seasonal change; it includes averages and departures from average. And because variation from seasonal norms, if not permanent, is included in the definition of climate, it is inherently defined over decades, while weather, though not instantaneous, is defined over hours or days.

The distinction, though useful, is inadequate for classifying some interventions. Seeding clouds is weather modification; there are no known or intended effects of more than short duration. Managing greenhouse gases that have long residence times, like carbon dioxide, is clearly climate modification, whether intended or not, and of course it could be intended. But two kinds of intervention are not so easily classified as weather or climate. One would be manipulating gases or particulates that affect incoming or outgoing radiation and have residence times of months or years, not days or decades. The other would be a sustained program of cloud seeding, if it could be reliably and predictably done, to affect permanently the probable rainfall or snowfall at certain times of year.

Our classification of modification techniques can take the "warming" due to CO_2 as point of departure: one category is affecting the global radiation balance.* According to the preceding chapters, there is now little doubt that this can be done on a huge scale. It is exactly what we are doing with CO_2. The fact that it is unintentional and that the consequences may not be welcome do not contradict that we know how, at some expense if necessary, to change the world's climate more than it has changed in the last 10,000 years.

Warming the atmosphere currently is more economical than cooling it, because it happens as a by-product of energy consumption that would be costly to reduce or terminate. If we were faced with a "little ice age" over the next centuries, we might be glad to get some of that CO_2 in the atmosphere at no cost and without having to negotiate climate change diplomatically. But we know that in principle cooling could be arranged. Volcanic eruptions have done it. Considering the development of nuclear energy in both its explosive and its controlled uses and the feat of landing a team on the moon and returning it safely, and that we now know how to warm the Earth's atmosphere, we should not rule out that technologies for global cooling, perhaps by injecting the right particulates into the stratosphere, perhaps by subtler means, will become economical during coming decades.

Next are nonradiative climate interventions. Possibilities include changing the courses of rivers (discussed in the Soviet Union), damming seas, and opening sea-level canals. Some would be within the jurisdictions of nations, some international.

Weather modification should also be assessed with a long-term perspective. Techniques for precipitating rain and snow, acquisitively because the precipitation is desired or pre-emptively to dump it elsewhere, could have a mitigating effect on undesired climate changes even if the time to develop them to the point of economic utilization were as long as the time that has elapsed since the first communication by electric spark.

Some control of hurricanes may be achievable. Attention is usually on their destructiveness, but in many places hurricanes are a determinant of rainfall over large areas. Systematic suppression would create a significant change in regional climate.

One more illustration that climate and weather modification are not an empty or uninteresting category is the possibility that the floating arctic ice cap could be made to disappear during part of the summer by depositing some substance, like soot, that would absorb the sun's

---

*"Warming" is in quotation marks because the greenhouse effect is a global average temperature phenomenon but may result in colder as well as warmer temperatures in particular localities, or during particular seasons, and the "warming" is, furthermore, likely to be less important as a temperature change than as a determinant of winds, clouds, precipitation, sunshine, and other components of weather and climate.

heat. Whether an ice-free Arctic itself should be called "weather," the absence of ice would affect the climate of the surrounding areas. (The word "climate" is awkward here, as it is in connection with the prevalence of hurricanes; the result is a permanent part of the seasonal cycle only if the activity is sustained or repeated; and while an individual hurricane would count as "weather," the effect of 2 months' ice-free Arctic would not fit the short-range definition of weather.)

Before going on to Category 4, it is important to recognize the distribution of incentives for climate or weather modification. It is characteristic of the "CO_2 problem" that though it may be in the interest of the world economy to restrict, at some cost, the use of fossil fuels, it is probably not in the interest of any single nation to incur on its own the cost of any reduction in global CO_2. Whether any of the climate and weather modifications that might be undertaken, aside from suppression of CO_2, would be unilaterally attractive to individual nations, is more problematic. If the capacity to affect the radiative balance at nonprohibitive cost were acquired by several nations that disagreed about the optimum balance, that technology could be a source of conflict. More obviously, countries that view hurricanes as disasters and countries that view them as water for their crops would have different preferences about which if any hurricanes to suppress or to modify, or to generate. (The possibility of unilateral action, especially if it could be surreptitious and unverifiable, could cause trouble.)

This section can be summarized by three points. (1) We know that in principle weather and climate modification are feasible; the question is only what kinds of advances in weather and climate modification will emerge over the coming century. (2) Interest in CO_2 may generate or reinforce a lasting interest in national and international means of climate and weather modification; once generated, that interest may flourish independently of whatever is done about CO_2. (3) Weather and climate modification may be more a source of international tension than a relief; and CO_2 may not in the future dominate discussion of anthropogenic climate change as it does now.

9.2.6 Adaptation

Adapting to a change in climate is often thought of as an alternative to preventing it. For reasons that are often implicit, but not necessarily illegitimate or unpersuasive on that account, prevention is usually the preferred alternative. Prevention will certainly not, however, be absolute. Such deliberate reduction in emissions as may be achieved will be costly--and increasingly costly as lower emission levels are approached. So there is an inescapable question of how much, if any, restriction or removal will appear worthwhile to those who have to take the decisions--or to pay for them. Deciding how much CO_2 suppression makes sense requires an assessment of adaptation in the aggregate.

Most adaptation will be undertaken by units the size of a nation or smaller--families, firms, ministries and departments, cities, counties, and states or provinces. Even participation in an international program of CO_2 suppression would be a national decision. The costs and benefits of adapting rather than preventing would thus be most meaningful at the level of each nation.

Thus, except for a comparison with CO_2 suppression or climate modification, adapting to change is not an aggregate process. It is largely a multitude of decentralized, unconnected actions. And most of the adaptation to change that will take place in most societies over the next hundred years will not be adaptation to climate change. Migration, for example, can be motivated by changing climate; but migration within and between countries may still be responding a hundred years from now, as it is now, to political conditions and economic opportunities, and changing climate would be only one element in the politics or the economics. If we think how life has changed in our own country in 75 years during which the population changed from rural and small town to urban--real income per capita doubled three times, the population 65 and older increased from 4 to 12%, and life expectancy at birth grew from 50 to 75 years--and suppose that comparably dramatic changes may occur at the same rate in the future (although not necessarily along the same dimensions), it is evident that in the way they live and earn their livings people will be making multitudinous adaptations. Furthermore, the microclimates of many urban areas (Los Angeles, Mexico City, Tokyo) have changed drastically in the last 50 or 75 years. And changing technology and changing incomes always entail continual adaptation to one's local climate, even when that climate is not changing.

Adaptation can be subdivided into governmental policies and private actions. Government policies can be further subdivided into those taken at a national or subnational level and those that are necessarily or preferably international.

And there are two aspects of a CO_2-based climate change that are so different that they need to be kept separate, especially as they may not seem to be what is usually meant by "climate." One is changing gaseous composition of the air we breathe. The other is changing sea level.

9.2.7 Breathing CO_2

Little attention is paid, by those who study carbon dioxide and climate change, to any possible direct effects of CO_2 in the air we breathe on human health or on the animal population. Any natural anxiety about the health effects of a doubling or quadrupling of an important gas in the air we breathe--the substance that actually regulates our breathing rate--is relieved by the observation that for as long as people have been living indoors, especially burning fuel to heat themselves, people have been spending large parts of their lives--virtually all of their lives for people who work indoors and travel in enclosed vehicles--in an atmosphere of elevated CO_2. Doubling or even quadrupling CO_2 would still present a school child with a lesser concentration during outdoor recess than in today's classroom.

There is, furthermore, no documented evidence that CO_2 concentrations of 5 or 10 times the normal outdoor concentration damage human or animal tissue, affect metabolism, or interfere with the nervous system. (Industrial safety limits for chronic exposure are currently 5000 ppm--about 15 times the atmospheric concentration.) Nor is there theoretical basis for expecting direct effects on health from the kinds of CO_2 concentrations anticipated.

But even though the answer is easy and reassuring, the question has to be faced. It will occur to people who hear about changes in the atmosphere that their grandchildren are going to breathe. And there have not been experiments with either people or large animals that spent their whole lives, including prenatal life, in an environment that never contained less than, say, 700 parts per million of carbon dioxide. So the question deserves attention, even though there is no known cause for alarm.

9.2.8 Change in Sea Level

A dramatic possible consequence of CO_2-induced climate change is a significant rise in sea level. The three phenomena related to climate that may raise the sea level are thermal expansion of the water, reduced land ice and snow due to melting and wind erosion, and enhanced flow of glacial ice. The West Antarctic ice sheet is grounded below sea level and possibly susceptible to a rise in ocean temperature, which might cause faster glacial flow or "collapse." The estimated volume susceptible to such a process is equivalent to a rise in ocean levels of 5 or 6 m. (Floating ice, of course, does not affect the water level.) Large portions of the Earth's population, territory, and capital structures are within 5 or 6 vertical meters of today's high-water mark.

Thermal expansion of the water and melting of land ice have possible effects, as explained in Chapter 8, of 60-70 cm if CO_2 should double over the coming century. Many serious shoreline problems are sensitive to sea-level changes on the order of tens of centimeters, and 65 cm, though modest sounding on a calm day at the seashore, could produce a variety of profound environmental changes. It is the West Antarctic ice sheet, however, that ranks as the major potential threat. Obviously speed and warning time are crucial to any assessment of costs and damages.

It is concluded in Chapter 8 that any disappearance of the West Antarctic ice sheet would take centuries rather than decades and would be progressive rather than sudden.

There are three principal ways that human populations can adapt to a rising sea level. One is retreat and abandonment. A second is to build dams and dikes. A third is to build on piers and landfill. The basic division is between abandonment and defense.

9.2.9 Defenses against Rising Sea Level

Defense against a sea-level rise of several meters has received little attention in the United States. The threat is unfamiliar and only recently recognized. (Actual defenses against high seawater are extremely localized in the United States.) Most people recall, when reminded, that parts of Holland have been below sea level for centuries. (More than half of the 14 million Dutch live below sea level.) People recall, when reminded, that large areas of many cities, including familiar airports, are built on landfill. The professions that defend against river and tidal floods, storm surges, and land subsidence have only very recently been drawn into the CO_2 discussion.

It is therefore worth emphasizing that there are ways to defend against rising sea levels. For many built-up and densely populated areas they could probably be cost-effective for a rise of 5 or 6 m. Even where defending against 5 m were not cost-effective, defending against a meter or two could make sense for a century or two. Defense is not an empty hypothetical or purely speculative option.

The economics of dikes and levees depends on the availability of materials (sand, clay, rock); on the configuration of the area to be protected; on the differential elevation of sea level and internal water table; on the depth of the dike where it encloses a harbor or estuary; on the tide, currents, storm surges, and wave action that it must withstand; and on the level of security demanded for contingencies like extreme ocean storms, extreme internal flooding, earthquakes, military action, sabotage, and uncertainties in the construction itself. The Dutch build their seaward dikes with sand from the nearby ocean bottom; other structural materials are used for the base of the dike (under the sand), to cover and enclose the sand, and to provide facing against waves and currents. The dikes are not impermeable; there is permanent seepage requiring some pumping. It is noteworthy that with current technology the Dutch have found that 5 or 6 m is about the mean-sea-level difference that they can safely and economically build against with the plentiful and nearby sea sand as the primary material. Much of the coastal diking does enclose land 5 or 6 m below sea level. (The dikes are actually built to 11 or 12 m to guard against heavy seas during an extreme spring tide augmented by a storm surge--the piling up of water by the force of storm winds acting on the funnel-shaped North Sea.)

Besides keeping the ocean out, there have to be dikes--we call them levees--for conducting rivers to the sea. If a river runs through a city and the final 50 km of the river is no more than 5 m above sea level, the surface of the river must rise by 5 m if the sea level does, and 50 km of levee are required. The dikes in Rotterdam hold back not the ocean but the waters of the Rhine and other rivers, the surfaces of which cannot be below sea level.

On the economics of diking it can be kept in mind that the Dutch for centuries have found it economical to reclaim the bottom of the sea, at depths of several meters, for agricultural, industrial, and residential purposes. This is true not only of inland seas of which only the mouth needs to be diked but of extensions directly into the ocean where half or more of the perimeter of the newly acquired land needs to be diked against the full fury of North Sea storms.

A rudimentary illustration of the economics can be based on the Boston area. A full 5 m would cover most of downtown Boston. Beacon

Hill, containing the State House, would be an island separated by about 3 km from the nearest mainland. Most of adjacent Cambridge would be awash. But it would take only 4 km of dikes, mostly built on land that is currently above sea level, to defend the entire area. Perhaps even more economical, because it would avoid the political costs of choosing what to save and what to give up and of condemning land for right-of-way, would be a dike 8 or 10 km in length to enclose all of Boston Harbor.

If that were done, new deep-water port facilities would have to be constructed outside the enclosed harbor; locks would permit small boats in and out. The Charles and Mystic Rivers would have to be accommodated. Whether in a couple of hundred years there would be any significant flow in those rivers would depend on changing climate and increasing demand for water. Levees, a diversion canal, or pumping could be compared for costs and ecological impacts.

We have no professional estimate of what such a system would cost. Some professional guesswork suggests that at today's values the cost of defending against even the full 5-m rise is less, perhaps by an order of magnitude, than the value preserved.

Actually, to judge by the Dutch experience, reclaiming tidelands and harbor bottoms and even land in the open sea could become irresistably attractive during the coming centuries, either by landfill or by dikes.

The situation is totally different for an area like the coast of Bangladesh. If we imagine the sea level rising by a meter or two per century for enough centuries to reach 5 or 6 m, Bangladesh differs from Boston in having a huge coastal area subject to inundation, rather than a concentration of capital assets that can be enclosed by a few miles of dikes, and in being so susceptible to internal flooding with freshwater that the levees required to protect the country would be many times greater than the length of the shoreline. Thus the example of Boston cannot demonstrate that protection would be the preferred course everywhere and that it would reduce damage everywhere by an order of magnitude. The example only demonstrates that defense may compete favorably with slow retreat from the sea for densely populated urban areas.

Where defense is not practicable, retreat is inevitable, at least selectively. In urban concentrations, where buildings may last a century, good hundred-year predictions of sea-level change (including likely erosion and storm damage) should permit orderly evacuation and demolition of buildings. Urban renewal and interstate highways have already had such effects. (In the United States, the law and the politics of real-estate development would determine whether recently evacuated areas could selectively be rebuilt using landfill or piers.)

The most severe dangers appear to be in areas, like Bangladesh, where dense populations dependent on agriculture occupy low coastal plains already subject to freshwater or seawater flooding.

9.2.10 Food and Agriculture

The only readily identified potential impact of significant magnitude on future living standards is on agriculture. Virtually all agriculture everywhere is outdoors. It is dependent on sun and rain or irrigation from remote rain and snow; it is sensitive to temperature, especially frost; it is subject to beneficial and harmful activities of insects, worms, and microorganisms, and to weeds, all of which in turn are affected by weather and climate. Rainfall affects leaching and salinity and erosion. Animals are dependent on weather and climate indirectly through the crops they feed on and the pests that afflict them and directly through their sensitivity to temperature, humidity, or wind and snow. Most farm labor is applied outdoors, in unconditioned climate. Much the same can be said for fishing.

Poultry has moved indoors in many parts of the world; not only is the climate conditioned but the length of the artificial day is controlled. Plastic has helped greenhouses to multiply. If genetic engineering can increase the efficiency of photosynthesis or facilitate the nonagricultural manufacture of protein, food production may become much less dependent on natural climate during the coming century. Still, in most of the world, a land-intensive agriculture, outdoors, seems almost sure to be the dominant form for at least most of the coming century.

Despite agriculture's almost certain substantial dependence on climate throughout the future, it is nevertheless not yet possible, with today's crops and today's technologies and today's distribution of agricultural activity over the Earth, to assess the aggregate impact accurately enough even to be sure of its algebraic sign. As mentioned earlier, there is a presumption that climate change, independent of what the change is, being costly or difficult to adapt to, has a disadvantageous expectation. There is a presumption that the direct effect of CO_2 on photosynthesis, though it could affect weeds, too, might increase yields somewhat. A warming in northern latitudes could bring additional land under cultivation, although the quality of such land for crops is not promising. The effects of rainfall and snowfall are too mixed and uncertain to allow a prediction of their net effect. It is prudent, however, to expect that if climate change occurs rapidly, the costs in the aggregate will be positive and for some countries severe.

A fair guess seems to be that any likely rate of change of climate due to CO_2 over the coming century would reduce per capita global Gross National Product by a few percentage points below what it would otherwise be. If global economic production were to stagnate and population to grow, so that food production became an even greater portion of world output, and so that large parts of the world's population continued poor and largely dependent on the production of food for their own livelihoods, the projected climate changes could be exceedingly bad news. If world productivity improves at a reasonable rate and population growth dampens over the coming century, food production should be a diminishing component of world income. A rise of 10% or even 20% in the cost of producing food would be a few percent of world income at the outside. That lost income should occur in a world of appreciably higher living standards than those that prevail today. A curve of world per capita income plotted over time would be set back probably less than half a decade. That is, the living standards that might have been achieved by 2083 in the absence of climate change would be achieved instead in the late 2080s.

Adaptation to changing climate would take several forms. The development of water resources, typically an enterprise with lead times of several decades, becomes more urgent. Novel sources of freshwater, like large-scale desalinization or the mining of icebergs, becomes relevant to accessible coastal areas. Crops resistant to saline soil, or irrigable with brackish water, could be at a premium in plant genetics. Cultivation technologies that do not turn or disturb the soil should be increasingly improved. Water storage and transport, and inhibition of evaporation, would receive continued attention.

A development that is difficult to assess is change in dietary habits. Besides reflecting changed levels of income, working arrangements, and climates themselves, these may be affected by food-preparation technologies that are now even beyond speculation.

9.2.11 Global Warming and Energy Consumption

The pure temperature feedback on the use of energy, both as a cost-saving (heating) and an additional cost (cooling) and as a consequent damper or booster to fuel consumption, is of obvious relevance. Such estimates as there are do not indicate that any overall reduction or increase in energy use, due solely to temperature change, would be of major significance, whichever way the net effect goes. It is curious that the most immediate consequence to come to mind in connection with a global warming is not a major element in the overall energy assessment.

In understanding such an estimate, it can be noted that there is little one can do except to imagine the difference it would make in present society, living as it does and where it does now, to have temperatures of the kind projected in relation to CO_2 to which we were already adapted. There have been no efforts to imagine the distribution of the population by climates a hundred years from now, the structures in which people would live and work, the technologies for heating or cooling air or human skin, the energy sources for heating and cooling, the effects of local population densities, or any of the other elements that would go into a cost comparison. Estimates also have had no basis in wind or humidity. This estimate, that temperature itself will matter little in overall energy use, is therefore a first approximation that can be straightforwardly arrived at by imagining summer and winter thermostat changes equivalent to the forecast change in ground-level atmospheric temperature in different regions or latitudes. Degree days of heating and cooling yield a result that is not an impressive fraction of the current cost of heating and cooling.

What this estimate does is to remind us that the most important effects of increases in global average atmospheric temperature, or even regional temperatures at different latitudes, are not temperatures per se but changes in the other dimensions of climate, especially precipitation, driven both by the average warming and by the changing temperature gradients between equator and pole.

9.2.12 Distributional Impact

The most likely possibility to emerge from the work done so far in relation to CO_2 is that the impact of climate change on global income and production, and specifically the agricultural component of it, would not be of alarming magnitude. Particular regions or countries, especially those dependent on agriculture for a large part of their earnings, could be severely affected. The result would look more like a redistribution of global income than a large subtraction from it. The absolute amount of such redistribution need not be large for it to affect some areas very adversely. Poorer countries would be especially at risk, if the global distribution of income bears close resemblance to that currently prevailing, mainly because it is in the poorer countries that food production is a large part of total income and the capacity to adapt would be the smallest. An important international means of adaptation would therefore be compensatory transfers of income, capital, and technical assistance.

Saying so does not make it politically feasible, but after 35 years of bilateral and multilateral foreign aid programs the compensatory approach to global imbalances and maldistributions appears to be a permanent part of the institutional landscape. In any event, in considering the disaggregated gross changes in production and income rather than the global net changes, we need not be committed to letting the chips lie where they fall. Certainly if global income were predicted to rise substantially in the aggregate at some level of CO_2, but particular areas were to be disastrously affected, some system of compensation would be optimal in a purely income-maximizing sense.

A distinction should be made between compensation arrangements oriented toward CO_2 or climate and transfers from richer to poorer countries not tied to the particular alleged origins of poverty or hardship. If CO_2 becomes the focus of concerted international action over a prolonged period, the predictions and measures of climate change and their attributions to CO_2 and other causes, together with national "contributions" to CO_2 through combustion of fossil fuels, will become articles in diplomatic commerce. In that case, claims for compensation as a matter of right may emerge as a redistributive basis, along with judicial-like procedures for assessing claims and obligations. If instead the CO_2 accumulation gradually changes climates but there is no internationally organized regime of fossil fuel restriction, compensation would be less like "categorical aid" and more like general welfare or income support--always subject, of course, to international political divisions and disputes.

Table 9.3 summarizes possible background changes and societal response strategies to climate change.

9.3 SUMMING UP

When the work leading to the current report was begun, the energy crisis that began in the winter of 1973-1974 was barely 5 years old and reactions to the rising energy prices of the 1970s were barely

TABLE 9.3 CO_2-Induced Climatic Change: Framework for Policy Choices

Possibly Changing Background Factors	Policy Choices for Response[a]: (1) Reduce CO_2 Production	(2) Remove CO_2 from Effluents or Atmosphere	(3) Make Countervailing Modifications in Climate, Weather, Hydrology	(4) Adapt to Increasing CO_2 and Changing Climate
Natural warming, cooling, variability			Weather Enhance precipitation Modify, steer hurricanes and tornadoes	Environmental controls: heating/cooling of buildings, area enclosures Other adaptations: habitation, health, construction, transport, military
Population global, distribution: nation, climate zone, elevation (sea level), density				Migrate--internationally, intranationally
Income global average distribution				Compensate losers--intranationally, internationally
Governments				
Industrial emissions Non-CO_2 greenhouse gases Particulates			Climate Change production of gases, particulates Change albedo ice, land, ocean Change cloud cover	

Energy Per capita demand Fossil versus nonfossil	Energy management Reduce energy use Reduce role of fossil energy Increase role of low-carbon fuels	Remove CO_2 from effluents Dispose in ocean, land Dispose of by-products in land, ocean		
Agriculture, forestry, land use, erosion Farming and other dust Agricultural emissions (N_2O, CH_4)	Land use Reduce rate of deforestation Preserve undisturbed carbon-rich landscapes	Reforest Increase standing stock, fossilize trees		Change agricultural practices: cultivation, plant genetics Change demand for agricultural products, diet Direct CO_2 effects Change crop mix Alter genetics
Water supply, demand, technology, transport, conservation, exotic sources (icebergs, desalinization)			Hydrology Build dams, canals Change river courses	Improve water-use efficiency

[a]Responses may be considered at individual, local, national, and international levels.

visible. Projecting a continued worldwide increase in the use of fossil fuels of 4% per year, reflecting historical experience, gave estimates of several thousand gigatons of additional carbon in the atmosphere by the end of the twenty-first century, something between two and three doublings over current levels. Such a projection was both alarming and unreasonable--alarming not only because of climate change but of sheer atmospheric contamination, and unreasonable in implying that coal consumption might actually increase fiftyfold or 100-fold. That was not an estimate, merely an illustrative projection of historical trend.

In the interim, the impact of rising fuel prices on fuel consumption has become more visible, and fuel prices have doubled again; current projections for fossil fuels are commonly closer to 2% growth. At that more moderate rate the accumulated atmospheric carbon at the end of another 100 years would still be at least double the current quantity, but far less than the several thousand gigatons of the earlier projection. The depressed rate of increase compared with the first three quarters of this century reflects higher fuel prices and national energy policies.

The prospects are therefore much less alarming than some of those earlier calculations made them appear. And if successive increments in CO_2 were going to be increasingly harmful to climate, the lowered estimates of CO_2 correspond to even more significant reductions in the impact of climate change. Nevertheless, there is still the prospect of climate change within the coming century that takes us outside the boundaries experienced within the past 10,000 years.

There is no reason for believing that that development is to be welcome, and there are many reasons for the contrary. The first policy question that usually suggests itself is what can be done through national and international efforts to reduce the combustion of fossil fuel. Next there comes a complex of questions about how to respond and adapt, internationally, nationally, locally, and individually, to the varieties of climate changes that are to be expected in the different places around the globe where people live.

For many reasons it is unlikely in the foreseeable future that national governments will embark on serious programs to reduce further their dependence on fossil fuels to protect the Earth's climate against change. One reason is that governments are already saturated with reasons to reduce their dependence on fossil fuels, having made what many of them believe are heroic adaptations to high and uncertain fuel prices during most of the decade that began 10 years ago. If the rate of increase of carbon emissions can be held to 2% instead of the 4% that might have been projected from earlier experience, the doubling time for atmospheric CO_2 is increased from about two thirds of a century to a full century. Had a universal tax on fossil fuels been proposed a decade ago to discourage the burning of fuel and to mitigate the coming climate changes, it is inconceivable that anyone would have proposed raising fuel prices as much as they have been raised. (Of course, no one would have proposed that the proceeds accrue as profits to the producers of fuels.) While some supply response to the current high price of fuels can be expected, the prospects are for continued high and even rising fuel prices despite some current softness in the market for crude oil. The appetite for still higher prices, or still further restrictions on fuel consumption, as a means of further stretching out future additions to atmospheric carbon dioxide will not be forthcoming.

As emphasized earlier, any single nation that imposes on its consumers the cost of further fuel restrictions shares the benefits globally and bears the costs internally. For only the very largest fuel-consuming nations, probably for only the Soviet Union and the United States, might it be in the national interest unilaterally to suppress further the use of fossil fuels in the interest of mitigating climate change. And even that trade-off is certain to look unpersuasive to consumers paying current fuel prices. Some global rationing scheme that enjoyed the participation of the major producers and consumers of fossil fuels would be required if there were to be severe action at the national level.

In the current state of affairs the likelihood is negligible that the three great possessors of the world's known coal reserves--the Soviet Union, the People's Republic of China, and the United States of America--will consort on an equitable and durable program for restricting the use of fossil fuels through the coming century and successfully negotiate it with the world's producers of petroleum and with the fuel-importing countries, developed and developing. It makes sense therefore to anticipate changing climates. In any event, no regime for further restricting fossil fuels would hold emissions constant, so climate change is what we should expect.

Evidently, the value of developing the nonfossil sources of energy is at least as great as it would have been under a regime of fossil-fuel restriction and, if anything, more valuable.

Are there long lead-time projects or policies that need now to be adapted to the prospects of changing climates? Water resources and related technology may have lead times of half a century; water is therefore a candidate for planning in a context of potential climate change. As forecasts for climate change become clearer, there may be strong indications for research and development related to agriculture, fisheries, and pests. Military planning will probably be alert to changes on land and sea. Certainly coastal planning should be affected by forecasts of rising sea levels. But nothing urgent is foreseeable yet.

The foreseeable consequences of climate change are no cause for alarm on a global scale but could prove to be exceedingly bad news for particular parts of the world. Generally, the more well-to-do countries can take in stride what may prove to be a reduction (probably not noticeable as such) by a few percent in living standards that will likely be greater per capita by more than 100% over today's. But one has to question whether this relatively calm assessment can be applied to a country, say Bangladesh, where food production is already at the margin of subsistence and coastal flooding is already serious. Is it an especial hardship for the people of Bangladesh that the nations on whom would depend any permanent regime for globally rationing the use of fossil fuels in the interest of stable climate are unable to take a long economic view and to reconcile intense political differences?

Those concerned with the future welfare of Bangladesh have more reason to be concerned with population growth than the growth of CO_2. They have to be more concerned about the floods that will occur during the next 20 or 30 years than the floods that may occur during the 20 or 30 years after the middle of the next century. If the developed countries were prepared to make substantial economic sacrifices now to help to provide a more benign climate for Bangladesh a hundred years from now, anyone responsible for Bangladesh would probably prefer to have those economic sacrifices take the form of more immediate economic contributions to the country's standard of living and economic growth. Specifically, if a global tax on carbon fuels were used to depress the trajectory of future carbon in the atmosphere, a country like Bangladesh would be far more interested in an immediate and continuing claim on some of the proceeds of such a tax than on the future climatic effects of the tax. It is unlikely that countries currently as poor as Bangladesh would elect to join the rest of the world in paying higher fuel prices just to suppress carbon dioxide.

I earlier referred to this "calm" assessment of the "foreseeable" consequences of climate change. As remarked earlier, there is probably some positive association between what we can predict and what we can accommodate. To predict requires some understanding, and that same understanding may help us to overcome the problem. What we have not predicted, what we have overlooked, may be what we least understand. And when it finally forces itself on our attention, it may appear harder to adapt to, precisely because it is not familiar and well understood. There may yet be surprises. Anticipating climate change is a new art. In our calm assessment we may be overlooking things that should alarm us. But it is difficult to know what will still look alarming 75 years from now. It will be a while before the subject settles down.

Annex 1
Report of Informal Meeting on CO_2 and the Arctic Ocean

Roger R. Revelle

On June 1-2, 1982, a group of experts was informally convened by the Carbon Dioxide Assessment Committee to discuss the implications of CO_2-induced climatic changes for the floating sea-ice coverage of the Arctic Ocean. The meeting was held at the Philadelphia Centre Hotel, Philadelphia, Pennsylvania, in conjunction with a meeting of the American Geophysical Union (AGU). The assistance of the AGU is gratefully acknowledged. A list of participants is appended to the following notes that summarize the views expressed at the meeting.

MAJOR POINTS OF DISCUSSION

1. The Arctic ice has been a stable climate feature. There is quite good evidence for persistence of the ice cover all year round for the last 700,000 years and perhaps for the past 3,000,000 years, although there is debate about whether the Arctic may have been open in summer from 700,000 to 3,000,000 years ago. The existence of glacial marine sediments in the Arctic basin shows that ice rafting occurred during the past 5,000,000 years. Longer ago than 5,000,000-15,000,000 years, the Arctic may have been open year round. Global cooling patterns are such that an initial freeze-up of the Arctic may have occurred 15,000,000 years before present, although there is no direct evidence. The physical reasons for the persistence of the Arctic ice are not well understood but may reflect both dynamic and thermodynamic processes such that when little (excess) ice exists, correspondingly more (less) ice is produced the next winter.

2. Studies on whether the Arctic sea ice will completely melt in summer, and if so, whether the ice will remain melted in winter, as suggested by Flohn (1982), have produced ambiguous results.

Ewing and Donn (1956) proposed an explanation for-ice sheet cycling (i.e., ice ages) based on an intermittently open Arctic. Quantitative analyses have not confirmed this hypothesis. MacCracken (1968) developed a simulation model, which indicated that an open Arctic is not stable; under present climatic and geographic conditions, there is such strong cooling in winter that a <u>year-round</u> open Arctic could not persist. MacCracken's result does not preclude an Arctic open only in summer. Energy balances worked out by Fletcher (1965, 1973) and Rakipova (1966) indicate that an open Arctic would be stable. Fletcher finds that during summer (April to August) atmospheric heat loss over an ice-free Arctic would be greater than at present. This implies more vigorous circulation and cooler lower latitudes than when the ice is present in summer.

Maykut and Untersteiner (1969) applied one-dimensional sea-ice models with prescribed atmosphere conditions. They suggest it would require an additional 6 kcal cm^{-2} yr^{-1} (~8 W m^{-2}) oceanic flux to melt the Arctic ice in summer. This amount is equivalent to a warming of the atmosphere of roughly 8°C during the summer only or a year-round warming of a few degrees greater than the Arctic surface warming that is projected by Manabe and Wetherald (1975) for a doubling of atmospheric CO_2 concentrations.

Parkinson and Kellogg (1979) developed and applied a regional ocean sea-ice model. They claimed quite good verification, although Hibler (1979) disputes whether verification was adequate. When the Parkinson-Kellogg model is run with either an annual average 5°C warmer Arctic atmosphere (Manabe and Wetherald, 1975, 2 x CO_2 result) or a 6°C summer, 9°C winter warmer atmosphere (Budyko, 1974, seasonal pattern), the result is that Arctic ice melts in summer and returns in winter. One caveat should be noted in the Parkinson-Kellogg analysis, namely, that their temperature change was applied uniformly in the vertical (instead of being assumed to occur primarily in the subinversion layer) to calculate the change in downward flux, which may overestimate the change in the flux in regions without clouds. Hibler has not yet run his model on a CO_2 study but doubts that the ice will melt with only a 5°C warming of air temperatures.

3. Oceanographic studies are still quite limited for the case of an open Arctic. There is now a very strong, salinity-induced, density stratification, the causes of which are not fully understood. If this stratification can be broken and does not reform, then the Arctic might be able to remain open through the winter. This possibility is not considered likely. Parkinson and Kellogg (1979) found that even an upward heat flux from the ocean ten times greater than the present flux would not prevent return of ice in winter. The stratification could be reduced in several ways, for example, by changes in wind mixing or by limiting sources of freshwater, which include sea-ice melt and river runoff. The latter, in turn, depends on precipitation in the upper mid-latitude watersheds of the major Arctic rivers and, potentially, on water-management policies.

4. There have been few studies of the effect of less ice or no ice in summer (independent of a CO_2 change) on atmospheric circulation. One modeling study (Herman and Johnson, 1978) indicates that with less ice, winter storm tracks shift poleward, presumably because the intensity of the winter polar high is reduced.

Since the surface temperature (and the temperature at the top of the intensified Arctic stratus layer) are not projected to change significantly in summer, there is little reason to expect significant changes in the 300-700-mbar temperature, the gradient of which is probably a major factor in determining meridional eddy fluxes of heat and water vapor. Thus, with an open Arctic in summer, there could be little hemispheric effect, despite the implications of Fletcher's (1965) energy-balance analysis.

With respect to the local meteorology around the basin, there are suggestions that the warmer land would induce a stronger sea-breeze regime. It is not agreed whether the open ice condition would lead to a minimonsoonal-induced increase in precipitation on surrounding lands or just to an intensification of the existing stratus regime (in the style of the U.S. West Coast) in which the clouds move inland and evaporate rather than precipitate. It is agreed, however, that the present summerlike Arctic conditions would last longer each year, winter being shorter.

During the winter, the delayed and less-extensive freezing that would result if the Arctic were open in summer probably implies a reduction in the intensity of the polar high and a poleward shift of 5-10° latitude of wintertime storm tracks. That the mid-latitude effect in the winter would be greater than the summertime effect is rather a new suggestion. It may well be, however, that as the duration of wintertime climate is shortened and the summertime lengthened by the warming, the most significant climatic effects will instead be during the transition seasons. If storm tracks are shifted 5-10°, a significant precipitation pattern change is likely to occur in mid-latitudes.

TENTATIVE CONCLUSIONS

1. Given the apparent long-term stability of Arctic ice, one must be cautious in projecting a melting due to prospective warming from increasing CO_2 concentrations. A number of climate and ice models suggest that the Arctic ice may melt in summer with a warming of about the magnitude that may be induced by a doubling of CO_2 and increase of other greenhouse gases, but this conclusion must be viewed as still tentative. The representations of the Arctic in energy balance and most climate models that have melted Arctic ice with a CO_2 warming usually do not include changes in cloud cover, ice dynamics, or the effects of open leads and salinity stratification.

   Owing to dynamic and thermodynamic processes, ice <u>thickness</u> may respond more readily to temperature increases than ice extent. However, verification of ice extent and thickness estimates from climate models is not yet adequate.

2. While atmospheric effects of reduction in Arctic ice remain highly speculative, some poleward shift of storm tracks seems likely, and most significant climatic effects may occur during transition seasons.

PROSPECTS FOR PROGRESS

There are a number of research efforts that should bring progress in understanding effects of a greenhouse warming on the Arctic. Specifically, efforts should be made to

1. Improve general circulation models and other models (e.g., sea ice, Arctic stratus, ocean dynamics, and radiation balance) and use

them in studies focused on Arctic response. Proper handling of cloud cover in the Arctic merits special attention, as do sensitivity studies using improved sea-ice models.

2. Study stability of the Arctic Ocean density stratification and the potential for its destruction.

3. Obtain long central Arctic sediment cores that could improve the record of the Arctic Ocean for the period 10,000-15,000,000 years ago.

REFERENCES

Budyko, M. I. (1974). Climate and Life. English edition, D. H. Miller, ed. International Geophysical Series, Vol. 18. Academic, New York.

Ewing, M., and W. L. Donn (1956). A theory of ice ages. Science 123:1061-1066.

Fletcher, J. O. (1965). The Heat Budget of the Arctic Basin and Its Relation to Climate. The RAND Corp., R-444-PR, Santa Monica, Calif.

Fletcher, J. O. (1973). Numerical simulation of the influence of Arctic sea ice on climate. Energy Fluxes over Polar Surfaces. WMO Publication No. 361, Geneva.

Flohn, H. (1982). Climate change and an ice-free Arctic Ocean. In Carbon Dioxide Review: 1982, W. C. Clark, ed. Oxford U. Press, New York, pp. 145-179.

Herman, G. F., and W. T. Johnson (1978). The sensitivity of the general circulation to Arctic sea ice boundaries: a numerical experiment. Mon. Wea. Rev. 106:1649-1664.

Hibler, W. D., III (1979). A dynamic thermodynamic sea ice model. J. Phys. Oceanog.

MacCracken, M. C. (1968). Ice age theory by computer model simulation. Ph.D. Dissertation. University of California, Davis/Livermore.

Maykut, G. A., and N. Untersteiner (1969). Numerical Prediction of the Thermodynamic Response of Arctic Sea Ice to Environmental Changes. The RAND Corp., RM-6093-PR, Santa Monica, Calif.

Manabe, S., and R. J. Wetherald (1975). The effects of doubling the CO_2 concentration on the climate of a general circulation model. J. Atmos. Sci. 32:3-15.

Parkinson, C. L., and W. W. Kellogg (1979). Arctic sea ice decay simulated for a CO_2-induced temperature rise. Clim. Change 2:149-162.

Rakipova, L. (1966). The Influence of the Arctic Ice Cover on the Zonal Distribution of Atmospheric Temperature. The Rand Corp, RM-5233-NSF, Santa Monica, Calif.

Informal Meeting on CO_2 and the Arctic Ocean
June 1-2, 1982
Philadelphia, Pennsylvania

Invited Participants

Roger R. Revelle, Chairman
University of California
at San Diego

Kirk Bryan, Jr.
Geophysical Fluid Dynamics
Laboratory/NOAA

David L. Clark
University of Wisconsin

Joseph O. Fletcher
National Oceanic and
Atmospheric Administration

William Hibler
Geophysical Fluid Dynamics
Laboratory/NOAA

William W. Kellogg
National Center for Atmospheric
Research

Michael C. MacCracken
Lawrence Livermore National
Laboratory

William Ruddiman
National Science Foundation (OCE)

John Walsh
University of Illinois

John S. Perry
National Research Council

David A. Katcher, Consultant
Chevy Chase, Maryland

Annex 2
Historical Note

Jesse H. Ausubel

The issue of carbon dioxide and climatic change has now been on the research agenda for more than a century. Indeed, by the 1980s it has acquired quite a distinguished scientific provenance. In the 1860s J. Tyndall began suggesting that slight changes in atmospheric composition could bring about climatic variations. The first precise numerical calculations about how much increased carbon dioxide concentrations would influence the Earth's surface temperature were made by Svante Arrhenius (1896, 1908). He estimated that a doubling of atmospheric CO_2 concentrations would produce a global warming of about 4-6°C. At the same time, T. C. Chamberlin (1899) was developing theories that the large variations in the Earth's climate, including periodic glaciation, could be attributable to changing carbon dioxide concentrations. C. F. Tolman (1899) provided the first major insights into the critical role of the oceans in the global distribution of carbon dioxide.

By the early decades of this century, there was a lively debate among scientists on the direction of future CO_2 concentrations. Some, like Arrhenius (1908), built their conception of future development on the expectation that the atmosphere is gaining in CO_2 under the present regime of "evaporating" our coal mines into the air. Others, like C. Schuchert (1919), stressed the volcanic origin of much CO_2 and worried that the ultimate extinction of the Earth's "plutonic fires" would bring in train the depletion of atmospheric CO_2 and the extinction of life. "Life and its abundance at any time are conditioned by the amount of this gas (CO_2) present in the atmosphere."

In his classic work, <u>Elements of Physical Biology</u>, A. J. Lotka (1924), stimulated by a general interest in the history of systems in the course of irreversible transformations, also explored the carbon cycle. Lotka offered one of the first eloquent formulations of the CO_2 issue:

> ...to us, the human race in the twentieth century [this phenomenon of slow formation of fossil fuels] is of altogether transcendent importance: The great industrial era is founded upon, and at the present day inexorably dependent upon, the exploitation of the fossil fuel accumulated in past geological ages.

> We have every reason to be optimistic; to believe that we shall be found, ultimately, to have taken at the flood this great tide in the affairs of men; and that we shall presently be carried on the crest of the wave into a safer harbor. There we shall view with even mind the exhaustion of the fuel that took us into port, knowing that practically imperishable resources have in the meanwhile been unlocked, abundantly sufficient for all our journeys to the end of time. But whatever may be the ultimate course of events, the present is an eminently atypical epoch. Economically we are living on our capital; biologically we are changing radically the complexion of our share in the carbon cycle by throwing into the atmosphere, from coal fires and metallurgical furnaces, ten times as much carbon dioxide as in the natural biological process of breathing. . .[T]hese human agencies alone would. . .double the amount of carbon dioxide in the entire atmosphere. . . .

Lotka estimated a doubling time of 500 years, based on continued usage of coal at 1920s levels. If he had used the logistic ("Lotka-Volterra") equations for which he was to become famous to calculate future emissions as a result of human activities, Lotka would have given a doubling time in the middle of the twenty-first century.

V. I. Vernadski (1926) was among the first to show the extent to which the Earth, its atmosphere as well as its hydrosphere and landscapes, is indebted to living processes, to the biota. The theoretical ecologist V. A. Kostitzin (1935) dealt extensively with the circulation of carbon in his monograph "Evolution de l'atmosphere, circulation organique, epoques glaciares." Kostitzin provides a general review of available information and some of the theories concerning the circulation of oxygen, carbon, and nitrogen and discusses the long-term changes in their abundance in the atmosphere and soil. He reviews the theories in light of a simple model, incorporating a system of linear and quadratic differential equations, one of the early formal attempts to model the cycles. In his concluding remarks Kostitzin warns against confusing the relative short-term stability of nature with the absolute, but misleading, long-term stability of mechanical systems "which does not, in fact, exist, either in mechanics or in biology."

By 1938 G. S. Callendar was focusing directly on the industrial production of carbon dioxide and its influence on temperature. He went on (1940, 1949) to speculate that a 10% increase in atmospheric CO_2 between 1850 and 1940 could account for the observed warming of northern Europe and northern America that had begun in the 1880s. G. Plass, of the Aeronutronic Division of the Ford Motor Company, was responsible during the 1950s for the development of surface energy balance approaches to climate sensitivity that yielded the first "modern" estimates of global surface temperature response to increased CO_2. R. Revelle and H. E. Suess, in the opening of their often-cited 1957 paper, dramatically emphasized the significance of the rise in atmospheric CO_2: "Human beings are now carrying out a large-scale geophysical experiment...." They also pointed out for the first time that most of the CO_2 produced by the combustion of fossil fuels would stay in the atmosphere and would not be rapidly absorbed by the ocean. Revelle was instrumental in incorporating accurate and regular measurements of the concentration of CO_2 into the program of the International Geophysical Year (IGY).

Meanwhile, the potential societal significance of climatic change had not gone unrecognized. J. von Neumann (1955), noting the effects of increasing atmospheric CO_2, anticipated that deliberate human modification of climate would become a major issue in world affairs.

> The most constructive schemes of climate control would have to be based on insights and techniques that would also lend themselves to forms of climatic warfare as yet unimagined... [U]seful and harmful techniques lie everywhere so close together that it is never possible to separate the lions from the lambs. This is known to all who have so laboriously tried to separate secret, "classified" science or technology (military) from the open kind; success is never more--nor intended to be more--than transient, lasting perhaps half a decade. Similarly, a separation into useful and harmful subjects in any technological sphere would probably diffuse into nothing in a decade. . .After global climate control becomes possible, perhaps all our present involvements will seem simple. We should not deceive ourselves: once such possibilities become actual, they will be exploited. It will, therefore, be necessary to develop suitable new political forms and procedures.

Broader public concern with the implications of rising CO_2 content of the atmosphere probably dates to a conference on the topic sponsored by The Conservation Foundation in March 1963. Conference participants included Plass and C. D. Keeling, who was responsible for the continuous monitoring program of atmospheric CO_2 begun at Mauna Loa in Hawaii and at the South Pole in 1957 during the IGY. The report of the conference states in part:

> It is known that the carbon dioxide situation, as it has been observed within the last century, is one which might have considerable biological, geographical and economic consequences within the not too distant future. . .It is estimated that a doubling of the carbon dioxide content of the atmosphere would produce an average atmospheric temperature rise of 3.8 degrees (Celsius). This could be enough to bring about an immense flooding of the lower portions of the world's land surface, resulting from increased melting of glaciers. . . .

The report goes on to recommend special emphasis on continuation of the CO_2 monitoring program and more exact quantitative knowledge of the biosphere, themes that have been maintained over the past two decades. The Conservation Foundation report also concluded:

> There is a need for a watchdog. The effects of the continuing rise in atmospheric CO_2 while not now alarming are likely to

become so if the rise continues. A committee of the National Academy of Sciences, National Research Council, might be charged with exploring the problem....

The CO_2 issue was subsequently raised as a national concern in Restoring the Quality of Our Environment, the Report of the Environmental Pollution Panel of the President's Science Advisory Committee, in 1965. Since that time, the CO_2 issue has been included in most lists of potentially serious environmental problems.

REFERENCES

Arrhenius, S. (1896). On the influence of carbonic acid in the air upon the temperature of the ground. Phil. Mag. 41:237.

Arrhenius, S. (1908). Worlds in the Making. Harper, New York.

Callendar, G. S. (1938). The artificial production of carbon dioxide and its influence on temperature. Q. J. Roy. Meteorol. Soc. 64:223.

Callendar, G. S. (1940). Variations in the amount of carbon dioxide in different air currents. Q. J. Roy. Meteorol. Soc. 66:395.

Callendar, G. S. (1949). Can carbon dioxide influence climate? Weather 4:310.

Chamberlin, T. C. (1899). An attempt to frame a working hypothesis of the cause of glacial periods on an atmospheric basis. J. Geol. 7:545.

Kostitzin, V. A. (1935). Evolution de l'atmosphere, circulation organique, epoques glaciares. Hermann, Paris.

Lotka, A. J. (1924). Elements of Physical Biology. Williams & Wilkins, Baltimore, Md. Reprinted 1956, Dover, New York.

Plass, G. (1956). Effect of carbon dioxide variations on climate. Tellus 8:140.

President's Science Advisory Committee (1965). Restoring the Quality of our Environment. Report of the Environmental Pollution Panel. The White House, Washington, D.C., November 1965.

Revelle, R., and H. E. Suess (1957). Carbon dioxide exchange between atmosphere and ocean and the question of an increase of atmospheric CO_2 during the past decades. Tellus 9:18.

Schuchert, C. (1919). In The Evolution of the Earth and Its Inhabitants, J. Barrell, C. Schuchert, C. Woodruff, R. Lull, and E. Huntington, eds. Yale U. Press, New Haven, Conn.

The Conservation Foundation (1963). Implications of rising carbon dioxide content of the atmosphere. The Conservation Foundation, New York.

Tolman, C. F., Jr. (1899). The carbon dioxide of the ocean and its relations to the carbon dioxide of the atmosphere. J. Geol. 7:585.

Tyndall, J. (1863). On radiation through the Earth's atmosphere. Phil. Mag. 4:200.

Vernadski, V. I. (1926). Biosphere, articles--first and second. Nautchtechizdat, Moscow (in Russian).

von Neumann, J. (1955). Can we survive technology. Fortune, June 1955.

Annex 3
Energy Security Act of 1980

94 STAT. 774 PUBLIC LAW 96-294—JUNE 30, 1980

SUBTITLE B—CARBON DIOXIDE

STUDY

42 USC 8911.

SEC. 711. (a)(1) The Director of the Office of Science and Technology Policy shall enter into an agreement with the National Academy of Sciences to carry out a comprehensive study of the projected impact, on the level of carbon dioxide in the atmosphere, of fossil fuel combustion, coal-conversion and related synthetic fuels activities authorized in this Act, and other sources. Such study should also include an assessment of the economic, physical, climatic, and social effects of such impacts. In conducting such study the Office and the Academy are encouraged to work with domestic and foreign governmental and non-governmental entities, and international entities, so as to develop an international, worldwide assessment of the problems involved and to suggest such original research on any aspect of such problems as the Academy deems necessary.

Report to Congress.

(2) The President shall report to the Congress within six months after the date of the enactment of this Act regarding the status of the Office's negotiations to implement the study required under this section.

Results, report to Congress.

(b) A report including the major findings and recommendations resulting from the study required under this section shall be submitted to the Congress by the Office and the Academy not later than three years after the date of the enactment of this Act. The Academy contribution to such report shall not be subject to any prior clearance or review, nor shall any prior clearance or conditions be imposed on the Academy as part of the agreement made by the Office with the Academy under this section. Such report shall in any event include recommendations regarding—

Recommendations.

(1) how a long-term program of domestic and international research, monitoring, modeling, and assessment of the causes and effects of varying levels of atmospheric carbon dioxide

Public Law 96-294, June 30, 1980; Title VII--Acid Precipitation Program and Carbon Dioxide Study; Subtitle B--Carbon Dioxide should be structured, including comments by the Office on the interagency requirements of such a program and comments by the Secretary of State on the international agreements required to carry out such a program;

(2) how the United States can best play a role in the development of such a long-term program on an international basis;

(3) what domestic resources should be made available to such a program;

(4) how the ongoing United States Government carbon dioxide assessment program should be modified so as to be of increased utility in providing information and recommendations of the highest possible value to government policy makers; and

(5) the need for periodic reports to the Congress in conjunction with any long-term program the Office and the Academy may recommend under this section.

(c) The Secretary of Energy, the Secretary of Commerce, the Administrator of the Environmental Protection Agency, and the Director of the National Science Foundation shall furnish to the Office or the Academy upon request any information which the Office or the Academy determines to be necessary for purposes of conducting the study required by this section.

(d) The Office shall provide a separate assessment of the interagency requirements to implement a comprehensive program of the type described in the third sentence of subsection (b).

Annex 4

Background Information on Committee Members

WILLIAM A. NIERENBERG, Chairman, is the Director of the Scripps Institution of Oceanography of the University of California at San Diego. Dr. Nierenberg was trained in physics and has performed and directed research extensively in nuclear physics and in physical oceanography. He is a Member of the National Academy of Sciences and the National Academy of Engineering and has served in numerous capacities as an advisor on national and international scientific affairs. He is a former Chairman of the National Advisory Committee on Oceans and Atmosphere and is currently a Member of the National Science Board as well as Chairman of the Peer Review Panel on Acid Rain of the President's Office of Science and Technology Policy.

PETER G. BREWER is a Senior Scientist at the Woods Hole Oceanographic Institution. For the past 2 years, Dr. Brewer has been on leave directing the Marine Chemistry program of the National Science Foundation. Dr. Brewer's research interests include measurement of CO_2 in the oceans and the use of tracers for ocean circulation studies.

LESTER MACHTA is Director of the Air Resources Laboratory of the National Oceanic and Atmospheric Administration. Dr. Machta's research interests focus on the measurement and analysis of secular and anthropogenic trends in atmospheric conditions. He has participated in several major studies of critical environmental problems and has recently served as Chairman of the Atmospheric Sciences Work Groups of the Joint U.S.-Canadian Task Force on Acid Rain.

WILLIAM D. NORDHAUS is Professor of Economics at Yale University and a staff member of the Cowles Foundation. Professor Nordhaus's research interests include energy economics, innovation and technological change, and a growing number of aspects of the carbon dioxide issue. During 1977-1979 Professor Nordhaus was a Member of the President's Council of Economic Advisers.

ROGER R. REVELLE is Professor of Science and Public Policy at the University of California, San Diego. Professor Revelle's research interests include physical and chemical oceanography, marine geology, water and energy resources, and population studies. He is one of the founders of the field of modern CO_2 studies, having begun his continuing work on the subject in the 1950s, established the monitoring of CO_2 at Mauna Loa during the International Geophysical Year (1957), and brought the subject to national attention while serving on the Panel on Environmental Pollution of the President's Science Advisory Committee in 1965. Professor Revelle is a Member of the National Academy of Sciences, a Past President of the American Association for the Advancement of Science, and the recipient of numerous awards and honors for his research and leadership in oceanography and resource analysis. Dr. Revelle chaired the NRC Committee that authored the 1977 report *Energy and Climate*.

THOMAS C. SCHELLING is Professor of Political Economy at the Kennedy School of Government of Harvard University. Professor Schelling has written extensively about conflicts between individual and collective behavior. His current interests include addictive behavior, energy policy, environmental policy, and arms control and national security. Professor Schelling is a Member of the Institute of Medicine and currently serves on the NRC Commission on Behavioral and Social Sciences and Education. Professor Schelling chaired the NRC Panel on Economic and Social Aspects of CO_2 Increase, the predecessor group of the Carbon Dioxide Assessment Committee.

JOSEPH SMAGORINSKY is a Visiting Senior Fellow in the Department of Geological and Geophysical Sciences at Princeton University and former Director of the Geophysical Fluid Dynamics Laboratory. Dr. Smagorinsky's interests include atmospheric general circulation, theory of climate, and atmospheric predictability. He has served on many NRC committees in atmospheric sciences and is Past Chairman of the Joint Organizing Committee for the Global Atmospheric Research Program and the Joint Scientific Committee for the World Climate Research Program.

PAUL E. WAGGONER is Director of the Connecticut Agricultural Experiment Station. Dr. Waggoner is trained in both meteorology and agricultural sciences. His research includes physiology of crop yield, pest management, and relationships between air quality and plant growth. Dr. Waggoner is a Member of the National Academy of Sciences and participated in the 1976 NRC study on *Climate and Food*.

GEORGE M. WOODWELL is Director of the Ecosystems Center at the Marine Biological Laboratory in Woods Hole. Dr. Woodwell's research has centered on the structure and function of natural communities and their role as segments of the biosphere. He has worked and written extensively on the ecological effects of toxic substances, especially ionizing radiation and persistent pesticides. He is a Past President of the Ecological Society of America, Fellow of the American Academy of Arts and Sciences, one of the founders of the Environmental Defense Fund, the Natural Resources Defense Council, and the World Resources Institute, and is currently Chairman of the World Wildlife Fund and the International Conference on the World after Nuclear War.